Molecular Associations in Biology

Contributors

S. Altman

Robert L. Baldwin

B. Beltchev

E. D. Bergmann

D. F. Bradley

Edward N. Brody

A. S. V. Burgen

Jacqueline Caillet

G. Cilento

Pierre Claverie

M. Cohn

A. Danchin

J. F. Danielli

Robert C. Davis

Michel Delmelle

L. Dimitrijevic

Paul Doty

P. Douzou

Jules Duchesne

E. Peter Geiduschek

M. Gilbert

M. Grunberg-Manago

Richard Harrison

J. G. Heathcote

Karst Hoogsteen

O. Jardetzky

S. Richard Jaskunas

D. O. Jordan

Ephraim Katchalski

H. Kersten

W. Kersten

L. S. Lerman

S. Lifson

Per-Olov Löwdin

Donald B. McCormick

Marie-José Mantione

J. C. Metcalfe

A. M. Michelson

H. A. Nash

Bernard Pullman

K. Rosenheck

Jean Salvinien

Sidney Shifrin

Meir Shinitzky

Oktay Sinanoğlu

M. A. Slifkin

M. N. Thang

Hugo Theorell

Ignacio Tinoco, Jr.

Gordon Tollin

Paul O. P. Ts'o

Olke Uhlenbeck

Gregorio Weber

David L. Wilson

K. Zinner

MOLECULAR ASSOCIATIONS
IN BIOLOGY

Proceedings of an International Symposium
Held in Celebration of the 40th Anniversary
of the Institut de Biologie Physico-chimique
(Fondation Edmond de Rothschild)

Paris, May 8–11, 1967

Edited by BERNARD PULLMAN

UNIVERSITÉ DE PARIS
INSTITUT DE BIOLOGIE PHYSICO-CHIMIQUE
PARIS, FRANCE

 1968

ACADEMIC PRESS   New York and London

ACADEMIC PRESS INC.
111 Fifth Avenue, New York, New York 10003

United Kingdom Edition published by
ACADEMIC PRESS INC. (LONDON) LTD.
Berkeley Square House, London W.1

LIBRARY OF CONGRESS CATALOG CARD NUMBER: 68-18679

PRINTED IN THE UNITED STATES OF AMERICA

List of Contributors

Numbers in parentheses indicate the pages on which the authors' contributions begin.

S. Altman[1] (271), Department of Biophysics, University of Colorado School of Medicine, Denver, Colorado

Robert L. Baldwin (145), Department of Biochemistry, Stanford University School of Medicine, Palo Alto, California

B. Beltchev[2] (183), Service de Biochimie, Institut de Biologie Physico-chimique, Paris, France

E. D. Bergmann (207), Department of Organic Chemistry, Hebrew University, Jerusalem, Israel

D. F. Bradley (137, 261), Laboratory of Neurochemistry, National Institute of Mental Health, Bethesda, Maryland

Edward N. Brody (163), Department of Biophysics, University of Chicago, Chicago, Illinois

A. S. V. Burgen (487), Harvard Medical School, Boston, Massachusetts

Jacqueline Caillet (217), Service de Biochimie Théorique, Institut de Biologie Physico-chimique, Paris, France

G. Cilento (309), Departamento de Quimica, Faculdade de Filosofia, Ciências e Letras, Universidade de São Paulo, São Paulo, Brazil

Pierre Claverie (115, 245), Service de Biochimie Théorique, Institut de Biologie Physico-chimique, Paris, France

M. Cohn[3] (183), Service de Biochimie, Institut de Biologie Physico-chimique, Paris, France

A. Danchin (183), Service de Biochimie, Institut de Biologie Physico-chimique, Paris, France

[1] Present address: Biological Laboratories, Harvard University, Cambridge, Massachusetts.
[2] Present address: Institute of Biochemistry, Bulgarian Academy of Sciences, Sofia, Bulgaria.
[3] Permanent address: Department of Biophysics and Physical Biochemistry, School of Medicine, University of Pennsylvania, Philadelphia, Pennsylvania.

v

J. F. Danielli (529), Center for Theoretical Biology and Department of Biophysics, State University of New York at Buffalo, Buffalo, New York

Robert C. Davis (77), Chemistry Department and Chemical Biodynamics Laboratory, University of California, Berkeley, California

Michel Delmelle (299), Department of Atomic and Molecular Physics, University of Liège, Cointe-Sclessin, Belgium

L. Dimitrijevic (183), Service de Biochimie, Institut de Biologie Physico-chimique, Paris, France

Paul Doty (107), Department of Chemistry, Harvard University, Cambridge, Massachusetts

P. Douzou (447), Service de Biospectroscopie, Institut de Biologie Physico-chimique, Paris, France

Jules Duchesne (299), Department of Atomic and Molecular Physics, University of Liège, Cointe-Sclessin, Belgium

E. Peter Geiduschek (163), Department of Biophysics, University of Chicago, Chicago, Illinois

M. Gilbert (245), Service de Biochimie Théorique, Institut de Biologie Physico-chimique, Paris, France

M. Grunberg-Manago (183), Service de Biochimie, Institut de Biologie Physico-chimique, Paris, France

Richard Harrison (107), Department of Chemistry, Harvard University, Cambridge, Massachusetts

J. G. Heathcote (343), Department of Pure and Applied Physics and Department of Chemistry, The University of Salford, England

Karst Hoogsteen (21), Department of Biophysics and Pharmacology, Merck Sharp and Dohme Research Laboratories, Division of Merck and Company, Rahway, New Jersey

O. Jardetzky[4] (487), Harvard Medical School, Boston, Massachusetts

S. Richard Jaskunas (77), Chemistry Department and Chemical Biodynamics Laboratory, University of California, Berkeley, California

D. O. Jordan (221), Department of Physical and Inorganic Chemistry, University of Adelaide, Adelaide, South Australia

[4] Present address: Department of Biophysics and Pharmacology, Merck Sharp and Dohme Research Laboratories, Division of Merck and Company, Rahway, New Jersey.

Ephraim Katchalski (361), Department of Biophysics, The Weizmann Institute of Science, Rehovoth, Israel

H. Kersten (289), Physiologisch Chemisches Institut, Universität Münster, Münster, Germany

W. Kersten (289), Physiologisch Chemisches Institut, Universität Münster, Münster, Germany

L. S. Lerman (271), Department of Molecular Biology, Vanderbilt University, Nashville, Tennessee

S. Lifson (261), The Weizmann Institute of Science, Rehovoth, Israel

Per-Olov Löwdin (539), Department of Quantum Chemistry, Uppsala University, Uppsala, Sweden, and Quantum Theory Project, University of Florida, Gainesville, Florida

Donald B. McCormick (377), Section of Biochemistry and Molecular Biology and Graduate School of Nutrition, Cornell University, Ithaca, New York

Marie-José Mantione (411), Service de Biochimie Théorique, Institut de Biologie Physico-chimique, Paris, France

J. C. Metcalfe (487), Department of Pharmacology, University of Cambridge, Cambridge, England

A. M. Michelson (93), Service de Biochimie, Institut de Biologie Physico-chimique, Paris, France

H. A. Nash (137), Laboratory of Neurochemistry, National Institute of Mental Health, Bethesda, Maryland

Bernard Pullman (1, 217), Service de Biochimie Théorique, Institut de Biologie Physico-chimique et Université de Paris, Paris, France

K. Rosenheck (517), Polymer Department, The Weizmann Institute of Science, Rehovoth, Israel

Jean Salvinien (461), Department of Chemistry, The Faculty of Sciences, Université de Montpellier, Montpellier, France

Sidney Shifrin (323), National Cancer Institute, National Institutes of Health, Bethesda, Maryland

Meir Shinitzky (361), Department of Biophysics, The Weizmann Institute of Science, Rehovoth, Israel

Oktay Sinanoğlu (427), Sterling Chemistry Laboratory and Department of Molecular Biophysics, Yale University, New Haven, Connecticut

M. A. Slifkin (343), Department of Pure and Applied Physics, and Department of Chemistry, The University of Salford, England

M. N. Thang (183), Service de Biochimie, Institut de Biologie Physicochimique, Paris, France

Hugo Theorell (471), Department of Biochemistry, Nobel Medical Institute, Stockholm, Sweden

Ignacio Tinoco, Jr. (77), Chemistry Department and Chemical Biodynamics Laboratory, University of California, Berkeley, California

Gordon Tollin (393), Department of Chemistry, The University of Arizona, Tucson, Arizona

Paul O. P. Ts'o (39), Department of Radiological Sciences, The Johns Hopkins University, Baltimore, Maryland

Olke Uhlenbeck (107), Department of Chemistry, Harvard University, Cambridge, Massachusetts

Gregorio Weber (499), Department of Chemistry and Chemical Engineering, University of Illinois, Urbana, Illinois

David L. Wilson (163), Department of Biophysics, University of Chicago, Chicago, Illinois

K. Zinner (309), Departamento de Quimica, Faculdade de Filosofia, Ciências e Letras, Universidade de São Paulo, São Paulo, Brazil

Preface

This volume contains the proceedings of an International Symposium held at the Institut de Biologie Physico-chimique, Fondation Edmond de Rothschild, in Paris in May 1967, to celebrate the fortieth anniversary of its foundation.

Forty years have indeed elapsed since Baron Edmond de Rothschild, the celebrated philanthropist and Maecenas, inspired by the daring thinking of Jean Perrin, decided to found a research institute devoted to the study and elucidation of the physicochemical aspects of life. By doing so he established what probably was one of the first institutes of molecular biology in the world. It seemed particularly appropriate to commemorate this anniversary with a symposium on one of the major themes of molecular biology, which, at the same time, is a subject essentially related to physical chemistry.

It is my pleasant duty to thank all those whose efforts made this memorable meeting possible. Our deepest thanks are due to the present Baron Edmond de Rothschild, the grandson of the founder, not only for his generosity, which made this meeting possible—generosity is a continuous tradition in his family—but still more for his complete understanding of our goals and for the sharing of our preoccupation. Our thanks are also due to the Administrative Councils of our institute for their enthusiastic support of the idea of the symposium and, in particular, to Professors Francis Perrin and René Wurmser for their invaluable help in planning the details of the meeting. I would like to acknowledge the efficient handling of secretarial and other problems by Mrs. de Hauss and Mrs. Landez.

Finally, I would like to express our gratitude to all our distinguished guests, speakers, and discussants, especially those who came from distant places in order to share their knowledge with us. Their contributions made this meeting successful. We look forward to seeing them all again at the fiftieth anniversary of our Institute.

December, 1967   BERNARD PULLMAN

Contents

Some Aspects of RNA Transcription   **163**

E. Peter Geiduschek, Edward N. Brody, and David L. Wilson

Influence of the Structure of Transfer RNA on Its Interaction with Enzymes and Divalent Cations   **183**

M. Grunberg-Manago, M. Cohn, M. N. Thang, B. Beltchev,
A. Danchin and L. Dimitrijevic

The Interaction of Aromatic Hydrocarbons with Nucleic Acids and Their Constituents   **207**

E. D. Bergman

On the Solubilization of Aromatic Carcinogens by Purines and Pyrimidines   **217**

Jacqueline Caillet and Bernard Pullman

The Interaction of Heterocyclic Compounds with DNA   **221**

D. O. Jordan

Molecular Associations
in Biology

Associations Moléculaires en Biologie: Théorie et Expérience. Propos d'Introduction*

BERNARD PULLMAN

Institut de Biologie Physico-Chimique
Laboratoire de Biochimie Théorique
Paris, France

En choisissant comme sujet de notre colloque les associations moléculaires en biologie, nous avons le sentiment d'avoir effectivement placé cette réunion au centre des préoccupations de la biologie moléculaire d'aujourd'hui. En effet, maintenant que la structure de la majorité sinon de la totalité des biomolécules simples est en grande partie déterminée, que la structure primaire des biopolymères devient également de plus en plus accessible, l'intérêt des chercheurs s'oriente nettement vers la détermination de la configuration spatiale de ces macromolécules, avec naturellement l'ambition d'élucider la nature des forces responsables de leurs caractéristiques et en particulier de l'existence et de la *stabilité* des structures ordonnées observées. Parallèlement de nombreuses et importantes recherches sur des processus biologiques fondamentaux—processus tels que transmission de l'information génétique, codage, mutagénèse, carcinogénèse—mettent en évidence le rôle prépondérant des interactions entre différents types de ces biopolymères, ou entre ces polymères et des entités plus petites.

Or il s'avère que ces différents groupes de phénomènes, et il y en a aussi d'autres non moins importants qui apparaîtront dans ce volume, que je n'énumère pas ici, mettent en jeu, souvent, le type d'interactions que l'on peut désigner sous la dénomination générale d'*associations moléculaires*. Leur caractéristique principale est qu'elles ne comportent pas la formation de véritables liaisons chimiques, fixes et fortes, mais impliquent comme élément moteur des forces nettement plus lâches et faibles dites *forces intermoléculaires*. Il en résulte un aspect beaucoup plus dynamique, plus facilement modifiable sous l'effet des actions ou perturbations extérieures, de telles structures ou mécanismes.

Si l'étude des associations moléculaires en biologie est par excellence un problème de biologie physico-chimique et, en tant que tel est l'objet de recherches dans plusieurs services de notre Institut, c'est en plus pour le théoricien que je suis un sujet particulièrement tentant et cela pour plusieurs raisons:

* Ce travail a été exécuté dans le cadre de la Convention 67-00-532, de la Délégation Générale à la Recherche Scientifique et Technique, Comité de Biologie Moléculaire.

1. *Les forces intermoléculaires sont moins bien connues, moins bien précisées que ne le sont les forces chimiques associées avec les liaisons essentiellement ou fortement covalentes.* Elles comportent plusieurs composantes parmi lesquelles les plus souvent citées sont les liaisons hydrogènes, les forces de Van der Waals-London (elles-mêmes subdivisées en général en forces électro-statiques, forces d'induction ou de polarisation et forces de dispersion) et les forces de transfert de charges. Les rapports entre ces différentes compo-santes, les valeur de leurs contributions relatives dans des circonstances déterminées sont en général difficiles à établir. En fait très souvent de regret-tables confusions règnent même dans les esprits sur la signification de ces différentes formes d'interaction. Ainsi par exemple lorsque fut établie la structure en hélice double de l'acide désoxyribonucléique (ADN) la majorité des biologistes ont cru que la stabilité de cet édifice provenait essentiellement des liaisons hydrogènes entre les paires de bases complémentaires. Lorsque plus tard divers arguments, sur lesquels je n'insiste d'ailleurs pas ici, ont indiqué qu'une telle conception était défectueuse ou pour le moins insuffisante, beaucoup d'auteurs ont avancé que la stabilité des acides nucléiques était due surtout aux interactions Van der Waals-London entre les bases ou les paires de bases empilées. Encore aujourd'hui on voit souvent posée la question de savoir si la stabilité des acides nucléiques est due principalement aux liaisons hydrogéné entre les bases horizontales ou aux forces Van der Waals-London entre les bases superposées. Or posée de cette façon la question est surtout mal posée car elle laisse implicitement supposer qu'il existe une différence fondamentale entre les forces opérant entre les bases horizontales et celles opérant entre les bases verticales; elle laisse supposer en particulier que les forces de Van der Waals-London n'opèrent pas entre les bases horizontales ou qu'elles y sont négligeables. Or c'est là une conception erronée, comme cela a été explicitement montré sur l'exemple même des interactions entre les bases puriques et pyrimidiques par DeVoe et Tinoco en 1962 et amplement confirmé et précisé depuis par d'autres. Anticipant sur la démonstration que j'en donnerai plus loin je peux déjà annoncer que les calculs démontrent que ce sont les forces de Van der Waals-London s'exerçant à la fois entre les bases liées horizontalement et les bases empilées verticalement et cela dans des contributions sensiblement comparables qui peuvent être considérées comme étant responsables en grande partie de la stabilité de la double hélice.

Comme autre exemple de confusion j'ajouterai l'imprécision des idées règnant sur le rôle des complexes de transfert de charges dans la détermination de structures et réactions biochimiques (Pullman et Pullman, 1966). Depuis que Mulliken a développé la théorie quantique des complexes de transfert de charges et que Szent-Gyorgyi (1960) a envisagé leur rôle possible dans des phénomènes biochimiques, beaucoup de malentendus se sont propagés à leur sujet. Ainsi, nombreux sont ceux pour qui la manifestation de l'existence d'un complexe de transfert de charges par l'apparition d'une bande d'absorption nouvelle implique nécessairement une contribution appréciable des forces de transfert de charges à la stabilisation de l'état fondamental du complexe, et certains postulent même un parallélisme entre la position de cette bande et la stabilisation du complexe, conception qui du point de vue théorique est gratuite et de ce fait s'avère souvent erronée. Dr. Tollin a été l'un des premiers, je crois, tout au moins en biochimie, à le démontrer sur l'exemple des associations moléculaires entre les flavines et les phénols (Fleischman et Tollin, 1965; voir aussi Dewar et Thompson, 1966). D'autres erreurs dans ce domaine concernent la surestimation générale de la valeur des forces de transfert de charges; j'entends par là la surestimation de la stabilisation de l'état fondamental grâce au transfert fractionnaire d'électrons entre les constituants du complexe. Il n'est pas trop difficile de montrer que dans de tels complexes d'autres forces intermoléculaires, en particulier encore les forces de Van der Waals-London, peuvent jouer, en ce qui concerne leur stabilisation, un rôle nettement plus important que les forces de transfert de charges. Mme Mantione de notre laboratoire illustrera plus loin dans ce volume cet état de choses par des exemples précis. Ces difficultés sont naturellement centuplées lorsque le complexe de transfert de charges est, comme cela arrive souvent, soupçonné mais non démontré.

2. *L'évaluation des forces intermoléculaires suppose la connaissance de certaines caractéristiques physico-chimiques des biomolécules.* Ainsi les composantes électrostatiques et d'induction des forces de Van der Waals-London sont, en général, évaluées dans l'approximation dipôle-dipôle et dipôle-dipôle induit. Leur calcul nécessite donc la connaissance des moments dipolaires et des polarisabilités des molécules interagissantes. L'évaluation de forces de dispersion nécessite en plus la connaissance de leurs potentiels d'ionisation qui jouent aussi un rôle important dans les transferts de charges. Or, très souvent ces caractéristiques physico-chimiques sont inconnues à l'heure actuelle et, en outre, très difficiles à mesurer. Ainsi par exemple, si l'on connaît expérimentalement le moment dipolaire des dérivés simples de l'adénine et de la thymine, on ignore ceux de la guanine et de la cytosine sans parler de ceux d'autres bases puriques et pyrimidiques. On ignore complètement expérimentalement la direction de localisation de ces moments. De même on manque presque complètement de toute donnée expérimentale sur les potentiels d'ionisation de biomolécules. La théorie peut pallier ces déficiences, en évaluant, aujourd'hui avec une garantie d'exactitude raisonnable, ces types de quantités. En fait des calculs très perfectionnés et difficiles ont été effectués dans ce domaine dans notre laboratoire durant ces dernières années, en particulier par Berthod *et al.* (1966a, b, 1967; Denis et Pullman, 1967), qui nous fournissent d'abondantes informations sur ces grandeurs physico-chimiques inconnues. Le Tableau I montre à titre d'exemple de telles évaluations des moments dipolaires des purines et des pyrimidines.

TABLEAU I

MOMENTS DIPOLAIRES DE PURINES ET PYRIMIDINES

Direction de localisation	Moment théorique (en D)	Moment expérimental (en D)	Direction de localisation	Moment théorique (en D)	Moment expérimental (en D)
Purine	4.15	4.3 dans 9-méthyl-purine	Uracile	3.86	3.9 dans 1,3-diméthyl-uracile
Adénine	3.16	3.0 dans 9-méthyl-adénine	Thymine	3.58	
Guanine	6.76		Cytosine	7.10	

3. A ce type de considérations on peut ajouter les tentantes *perspectives de perfectionnement des approximations impliquées dans la représentation de différentes forces intermoléculaires*. Ainsi, par exemple, j'ai dit tout à l'heure que les forces de Van der Waals-London sont évaluées *en général* dans l'approximation dipolaire. Or, une telle approximation n'est ni justifiable ni appropriée, en fait, que lorsque les systèmes interagissants sont séparés par des distances relativement grandes, supérieures nettement à leurs dimensions propres. Ce n'est évidemment pas le cas par exemple pour les interactions entre les purines et pyrimidines des acides nucléiques. Dans de tels cas il convient d'abandonner cette approximation et d'utiliser à sa place, comme l'ont indiqué explicitement parmi les premiers, Bradley *et al.* (1964) et Hirschfelder (1965), l'approximation des monopôles dans laquelle les interactions électrostatiques s'exercent entre les charges nettes atomiques elles-mêmes. Or cette distribution de charges ne peut pour l'instant être atteinte par aucune méthode expérimentale. Le poids d'une telle détermination repose donc entièrement sur la théorie. De sorte que même si de telles évaluations sont nécessairement toujours approximatives, elles sont néanmoins d'une utilité essentielle.

Ainsi ce passage de dipôles aux monopôles dans le calcul des forces de Van der Waals-London a des conséquences hautement significatives. L'illustration la plus claire de cette situation me paraît être contenue dans l'exemple suivant que je tire d'un travail de Nash et Bradley (1966). Les auteurs ont recherché les minima d'énergie potentielle susceptibles d'apparaître lorsqu'on promène un uracile autour d'une adénine dans le plan de celle-ci. Or, lorsque de tels calculs sont effectués dans l'approximation des monopôles, des minima bien caractérisés apparaissent qui correspondent à des arrangements mutuels tels qu'ils existent dans les liaisons hydrogène. Ce résultat est dû à la grande contribution à l'énergie électrostatique des atomes rapprochés dans les configurations correspondantes aux liaisons hydrogène et peut être considéré comme confirmant explicitement le caractère essentiellement électrostatique de telles liaisons. Rien de pareil n'est visible dans l'approximation dipolaire. En réalité du fait de la *faible distance* des associations moléculaires intervenant en biologie par rapport aux dimensions des composés impliqués de nombreuses autres approximations classiques de procédés d'évaluations des forces intermoléculaires, adaptées en général aux interactions à travers des distances plus grandes, sont à réviser et à perfectionner. Nous travaillons beaucoup en ce moment, comme le font d'ailleurs aussi d'autres laboratoires, sur ces perfectionnements et M. Claverie indiquera dans sa contribution quelques uns de tels perfectionnements, actuellement en cours d'élaboration dans notre laboratoire.

Ainsi toutes ces différentes considérations me font croire que le domaine des associations moléculaires est un domaine de choix pour une étroite collaboration entre le théorie et l'expérience. Il ne reste qu'a montrer par un exemple qu'il peut en être effectivement ainsi. On n'a pas de peine d'ailleurs pour trouver un tel exemple. En effet, il suffit de considérer à ce point de vue, le sujet général des interactions et des associations entre les purines et pyrimidines qui sera amplement décrit dans ce volume. Ainsi, bien que ce soit naturellement leur signification pour la stabilité de la structure ordonnée des acides nucléiques qui est l'objet final des recherches dans ce domaine, celles-ci ont mis en évidence et posé toute une série de problèmes connexes. Parmi ceux-ci les plus frappants me paraissent associés aux observations suivantes:

1. La découverte par Hoogsteen (1959) que la cocristallisation de l'adénine et de la thymine, substituées toutes deux sur leurs azotes glycosidiques, conduit à une association par liaison hydrogène, qui n'est pas conforme au modèle de Watson-Crick: la thymine est liée à N_7 de l'adénine et non pas à N_1. Ce type de cristaux mixtes s'est montré assez général dans les associations entre les dérivés de l'adénine et de l'uracile. En revanche c'est la configuration Watson-Crick qui paraît la seule observée dans les cocrystallisations des dérivés de la guanine et de la cytosine (toujours substitués sur leurs azotes glycosidiques).

2. L'exclusivité des associations par liaisons hydrogène (que ce soit dans des cocrystallisations ou en solution dans des solvants non aqueux) entre les bases complémentaires dans le sens de Watson-Crick (A—T ou A—U et G—C), aucune association ne paraissant s'établir entre les bases non complémentaires en ce sens (A—G, C—T, A—C ou G—T). C'est un phénomène qui, à première vue, a une allure un peu magique, car *chimiquement* rien ne paraît s'opposer à ce que de telles associations non complémentaires se forment en dehors des acides nucléiques. Je précise que cette exclusivité concerne les bases présentes dans les acides nucléiques et portant dans ces expériences un substituant simple sur leur azote glycosidique.

Cette exclusivité dans les cocrystallisations ou les associations en solution ne préjuge pas naturellement de la possibilité d'établissement de tels couplages non complémentaires dans d'autres circonstances plus particulières, où ils pourraient être imposés par des facteurs extérieurs. En fait la possibilité de couplages non complémentaires a été envisagee par exemple dans la "wobble" hypothèse de Crick (1966) à propos des interactions codon–anticodon, ou comme pouvant intervenir dans la structure de certains acides ribonucléiques (ARN) (Warshaw et Tinoco, 1966; Traub et Elson, 1966).

3. Lorsque au lieu d'utiliser les bases substituées sur leur azote glycosidique on fait appel à des bases entièrement libres, *aucune* association ne paraît plus s'établir entre elles.

4. Des phénomènes analogues des associations préférentielles, à géométrie probablement choisie, se produisent aussi pour des interactions verticales (en sandwich) entre les purines et pyrimidines, telles qu'elles se manifestent

dans les associations entre ces bases en solution dans l'eau, si abondamment et magistralement étudiées par Ts'o et ses collaborateurs et par Jardetzky et dans les interactions entre les bases dans les dinucléotides, oligonucléotides, et polynucléotides, étudiées par toute une phalange de chercheurs representée dignement dans ce volume par MM. Tinoco, Doty, Michelson et Brahms, et qui peuvent se résumer dans cette question essentielle que doivent se poser les nucléotides: *To stack or not to stack*?

Ce vaste ensemble d'observations est-il susceptible d'une interprétation homogène? Les calculs quantiques paraissent fournir une réponse positive à cette question. Ainsi, le Tableau II résume tout un ensemble des résultats de

TABLEAU II

ENÉRGIE D'INTERACTION (KCAL/MOLE) DANS LES PAIRES DE BASES LIEES PAR LIAISONS HYDROGENE

A—A T—T	−5.8 −5.2	A—T	−7·0
G—G C—C	−14.5 −13	G—C	−19·2
A—A C—C	−5.8 −13	A—C	−7·8
G—G T—T	−14.5 −5.2	G—T	−7·4
C—C T—T	−13 −5.2	C—T	−6·5
A—A G—G	−5.8 −14.5	A—G	−7·5

$$A—T > \frac{A—A}{T—T} \qquad \begin{array}{l} A—C < C—C \\ G—T < G—G \end{array}$$

$$G—C > \frac{G—G}{C—C} \qquad \begin{array}{l} C—T < C—C \\ A—G < G—G \end{array}$$

calculs effectués dans notre laboratoire (Pullman *et al.* 1966a, b, c) sur les interactions horizontales entre les purines et pyrimidines nucléiques correspondant à la formation des associations par liaisons hydrogènes. Les résultats figurant sur le Tableau II correspondent aux calculs effectués dans l'approximation des monopôles, le nombre figurant sur le tableau représentant l'énergie totale de l'interaction, somme des énergies électrostatiques, d'induction et de dispersion. Ces nombres correspondent dans chaque cas à l'association la plus forte obtenue dans l'hypothèse où les azotes glycosidiques ne sont pas disponibles pour l'association. Comme nous le verrons dans le cliché suivant

TABLEAU III
Energie d'interaction[a](kcal/mole) dans les Différentes Configurations Adénine-Thymine et Guanine-Cytosine

Configuration	$E_{\rho\rho}$	$E_{\rho\alpha}$	E_L	E_{totale}
	−4.61	−0.27	−0.77	−5.65
	−5.85	−0.22	−0.98	−7.05
	−5.64	−0.18	−1.03	−6.85
	−15.91	−2.02	−1.25	−19.18

TABLEAU III (suite)

Configuration	$E_{\rho\rho}$	$E_{\rho\alpha}$	E_L	E_{totale}
	-3.98	-1.33	-0.44	-5.75

[a] $E_{\rho\rho}$ = energie monopôle-monôpole; $E_{\rho\alpha}$ = energie monôpole-dipôle induit; E_L = energie de dispersion.

ce sera, par exemple, la configuration de Hoogsteen pour la pair A—T et celle de Watson-Crick pour la paire G—C. Comme vous pouvez le remarquer, nos calculs couvrent à la fois les associations existantes et celles qui n'existent pas. C'est évidemment un avantage de théoricien sur l'expérimentateur que de pouvoir étudier ce qui n'existe pas.

Remarquons tout d'abord, et cela sera le point de départ de la réponse à certains problèmes qui nous intéressent, qu'en ce qui concerne les autoassociations de bases on peut diviser celles-ci en deux groupes: d'une part les autoassociations G—G et C—C correspondant aux énergies d'interactions élevées et d'autre part A—A et T—T auxquelles correspondent des énergies d'interaction nettement plus faibles.

Si nous regardons maintenant les différentes associations mixtes que l'on peut construire à partir de ces mêmes bases, on constate que celles-ci aussi se divisent en deux groupes: d'une part les paires complémentaires A—T et G—C dont les énergies d'interaction sont supérieures aux énergies d'auto-association de leur deux constituants (ou a la moyenne de ces énergies) et, d'autre part, toutes les paires non complémentaires pour lesquelles les énergies d'interaction seraient en elles-mêmes appréciables mais toujours inférieures à l'une des énergies d'autoassociation de l'un de leur constituant (G ou C). Cette situation suggère par elle-même l'explication de l'exclusivité "magique" des associations complémentaires A—T et G—C, seules suffisamment stables par rapport aux autoassociations pour pouvoir se former à leur dépens.

Le Tableau III indique avec un peu plus de détails les résultats de calculs pour les différents modes de couplage possibles dans les associations A—T et G—C. Soulignons la prépondérance du couplage Hoogsteen pour la paire A—T et du couplage Watson-Crick pour la paire G—C. Remarquons également que la partie essentielle de l'énergie d'interaction provient dans ces

TABLEAU IV
RESULTATS EXPERIMENTAUX SUR LES INTERACTIONS PAR LIAISONS HYDROGÈNE
ENTRE LES BASES DES ACIDES NUCLÉIQUES

Résultat	Solvent	Méthode	Références
A—U > A—A ou U—U	$CDCl_3$	IR	Hamlin et al. (1965)
	$CDCl_3$	IR	Kyogoku et al. (1967)
	$CDCl_3$	IR	Miller et Sobell (1967)
G—C > G—G ou C—C	$CDCl_3$	IR	Katz et Penman (1966); Kyogoku et al. (1966)
G—C > A—T ou A—U	$DMSO + CHCl_3$	NMR	Katz et Penman (1966);
	$CHCl_3$	IR	Bitha et al. (1966)
	DMSO	NMR·	Shoup et al. (1966).
G—G > C—C	$CHCl_3$	IR	Kyogoku et al. (1966)
U—U > A—A	$CHCl_3$	IR	Kyogoku et al. (1967)
Aucune interaction entre les bases non complementaires	Pas de cocrystallisation		Haschemeyer et Sobell (1964); Shoup et al. (1966); Kyogoku et al. (1966);
Aucune interaction entre les bases non substituées sur l'azote glycosidique			Miller et Sobell (1966); Sobell (communications privée)

couplages de la composante électrostatique. Signalons aussi que, comme l'indique le Tableau IV, presque toutes les prédictions de détails contenues dans les calculs se trouvent vérifiées actuellement par l'expérience. Le seul désaccord avec celle-ci concerne la stabilité relative des autoassociations U—U et A—A. Signalons aussi que ces calculs ont permis de rendre compte de la structure cristalline de différentes purines et pyrimidines (Pullman et al., 1966a).

Jusqu'ici nous avons considéré, en accord avec la réalité expérimentale la plus courante, des interactions entre les bases substituées sur les azotes glycosidiques. La question peut être posée de savoir ce qui se passerait si l'on mettait en présence des bases entièrement libres. Dans ces cas il convient d'envisager un type complémentaire d'autoassociation et d'association mixte de bases mettant en jeu le proton attaché à l'azote glycosidique. Les résultats de calculs obtenus dans cette nouvelle hypothèse ou plutôt les modifications que cette nouvelle hypothèse entraîne pour les résultats antérieurs sont illustrés sur le Tableau V. La modification essentielle concerne l'énergie (maximum) d'autoassociation de l'adénine qui dans ce nouveau mode d'interaction est supérieure à toutes celles prévues pour les associations mixtes possibles entre

TABLEAU V
ENERGIES D'INTERACTIONS (KCAL/MOLE) ENTRE LES BASES NON SUBSTITUÉES

A—A	−~~5.8~~ 8.13	A—T	−7·0
T—T	−5.2		
G—G	−14.5	G—C	−19·2
C—C	−13		
A—A	−~~5.8~~ 8.13	A—C	−7·8
C—C	−13		
G—G	−14.5	G—T	−7·4
T—T	−5.2		
C—C	−13	C—T	−6·5
T—T	−5.2		
A—A	−~~5.8~~ 8.13	A—G	−7·5
G—G	−14.5		

$$A—T \not< \begin{matrix} < & A—A \\ > & T—T \end{matrix} \qquad \begin{matrix} A—C < C—C \\ G—T < G—G \end{matrix}$$

$$G—C > \begin{matrix} G—G \\ C—C \end{matrix} \qquad \begin{matrix} C—T < C—C \\ A—G < G—G \end{matrix}$$

l'adénine et la thymine. Par conséquent, en accord avec la règle précédente et à la différence de ce qui se produit pour les dérivés substitués, l'adénine et la thymine libres ne devraient pas s'associer. En revanche, le nouveau mode d'autoassociation de la guanine et de la cytosine correspond toujours à une énergie d'interaction inférieure à celle des associations mixtes G—C et ces deux bases libres devraient pouvoir s'associer. L'expérience indique que jusqu'ici aucune association entre les bases nucléiques libres n'a pu être mise en évidence. Toutefois, l'expérimentation avec la guanine et la cytosine ne saurait être considérée comme décisive du fait de l'insolubilité presque totale de la guanine dans les solvants utilisés.

Avant de quitter ce domaine d'associations par liaisons hydrogène, signalons que ce type de calculs a été étendu récemment à l'étude des *trimères* de bases, comme il en apparaît par l'exemple dans poly I ou poly (A + 2U) (Pullman *et al.*, 1967), aux associations erronées (*miscouplings*) impliquant les bases dans les formes tautomères rares (Pullman et Caillet, 1967a), et aux associations entre les analogues de bases nucléiques (Pullman et Caillet, 1967b). En relation avec la discussion précédente il peut être utile de dire quelques mots à propos des résultats concernant la configuration du trimère A + 2U. Deux configurations, I et II du Tableau VI, différentes entre elles par l'orientation du noyau d'uracile lié à N_7 de l'adénine ont été envisagées. Une étude minutieuse du déplacement des fréquences de vibration

TABLEAU VI

ENERGIE D'INTERACTION (KCAL/MOLE) DANS DEUX MODÈLES DU TRIPLET A + 2U

Modèle	Interaction	$E_{\rho\rho}$	$E_{\rho\alpha}$	E_L	E_M
I. Poly (A + 2U)	A—U₁	-4.64	-0.25	-0.69	-5.58
	A—U₂	-5.63	-0.17	-0.94	-6.74
	U₁—U₂	0.57	-0.01	-0.03	0.53
	Totale	-9.70	-0.43	-1.66	-11.79
II. Poly (A + 2U)	A—U₁	-4.64	-0.25	-0.69	-5.58
	A—U₂	-5.86	-0.22	-0.88	-6.96
	U₁—U₂	0.96	-0.02	-0.03	0.91
	Totale	-9.54	-0.49	-1.60	-11.63

infrarouges lors de l'établissement de la triple hélice ont permis de démontrer (Miles, 1964) que la configuration du trimère dans poly (A + 2U) est celle représentée par I. La liaison entre l'adénine et l'uracile lié à son N_7 *n'y est donc pas* celle observée dans le cristal mixte adénine-thymine, mais représente encore une autre possibilité de jonction (présente par ailleurs dans le cristal mixte adénine-6-bromouracil). Les calculs fournissent une interprétation possible de ce phénomène. Ainsi, la décomposition de l'interaction totale en composantes correspondantes aux interactions partielles entre les différentes bases du trimère (Tableau VI) montre que l'interaction entre les deux uraciles non liés introduit une répulsion. Or c'est cet élément de répulsion, bien que numériquement relativement faible, qui paraît en fait responsable de l'orientation de l'uracile lié à N_7 de l'adénine. En effet on constate qu' en l'absence de cette répulsion la configuration II serait plus stable que la configuration I.

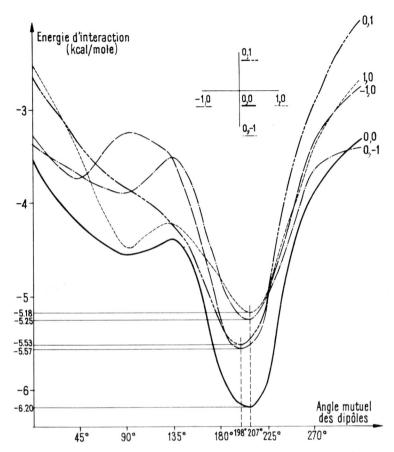

FIG. 1. Interaction adénine-uracile en stacking (sans retournement).

Toutefois la valeur de la répulsion entre les deux uraciles est plus forte dans II que dans I et produit le renversement des stabilités totales en faveur de I. Bien sûr le fait que nous ayons affaire en réalité à un trimère dans un polynucléotide en solution ne nous permet pas d'affirmer que nous tenons l'explication complète du phénomène. Mais il paraît probable que le facteur considéré ici joue un rôle significatif.

Des calculs analogues peuvent également être effectués et l'on été (Claverie et al., 1966; Nash et Bradley, 1965) pour les empilements de bases et cela, naturellement, que ce soit entre les bases libres en solution ou des bases liées comme c'est le cas pour les di-, oligo- ou polynucléotides. Les forces prises en considération sont les mêmes que précédemment, les calculs indiquant toutefois que dans ce type d'interaction ce n'est plus la composante électrostatique mais plutôt la composante de dispersion qui risque de prédominer,

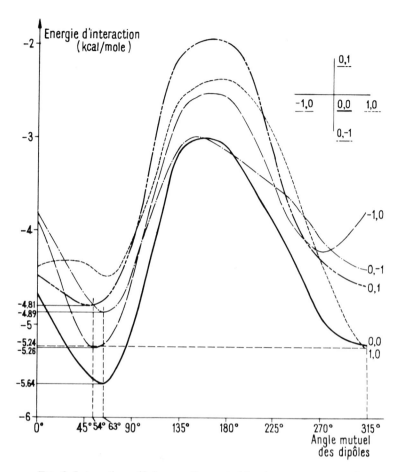

FIG. 2. Interaction adénine-uracile en stacking (avec retournement).

bien que dans l'ensemble la contribution des trois composantes soit plus équilibrée. D'une façon générale, ces calculs des énergies d'empilement sont plus difficiles à mener à bien que les calculs des associations par liaisons hydrogène car ne correspondant pas aux géométries connues d'avance, ils impliquent la recherche des positions du maximum ou des maxima d'interaction. Or cela peut être un processus long et fastidieux. Les Figs. 1 et 2 indiquent à titre d'exemple, les résultats d'une partie des calculs qu'il est nécessaire d'effectuer en vue d'établir les énergies d'interaction verticale entre une adénine et un uracile et la Fig. 3 montre la configuration apparemment la plus stable. D'actifs travaux sont poursuivis dans notre laboratoire dans ce domaine, en particulier par Mme Caillet, à qui nous devons d'ailleurs les courbes reproduites ici et je signale que ce type de recherches vient de progresser sensiblement tout récemment par l'introduction dans le calcul des forces de *répulsion* à courte distance ce qui rend possible *la détermination de la distance d'équilibre* entre les bases empilées.

Les Figs. 1–3 nécessitent quelques mots d'explication. Pour la commodité des calculs sur ces interactions «en stacking» les axes des molécules sont fixés par rapport à leur moment dipolaire. L'origine de ces axes se trouve au centre du moment. L'axe $0y$ est orienté du pôle négatif vers le pôle positif de ce moment, l'axe $0x$ lui est perpendiculaire. Les axes $0x$ et $0y$ ne sont représentés sur nos figures que pour l'adénine, par rapport à laquelle on envisage les déplacements de l'uracile.

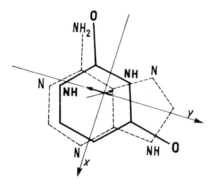

FIG. 3. Position du maximum d'interaction dans le stacking entre l'adénine et l'uracile.

Les différentes positions relatives des molécules sont obtenues par deux mouvements successifs:

1. En partant d'une position initiale correspondant au parallélisme des moments, notée $(0, 0)$, on effectue des translations de l'uracile de 1 Å suivant les axes de l'adénine, translations notées (voir Fig. 4): $(1, 0)$, $(0, -1)$, $(-1, 0)$ et $(0, -1)$. On noterait de même $(2, 0)$ etc. des translations de 2 Å etc. On peut considérer aussi des translations de 1 Å suivant deux axes simultanément, c'est-à-dire de 1.414 Å suivant les bissectrices que l'on note $(+1, -1)$ etc., le déplacement selon l'axe $0x$ étant toujours indiqué le premier. Toutes ces translations sont suivies d'une translation verticale de l'uracile, prise dans notre cacul égale à 4 Å.

2. On effectue ensuite des rotations de 45 degrés en 45 degrés autour d'un axe vertical passant par le milieu du moment dipolaire de l'uracile.

De plus on envisage les même translations et rotations dans la configuration dite «avec retournement» (Fig. 2) ou l'on prend comme position initiale l'antiparallélisme des moments dipolaires obtenu par une rotation de l'uracile de 180 degrés autour de l'axe des $0x$.

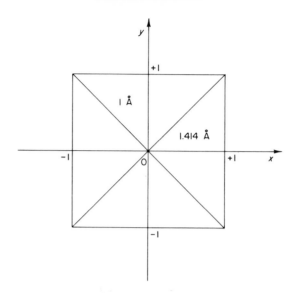

FIG. 4. Les axes pour l'étude des translations dans le stacking.

Les courbes d'énergie d'interaction des Figs. 1 et 2 sont tracées par chaque translation horizontale fonction des angles relatifs des moments dipolaires. Elles le sont pour les deux arrangements, sans (Fig. 1) et avec (Fig. 2) retournement, pour le cas sans translation et les cas avec une translation de 1 Å suivant les axes. Les énergies correspondant aux translations suivant les bissectrices, non représentées ici, sont inférieures aux précédentes.

La position de maximum d'interaction représentée dans la Fig. 3 correspond au minimum de la courbe (0, 0) de la Fig. 1. Elle est très proche d'une configuration correspondant à l'antiparallélisme des moments (rotation de 207 degrés au lieu de 180 degrés) dans l'arrangement sans retournement.

Je n'ai malheureusement pas le temps de m'attarder ici sur les résultats obtenus dans différents cas particuliers d'autant plus qu'ils nécessitent parfois une discussion plus poussée, mais je peux signaler que ces résultats permettent de rendre compte de certaines observations importantes dans ce domaine comme par exemple de la tendance plus grande à l'empilement dans les dinucléotides et polynucléotides de la guanine, de l'adénine et de la cytosine par rapport à l'uracile, tout en indiquant, en accord en particulier avec les travaux de Michelson (Michelson et Monny, 1966) un gain d'énergie non négligeable associé avec le stacking même dans ce dernier cas.

En revanche, je voudrais indiquer encore les résultats auxquels ce genre de calcul conduit dans le cas de couples de paires de bases telles qu'elles sont présentes dans l'ADN lui-même, dans lequel s'ajoutent les deux types d'interaction, par liaison hydrogène et par empilement. Ces résultats sont visibles sur le Tableau VII qui ne fait d'ailleurs que traduire, dans l'approximation monopolaire, les résultats obtenus antérieurement par DeVoe et Tinoco dans l'approximation dipolaire. Les conclusions auxquelles l'examen de ce tableau conduit sont évidentes.

Il apparaît ainsi nettement que les interactions horizontales et les interactions verticales contribuent toutes deux et cela d'une façon assez comparable à la stabilité de la structure à double hélice. On constate que les différentes combinaisons de paires de base se divisent, au point de vue de l'énergie totale d'interaction, en trois groupes: les combinaisons les plus stables

TABLEAU VII

ENERGIE D'INTERACTION (KCAL/MOLE) ENTRE PAIRES DE BASES VOISINES
DANS DNA DANS LE VIDE

Paires adjacentes[a]	Interactions verticales			Energie totale d'empile- ment	Contribution moyenne des interactions horizontales	Energie totale d'inter- action
	$E_{\rho\rho}$	$E_{\rho\alpha}$	E_L			
↑C—G| G—C↓	+0.9	−2.0	−10.2	−11.3	−19.2	−30.5
↑G—C| C—G↓	−1.6	−2.5	−4.0	−8.5	−19.2	−27.7
↑G—C| G—C↓	+2.6	−2.0	−8.3	−7.7	−19.2	−26.9
↑A—T| G—C↓	+1.2	−0.8	−10.3	−9.9	−12.2	−22.1
↑A—T| C—G↓	−0.6	−1.7	−4.9	−7.2	−12.2	−19.4
↑T—A| C—G↓	−0.1	−1.7	−5.2	−7.0	−12.2	−19.2
↑T—A| G—C↓	+1.8	−1.0	−7.8	−7.0	−12.2	−19.2
↑A—T| A—T↓	+0.5	−0.5	−7.4	−7.4	−5.5	−12.9
↑A—T| T—A↓	+0.4	−0.3	−6.2	−6.1	−5.5	−11.6
↑T—A| A—T↓	+1.5	−0.7	−5.8	−5.0	−5.5	−10.5

[a] Les flèches qui désignent la direction de la chaîne sont dirigées du carbone 3′ sur un sucre vers le carbone 5′ sur le sucre adjacent. Exemple: T̲A̲⃗ représente: T—sucre-3′-phosphate-5′-sucre—A.

s'établissent entre deux paires G—C, les moins efficaces ont lieu entre deux paires A—T, et les différentes combinaisons $\begin{pmatrix} G—C \\ A—T \end{pmatrix}$ correspondent à une stabilité intermédiaire. On peut remarquer que si ces résultats, qui de toute évidence peuvent être rapprochés de l'accroissement de la stabilité thermique des acides nucléiques en fonction de leur contenu en G—C, mise en évidence par Marmur et Doty (1959), correspondent également à l'ordre d'interactions horizontales, ils ne correspondent pas au seul ordre des interactions verticales.

Cet ensemble de résultats me paraît démontrer clairement le rôle prépondérant des forces de Van der Waals-London dans la détermination de la stabilité des acides nucléiques et de leurs analogues. Ils indiquent que ces *mêmes* forces s'exercent *d'une façon comparable* entre les bases associées horizontalement et celles associées verticalement. Ils démontrent donc le caractère irrationel des querelles sur la prépondérance des liaisons hydrogène ou des forces de stacking. Ce sont les mêmes forces mais qui, n'ayant pas le caractère directionel exclusif des liaisons chimiques, s'exercent plus librement dans des directions multiples. Compte tenu de la remarque que j'ai faite précédemment sur la contribution importante des forces de Van der Waals-London aussi à la stabilité des complexes dits de «transfert de charges,» de leur rôle essentiel (qui sera précisé dans ce volume par M. Gilbert de notre laboratoire) dans l'intercalement éventuel des hétérocycles tels que les aminoacridines dans l'ADN, de leur rôle essentiel que nous avons, indiqué ailleurs (Pullman *et al.*, 1965; voir aussi Caillet et Pullman, ce volume) dans les interactions *physiques* entre les bases puriques et les hydrocarbures aromatiques telles qu'elles se manifestent dans la solubilisation de ceux-ci par celles-ci, il apparaît que c'est donc ce type de forces qui doit jouer un rôle prédominant dans l'établissement des associations moléculaires en biologie.

Avec une restriction importante toutefois. J'ai parlé souvent des résultats des expériences en solution; tous les calculs que je vous ai présentés correspondent toutefois en principe aux phénomènes étudiés dans le vide. Il se fait que pour les phénomènes que j'ai signalés ici, des réponses tout au moins qualitativement satisfaisantes sont obtenues même dans cette approximation. Il n'est évidemment pas du tout certain qu'il en sera toujours ainsi et de toute façon le rôle des *solvants* et en particulier de *l'eau* sur les phénomènes évoqués est appréciable et son introduction dans le calcul risque de modifier sensiblement les valeurs numériques des énergies d'interactions précédemment citées. Si je n'ai pas parlé de cet important effet, c'est parce que son rôle précis est jusqu'ici mal défini, et aussi parce que ce problème sera évoqué avec plus de détails par le Professeur Sinanoğlu qui est l'un de ceux qui ont le plus contribué au développement de cet aspect du problème et qui aura certainement beaucoup plus à en dire à ce sujet que je ne saurais le faire.

BIBLIOGRAPHIE

Berthod, H., Giessner-Prettre, C., et Pullman, A. 1966a. *Theoret. Chim. Acta* **5**, 53.

Berthod, H., Giessner-Prettre, C., et Pullman, A. 1966b. *Compt. Rend.* **262**, 2657.

Berthod, H., Giessner-Prettre, C., et Pullman, A. 1967. *Intern. J. Quantum Chem.* **1**, 123.

Bradley, D. F., Lifson, S., et Honig, B. 1964. Dans "Electronic Aspects of Biochemistry" (B. Pullman, ed.), p. 77. Academic Press, New York.

Claverie, P., Pullman, B., et Caillet, J. 1966. *J. Theoret. Biol.* **12**, 419.

Crick, F. H. C. 1966. *J. Mol. Biol.* **19**, 548.

Denis, A., et Pullman, A. 1967. *Theoret. Chim. Acta* **7**, 110.

DeVoe, H., et Tinoco, I., Jr. 1962. *J. Mol. Biol.* **4**, 500.

Dewar, M. J. S., et Thompson, C. C., Jr. 1966. *Tetrahedron Suppl.* **7**, 97.

Fleischman, D. E., et Tollin, G. 1965. *Proc. Natl. Acad. Sci. U.S.A.* **53**, 38.

Hamlin, R. M., Lord, R. C., et Rich, A. 1965. *Science* **148**, 1734.

Haschemeyer, A. E. V., et Sobell, H. M. 1964. *Nature* **202**, 969.

Hirschfelder, J. O., 1965. Dans "Molecular Biophysics" (B. Pullman et M. Weissbluth, eds.), p. 325. Academic Press, New York.

Hoogsteen, K. 1959. *Acta Cryst.* **12**, 822.

Katz, Z., et Penman, S., 1966. *J. Mol. Biol.* **15**, 220.

Kyogoku, Y., Lord, R. C., et Rich, A. 1966. *Science* **154**, 518.

Kyogoku, Y., Lord, R. C., et Rich, A. 1967. *J. Am. Chem. Soc.* **89**, 497.

Marmur, J., et Doty, P. 1959. *Nature* **183**, 1427.

Michelson, A. M., et Monny, C. 1966. *Proc. Natl. Acad. Sci. U.S.A.* **56**, 1528.

Miles, H. T. 1964. *Proc. Natl. Acad. Sci.* **51**. 1105.

Miller, J. H., et Sobell, H. M. 1966. *Proc. Natl. Acad. Sci. U.S.A.* **55**, 1201; Sobell, H. M., communication privée.

Miller, J. H., et Sobell, H. M. 1967. *J. Mol. Biol.* **24**, 345.

Nash, H. A., et Bradley, D. F. 1965. *Biopolymers* **3**, 261.

Nash, H. A., et Bradley, D. F. 1966. *J. Chem. Phys.* **45**, 1380.

Pitha, J., Norman Jones, R., et Pithova, P. 1966. *Canad. J. Chem.* **44**, 1045.

Pullman, A., et Pullman, B. 1966. Dans "Quantum Theory of Atoms, Molecules and the Solid State" (P.O. Löwdin, ed.), p. 345. Academic Press, New York.

Pullman, B., et Caillet, J. 1967a. *Compt. Rend.* **264**, 1900.

Pullman, B., et Caillet, J. 1967b. *Theoret Chim. Acta* **8**, 223.

Pullman, B., Claverie, P., et Caillet, J. 1965. *Science* **147**, 1305.

Pullman, B., Claverie, P., et Caillet, J. 1966a. *Proc. Natl. Acad. Sci. U.S.A.* **55**, 905.

Pullman, B., Claverie, P., et Caillet, J. 1966b. *J. Mol. Biol.* **22**, 373.

Pullman, B., Claverie, P., et Caillet, J. 1966c. *Compt. Rend.* **263**, 2006.

Pullman, B., Claverie, P., et Caillet, J. 1967. *Proc. Natl. Acad. Sci. U.S.A.* **57** 1663.

Shoup, R. R., Miles, H. T., et Becker, E. D. 1966. *Biochem. Biophys. Res. Communs.* **23**, 194.

Szent-Gyorgyi, A. 1960. "Introduction to a Submolecular Biology." Academic Press, New York.

Traub, W., et Elson, D. 1966. *Science* **153**, 3732.

Warshaw, M. M., et Tinoco, I., Jr. 1966. *J. Mol. Biol.* **20**, 29.

Hydrogen Bonding between Purines and Pyrimidines

KARST HOOGSTEEN

Department of Biophysics and Pharmacology
Merck Sharp and Dohme Research Laboratories
Division of Merck and Co., Rahway, New Jersey

I. Introduction

The specific association of the purine and pyrimidine bases in the nucleic acids is a property which provides the basis for the storage, transmission, and expression of genetic information. This fundamental rule, first proposed by Watson and Crick (1953), is explained by the specific complementarity of the hydrogen-bonding geometry of adenine with thymine or uracil and of guanine with cytosine. In this manner, the base sequence of DNA is replicated and the genetic information, expressed as this base sequence, is transferred to the process of protein biosynthesis via messenger RNA.

Donohue (1956) and Donohue and Trueblood (1960) have shown that with the geometrical requirements for the formation of single hydrogen bonds between nitrogen and oxygen atoms, i.e., approximate colinearity of the three atoms A—H $\cdots$ B involved and interatomic distances between A and B ranging from about 2.80 to 3.00 Å as the only restrictions, 29 base pairs connected by two or three hydrogen bonds could be formed between the four nucleosides present in the nucleic acids. The reason for the specificity was, therefore, initially thought to reside largely in the stereochemical constraints imposed on the purines and the pyrimidines because of the double helical structure of DNA.

After it became known that N-methylated derivatives of these purines and pyrimidines could be cocrystallized from solution to form coplanar hydrogen-bonded base pairs, a large amount of information, experimental and theoretical, became available pointing out that the physical basis of specific complementarity resides in the structure and the cooperation of all intermolecular forces, i.e., the Coulombic, van der Waals, and hydrogen-bonding forces, between the constituent bases of the nucleic acids.

In this article a review will be given of some of the experimental results of the structural studies of coplanar base pairs in the crystalline state, together with some spectroscopic data about their interactions in solutions. This survey will be largely restricted to those purine and pyrimidine derivatives for which the configuration of the hydrogen donors and acceptors resemble those of the four bases adenine, thymine, cytosine, and guanine.

II. Base Pairs in the Crystalline State

The information concerning the structure of base pairs in the crystalline state has been obtained by means of single-crystal X-ray structure determinations. For each of the structures mentioned a large number of diffraction intensities, ranging from 1000 to 2000, have been measured. Each of these intensities $I(hkl)$, identified by three indices, is a function of four to nine parameters—three positional parameters together with one to six parameters describing the thermal vibrations of the atoms in the crystal for each of the atoms in the "asymmetric unit" of the crystallographic unit cell.

The structures under consideration here contain from twenty to forty atoms, and the unambiguous and accurate determination of all these parameters necessitates the large number of observations. The quality of a proposed structure is usually assessed with the help of a disagreement index R. The structure factor, F, equals the square root of the observed diffraction intensity,

$$R = \frac{\sum \left| |F_c| - |F_o| \right|}{\sum |F_o|}$$

corrected for some experimental constants; F_c is the value of the structure factor as calculated with the parameters of the structure, while F_o is the observed value.

After a trial structure has been proposed, proof for the correctness of the structure and an evaluation of the accuracy of the atomic parameters is usually obtained by means of least-squares refinement until the differences between the observed and the calculated values of the structure factors are close to the observed errors.

It is important to realize that, although initially chemical information is being used to determine the structure, the subsequent refinement and proof of correctness is largely based on the randomness of observed errors and, with a few exceptions, independent of chemical constraints. For the structures considered here the R factor usually is about 0.10 which can be considered a reasonable agreement.

A. Adenine-Thymine Base Pairs

The first evidence that combinations of purine and pyrimidine components of nucleic acids could be crystallized from solution to form coplanar hydrogen-bonded base pairs in the crystalline state was found with 1-methylthymine and 9-methyladenine (Hoogsteen, 1959, 1963). The crystals, grown from an aqueous solution containing equimolecular quantities of these two compounds, are monoclinic with the apparent space group P_{2_1}/m. The structure

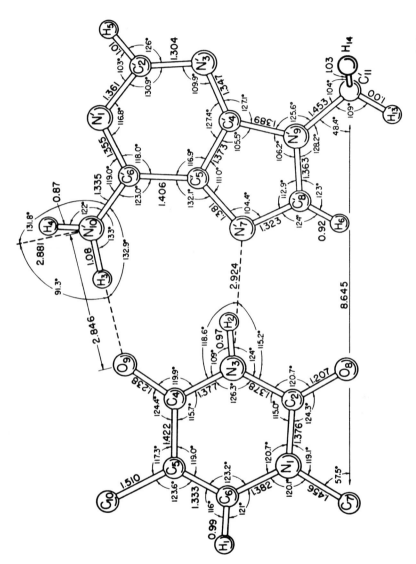

Fig. 1. Molecular dimensions of the 1-methylthymine-9-methyladenine base pair (Hoogsteen, 1963).

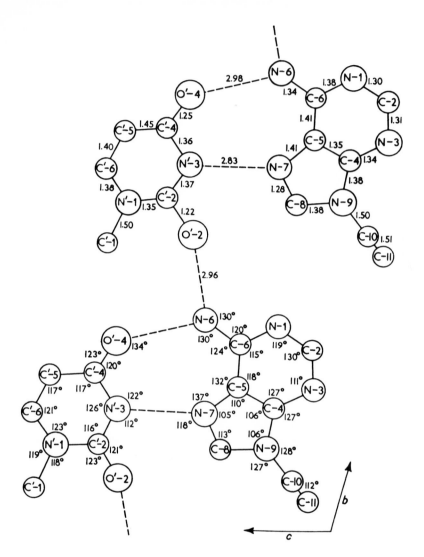

Fig. 2. Bond distances and angles for *N*-ethyladenine and *N*-methyluracil (Mathews and Rich, 1964).

was refined to $R = 0.08$ and the calculated standard deviations in the bond lengths and bond angles are 0.005 Å and 0.2°. The configuration of the base pair together with the molecular dimensions is shown in Fig. 1. The amino nitrogen of 9-methyladenine forms a hydrogen bond with the oxygen atom at C-4 of 1-methylthymine (2.85 Å), while N-7 of 9-methyladenine is hydrogen-bonded to N-3 of 1-methylthymine (2.92 Å). Since the complex is located in a mirror plane, deviations of the atoms from the plane of the base pair were not observed. The molecular dimensions of the two molecules in the base pair do not deviate largely from the dimensions as found in the crystal structures of the separate compounds. For 9-methyladenine the largest discrepancies with the dimensions found in the crystal structure of this

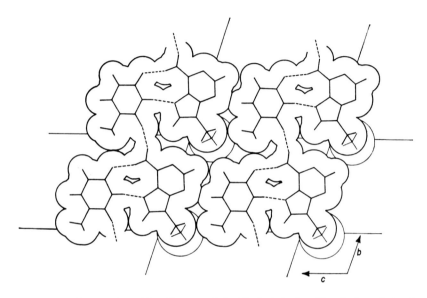

Fig. 3. A van der Waals packing diagram of the N-ethyladenine-N-methyluracil structure (Mathews and Rich, 1964).

compound (Stewart and Jensen, 1964) are: 0.030 Å and 4.4°. A notable feature in this structure is the small deviation from linearity of the atoms involved in the base-pair hydrogen bonds. The positions of the hydrogen atoms, indicated in Fig. 1, were obtained experimentally by means of a Fourier synthesis. The largest deviation from colinearity is 9° for the N—H $\cdots$ N hydrogen bond which is within the limits of 15°, allowed on the basis of a study of published data by Donohue (1956).

An intermolecular, coplanar complex between 9-ethyladenine and 1-methyluracil was isolated by Mathews and Rich (1964). Crystals, triclinic with the space group $P_{\bar{1}}$, could be obtained from a solution of equimolecular quantities of these two compounds in either dimethyl sulfoxide or tetrahydrothiophene sulfone. The structure was refined with 2700 reflections to $R = 0.14$. The errors in bond lengths and bond angles were estimated at 0.008 Å and 0.6°. Figure 2 shows the geometry of the base pair. It is interesting that again the hydrogen bonding of the base pair does not conform to the Watson-Crick configuration, but that the imidazole nitrogen atom N-7 of 9-ethyladenine forms a hydrogen bond with N-3 of 1-methyluracil. There is a remarkable similarity between this structure and the one mentioned above, despite the differences in crystal symmetry and solvents from which the crystals were grown. Figure 3 shows a van der Waals packing diagram of the structure of the 9-ethyladenine-1-methyluracil complex. The molecules are lying in a plane almost parallel to the bc plane and the packing in this plane closely resembles the packing of the 1-methylthymine-9-methyladenine complexes in the ac plane of the first structure. Apparently, C-11 of the N-9 ethyl substituent can be exchanged with the hydrogen atom at C-5 of the uracil molecule (Fig. 3), converting this structure to the 1-methylthymine-9-methyl-adenine structure with only minor shifts in the positions of the molecules within a layer. This will cause the packing of consecutive layers of molecules to change, however, giving rise to the different symmetries of the two struc-tures. The molecular dimensions shown in Fig. 2 are in reasonable agreement with those shown in Fig. 1. Although the positions of the hydrogen atoms were not determined, large deviations from linearity for the atoms involved in the base-pair hydrogen bonds are not to be expected. The 9-ethyladenine (C-11 excepted) and 1-methyluracil are planar molecules and the two mole-cules are nearly coplanes; the dihedral angle between the two planes being 4.4°. It is interesting to note that in this structure the N—H $\cdots$ O distance (2.98 Å) is 0.14 Å longer and the N—H $\cdots$ N distance (2.83 Å) is 0.10 Å shorter than the corresponding distances in the first structure. It seems that the changes in the Coulombic and van der Waals interactions can easily cause shifts of about 0.1 Å or more in the distances between hydrogen-bonded atoms.

   Another hydrogen-bonded base pair, interesting in many aspects, was prepared by Haschemeyer and Sobell (1963, 1965a) by cocrystallization of adenosine and 5-bromouridine from an aqueous solution. The structure, orthorhombic with space group $P_{2_12_12_1}$, was refined with 1986 reflections to $R = 0.138$, with 0.03 Å and 2° as the estimated standard deviations in the bond lengths and angles of the carbon, nitrogen, and oxygen atoms.

   This is the first base pair between complementary nucleosides, and the presence of the two ribose sugars together with one molecule of water, distributed statistically over at least two different sites, gives rise to a crystal structure with a very complex system of hydrogen bonds. Figure 4 presents a view of this structure in which the broken lines indicate the probable hydrogen bonds. Again, the base pair has not assumed the Watson-Crick configuration, while it also differs from the configuration of the two previous structures. The N-7 nitrogen atom of adenosine is hydrogen-bonded to N-3 of 5-bromouridine, but the amino nitrogen atom of adenosine is now

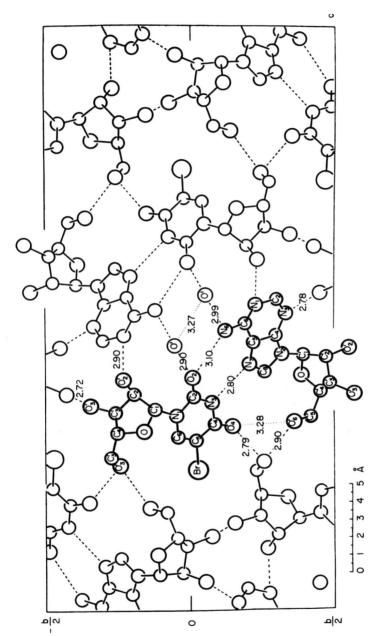

FIG. 4. Packing diagram of the adenosine-bromouridine crystal structure (Haschemeyer and Sobell, 1965a).

hydrogen-bonded to O-2 rather than O-4 of the bromouridine molecule. The N—H $\cdots$ N distance (2.80 Å) is rather short, whereas the N—H $\cdots$ O distance (3.10 Å) is surprisingly long. The base pair is approximately planar, the dihedral angle between the planes of the two bases is 4°55'. The bond distances and angles do not differ appreciably from those found in crystals of the separate closely related compounds, indicating that the interaction between the two nucleosides does alter their molecular configuration.

The same "reversed" configuration was found by Katz et al. (1965) in crystals containing the 1 : 1 complex of 9-ethyladenine and 1-methyl-5-bromouracil. These crystals, grown from a solution in dimethyl sulfoxide, are triclinic with the space group $P_{\bar{1}}$. The atomic parameters were refined to $R = 0.127$. Figure 5 shows the configuration together with the molecular dimensions of this base pair. The bond distances and angles are in good

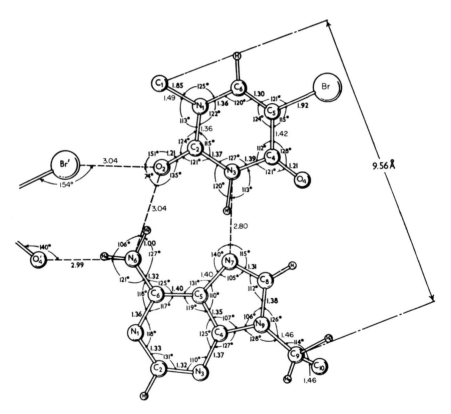

Fig. 5. Molecular dimensions of the 9-ethyladenine-1-methyl-5-bromouracil complex (Katz et al., 1965).

agreement with those found in the other complexes. The N—H $\cdots$ O distance of 3.04 Å is long and the N—H $\cdots$ N distance with 2.80 Å is rather short which is also the case in the adenosine-5-bromouridine complex. The base pair is nearly planar with a dihedral angle of 6.3° between the planes of the adenine and the uracil molecules.

The crystal structure, except for the reversed orientation of the uracil molecule with respect to the adenine, closely resembles the structures of the 9-ethyladenine-1-methyluracil and the 9-methyladenine-1-methylthymine complexes. Aside from the two base-pair hydrogen bonds, the only other hydrogen bond occurring in these structures is formed between the amino nitrogen atom of adenine to the O-4 or O-2 oxygen atom of the uracil or thymine molecule of an adjacent base pair.

A rather interesting feature in this structure is the occurrence of molecular disorder. The authors have found that in approximately 6% of the base pairs the bromouracil molecules have assumed the configuration in which O-4 rather than O-2 is hydrogen bonded to the adenine amino group, which is the configuration of the 1-methylthymine-9-methyladenine and the 1-methyluracil-9-ethyladenine base pairs. Haschemeyer and Sobell (1963) have made the suggestion that the inductive effect of the electronegative bromine atom would make O-2 more electronegative than O-4, making the "reversed" configuration the preferred one, although the crystal forces and the ribose hydroxyl hydrogen bonds could be responsible for this phenomenon.

The closer similarity of the ethyladenine-bromouracil structure to the structures of the base pairs that do not contain the 5-bromo substituent would tend to favor the theory concerning the inductive effect. The recent structure determination of the 1-methyl-5-bromouracil-9-methyladenine base pair by Baklagina et al. (1966), however, has shown that in this structure the configuration of the complex is the same as that of the nonbrominated base pairs. The crystals of this complex, grown from an aqueous solution, belong to the triclinic system with the space group P$\bar{1}$. The amino nitrogen atom of 9-methyladenine is hydrogen-bonded to O-4 of the bromouracil with a length of 2.97 Å, whereas the N-7 $\cdots$ N-3 hydrogen bond is 2.86 Å.

B. Guanine-Cytosine Base Pairs

Initial attempts to cocrystallize guanine and cytosine derivatives were unsuccessful because of the low solubility of the guanine derivatives in water. Crystals containing the bases 9-ethylguanine and 1-methylcytosine, however, could be grown from a solution in dimethyl sulfoxide and the structure was determined by O'Brien (1963, 1967). The structure, triclinic with space group P$\bar{1}$, was refined to $R = 0.112$. The positions of most of the hydrogen atoms were determined and it was established that both the ethylguanine

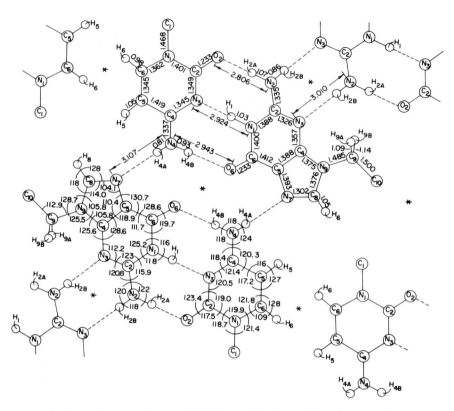

FIG. 6. Molecular dimensions of the 9-ethylguanine-1-methylcytosine base pair (O'Brien, 1967).

and the 1-methylcytosine molecules assume the keto-amino tautomeric forms. Figure 6 shows the molecular dimensions of the base pair. The configuration of the base pairs is the one proposed by Watson and Crick (1953) and the presence of three relatively short hydrogen bonds, linking the two molecules together, has been established. The positions of the hydrogen atoms show the hydrogen bonds all to be linear, the deviations from linearity being smaller than the estimated errors. The angle between the guanine and cytosine planes is 6.5°.

The same author also determined the structure of the complex between 9-ethylguanine and 1-methyl-5-fluorocytosine (O'Brien, 1966, 1967). This structure is isomorphous with the previous one and close agreement was found in the corresponding bond distances and angles of the two structures. The same base-pair configuration was discovered by Sobell et al. (1963) in the complex between 9-ethylguanine and 1-methyl-5-bromocytosine. The crystals of this base pair, grown from a solution in dimethyl sulfoxide, are monoclinic with the space group P_{2_1}/c. Figure 7 shows the configuration and the dimensions of this complex as it occurs in this crystal. Two of the hydrogen bonds in this pair differ in length from the two previous complexes. N-2 ··· O-2 is 0.1 Å longer and O-6 ··· N-8 is 0.08 Å shorter in the brominated base pair. The close agreement in the distance between the two structures determined by O'Brien shows that the crystal packing forces rather than the inductive effect of the halogen on the pyrimidine ring are responsible for these differences.

The consistency with which, at least to date, the Watson-Crick configuration occurs in the guanosine-cytosine base pairs is underlined by the structure of the complex between the nucleosides deoxyguanosine and 5-bromo-deoxycytidine, determined by Haschemeyer and Sobell (1964, 1965b).

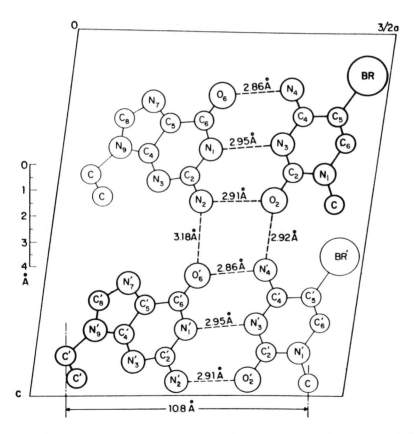

FIG. 7. The structure and molecular dimensions of the 9-ethylguanine-1-methyl-5-bromocytosine (Sobell et al., 1963).

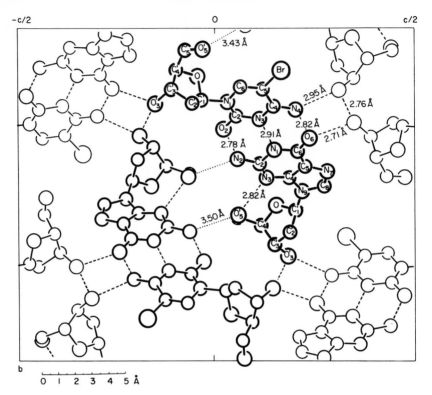

FIG. 8. A diagram of the deoxyguanosine-5-bromodeoxycytidine crystal structure. The presumed hydrogen bonding contacts are shown with dashed lines (Haschemeyer and Sobell, 1964).

The crystals were grown from an aqueous solution and they are orthorhombic with the space group $P_{2_1 2_1 2_1}$. In this structure, which is shown in Fig. 8, the guanine and cytosine rings are again approximately coplanar. The lengths of the hydrogen bonds connecting the two bases, indicated in the figure, were found to differ by as much as 0.12 Å from the values determined in the 9-ethylguanine-1-methylcytosine base pair.

C. Discussion

The main feature these structures, mentioned before, have in common is the formation of complementary, coplanar base pairs by crystallization from various polar solvents. Although it is impossible to indicate how the intermolecular forces determine the orientation of the molecules to each other in the crystalline state, the consistent presence of these base pairs in a number of structures of different symmetries and different complexities indicates that coplanar base-pair formation is an intrinsic property of the purine and pyrimidine derivatives. The question whether significant complex formation already exists in solution prior to crystallization will be treated in the next section.

Numerous attempts have been made to cocrystallize purine and pyrimidine derivatives in all possible combinations. The failure of these experiments so far to produce in polar solvents coplanar complexes in combinations other than the Watson-Crick complementary pairs is indicative for the specificity of their interactions in these solvents, but does not constitute proof in itself.

In the structures of the five complexes containing the adenine and the thymine (or uracil) derivatives the configuration postulated by Watson and Crick to occur in the DNA molecule has not been found. In spite of the variation in crystal structure, symmetry, and solvent of crystallization, the N-3 atom of thymine (uracil) is always hydrogen-bonded to N-7 rather than N-1 of adenine. This is rather surprising, since theoretical calculations of the energy of interaction between the adenine and the uracil molecules *in vacuo* have indicated that there is very little difference between the stability of the Watson-Crick pairing and the alternate schemes (Pullman *et al.*, 1965; Pollak and Rein, 1966; Nash and Bradley, 1966).

The variations in the hydrogen-bond distances are rather large. The average distance for the first five structures mentioned is 2.99 Å with deviations of -0.14 to $+0.10$ Å for the N—H $\cdots$ O and 2.84 Å with -0.04 to $+0.08$ Å, respectively, for the N—H $\cdots$ N distances. These large variations and the fact that the distances are not abnormally short, sometimes rather long, point out that the interaction between these base pairs can hardly be explained by virtue of these hydrogen bonds per se.

The last four structures, all of the guanine-cytosine type, show the Watson-Crick configuration, with somewhat smaller variations in the base-pair hydrogen-bonding distances.

III. Base Pairs in Solution

So far only the crystallographic observations for the occurrence of some of the purine-pyrimidine base pairs in the crystalline state has been reviewed. Although this has given information on the precise molecular geometry of the complexes in the solid state, it has not provided any evidence if and to what extent base pairing occurs in solution. Recently, however, important information has become available showing that significant pairing exists in solution and that these interactions show a specificity similar to that in the DNA molecule.

The methods used for the detection of hydrogen bonding in solution has been reviewed by Pimentel and McClellan (1960). In the work quoted here infrared and nuclear magnetic resonance spectroscopy has been used. The first method detects hydrogen bonding by the appearance of new absorption bands in the region of 3000–3500 cm^{-1}, due to the N—H stretching frequencies of the hydrogen-bonded atoms, while in the second method the decrease in magnetic shielding of the proton involved in hydrogen bonding gives rise to a downfield shift in the magnetic resonance spectrum.

Chan *et al.* (1964) found that in aqueous solution the C—H proton resonances of purine are all shifted to higher field, which they attributed to parallel stacking of the planar molecules. Jardetzky (1964) found the same effect, but also noted that in a series of purine and pyrimidine derivatives the magnitude of the upfield shifts decreased with an increase in the number of polar substituents. Although formation of hydrogen bonds, in addition to stack formation, on account of this, could not be ruled out, a study of a 1 : 1 mixture of deoxyadenosine monophosphate and thymidine failed to show any evidence of base-pair formation.

Tuppy and Kuechler (1964) have provided direct evidence for the specific interaction in water by studying the elution diagrams of mixtures of nucleosides on columns of Amberlite to which nucleosides were attached covalently. In this manner a relative retardation of deoxycytidine and cytidine on guanosine-Amberlite and a retardation of thymidine and uridine on adenosine-Amberlite was observed. This effect could not be detected on untreated Amberlite. Since in solutions containing 7 *M* urea these specific retardation effects were not observed, hydrogen-bond formation is responsible for this phenomenon.

Much more detailed information about base-pair formation in solution through hydrogen bonding was provided by a series of infrared spectroscopy experiments in nonaqueous solvents. In these studies purine and pyrimidine compounds were used that were suitably derivatized to enhance their solubility in nonaqueous solvent. Hydrogen-bonding forces should predominate in these nonpolar solvents, whereas the stacking interactions between the planar bases should be considerably weakened. Hamlin *et al.* (1965) provided the first evidence by studying the interaction between 9-ethyladenine and 1-cyclohexyluracil in deuterochloroform. In a solution of 0.022 *M* 1-cyclohexyluracil there is no evidence of association by hydrogen bonding, while 9-ethyladenine at the same concentration shows only very little self-association by the appearance of weak absorption bands which become more pronounced in a saturated solution. When these solutions are mixed pronounced absorption bands, due to hydrogen bonding, appear at 3490 and 3330 cm^{-1} together with some weaker bands that are not present in the spectra of the separate compounds. Figure 9 shows the difference in optical density between the mixtures and the separate components for three association bands as a function of the molar ratio of the two components at a constant molarity of total solute. There is a maximum in optical density for the 1 : 1 mixture, and

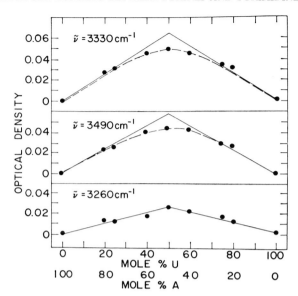

FIG. 9. The change in optical density of three bands due to hydrogen bonding as a function of the molar ratio of ethyladenine and cyclohexyluracil in deuterochloroform (Hamlin *et al.*, 1965).

the association constant in this solvent for the 1 : 1 dimer was estimated at 10^3 liters per mole.

Similar results were reported by Kuechler and Derkosch (1966) who studied the concentration dependency of the infrared spectrum for the 2', 3'-isopropylidene-5'-trityl derivatives of uridine and adenosine in carbon tetrachloride. A careful quantitative analysis of the data showed that the association constant at 20°C was 5 to 10 times higher than the self-association constants, and that up to a concentration of 10^{-2} mole per liter the association is first-order in both components giving a 1 : 1 complex. Direct information about the geometry of the adenine-uracil base pairs could not be obtained from these spectroscopic data.

The interaction between guanine and cytosine derivatives has been studied by Pitha *et al.* (1966). This work has shown that, although guanosine interacts strongly with itself, an even stronger interaction takes place between the guanosine and cytosine derivatives. Concentration variations show that the absorption bands due to the association increase to a maximum, whereas the absorption bands due to the free N—H groups decrease to a minimun in the 1 : 1 composition. The absorption bands due to the unassociated N—H stretching vibrations of both the guanosine and the cytosine moiety of the complex decrease. This indicates that the geometry of this base pair in solution is the Watson-Crick pairing and that three hydrogen bonds are involved in its formation.

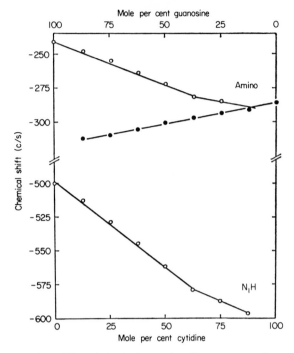

Fig. 10. The chemical shifts of the hydrogen-bonding protons of guanosine (O) and cytidine (●) in mixtures of the two species in dimethyl sulfoxide. The measurements were made at 16°C on solutions containing a total nucleoside concentration of 0.5 M (Katz and Penman, 1965).

Kyogoku *et al.* (1966) have reported similar results with the use of 2', 3'-benzylidene-5'-trityl of guanosine and cytidine in deuterochloroform. By studying the infrared spectrum of all possible pair combinations of derivatives of the four bases adenine, uracil, cytosine and guanine in deutero-chloroform solution, they were able to establish that interactions in solution show the same selectivity that is also found in the nucleic acids. The spectra of the combinations of adenine and uracil derivatives and those of guanine and cytosine show a substantial difference from additivity, whereas all other combinations in up to 0.008 M concentration do not. These authors also determined the association constants of several 1-cyclohexyluracil and 9-ethyladenine derivatives (Kyogoku *et al.*, 1967). Many interesting correlations were detected between the relative association constants of bases, modified with ring substituents of different nature.

Methylation of the hydrogen-bonding sites abolishes association; interaction between 9-ethyladenine and 1-cyclohexyl-3-methyluracil and between 9-ethyl-6-dimethylaminopurine and 1-cyclohexyluracil could not be detected. For the nonmethylated compounds, the absorption bands due to the free

NH_2 group in adenine and the free NH group in uracil decreased at constant intensity ratio with increasing concentration, indicating the formation of a cyclic dimer in which both groups are involved in hydrogen bonding.

Many of the pertinent results of the infrared spectroscopy measurements, mentioned above, have also been found by magnetic resonance experiments. Identification of most or all the protons and the possibility of working with solutions of higher concentrations are two of the many advantages of this method. As previously indicated, the use of nonaqueous solvents is necessary for the interaction studies.

Katz and Penman (1965) studied the interaction of guanosine and cytidine in dimethyl sulfoxide. Figure 10 shows the chemical shifts of the N—H protons for various guanosine-cytidine ratios at 0.5 M total nucleoside concentration and 16°C. An increasing downfield shift, linear over a wide range, is observed for the amino and the N-1 protons of guanosine with increasing cytidine content, and similarly the cytidine amino proton resonances shift downfield as the proportion of guanosine increases. The linear dependency suggests a 1 : 1 complex formation (since all hydrogen-bonding protons are affected) in the triple-bonded Watson-Crick configuration.

No interaction was found between adenine and uracil in dimethyl sulfoxide, but by making use of the system of Hamlin et al. (1965), the authors were able to observe base-pair interaction between 9-ethyladenine and 1-cyclohexyl-uracil in chloroform solution at 0.1 M total concentration. The concentration dependency of the shifts of the adenine amino proton displays a marked deviation from linearity at high uracil-adenine ratio, indicating that the association between these two compounds is not a simple one-to-one complex at higher uracil concentrations.

With the use of an equal-volume mixture of dimethyl sulfoxide and benzene as solvent, in which solvent the adenine-uracil hydrogen bonding could be observed, the authors also could obtain evidence for the specificity of the interactions between the four nucleosides—adenosine, guanosine, cytidine, and uridine. The shifts in the resonance frequency of the N-1 proton of guanosine in the presence of the four afore-mentioned nucleosides are -0.01, -7.1, -134.7, and -1.2 cps, respectively, while those for the N-3 proton of uridine are -8.2, 0, -0.6, and 0 at -4°C and 0.05 M total nucleoside concentration.

Shoup et al. (1966) have also made a nuclear magnetic resonance study of the four purine and pyrimidine components of nucleic acids and they have obtained similar results.

IV. Conclusion

This review has been limited to those purine and pyrimidine derivatives for which the configuration of the hydrogen donors and acceptors is the same as that of the four main constituents of the nucleic acids. For this reason, many interesting results in the works quoted have not been brought forward, especially in regard to "abnormal" bases. Crystallographic and spectroscopic studies of "abnormal" bases and normal bases in combinations violating the Watson-Crick specificity principle are actively being pursued (Sobell, 1966).

REFERENCES

Baklagina, Ju. G., Volkenstein, M. V., and Kondraschev, Ju. D. (1966). *Zh. Strukt. Khim.* **7**, 399.

Chan, S. I., Schweizer, M. P., Ts'o, P. O. P., and Helmkamp, G. K. (1964). *J. Am. Chem. Soc.* **86**, 4182.

Donohue, J. (1956). *Proc. Natl. Acad. Sci. U.S.* **42**, 60.

Donohue, J., and Trueblood, K. (1960). *J. Mol. Biol.* **2**, 363.

Hamlin, R. M., Jr., Lord, R. C., and Rich, A. (1965). *Science* **148**, 1734.

Haschemeyer, A. E. V., and Sobell, H. M. (1963). *Proc. Natl. Acad. Sci. U.S.* **50**, 782.

Haschemeyer, A. E. V., and Sobell, H. M. (1964). *Nature* **202**, 969.

Haschemeyer, A. E. V., and Sobell, H. M. (1965a). *Acta Cryst.* **18**, 525.

Haschemeyer, A. E. V., and Sobell, H. M. (1965b). *Acta Cryst.* **19**, 125.

Hoogsteen, K. (1959). *Acta Cryst.* **12**, 822.

Hoogsteen, K. (1963). *Acta Cryst.* **16**, 907.

Jardetzky, O. (1964). *Biopolymers, Symp.* **1**, PP 501.

Katz, L., and Penman, S. (1965). *J. Mol. Biol.* **15**, 220.

Katz, L., Tomita, K., and Rich, A. (1965). *J. Mol. Biol.* **13**, 340.

Kuechler, E., and Derkosch, J. (1966). *Z. Naturforsch.* **21b**, 209.

Kyogoku, Y., Lord, R. C., and Rich, A. (1966). *Science* **154**, 518.

Kyogoku, Y., Lord, R. C., and Rich, A. (1967). *Proc. Natl. Acad. Sci. U.S.* **57**, 250.

Mathews, F. S., and Rich, A. (1964). *J. Mol. Biol.* **8**, 89.

Nash, H. A., and Bradley, D. F. (1966). *J. Chem. Phys.* **45**, 1380.

O'Brien, E. J. (1963). *J. Mol. Biol.* **7**, 107.

O'Brien, E. J. (1966). *J. Mol. Biol.* **22**, 377.

O'Brien, E. J. (1967). *Acta Cryst.* **23**, 92.

Pimentel, G. C., and McClellan, A. L. (1960). "The Hydrogen Bond." Freeman, San Fransisco, California.

Pitha, J., Jones, R. N., and Pithova, P. (1966). *Can. J. Chem.* **44**, 1045.

Pollak, M., and Rein, R. (1966). *J. Theoret. Biol.* **11**, 490.

Pullman, B., Claverie, P., and Caillet, J. (1965). *Proc. Natl. Acad. Sci. U.S.* **55**, 904.

Shoup, R. R., Miles, H. T., and Becker, E. D. (1966). *Biochem. Biophys. Res. Commun.* **23**, 194.

Sobell, H. M. (1966). Private communication.

Sobell, H. M., Tomita, K., and Rich, A. (1963). *Proc. Natl. Acad. Sci. U.S.* **40**, 885.

Stewart, R. F., and Jensen, L. H. (1964). *J. Chem. Phys.* **40**, 2071.

Tuppy, H., and Kuechler, E. (1964). *Biochim. Biophys. Acta* **80**, 669.

Watson, J. D., and Crick, F. H. C. (1953). *Nature* **171**, 737.

The Physicochemical Basis of Interactions of Nucleic Acid

PAUL O. P. TS'O

Department of Radiological Sciences
The Johns Hopkins University
Baltimore, Maryland

Several decades of research on nucleic acids have gradually provided us with the necessary information about the nature and the magnitude of the forces to determine the secondary structures, the interactions, and perhaps even the replication processes of nucleic acids. In this chapter we wish to report briefly about some of the contribution from our laboratory to this problem.

During the early 1950's, when the Watson-Crick model for the DNA double helix had gained popular acceptance, the prevailing view was that the hydrogen bonding between the base pairs is responsible for holding the two strands together in the double helix. This view was not derived from compelling experimental observations, but was a logical presupposition, since the only other kind of force operating in nucleic acids that we knew at that time was the electrostatic repulsion of the charged phosphate groups along the strands. Thus, hydrogen bonding between the pairing bases was assumed to be the major stabilizing force in maintaining the conformation of the double helix.

As early as 1958, it was suggested that hydrogen bonding was probably not the sole source of stability for the DNA helix (Rice *et al.*, 1958; Sturtevant *et al.*, 1958). This suggestion came from the observation that at sufficiently low temperature the DNA double helix can be kept intact at a pH low enough to break most of the hydrogen bonding by protonation. The nature of other contributions to the stability, however, was not certain at that time. Subsequent work in the early 1960's on the properties of nucleic acids in organic solvents including those contributions from our laboratory supported the above suggestion and further proposed that hydrophobic interaction of bases contributes significantly to the stability of the helix (Herskovits *et al.*, 1961; Herskovits, 1962; Helmkamp and Ts'o, 1961; Marmur and Ts'o, 1961; Ts'o *et al.*, 1962, 1963a).

This point of view was originated from the observations that organic solvents are effective denaturants for the helical nucleic acids. It was reasoned at that time that the hydrogen bonding of the base pairs should be strengthened in organic solvents as compared to the hydrogen bonding of the base pairs in water. This reasoning is borne out now completely by the recent studies of the properties of the bases in organic solvents (Hamlin et al., 1965; Pitha et al., 1966; Kyogoku et al., 1966, 1967; Katz and Penman, 1966; Shoup et al., 1966; Küchler and Derkosch, 1966). For example, base pairing can be observed in dimethyl sulfoxide (Katz and Penman, 1966; Shoup et al., 1966), but not in water (see below); base stacking can be observed in water, but not in dimethyl sulfoxide (Chan et al., 1964). Nevertheless, dimethyl sulfoxide was found to be a very powerful denaturant for nucleic acids (Helmkamp and Ts'o, 1961). Therefore, it is apparent that the hydrophobic stacking interaction of bases in water must exert a pronounced influence on the conformation of nucleic acids.

Our results will be described in three sections. Section I concerns the properties of the monomers (bases and nucleosides) in aqueous solutions. This study clearly indicates the hydrophobic stacking properties of the base moiety in water. Section II is concerned with the specific and cooperative interaction of the nucleosides with the complementary polynucleotides. Here, we can see how the hydrophobic stacking forces and the hydrogen-bonding forces work together resulting in the formation of an unique secondary structure of the nucleoside-polynucleotide complex. Finally, in Section III, some of the results which indicate the influence of the 2'-OH group of the ribosyl moiety on the secondary structure of the homopolynucleotides will be described.

I. Association of Bases and Nucleosides in Aqueous Solutions

Vapor pressures of solutions of purine, 6-methylpurine, and 14 pyrimidine and purine nucleosides have been measured thermoelectrically at varying concentrations (Ts'o et al., 1963b; Ts'o and Chan, 1964; Broom et al., 1967). Osmotic coefficients, ϕ, were calculated from the data and these values are given in Table I. Compounds such as purine, 6-methylpurine, uridine, 5-bromouridine, and cytidine are more soluble than others and, therefore, they can be studied in a larger range of concentration. Activity coefficients for these more soluble compounds at 25°C were calculated from the osmotic coefficients by the Gibbs-Duhem relationship using a computer which performed a numerical integration on the fitted polynomials and related molal concentration to ϕ (Table II). The data clearly indicated that the properties of these bases and nucleosides in solutions are far from ideal. Values of both osmotic coefficients and activity coefficients are well below unity. These results establish the concept that purine and pyrimidine nucleosides do interact extensively in aqueous solution.

TABLE I
MOLAL OSMOTIC COEFFICIENTS (ϕ) AT 25°C

Compound	Molal concentration															
	0.025	0.05	0.10	0.15	0.20	0.25	0.30	0.35	0.40	0.45	0.50	0.60	0.70	0.80	0.90	1.00
Uridine[a]	—	0.969	0.943	0.921	0.901	0.883	0.866	0.849	0.833	0.817	0.801	0.773	0.775	—	—	—
Cytidine[a]	—	0.967	0.935	0.905	0.876	0.850	0.826	0.804	0.785	0.768	0.752	0.724	0.695	—	—	—
Thymidine[b]	—	—	0.905	—	0.865	—	—	—	—	—	—	—	—	—	—	—
5-Bromouridine[c]	—	0.902	0.811	0.732	0.666	—	0.569	—	—	—	—	—	—	—	—	—
Inosine[b]	0.994	0.957	0.888	0.830	—	0.766	—	0.715	0.502	—	—	—	—	—	—	—
1-Methylinosine[b]	0.962	0.926	0.860	0.800	0.750	—	—	—	—	—	—	—	—	—	—	—
Ribosylpurine[b]	0.965	0.930	0.860	0.810	0.770	—	0.710	—	—	—	—	—	—	—	—	—
Purine[a]	—	0.917	0.849	0.794	0.749	0.714	0.685	0.662	0.643	0.627	0.614	0.590	0.567	0.544	0.522	0.505
Adenosine[b]	0.915	0.836	0.740[d]	—	—	—	—	—	—	—	—	—	—	—	—	—
2'-O-Methyladenosine[b]	0.908	0.828	0.723	0.658	0.611	—	0.550	—	—	—	—	—	—	—	—	—
2'-Deoxyadenosine[b]	0.900	0.800	0.668	0.598	—	—	—	—	—	—	—	—	—	—	—	—
6-Methylpurine[c]	—	0.786	0.682	0.624	0.582	0.544	0.510	0.484	0.469	0.461	0.456	0.427	0.410	—	—	—
N-6-Methyladenosine[b]	0.805	0.685	0.558	0.480	—	—	—	—	—	—	—	—	—	—	—	—
N-6-Methyl-2'-deoxyadenosine[b]	0.790	0.680	0.540	0.468	—	—	—	—	—	—	—	—	—	—	—	—
N-6-Dimethyladenosine[b]	0.712	0.608	0.470	0.408	0.378	—	—	—	—	—	—	—	—	—	—	—

[a] From Ts'o et al. (1963b).
[b] From Broom et al. (1967).
[c] From Ts'o and Chan (1964).
[d] At solubility limit of 0.085 M.

TABLE II

MOLAL ACTIVITY COEFFICIENTS[a] AT 25°C COMPUTED
FROM THE FITTED OSMOTIC COEFFICIENTS[b]

Molal concentration	Purine	6-Methyl-purine	Uridine	5-Bromo-uridine	Cytidine
0.05	0.844	0.626	0.939	0.902	0.936
0.10	0.728	0.469	0.888	0.811	0.878
0.15	0.641	0.385	0.845	0.732	0.824
0.20	0.575	0.329	0.808	0.666	0.776
0.25	0.522	0.287	0.775	0.613	0.733
0.30	0.480	0.255	0.744	0.569	0.695
0.35	0.446	0.230	0.716	0.533	0.661
0.40	0.418	0.211	0.690	0.502	0.631
0.45	0.394	0.196	0.665	—	0.604
0.50	0.374	0.185	0.641	—	0.580
0.55	0.355	0.173	0.620	—	0.558
0.60	0.339	0.162	0.600	—	0.537
0.65	0.324	0.152	0.582	—	0.518
0.70	0.311	0.146	0.568	—	0.499
0.75	0.297	—	—	—	—
0.80	0.286	—	—	—	—
0.85	0.275	—	—	—	—
0.90	0.264	—	—	—	—
0.95	0.255	—	—	—	—
1.00	0.247	—	—	—	—
1.05	0.240	—	—	—	—
1.10	0.235	—	—	—	—

[a] Data from Ts'o et al. (1963b) and Ts'o and Chan (1964).
[b] See Table I.

After further analysis for their congruence to different models for multiple equilibria, the thermodynamic data were found to be incompatible with the model which assumes that only dimers are formed (Ts'o et al., 1963b). Thus, the degree of association of these compounds may go beyond the dimer stage to a higher degree of polymerization. Most of the results are consistent with the model which assumes that the association process continues through many successive steps (at least more than five steps) with the same equilibrium constant (Ts'o et al., 1963b; Ts'o and Chan, 1964). The apparent equilibrium constant, K, for the association at various steps can be obtained by the following equation.

$$K = \frac{1 - \phi}{m\phi^2} \tag{1}$$

where m is the concentration in molality. For more complicated situations, the knowledge of the activity coefficient is also required (Ts'o and Chan, 1964). Comparison of the equilibrium constant and, thus, the standard free energy changes of these nucleosides and urea (Schellman, 1956) are given in Table III. One immediate general conclusion is that the tendency of purine to associate is much greater than that of pyrimidine nucleosides, which in turn is greater than that of urea.

TABLE III

EQUILIBRIUM CONSTANTS AND THE STANDARD FREE ENERGY CHANGE OF THE ASSOCIATION FOR THE FOLLOWING COMPOUNDS AT 25°C

Compound	K (molal^{-1})	$F°$ ($RT \ln k$, cal)
Urea[a]	0.041	1190
Uridine[b]	0.61	290
Cytidine[b]	0.87	80
Thymidine[c]	0.91	60
5-Bromouridine[d,e]	$K_1 = 1.0$	0
	$\overline{K} = 2.9$	−630
Ribosylpurine[f]	1.9	−380
1-Methylinosine[f]	1.8–2.0	−360 to −410
Purine[f]	2.1	−440
Adenosine[f]	4.5	−900
2'-O-Methyladenosine[f]	5.1	−970
6-Methylpurine[f]	6.7	−1120
2'-Deoxyadenosine[f]	7.5–4.7	−1195 to −920
N-6-Methyladenosine[f]	14.9–11.8	−1600 to −1460
N-6-Methyl-2'-deoxyadenosine[f]	15.9	−1640
N-6-Dimethyladenosine[f]	22.2	−1840

[a] From Schellman (1956).
[b] From Ts'o et al. (1963b).
[c] From Solie (1965).
[d] From Ts'o and Chan (1964).
[e] The treatment of multiple equilibria for this compound requires two equilibrium constants. K_1 is for the first step and $\overline{K}$ is for the successive steps. See original paper.
[f] From Broom et al. (1967).

What is the mode of association of these molecules in aqueous solution? Do they associate with each other vertically through hydrophobic and stacking interactions, or do they associate horizontally through hydrogen bonding? The thermodynamic data on these compounds conclusively do not support the hypothesis of horizontal association through hydrogen bonding because of the following reasons:

1. All these bases and nucleosides associate much more extensively than urea which is already known to be one of the best hydrogen-bonding agents in water.

2. Methylation enhances association. From the values of K of association (Table III) and ϕ (Table I), the order of association tendencies with respect to the degree of methylation can be listed as follows: 2'-deoxyadenosine < N-6-methyl-2'-deoxyadenosine; adenosine < N-6-methyladenosine < N-6-dimethyladenosine; and inosine < 1-methylinosine. In every case examined, substitution of a hydrogen of the base by a methyl group removes a hydrogen-bond donor and the association tendency is enhanced significantly. 1-Methyl-inosine and N-6-dimethyladenosine, in spite of the fact that the hydrogen-bond donor sites of those two compounds have been completely removed by methylation, do associate substantially more than inosine or adenosine, respectively. It is interesting to note that in order to obtain the promoting effect of methylation, the methylation has to take place at the base and not at the pentose. Thus, the association tendency of 2'-O-methyladenosine is about the same as that of adenosine and may be slightly less than that of 2'-deoxy-adenosine.

More direct information about the mode of association of the bases and nucleosides in solution can be obtained by the study of nuclear magnetic resonance. It is well known that nuclear magnetic shielding is a very sensitive probe of inter- and intramolecular interactions. In this case, vertical stacking interactions are easily distinguished from hydrogen-bonding interactions since these interactions manifest themselves differently in NMR. It is therefore hoped that the concentration dependence of the NMR spectra in aqueous solutions of purine and nucleosides will shed some light on the association mechanism. NMR spectra of purine have been studied over the concentration range of 0.05 to 1 M (Chan et al., 1964). Chemical shifts of the three protons in purine vs. the concentration are shown in Fig. 1. A pronounced concentration effect has been observed. Proton resonances in purine are all shifted to higher fields as the solute concentration is increased. Shifts to high fields with concentration are well known for aromatic systems and are generally attributed to the magnetic anisotropy associated with the ring currents in neighboring molecules. Because of the mobile electrons, a large diamagnetic current is induced in the plane of the ring by an external magnetic field when the field is perpendicular to the plane of the molecule. This ring current gives rise to a small secondary magnetic field which reinforces the primary field at the peripheral protons in the plane of the ring. In the region directly above and below the molecular plane, however, the two fields are opposed. As the concentration of a solution of aromatic molecules is increased, the average distance between molecules decreases and the protons of a given molecule will

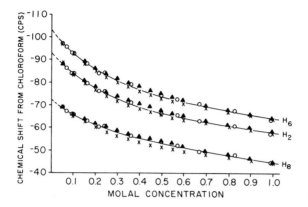

FIG. 1. Concentration dependence of the proton chemical shifts for purine in aqueous solution at 25°C (corrected for bulk susceptibility); shifts measured from external chloroform reference: —O—, experimental values; —×—, calculated values from overall average model; —▲—, calculated values from statistical partial-overlapping model.

feel the secondary magnetic fields produced by the ring current of neighboring molecules. Since it is much more probable to find the molecules somewhere above or below the molecular plane of another aromatic molecule due to the dish-shaped nature of the aromatic molecules, this magnetic anisotropy of the ring current effect will lead to a high-field shift with concentration or to a low-field shift upon dilution. At higher temperatures, or when the purine is dissolved in an organic solvent such as dimethylformamide, such concentration-dependent chemical shifts for the purine protons are greatly reduced. Furthermore, when the purines are protonated by hydrochloride so that they cannot associate because they carry a positive charge, such concentration-dependent chemical shifts are again practically eliminated. These data clearly suggest that the mode of association of purine is by the vertical stacking of rings in a partially overlapping fashion. As described above, the osmotic and activity coefficients of purine have been interpreted in terms of multiple equilibria and, on this basis, populations of various associated species at varying concentrations were computed. Based on these population distributions of the associated species, we can calculate the concentration dependence of the chemical shifts which is also given in Fig. 1 (Chan *et al.*, 1964). It can be seen that the calculated value and the experimental value are in satisfactory agreement. Therefore, a numerical correlation between the NMR data and osmotic data has been successful in the sense that they reinforce and support the interpretations of each other.

Similar results have been obtained from the purine nucleosides, especially the adenine nucleoside series (Table IV). The close correlation between the

TABLE IV

CONCENTRATION DEPENDENCE OF CHEMICAL SHIFTS FOR 11 PURINE NUCLEOSIDES
(0.0–0.2 M IN D_2O)

Compound	Temp. (°C)	$\Delta\delta$ (cps)				
		H-2	H-8	H-6	H-1'	-CH₃
Inosine	32	6.4	5.3	—	7.1	—
1-Methylinosine[a]	33	8.9	6.4	—	6.8	5.3
Ribosylpurine	30	10.7	6.4	13.1	8.8	—
Purine[b]	25–27	12.6	9.6	14.2	—	—
2'-O-Methyladenosine	31	13.7	7.5	—	8.8	—
6-Methylpurine[b]	25–27	19.4	13.3	—	—	17.0
2'-Deoxyadenosine	30	19.8	13.0	—	13.6	—
N-6-Methyl-2'-deoxy-adenosine	32	26.0	15.8	—	14.0	15.2
N-6-Dimethyladenosine	28	27.2	14.5	—	14.4	25.5
N-6-Methyladenosine	26	32.6	17.5	—	12.6	18.1
2'-Deoxyadenosine[c]	30	14.8	10.0	—	9.8	—
Adenosine[c]	32	14.8	8.3	—	6.9	—
3'-Deoxyadenosine[c]	25	15.8	9.0	—	9.6	—

[a] Peak positions of H-8 and H-2 are reversed with respect to the other 6-substituted nucleosides studied.

[b] From Chan et al. (1964).

[c] Differences measured over the concentration range 0–0.10 M because of solubility limitation.

magnitude of $\Delta\delta$, the concentration dependence of the chemical shifts, and the values of ϕ and K (Tables I and II) for this series of purine nucleosides in solution not only establishes that stacking is the mode of association of these solutes, but also verifies the usefulness of NMR as a tool for studies of association. The differentials in the magnitude of $\Delta\delta$ for various protons of a given nucleoside may provide additional information about the average geometry of the stacks in solutions. For instance, in the case of 2'-deoxyadenosine, the concentration dependence of the chemical shift is even larger than that of the purine (Fig. 2). In all these cases, the H-2 proton of the 6-membered ring of the adenine is shifted to a higher field than the H-8 proton in the 5-membered ring. This indicates that the 6-membered ring of the adenine does participate to a greater extent in the stacks than the 5-membered ring of the adenine nucleosides. The pentose protons of H-1' are also considerably shifted to higher fields when concentration is increased, whereas the pentose protons of the H-5' are hardly affected. As one proceeds around the pentose ring from the C-1' to the C-5', there is a progressive drop or decrease in the magnitude of these concentration-dependent chemical shifts. This indicates that adenine

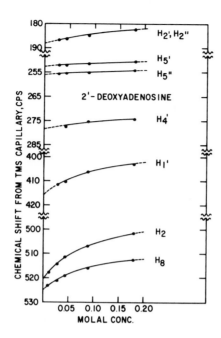

FIG. 2. Concentration dependence of the proton chemical shifts for 2′-deoxyadenosine.

nucleoside interaction is preferentially localized at the purine base of the nucleoside, so that the ring current magnetic anisotropy is principally felt by the base protons. From this type of study, therefore, not only can we obtain the general picture about the mode of association, but we can even get down to the detailed molecular structure of the stacks. Two models (Fig. 3a,b) for the preferred average orientation of the nucleoside bases in the stacks have been proposed to account for the results observed (Broom *et al.*, 1967). In these dimer models, one can see that the H-2 proton will be shielded strongly by the 6-membered ring of the neighboring bases most of the time. The H-8 and the H-1′ protons will be shielded mainly by the 5-membered ring. In addition, according to these models, H-8 and H-1′ will spend less time on the average in the proximity of the 5-membered ring than H-2 will in the proximity of the 6-membered ring. These two models are related to each other by symmetry considerations. The base-stacking arrangement for model Fig. 3a is face-to-back; and that for Fig. 3b is face-to-face (or back-to-back), one of the nucleosides being rotated 180° along an axis in the plane bisecting the C-4–C-5 bond. Thus, in model Fig. 3a the two ribosyl substituents at N-9 will be on the *same side* of the dimer (straight stack). On the other hand, in model Fig. 3b, the two ribosyl moieties at N-9 will be *opposite* to each other in the dimer resulting in an alternating arrangement (alternate stack). The alternate stack

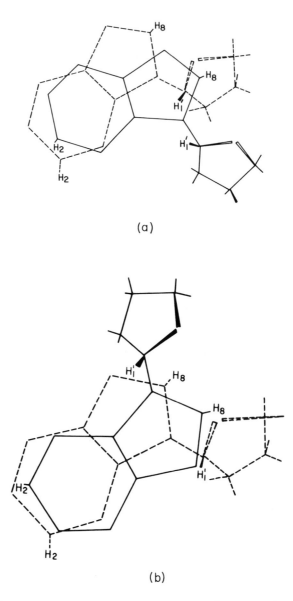

(a)

(b)

FIG. 3. Models illustrating the proposed arrangements for two nucleosides in stacks. (a) Straight stack (face-to-face); the ribosyl substituents at N-9 are on the same side of the stack. (b) Alternate stack (face-to-face or back-to-back); one of the nucleosides is rotated by 180° along an axis in the plane bisecting the C-4–C-5 bond. The ribosyl substituents are opposite each other in the stack.

(Fig. 3b) may be favored based on two considerations: (1) the steric hindrance of the ribosyl group is reduced and (2) this arrangement reduces the repulsion between the dipole moments of the adjacent bases expected in the straight stack. No critical evaluation of these two models can be pursued further at this time. These models are similar to the arrangement in 8-azaguanine monohydrate crystals (MacIntyre, 1965) and to one of the stacking modes of purine or 6-methylpurine suggested by Pullman et al. (1958).

Two qualifying statements should now be mentioned in the discussion of these models. (a) The standard free energy change for the association is in the order of the thermal energy kT, and no significant line broadening or separate shifts for stacked and free species are observed in the NMR studies; therefore, these stacks must break and reform rapidly. What we observe and describe is an overall average phenomenon. (b) The preferred orientation of the bases in the stack must vary to a certain degree depending on the nature of the bases in the nucleosides. Substitution of an oxo, amino, methyl, or dimethylamino

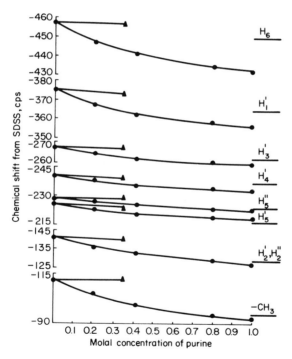

FIG. 4. Chemical shift dependence of thymidine protons upon thymidine (—▲—) and purine concentration (—●—) at 35°C in D_2O. Only two thymidine concentrations were measured: 0.1 M (placed at origin of the abscissa) and 0.35 m (placed at the 0.35 m scale of the abscissa). Shifts measured from external SDSS. Magnetic field increases from top to bottom along the ordinant. Spectra obtained at 60 Mc.

group, should exert an influence on the geometric arrangement of the stacks. Nevertheless, these stacks will have a partially overlapped orientation so that the H-8 and H-1' can both be shielded to a similar extent.

The chemical shifts of the base protons (H-5, H-6, or 6-CH$_3$) of the pyrimidine nucleosides are not concentration dependent even though these compounds have all been shown to associate in aqueous solutions by a lowering of osmotic pressure (Fig. 4). Therefore, we should interpret this result on the basis that these pyrimidine nucleosides do not support ring currents as the aromatic purine bases do, so that their interactions cannot be monitored by proton magnetic resonance via the effect of the ring-current magnetic anisotropy (Schweizer et al., 1965). Interaction of pyrimidine nucleosides with purine or purine nucleoside, however, can be monitored by NMR as shown in Fig. 4 for the interaction of thymidine and purine. Marked upfield shifts are noticed, particularly for the thymidine base protons (6-CH$_3$ and H-5) and the anomeric proton H-1'. As one proceeds around the deoxyribose ring from C-1' to C-5', however, there is a progressive drop off or decrease in the magnitude of these upfield shifts. This phenomenon, which is similar to that observed for the self-interaction of purine nucleosides, again indicates that the purine-pyrimidine nucleoside interaction is preferentially localized at the pyrimidine base of the nucleoside and the mode of interaction is vertical ring stacking of the bases. The same effect can be observed when a purine nucleoside is used in place of purine. For example, a mixture of 0.2 molal 2'-O-methyladenosine and 0.1 molal thymidine results in the following upfield shifts (in cps) of the thymidine protons: 8.2 (-CH$_3$), 8.9 (H-6), 7.1 (H-1'), 4.3 (H-2', H-2''). Similar effects were noted for N-6-methyladenosine and 2'-deoxyadenosine in thymidine solutions, whereas ribosylpurine is only 70% as effective as the adenosine derivatives.

What are the forces involved in this type of stacking interaction? It is interesting to note that the substitution of the 6-amino group to the ribosylpurine to give adenosine substantially enhances the association of the nucleoside as studied by osmometry or NMR. The substitution of a polar group is not likely to reduce the solvation properties of the nucleoside; therefore, one cannot explain this enhancement on general hydrophobic terms. Another interesting example of this type is that the extent of association of 5-bromouridine is higher than that of the thymidine. The increase in association due to the bromo substitution for a methyl group again required further reflection on the driving force of this association. In Table V, the correlation among the osmotic coefficients (ϕ) of the nucleosides, the polarizabilities (α) and the dipole moments (μ) of the bases is listed. The values of α and μ are those calculated by Pullman (1965a,b). The α values for adenine and cytosine from the paper by Pullman are identical to those previously published by DeVoe and Tinoco (1962). The μ values from MO calculations for the methylated bases obtained by DeVoe and Tinoco (1962) are also similar to those by Pullman (9-methyl-

TABLE V

CORRELATION BETWEEN THE OSMOTIC COEFFICIENTS (ϕ) OF THE NUCLEOSIDES
AND THE POLARIZABILITIES (α)[a] AND DIPOLE MOMENTS (μ)[a] OF THE BASES

Compound	ϕ (0.1 molal, 25°C)	α (Å)³	μ (D)
Uridine[b]	0.943	10.2	(3.9)[c]
Cytidine[b]	0.935	11	7.2
Thymidine[d]	0.905	12	3.6
Inosine[d]	0.888	13.0	5.2
Ribosylpurine[d]	0.860	12.5	4.3(4.2)[c]
2'-O-Methyladenosine[d,e]	0.723	13.9	3.0(3.2)[c]
2'-Deoxyadenosine[d,e]	0.688		

[a] From Pullman (1965a,b).
[b] From Ts'o et al. (1963b).
[c] Experimental value of DeVoe and Tinoco (1962).
[d] From Broom et al. (1967).
[e] Adenosine is not sufficiently soluble to give a 0.1 M solution.

adenine, 2.8 μ; 3-methylcytosine, 8.0 μ; and 9-methylpurine 3.6 μ). These small differences do not affect the conclusion presented later in this paragraph. It is also gratifying to note that the experimental values of μ for methyl-adenine, dimethyluracil, and methylpurine are in good agreement with the calculated values (DeVoe and Tinoco, 1962). Comparison of the order of ϕ and μ among the four purine nucleosides clearly indicates that these two quantities relate to each other in a reverse manner. Comparison of the order of ϕ and μ among the three pyrimidine nucleosides again shows no correlation between these two quantities. Thymine has the smallest dipole moment, but thymidine has the greatest tendency to associate, whereas cytosine has the greatest dipole moment among the six compounds in Table V, but cytidine has the next to the lowest tendency to associate. Therefore, it can be concluded with a considerable degree of certainty that as far as these nucleosides are concerned, permanent dipole moment attraction is not the most important driving force for their self-association. This conclusion is not surprising in view of the high dielectric constant of the solvent, water. It is also in accord with the model proposed in later sections for the mode of stacking of these purine nucleosides based on the NMR studies. In these models, the permanent dipole moment of the bases in stacks is likely to exert a negative influence rather than a positive promotion on association.

On the other hand, the correlation between the order of ϕ and the order of α, the polarizability, is good (Table V). The only exception is that although hypoxanthine was calculated to have a larger polarizability than purine, the association tendency of inosine appears to be slightly less than that of ribosyl-purine. This anomaly may be due to the negative influence of the large permanent dipole moment of hypoxanthine which may hinder the stacking either directly or indirectly by more extensive hydration. Other than this exception, the rather good correlation here is in accord with the recent conclusion based on a correlation of polarizabilities and the effectiveness of a variety of denaturing agents, that London dispersion forces are responsible for the stability of the DNA helix (Hanlon, 1966). In summary, the order of nucleoside osmotic coefficients which is an index of association by stacking, is in good agreement with the reported polarizability values of the respective base and is not in agreement with the base dipole moment values.

Interesting and important information was provided by the comparative studies on the spectral position of the chemical shifts of the protons of various nucleosides at infinite dilution (Broom *et al.*, 1967). Here, we shall pay attention to only one aspect which is related to Section III. The chemical shifts of the H-2 and the H-8 protons (cps from TMS capillary) for adenosine, 2-O-methyladenosine, and deoxyadenosine, respectively, are 524, 528.5; 521.7, 525.6; and 521.2, 525.4. It is clear that the ring protons of adenine ribonucleoside are shifted downfield by about 0.05 ppm from the corresponding adenine 2'-deoxy or 2'-methoxyl nucleoside. This observation suggests the existence of hydrogen bonding from the 2'-OH group of the pentose to the N-3 of the purine ring. Preliminary potentiometric titration experiments done in our laboratory (0.05 M nucleoside, 0.15 N NaCl, 25°C) indicate that 2'-deoxyadenosine has a pK_a 0.1 pH unit higher than that of adenosine. We have also found that H-2 and H-8 of 2'-deoxyadenosine are shifted 10 cps downfield going from neutral solution to pH 3.6, whereas H-2 and H-8 of adenosine are shifted only 5 cps to lower field over the same pH range. Thus, the base of 2'-deoxyadenosine is more readily protonated, indicating a higher pK_a. These results again are in accord with the concept of intramolecular hydrogen bonding. This observation is pertinent to the experiments reported in Section III.

II. Specific and Cooperative Interaction of the Nucleosides with the Complementary Polynucleotides

The simple system of monomer-monomer interactions can be studied quantitatively by thermodynamic and spectroscopic methods. However, this system does not have the specificity exhibited at the level of polymer-polymer interaction. Consequently, we turn our attention to the nucleic acid interactions at the polymer-monomer level. A model system for this kind of study should have the following characteristics. (1) The polymer should have minimal degree of self-interactions. (2) Solubility of both polymer and monomer should be sufficiently high. (3) The electrostatic forces should be minimal. (4) Its properties are relevant to those of a well-characterized polymer-polymer interaction system.

The above criteria are apparently met by the system: polyuridylic acid and adenosine (Huang and Ts'o, 1966). The binding of adenosine to poly U was first studied by equilibrium dialysis at 5°C in 0.4 M NaCl. When the fraction of the occupied poly U binding sites is plotted as a function of free adenosines (Fig. 5), the resulting adsorption isotherm shows a very steep transition. No binding was detectable until a critical threshold concentration of adenosine was reached. This steep curve of the adsorption isotherm is analyzed by the following equation derived from lattice statistics based on the nearest-neighbor interaction (Hill, 1960).

$$\left(\frac{\partial \theta}{\partial \ln Y}\right)_{\theta=1/2} = \frac{\exp(-W/2kT)}{4} \tag{2}$$

where θ is the fraction of sites occupied, W is the interaction energy of the nearest neighbor, and Y is the function of absolute activity of the adsorbate $(\lambda = e^{u/kT})$ and the partition function for a molecule of bound adsorbate, q; λq at dilute solution is equivalent to $K_0 M$, where K_0 is intrinsic association constant for one molecule of adsorbate with a single site, and M is the molar concentration of free adsorbate (Steiner and Beers, 1960). Therefore,

$$\left(\frac{\partial \theta}{\partial \ln M}\right)_{\theta=1/2} = \frac{\exp(-W/2kT)}{4} \tag{3}$$

Estimation from the slope of the curve (Fig. 5, open circle) yields a value of 30–60. In Eq. (3) W is calculated to be -5 to 6 kcal/mole, which is the stacking

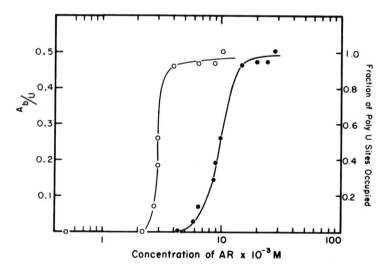

FIG. 5. Adenosine bound per UMP of the poly U (1.5×10^{-2} M) vs. adenosine input concentration at 5°C, 0.4 M NaCl, 0.01 M phosphate (HMP) (—●—). The fraction of poly U sites occupied vs. free adenosine concentration is also shown, —○—.

energy of adenosine upon pairing with 2U of poly U. This is comparable to the value of -4.8 or -7.5 kcal/mole calculated for the stacking energy of poly dAT and poly dI:dBC, respectively, by Crothers and Zimm (1964).

Similar experiments were also performed using cytidine or inosine as the dialyzable components. No detectable binding was found even at input nucleoside concentration as high as 2×10^{-2} M. Therefore, this interaction has the same specificity as the system of long-chain polymers, i.e., the base-pairing scheme of Watson-Crick. The stoichiometry of this binding reaction was studied using solubility measurements, and it was found that at low temperature the stoichiometry is 2U to 1A, while at 20°C the stoichiometry becomes 1A to 1U. The physical properties of this poly U-adenosine complex were further analyzed by sedimentation, viscosity, and optical rotation measurements.

The formation of the poly U-adenosine (AR) complex can be demonstrated by analytical ultracentrifugation. Sedimentation coefficients (S) of poly U in the absence (control) and presence of nucleosides are given in Table VI. When

TABLE VI

SEDIMENTATION OF POLY U IN NUCLEOSIDE SOLUTIONS[a]

Nucleosides	Buffer (M)	Temp. (°C)	S_{20} Control	S_{20} Complex	Increase (%)
Adenosine	0.4 NaCl	5	4.68	6.21	33
Adenosine	0.4 NaCl	10	4.03	5.78	43
Adenosine	0.4 NaCl	19	4.00	6.11	53
Adenosine	0.02 MgCl$_2$	5	4.70	6.45	37
L-adenosine	0.4 NaCl	5	4.77	6.69	40
N-6-Methyl-adenosine	0.4 NaCl	5	4.77	4.77	0
Cytidine	0.4 NaCl	5	4.06	4.06	0
Inosine	0.4 NaCl	5	4.68	4.68	0

[a] Poly U concentration, 1.5×10^{-2} M; nucleoside concentration, 1.5×10^{-2} M.

N-6-methyladenosine, cytidine, or inosine was mixed with poly U in equal amounts (1.5×10^{-2} M each) at 5°C and 0.4 M NaCl, no change in either the pattern or the S value was found. As adenosine was mixed with poly U under identical conditions, a 33% increase in S value and a sharpening of the boundary was observed as compared with the control. Similar results were obtained in 0.02 M MgCl$_2$ with the same mixture. In 0.4 M NaCl, as the temperature was raised, the percentage change in the S value also increased to 43% at 10°C and 53% at 19°C, but it was accompanied by a decrease in sharpness of the boundary. The specific viscosities of poly U (1.5×10^{-2} M)

and poly U-AR complex (1.5×10^{-2} M of each) in 0.4 M NaCl at 5°C were 0.602 and 1.05, respectively. As previously stated a parallel increase in S value (33%) has also been observed. The concurrent increase in both specific viscosity and the sedimentation coefficient of the poly U-AR complex as compared with those of poly U unambiguously showed that there is a molecular weight increase in the polymer resulting from the complex formation.

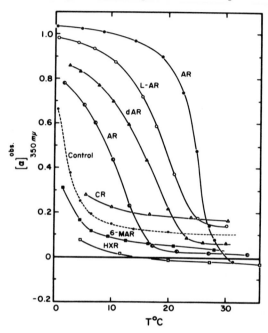

FIG. 6. Observed rotation of poly U (1.5×10^{-2} M)-nucleoside mixtures vs. temperature in 0.4 M NaCl, HMP. Concentrations of the nucleosides are: adenosine (AR), 1.5×10^{-2} M (—●—) and 7.3×10^{-3} M (—◑—); L-adenosine (L-AR), 9.3×10^{-3} M; deoxyadenosine (dAR) 7.8×10^{-3} M; cytidine (CR), 1.1×10^{-2} M; N-6-methyladenosine (6-MAR), 7×10^{-3} M; inosine (HXR), 1.5×10^{-2} M.

Optical rotation measurements at 350 mμ were used to determine the conformation and stability of the poly U-AR complex. Poly U in 0.4 M NaCl gave a small positive rotation at low temperature (at 1.5×10^{-2} M, the observed rotation was about 0.2 degree at 5°C). The rotation decreased with increasing temperature, finally becoming temperature insensitive beyond 12°C as shown in the control curve in Fig. 6. On the other hand, 1.5×10^{-2} M adenosine alone gave an observed rotation of $-0.09°$ calculated from the rotation at 1.2×10^{-2} M which was temperature independent. Nevertheless, when the two were mixed, a large increase in positive rotation was observed, $+1.03°$ at 5°C. At the temperature-insensitive region, the rotation of the mixture was the algebraic sum of its constituents. We took this to mean that the poly U-AR complex formed an ordered structure in 0.4 M NaCl and its stability was reflected by its melting behavior in response to the temperature variation. In 0.4 M salt solution, the optical rotation measurements remained essentially invariant with a temperature range from 0.5° to 20°C. When poly U is mixed with cytidine, inosine, or methylated adenosines no complex formation is observed (Fig. 6).

Formations of poly U-AR complex and its thermostability were highly dependent on adenosine concentration as illustrated in Fig. 7. When a constant amount of poly U (1.5×10^{-2} M) was allowed to interact with varying

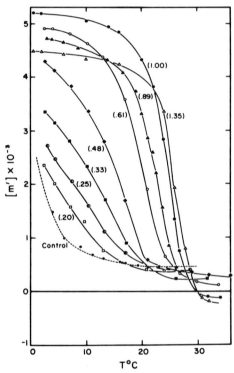

FIG. 7. The melting of poly U-AR complex in 0.4 M NaCl, HMP measured by rotation at 350 mμ. The poly U concentration is constant at 1.5×10^{-2} M. The parentheses indicate the input AR per UMP of poly U (A/U).

amounts of adenosine ranging from 3×10^{-3} to 2×10^{-2} M, a saturation phenomenon similar to that observed in the equilibrium dialysis was also found, i.e., the magnitude of the maximum rotation and apparent stability remained unchanged after the ratio of input adenosine per UMP of poly U (denoted by A/U) reached unity.

Various analogs of adenosine were also tested for their binding capacity to poly U using the optical rotation and sedimentation methods with the expectation of obtaining information about an involvement of binding sites and the role of the sugar moiety. The following compounds were tested: deoxyadenosine, L-adenosine (the pentose was L-ribose instead of D-ribose), (9-γ-hydroxypropyl)adenine and (9-hydroxypentyl)adenine (long-chain alcohols replacing the sugar moiety). Complexing with poly U was found for all these four compounds. When the point of attachment of the purine ring was changed from the 9-position to the 3-position as in the case of 3-isoadenosine, complex formation could still take place. All these observations indicate that the sugar moiety of the adenosine does not play an important role for the binding. Optical rotation studies of mixtures of poly U with N-6-methyladenosine, 1-methyladenosine, and tubercidin (A pyrrolopyrimidine-2,3-D riboside) revealed that no interaction took place. Therefore, the N-6-amino group and N-1 position of the adenine appears to be involved definitely in bonding with poly U. The other possible bonding site is the N-7 position of the adenine.

The two important aspects of the participation of adenosine in the interaction are its concentration dependence and specificity. The complex formation is undetectable in low nucleoside concentration. After a threshold concentration of adenosine is reached, the binding increases rapidly in a cooperative manner until saturation. The key to the understanding of the interaction resides in the properties of nucleosides in solutions of moderate concentration as detailed in the section of the monomer-monomer interaction. From these studies, we know that the stacking of adenosine occurs when the concentration increases. These stacks behave like the oligonucleotides and, therefore, have much greater affinity to poly U than the free adenosine. At moderate concentrations, these associated stacks may serve as initiators for the subsequent binding of the adenosine molecule to poly U by a cooperative mechanism. In fact, the stability of the completely interacting complexes measured in our experiments is comparable to that obtained for the poly U-trimer or -tetramer (oligonucleotides) interaction. The forces responsible for stacking energy are short-ranged. Calculations based on consideration of the nearest neighbor only gave an estimation of approximately 5 or 6 kcal/mole as the free energy of stacking for this poly U-AR system. The results clearly indicated that hydrogen bonding cannot be the sole force responsible for the binding, since in dilute solution no binding is detected, even though hydrogen bonding capacity is still present. On the other hand, hydrophobic stacking forces alone do not allow the interaction to occur. Inosine, methylated adenosines, and other adenine analogs probably all form stacks, yet they fail to bind to poly U. It appears, therefore, the hydrogen bonding and the hydrophobic stacking forces are both essential, with the former related to specificity and the latter related to stability. This system has been also investigated by the use of infrared spectroscopy and other techniques (Howard *et al.*, 1966).

The combined action of the hydrophobic stacking forces and the hydrogen bonding in the polymer-monomer system can be evaluated separately, and the results further indicate the importance of the stacking forces in this specific interaction. This is achieved in our laboratory by the introduction of guanosine into this well-characterized adenosine-poly U system in equilibrium dialysis experiments (Huang, 1967). In 0.4 M NaCl, Tris buffer (0.01 M, pH 7),

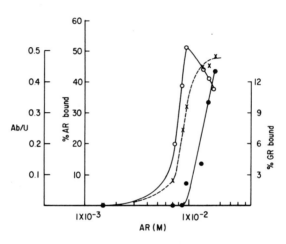

FIG. 8. The equilibrium dialysis of poly U (1.5×10^{-2} M) and mixed nucleosides at 5°C in 0.4 M NaCl, pH 7.0. The concentration of guanosine was fixed at 1.4×10^{-5} M and that of adenosine varied. (—×—), adenosine bound per UMP of poly U (Ab/U); (—O—), % adenosine bound; and (—●—), % guanosine bound.

and at 5°C, poly U (1.5×10^{-2} M) was dialyzed against a fixed amount of ^{14}C-guanosine (1.4×10^{-5} M) together with varying amounts of adenosine, ranging from 1.5×10^{-3} to 1.8×10^{-2} M. Figure 8 shows the distribution of adenosine and guanosine as the result of the equilibrium dialysis against poly U. The following observations are noted.

(1) The interaction of adenosine with poly U is not affected by the presence of guanosine at this concentration (1.4×10^{-5} M). A sigmoidal adsorption isotherm for adenosine with the same threshold concentration is again obtained as was previously described in Fig. 5.

(2) Guanosine shows no accumulation inside the dialysis tubing until the input concentration of adenosine reaches a threshold. This clearly indicates that guanosine does not bind to poly U by its own action. The accumulation of guanosine at high concentrations of adenosine is a result of adenosine binding to poly U. Since it is generally recognized that guanosine does not interact with poly U through complementary hydrogen bonding, the binding of

guanose into poly U in the presence of adenosine is then attributed to the co-stacking of the bases. However, the threshold concentration requirement in AR concentration for significant binding of guanosine is 9×10^{-3} M; higher than that for adenosine, which is 6×10^{-3} M. Consequently, the guanosine adsorption profile is correspondingly shifted toward higher input adenosine concentrations.

(3) The per cent binding of guanosine increases with the adenosine concentration. Such binding, however, continues to increase even when the adenosine binding to poly U reaches saturation ($A_b/U = 0.5$) at 1.5×10^{-2} M or higher. The interaction of adenosine and poly U is governed by the extent of the adenosine self-association as well as by the hydrogen bonding of the stacked species to the polymer. As soon as the binding sites of poly U are fully occupied, an increase in the input adenosine concentration can no longer change the difference in adenosine concentrations between the two sides of the dialysis tubing; yet the concentration of the free adenosine can still increase. Since the association of guanosine is expected to show a concentration dependence (Ts'o and Chan, 1964; Schweizer et al., 1965), the increase in free adenosine concentration will, therefore, increase the costacking between guanosine and adenosine. Since it is the stacked species that is binding to the polymer, the binding of guanosine to poly U continues to increase even after the binding of adenosine to poly U has reached saturation.

These experiments on the binding of the guanosine to the noncomplementary poly U by costacking with the adenosine again show the significance of the hydrophobic stacking force in the monomer-polymer system.

III. Contribution of the 2′-Hydroxyl Group to the Conformation and Interaction of Nucleic Acids

The existence of intramolecular hydrogen bonding between the 2-carbonyl group of the pyrimidine or the N-3 of the purine, and the 2′-hydroxyl group in nucleosides have been well supported by UV spectroscopy, pK_a determination, infrared spectroscopy of compounds in organic solvents, enzymatic catalysis, and NMR studies (see evidences summarized in Ts'o et al., 1966). Some of the evidences for intramolecular hydrogen bonding from the studies of the purine nucleosides by NMR have been presented above. This intramolecular hydrogen bonding should have the following effects on the pyrimidine nucleosides. (1) It should reduce the freedom of rotation between the glycosyl linkage of the N-1–C′-1 bond between the base and the pentose (Michelson, 1963). (2) The pK_a of the nucleosides and nucleotides should be affected. (3) It should significantly hinder intermolecular hydrogen-bond formation by the carbonyl group. The first two of these effects should also be true for the purine nucleosides, namely, reduction in freedom of rotation about the glycosyl linkage and lowering the pK_a of the bases. The N-3 position, however, is not expected to be involved in the intermolecular hydrogen-bonding scheme

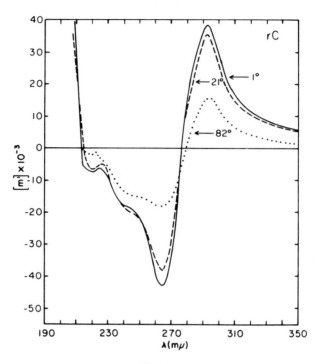

FIG. 9a

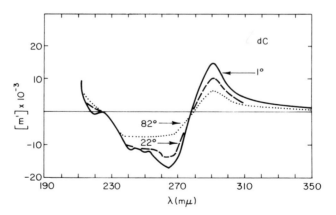

FIG. 9b

FIG. 9. ORD curves of r(C)$_n$ and d(C)$_n$. (a) r(C)$_n$ in 0.05 M NaPO$_4$, pH 8.4. At 1°C (—); 21°C (- - -); 82°C ($\cdots$). (b) d(C)$_n$ in 0.05 M NaPO$_4$, pH 8.4. At 1°C (—); 22°C (- - -); 82°C($\cdots$). (c) r(C)$_n$ in 0.05 M NaClO$_4$, 1 × 10^{-3} M Na-acetate; pH 4.4. At 20°C (—); 79.5°C (- - -); 84.5°C($\cdots$). (d) d(C)$_n$ in 0.05 M NaClO$_4$, 1 × 10^{-3} M Na-acetate; pH 5.1. At 20°C (—); 58°C (- - -); 79°C($\cdots$).

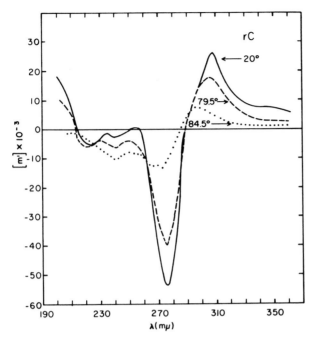

FIG. 9c

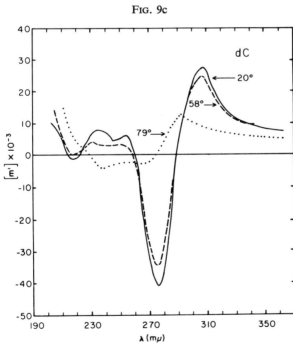

FIG. 9d

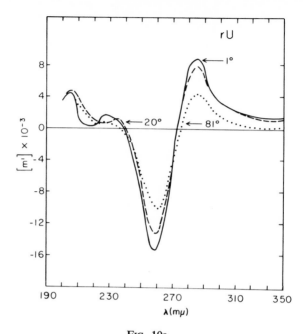

FIG. 10a

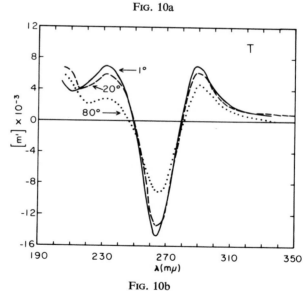

FIG. 10b

FIG. 10. ORD curves of r(U)$_n$ and d(T)$_n$. (a) r(U)$_n$ in 0.05 M NaClO$_4$, pH 7.0. At 1°C (—); 20°C (---); 81°C(···). (b) d(T)$_n$ in 0.05 M NaClO$_4$, pH 7.0. At 1°C (—); 20°C (---); 80°C (···). (c) r(U)$_n$ in 0.02 M Mg(ClO$_4$)$_2$, pH 6.4. At 1°C (—); 6°C (---); 10°C(···); 26°C(– · –). (d) d(T)$_n$ in 0.02 M Mg(Cl)$_2$, pH 6.4. At 2°C (—); 11°C (---); 32°–60°C (···).

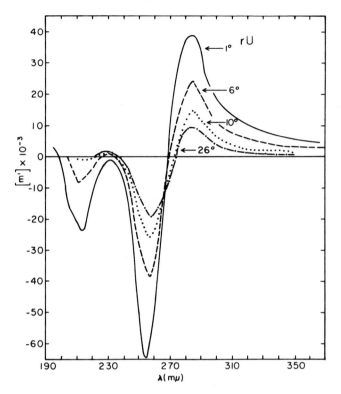

FIG. 10c

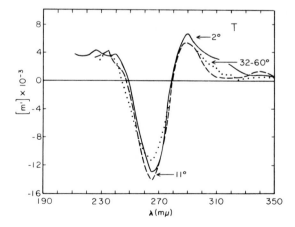

FIG. 10d

and should produce no effect in this regard. Currently, the occurrence of this intramolecular hydrogen bonding at the level of nucleotides and dinucleotides is being investigated in our laboratory. At present, this property of the monomer appears to provide an explanation for observed properties of the homopolymers. In the following paragraphs, the results from our laboratory concerning the influence of the 2'-OH group on the homopolymers of cytosine [$d(C)_n$ or $r(C)_n$], uracil, or thymidine [$r(U)_n$ or $d(T)_n$] and homopolymers of adenine [$r(A)_n$ or $d(A)_n$] will be briefly described (Ts'o et al., 1966).

The comparative study of the properties of $d(C)_n$ vs. those of $r(C)_n$ yields the following three conclusions:

1. In neutral or slightly alkaline medium $d(C)_n$ has much less stacking interaction than $r(C)_n$. This conclusion is arrived at as follows: (a) Between 23° and 90°C, $r(C)_n$ exhibits 15–16% hyperchromicity and less than 1% was observed for $d(C)_n$. Correspondingly, the maximum molar extinction coefficient of $d(C)_n$ (7.4×10^3) is about 15% higher than that of the $r(C)_n$ (6.5×10^3) at 23°C. (b) The ORD curves of $d(C)_n$ and $r(C)_n$ are generally similar to each other, but the absolute rotation value of the $d(C)_n$ curve at 1°C is about the same as that of $r(C)_n$ at 82°C (Fig. 9a,b), indicating a higher degree of secondary structure arising from stacking interaction in $r(C)_n$.

2. In acidic solution $d(C)_n$ and $r(C)_n$ form similar structures, as indicated by identical ORD curves (Fig. 9c,d) and a common requirement for protonation of the helix formation. There is, however, a major difference in the stabilities of the acid forms of $d(C)_n$ and $r(C)_n$. Titration studies at 25°C have indicated that the transition pH of $d(C)_n$ in 0.05 M Na ions is at pH 7.2 (Inman, 1964) and the transition pH of $r(C)_n$ in 0.1 M Na ions is at pH 5.7 (Hartman and Rich, 1965). Thus, at room temperature helix formation in $r(C)_n$ requires more protons in solution (1.5 pH unit) than $d(C)_n$ does.

3. The hydrogen-bonding scheme of the acid form of $r(C)_n$ has been shown to involve two pairs of interchain hydrogen bonds from the 2-carbonyl group to the 4-amino group, with a proton shared by two N-3 ring nitrogens from both chains (Akinrimisi et al., 1963; Langridge and Rich, 1963; Hartman and Rich, 1965). The hydrogen-bonding scheme may be the same for $d(C)_n$.

In neutral or slightly alkaline solution, the higher degree of secondary structure in $r(C)_n$ can be explained on the ground that the intramolecular hydrogen bonding of the 2'-hydroxyl group to the 2-carbonyl group greatly reduces rotational freedom around the N-1–C'-1 bond. This may enhance the stacking of bases along the chain. The lower stability of the helix of $r(C)_n$ can be explained by the following reasons: (a) that the intramolecular hydrogen bonds greatly hinder participation of the 2-carbonyl group in interchain hydrogen bonding and (b) that the pK_a of the ribosyl cytosine group is lowered by intramolecular hydrogen bonding. Both of these effects will tend to lower the transition pH of $r(C)_n$ as compared to $d(C)_n$.

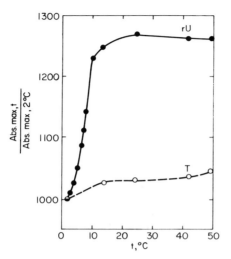

FIG. 11. The absorbance vs. temperature profile of r(U)$_n$ and d(T)$_n$ at 260 mμ in 0.02 M Mg(ClO$_4$)$_2$, pH 6.4.

The comparative study of d(T)$_n$ vs. r(U)$_n$ and r(T)$_n$ yields different conclusions. The ORD patterns (Fig. 10a,b) of the d(T)$_n$ and the r(U)$_n$ are much the same at room temperature and in the absence of Mg^{2+}. In the presence of Mg^{2+} ions (0.01–0.2 M) and at low temperature r(U)$_n$ acquires an ordered structure (Fig. 10c) having a T_m of about 8°C (Fig. 11) (Lipsett, 1960; Shugar and Szer, 1962). r(T)$_n$ in 0.01 M MgCl$_2$ was found to have a T_m of 36°C (Shugar and Szer, 1962) and the higher T_m can generally be explained on the basis of increase of hydrophobic stacking interaction, since thymidine associates to a greater extent than uridine in water (Ts'o *et al.*, 1963b; Broom *et al.*, 1967). On the other hand, both optical density vs. temperature profile (Fig. 11) and the ORD patterns vs. temperature studies (Fig. 10d) indicate that d(T)$_n$ has very little stacking interaction and secondary structure even at 1°C, 0.02 M Mg^{2+}. It appears that the presence of 2'-hydroxyl groups in the polymer contributes a major stabilizing influence on the secondary structure. This effect is opposite to the effect observed in r(C)$_n$, where the presence of 2'-hydroxyl groups destabilizes the helical structure of r(C)$_n$.

The hydrogen-bonding scheme of the ordered forms of r(U)$_n$ or r(T)$_n$ has not been established although the N-3 position is mostly likely to be involved (Szer and Shugar, 1961). The most plausible hydrogen-bonding scheme proposed (Donohue, 1956; Green *et al.*, 1962) is a double-strand helix with the two interchain hydrogen bonds formed between the N-3 (donor) from one chain and the O-4 (acceptor) from the other.

If this hydrogen bonding scheme for r(U)$_n$ or r(T)$_n$ is accepted, then the effect of 2'-hydroxyl group can be readily understood. Since the 2-carbonyl

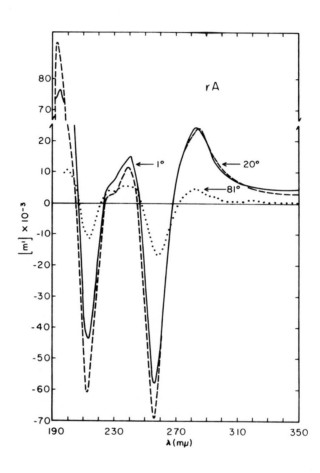

FIG. 12. ORD curves of r(A)$_n$ and d(A)$_n$. (a) r(A)$_n$ in 0.05 M NaClO$_4$, pH 7.0. At 1°C (—); 20°C (– – –); 81°C ($\cdots$). (b) d(A)$_n$ in 0.05 M NaClO$_4$, pH 7.35. At 3.3°C (—); 20°C (– – –); 85.3°C ($\cdots$). (c) d(A)$_n$ and r(A)$_n$ in acidic solutions; d(A)$_n$ at 20°C, 1 × 10^{-3} M acetic acid and at pH 3.4 (—); r(A)$_n$ at 20°C (– – –) and at 64°C ($\cdots$). pH 5.0 in 0.05 M NaClO$_4$, 1 × 10^{-3} M HAc–NaAc.

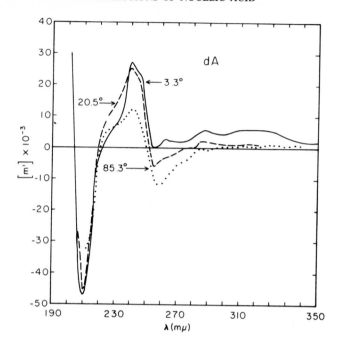

FIG. 12b

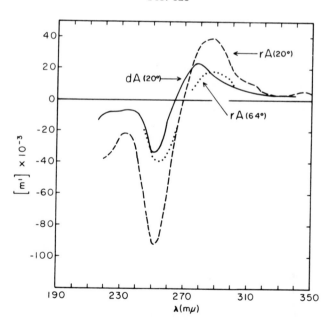

FIG. 12c

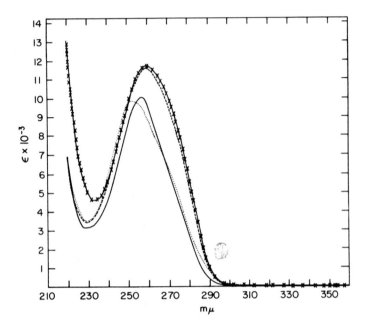

FIG. 13. The UV absorption spectra of d(A)$_n$: (—) pH 8.0, 1×10^{-3} M Tris; (---) pH 5.25, 1×10^{-3} M acetate; (—×—) pH 4.07, 0.2 M NaClO$_4$, 0.02 M NaAc; (···) pH 3.23, 1×10^{-3} M acetate.

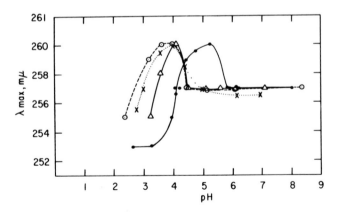

FIG. 14. The effect of salt concentration and pH on the λ_{max} of d(A)$_n$ UV absorption spectra: (—●—) 10^{-3} M buffers (Tris or acetate); (—△—) 0.02 M Na—PO$_4$ or acetate buffer plus 0.03 M NaClO$_4$; (---×) 0.02 M PO$_4$ or acetate buffer plus 0.09 M NaClO$_4$; (---○) 0.02 M Na—PO$_4$ or acetate buffer plus 0.2 M NaClO$_4$.

group does not participate in this hydrogen-bonding scheme, the intramolecular hydrogen bonding of the carbonyl group to the 2-hydroxyl group does not hinder the helix formation. A reduction of the rotational freedom along the axis of N-1–C'-1 bond may enhance base stacking and increase stability of the helix. Thus, we postulate that the helical form of $r(U)_n$ favors the hydrogen bonding of the 2'-hydroxyl group to the 2-keto group more than the coil form of $r(U)_n$, thereby gaining extra energy for stabilization. Therefore, the opposite effect of intramolecular hydrogen bonding on the stability of the helix or $r(C)_n$ vs. $d(C)_n$ occurs because here the 2-carbonyl group participates in the H-bonding scheme of the helical $r(C)_n$. The protonation requirements in the C polymers also contribute to this difference.

The differences between $d(A)_n$ vs. $r(A)_n$ can be discussed from three different aspects.

1. At neutral pH $d(A)_n$ and $r(A)_n$ have the same UV hyperchromicity upon heating. This would seem to indicate that the two polymers have the same degree of stacking interaction. The ORD patterns of $d(A)_n$ and $r(A)_n$ are, however, vastly different. The prominent peak and trough in the region of 250–210 mμ $r(A)_n$ (Fig. 12a) is absent in the ORD pattern of $d(A)_n$ (Fig. 12b). According to exciton theory the rotational strength (R_k) is a trigonometric function of the angle between a given transition moment in one base and the corresponding moment in the neighboring base along the polymer. This angle is called α in the notation of Van Holde et al. (1965) and Brahms et al. (1966) (denoted as $2\pi/P$ in the general theory of Bradley et al. (1963), where P is the number of residues per turn in a regular helix) in their calculation of the circular dichroism of the oligomers of riboadenylate. If α is $0°$ ($P = 1$, straight stack) or $180°$ ($P = 2$, alternating stack), the rotational strength arising from the nearest-neighbor interaction is zero, and no optical activity will be observed. The difference in ORD of $r(A)_n$ and $d(A)_n$ at 240–310 mμ suggest that the angle between this transition moment must be neither $0°$ nor $180°$ in the case of $r(A)_n$ and very close to $0°$ or $180°$ in the case of $d(A)_n$. The blue shifts of the polymer absorption maximum, 257 mμ as compared to 259 mμ for the monomer, indicate that the bases of the $d(A)_n$ are probably in a straight stack ($\alpha = 0$). Since these two polymers are identical in their primary structures, except for the 2'-OH group, it appears that an intramolecular hydrogen bond from the 2'-OH to N-3 of adenine may cause the angle α, formed between the neighboring bases in the stacks of $r(A)_n$, to be more oblique.

2. At acid pH the ORD of $d(A)_n$ is similar to that of $r(A)_n$ indicating resemblance of overall structure and intermolecular bonding in the helical form (Rich et al., 1961). The magnitude of the peak and the trough of $d(A)_n$ is less than that of the $r(A)_n$ in full helical form. This information suggests a slightly different arrangement of the bases in the helical $d(A)_n$ as compared to $r(A)_n$.

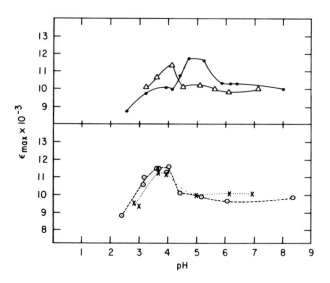

FIG. 15. The effect of salt concentration and pH on the E_{max} of $d(A)_n$ UV absorption spectra. Symbols are the same as in Fig. 14.

3. The spectrum of protonated $d(A)_n$ (Figs. 13–15) is greatly different from that of $r(A)_n$ (Helmkamp and Ts'o, 1961). Partial protonation (pH 5.0 or below) results in a 3-mμ bathochromic shift and a 17% increase in E_{max} in $d(A)_n$, whereas $r(A)_n$ exhibits a hypsochromic shift of 4 mμ and a lowering of 18% in E_{max}. ORD shows $d(A)_n$ to be in the helical form at this pH (comparison of Fig. 16a,b with Fig. 15). This remarkable difference between the spectra of the acidic forms of $r(A)_n$ and $d(A)_n$ suggests that all the protons may not go to the N-1 position as generally expected for the $r(A)_n$. Studies of the UV spectra of suitable model compounds upon protonation indicate that the reason for the red shift of λ_{max} and the enhancement of E_{max} in the spectrum of $d(A)_n$ upon protonation is that some of the protons go to N-3 instead of N-1 (Ts'o *et al.*, 1966). The possibility of protonation at the N-7 position is considered unlikely if the hydrogen-bonding scheme of the $d(A)_n$ helix is essentially that of the $r(A)_n$ helix—then the N-7 is in the middle of the helix, hydrogen-bonded to the 6-amino group. Protonation at this nitrogen will not allow the helix to form. Though at present we cannot quantitatively assess the proportion of the protons going to the N-3 vs. those going to N-1 in $d(A)_n$, all this information thus indicates that upon partial protonation the proportion of the protons going to N-3 in $d(A)_n$ is higher than that in $r(A)_n$.

When the pH of the solution is lowered further the spectrum exhibits a hypsochromic shift of λ_{max} and a lowering of E_{max}. The ORD pattern is essentially unchanged. Although there is aggregation at this pH we do not

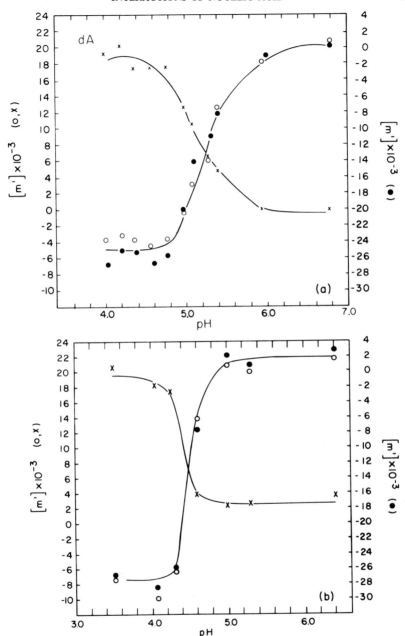

FIG. 16. The (m') vs. pH profile of $d(A)_n$. (a) In 1×10^{-3} M Na-acetate, titration starting from pH 4.0 with addition of small amount of NaOH; (—×—) at 279 mμ (left scale); (—●—) at 252.5 mμ (right scale); (—O—) at 240 mμ (left scale). Temperature at 22°C. (b) In 0.2 M NaClO$_4$, 0.02 M Na-acetate. Samples were prepared individually at different pH values; (—×—) at 279 mμ (left scale); (—●—) at 252 mμ (right scale); and (—O—) at 240 mμ (left scale). Temperature, 20°C.

believe these observations to be optical artifacts of aggregation. If our reasoning on this point is valid, then this is an indication that upon further increase in H^+ more protons now go to the N-1 position of $d(A)_n$. The reason for this is not immediately apparent.

Why is it that during early stages of protonation the proton tends to go to N-3 in the $d(A)_n$ helix, but goes to N-1 in the $r(A)_n$ helix? We propose that in the $r(A)_n$ helix there is a degree of hydrogen bonding between N-3 and the 2'-OH as in the case of the adenosine which cannot occur in the $d(A)_n$ helix. Support for this notion may be taken from alkylation studies on nucleic acids (Lawley and Brookes, 1963). No 3-methyladenine is found in RNA after the reaction (most of the alkylation takes place at N-1), but 3-methyladenine is the *major* product from the alkylation of native DNA and is a *minor* product for heat-denatured DNA.

4. The pH for the helix-coil transition at 20°C examined by ORD indicates that in 0.001 M salt this transition is at pH 5.3 and in 0.22 M salt it is at pH 4.4 (Fig. 16a,b). Similar results are also obtained from the UV spectrum (Figs. 14 and 15). The effect of salt on the pH transition for $d(A)_n$ is similar to $r(A)_n$ as studied by potentiometric titration. This confirms that the transition pH of $r(A)_n$ in 0.001 M KCl at 20°C is 6.8 and in 0.15 M KCl it is 6.0 (Holcomb, 1966). In 0.01 M KCl at 26°C Steiner and Beers (1957) found a transition pH of 6.4 and in 0.1 M KCl at 26°C it is 6.0. Under similar conditions $d(A)_n$ apparently requires a pH about 1.5 units lower than that required by the $r(A)_n$ to go into the helical form. The greater number of protons required for the formation of $d(A)_n$ helix indicates that the $r(A)_n$ helix is more stable. The greater stability of the $r(A)_n$ helix may be ascribed to the intramolecular hydrogen bonding of the 2'-OH group to the base.

In summary, the comparative studies of polyribo- vs. polydeoxyribo-nucleotides of cytosine, uracil (or thymine), and adenine yield the following conclusions.

1. Although both poly C and poly A require protonation for the formation of helix, under comparable conditions, poly dC is more stable than the poly rC by about 1.5 pH units; while the poly dA is less stable than the poly rA by about 1.5 pH units.

2. There is no indication that the hydrogen-bonding schemes of the double helices are different for the ribosyl- and the deoxyribosyl homopolymers. The UV spectrum study, however, indicated that the site of protonation is not exactly the same for $d(A)_n$ and $r(A)_n$. At the early phase of protonation, more protons appear to go to N-3 and less go to N-1 in the case of $d(A)_n$ as compared to those in the case of $r(A)_n$. Such a difference in sites of protonation is not noticed between $d(C)_n$ and $r(C)_n$.

3. In the nonhelical form, the $d(C)_n$ appears to have much less secondary structure than the $r(C)$; the $d(T)_n$ also appears to have much less secondary

structure than the $r(U)_n$ and $r(T)_n$. In fact, no helical state of $d(T)_n$ has yet been found at low temperature in the presence of Mg^{2+} ions. For poly A in the nonhelical form, the $r(A)_n$ and $d(A)_n$ appear to have the same UV absorption spectral properties. According to the hypochromicity measurement, these two polymers should have the same extent of secondary structure. However, their ORD patterns are very different from each other. This suggests that the geometry of stacking in the secondary structure is different between the $r(A)_n$ and $d(A)_n$. These two polymers, therefore, may have different forms of conformation. Such an incongruence between the results of ORD and UV absorbance has not been found for the comparison of optical properties of other polymers.

It can be concluded, therefore, that the effect of 2'-hydroxyl group on the polynucleotides is specific for different bases. At present, these effects appear to be explainable in many cases by the intramolecular hydrogen bonding found in the nucleosides. Though, it is also tempting to postulate about the hydrogen bonding between the 2'-hydroxyl group to the oxygen atom of the adjacent 3'-phosphate group (Spencer *et al.*, 1962; Langridge and Gamatos, 1963). Such a bonding may, indeed, exist in view of the formation of the cyclic phosphate during hydrolysis of polyribonucleotides. At present, however, it is very difficult to invoke such a bonding phenomenon of hydroxyl group to the phosphate in order to explain the different effects observed on the different bases of the polynucleotides.

IV. Final Remarks

Together with the recent studies on the properties of the bases and nucleosides in organic solvents, the studies reported in this chapter provide a physicochemical basis for the understanding of the interaction of nucleic acids. Not much is yet known, however, about the influence of the backbone of the polynucleotide chain, especially of the rigid furanose ring, on the conformation and interaction of nucleic acids. The research on the differences in the ribosyl- and deoxyribosyl-polynucleotides done in our laboratory and in others (Chamberlin, 1965; Chamberlin and Patterson, 1965) is only but a beginning toward this direction. More effort will be devoted to this direction from our laboratory.

The insight we gained about the physicochemical forces which govern the interaction and conformation of nucleic acids may also provide us with additional knowledge about the mechanism of replication or transcription of nucleic acids. Our laboratory is currently working toward this goal (Huang, 1967).

REFERENCES

Akinrimisi, E. O., Sander, C., and Ts'o, P. O. P. (1963). *Biochemistry* **2**, 340.

Bradley, D. F., Tinoco, I., Jr., and Woody, R. W. (1963). *Biopolymers* **1**, 239.

Brahms, J., Michelson, A. M., and Van Holde, K. E. (1966). *J. Mol. Biol.* **15**, 467.

Broom, A. D., Schweizer, M. P., and Ts'o, P. O. P. (1967). *J. Am. Chem. Soc.* **89**, 3612.

Chamberlin, M. J. (1965). *Federation Proc.* **24**, 144.

Chamberlin, M. J., and Patterson, D. L. (1965). *J. Mol. Biol.* **12**, 410.

Chan, S. I., Schweizer, M. P., Ts'o, P. O. P., and Helmkamp, G. K. (1964). *J. Am. Chem. Soc.* **86**, 4182.

Crothers, D. M., and Zimm, B. H. (1964). *J. Mol. Biol.* **9**, 1.

DeVoe, H., and Tinoco, I., Jr. (1962). *J. Mol. Biol.* **4**, 500.

Donohue, J. (1956). *Proc. Natl. Acad. Sci. U.S.* **42**, 60.

Green, D. W., Mathews, F. S., and Rich, A. (1962). *J. Biol. Chem.* **237**, 3573.

Hamlin, R. M., Jr., Lord, R. C., and Rich, A. (1965). *Science* **148**, 1734.

Hanlon, S. (1966). *Biochem. Biophys. Res. Commun.* **23**, 861.

Hartman, K. A., and Rich, A. (1965). *J. Am. Chem. Soc.* **87**, 2033.

Helmkamp, G. K., and Ts'o, P. O. P. (1961). *J. Am. Chem. Soc.* **83**, 138.

Herskovits, T. T. (1962). *Arch. Biochem. Biophys.* **97**, 433.

Herskovits, T. T., Singer, S. J., and Geiduschek, E. P. (1961). *Arch. Biochem. Biophys.* **94**, 99.

Hill, T. L. (1960). "Introduction to Statistical Thermodynamics," Chapter 14. Addison-Wesley, Reading, Massachusetts.

Holcomb, D. M. (1966). Unpublished results.

Howard, F. B., Frazier, J., Singer, M. F., and Miles, H. T. (1966). *J. Mol. Biol.* **16**, 415.

Huang, W. M. (1967). Ph.D. Thesis, Johns Hopkins University.

Huang, W. M., and Ts'o, P. O. P. (1966). *J. Mol. Biol.* **16**, 523.

Inman, R. B. 1964. *J. Mol. Biol.* **9**, 624.

Katz, L., and Penman, S. (1966). *J. Mol. Biol.* **15**, 220.

Küchler, E., and Derkosch, J. (1966). *Z. Naturforsch.* **21b**, 209.

Kyogoku, Y., Lord, R. C., and Rich, A. (1966). *Science* **154**, 518.

Kyogoku, Y., Lord, R. C., and Rich, A. (1967). *J. Am. Chem. Soc.* **89**, 496.

Langridge, R., and Gamatos, P. J. (1963). *Science* **141**, 649.

Langridge, R., and Rich, A. (1963). *Nature* **198**, 725.

Lawley, P. D., and Brookes, P. (1963). *Biochem. J.* **89**, 127.

Lipsett, M. N. (1960). *Proc. Natl. Acad. Sci. U.S.* **46**, 445.

MacIntyre, W. M. (1965). *Science* **147**, 507.

Marmur, J., and Ts'o, P. O. P. (1961). *Biochim. Biophys. Acta* **51**, 32.

Michelson, A. M. (1963). "The Chemistry of Nucleosides and Nucleotides," Chapter 8. Academic Press, New York.

Pitha, J., Jones, R. N., and Pithova, P. (1966). *Can. J. Chem.* **44**, 1045.

Pullman, B. (1965a). *J. Chem. Phys.* **43**, S233.

Pullman, B. (1965b). *In* "Molecular Biophysics" (B. Pullman and M. Weissbluth, eds.), pp. 154–157. Academic Press, New York.

Pullman, B., Claverie, P., and Caillet, J. (1958). *Compt. Rend.* **20**, 5387.

Rice, S. A., Wada, A., and Geiduschek, E. P. (1958). *Discussions Faraday Soc.* **25**, 130.

Rich, A., Davies, D. R., Crick, F. H. C., and Watson, J. D. (1961). *J. Mol. Biol.* **3**, 71.

Schellman, J. A. (1956). *Compt. Rend. Trav. Lab. Carlsberg, Ser. Chim.* **29**, 223.

Schweizer, M. P., Chan, S. I., and Ts'o, P. O. P. (1965). *J. Am. Chem. Soc.* **87**, 5241.

Shoup, R. R., Miles, H. T., and Becker, E. D. (1966). *Biochem. Biophys. Res. Commun.* **23**, 194.

Shugar, D., and Szer, W. (1962). *J. Mol. Biol.* **5**, 580.

Solie, T. (1965). Ph.D. Thesis, University of Oregon.

Spencer, M., Fuller, W., Wilkins, M. H. F., and Brown, G. L. (1962). *Nature* **194**, 1014.

Steiner, R. F., and Beers, R. F., Jr. (1957). *Biochim. Biophys. Acta* **26**, 336.

Steiner, R. F., and Beers, R. F., Jr. (1960). "Polynucleotides." Elsevier, Amsterdam.

Sturtevant, J. M., Rice, S. A., and Geiduschek, E. P. (1958). *Discussions Faraday Soc.* **25**, 138.

Szer, W., and Shugar, D. (1961). *Acta Biochim. Polon. (English Transl.)* **9**, 225.

Ts'o, P. O. P., and Chan, S. I. (1964). *J. Am. Chem. Soc.* **86**, 4176.

Ts'o, P. O. P., Helmkamp, G. K., and Sander, C. (1962). *Biochim. Biophys. Acta* **55**, 584.

Ts'o, P. O. P., Helmkamp, G. K., Sander, C., and Studier, F. W. (1963a). *Biochim. Biophys. Acta* **76**, 54.

Ts'o, P. O. P., Melvin, I. S., and Olson, A. C. (1963b). *J. Am. Chem. Soc.* **85**, 1289.

Ts'o, P. O. P., Rapaport, S. A., and Bollum, F. J. (1966). *Biochemistry* **5**, 4153.

Van Holde, K. E., Brahms, J., and Michelson, A. M. (1965). *J. Mol. Biol.* **15**, 467.

Base-Base Interactions in Nucleic Acids

IGNACIO TINOCO, JR., ROBERT C. DAVIS,
AND S. RICHARD JASKUNAS

Chemistry Department and
Chemical Biodynamics Laboratory
University of California, Berkeley, California

One way of learning about the three-dimensional structure of nucleic acids in solution is to study the properties of their individual components. We have been studying the structures of oligoribonucleotides and would like to summarize our knowledge to date. Our goal is twofold: (1) to be able to draw general conclusions about conformation simply from the sequence of bases in the polymer; and (2) from some physical measurements to determine a more precise conformation for the nucleic acid at a particular temperature, pH, ionic strength, etc.

I. Components

We have considered four bases (adenine, guanine, uracil, and cytosine) and D-ribose. If we knew how these molecules interacted when connected in a ribonucleic acid molecule, the first goal would nearly be reached. In addition, of course, we need to know the interactions involving the phosphate group.

A. Mononucleotides

The most important aspect of the conformation of a mononucleoside is the angle of rotation around the glycosidic link between the ribose and base. Donohue and Trueblood (1) concluded that the base could exist in two conformations differing by approximately 180°. The conformation which placed the C-6–H of cytosine and uracil and the C-8–H of adenine and guanine near the ribose ether oxygen was termed anti; the conformation with these away from the ribose is syn. An analysis of many crystal structures of mononucleosides and mononucleotides by Haschemeyer and Rich (2) shows that in these crystals adenine, cytosine, and uracil exist in the anti conformation, but guanine is syn in the one crystal structure studied (deoxyguanosine). These conformations are illustrated in Fig. 1. We have made calculations of the potential energy of interaction of each base with ribose as a function of rotation around the glycosidic bond. Each atom is assigned a charge, a polarizability, an ionization energy, and a steric repulsion parameter. The

77

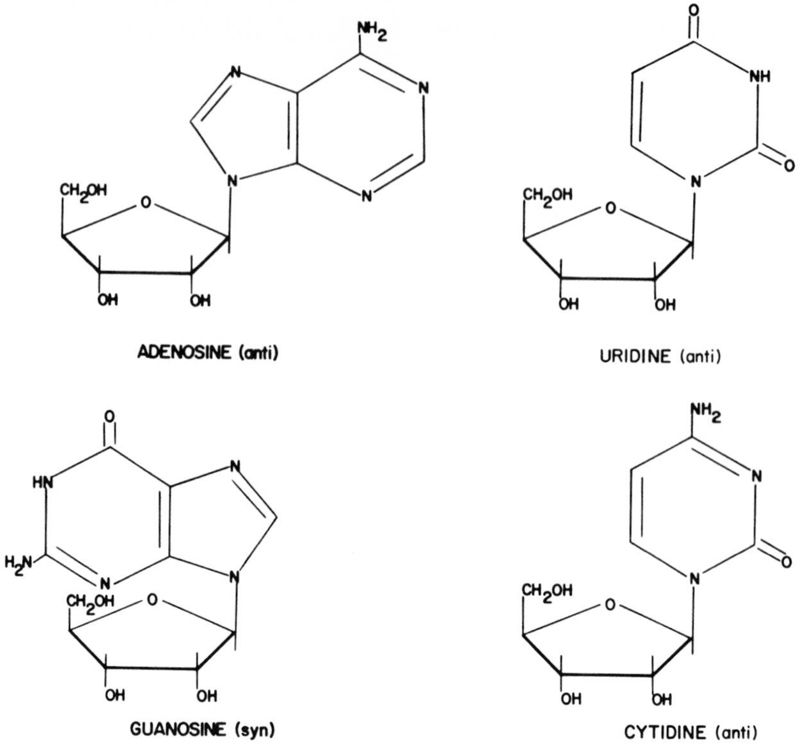

FIG. 1. Conformations found in crystals and assumed for solutions of four mononucleosides.

calculations are very similar to those done for polypeptides and amino acids by Ramachandran, Scheraga, and others (3, 4). The results can be represented as a probability of finding the base in a particular conformation at any temperature. The probability is defined as equal to $\exp(-E_i/RT)/\sum_i \exp(-E_i/RT)$ where E_i equals the energy of conformation i. Calculations were done for a number of possible ribose geometries (5); the conclusions do not depend on the choice of geometry. The results for the ribose conformation found in crystals of cytidine 3′-phosphate (6) (2′-endo) are shown in Fig. 2. The angle ϕ_{CN} is defined in Ref. 1. It is seen that adenosine, cytidine, and uridine are anti, whereas guanosine is syn. This implies that the conformation found in the crystal is retained in solution. Of course, specific solvent interaction or interaction with other molecules may very well change this. The calculated difference in energy between syn and anti is only 1 to 2 kcal/mole for adenosine and guanosine and 5 to 7 kcal for uridine and cytidine. We will assume that the conformation of the pyrimidine mononucleosides and mononucleotides are anti, but that of the purine monomers can be either syn or anti. Direct

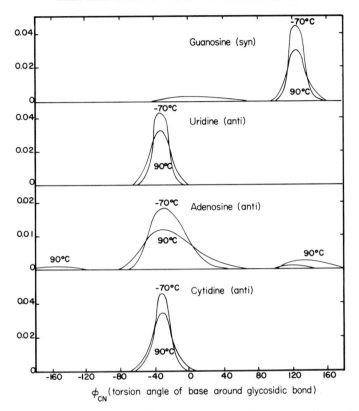

FIG. 2. Calculated probability of finding a mononucleoside in a particular conformation at two temperatures.

evidence that the pyrimidine mononucleosides are anti in solution has been obtained from studies of their optical rotatory dispersion (7).

Further calculations have shown that reasonable changes in geometry and parameters lead to syn as the most stable conformation for adenosine. This is consistent with the assumption that the purine nucleosides in solution can be either syn or anti.

B. Dinucleoside Phosphates

The only crystal structure reported for a dinucleoside phosphate is that of the 2′–5′-linked ApU (8). The bases are anti and essentially stacked one above the other. We have not made any potential energy calculations as yet, but there is much evidence that the bases also stack in aqueous solution (9–11). From earlier studies of the optical properties (11) we know that ApA and UpU show the most extreme properties among the set of 16 dinucleoside

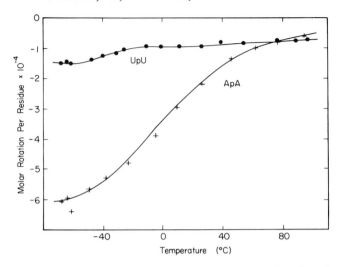

Fig. 3. The optical rotatory dispersion of ApA and UpU as a function of temperature in 25.2% LiCl at pH 7.

phosphates that can be formed from the four mononucleotides. We will discuss these two in detail here. Studies of all the dimers will be reported elsewhere (12).

The optical rotation of ApA and UpU indicates a low-temperature ordered form and a high-temperature disordered form for each molecule. This is illustrated in Fig. 3. It is also clear that the behavior of the two molecules is quite different. ApA is half-ordered at $+5°C$, whereas UpU may not reach this state until $-40°C$ or lower. If one assumes a simple two-state model for the equilibrium between ordered and disordered forms, standard thermodynamic parameters can be obtained. The temperature dependence can be fitted by a two-state model for ApA in 25.2% LiCl; it leads to a value of $\Delta H°$ for unstacking of 5.3 kcal/mole and a $\Delta S°$ of 19 eu. This is the same as estimated from the ORD temperature dependence from $0°$ to $90°C$ in $0.10\ M$ NaCl in water. For UpU in 25.2% LiCl the rotations are so nearly equal to the mononucleotide rotation that thermodynamic values cannot be obtained. The high concentration of LiCl evidently disorders uracil-containing dimers more than it does ApA. For the unstacking of bases in a dinucleoside phosphate we do not think that a two-state process is actually occurring. It is much more reasonable to expect a gradual increase in disorder as rotation about various bonds increases with temperature (see Fig. 2); i.e., we expect an increase in the average base-base distance and an increase in the fluctuations of this distance with temperature. Finally each base may become individually solvated and then we conclude the bases are unstacked. Evidence for this view is threefold. (i) The absorption and optical rotatory dispersion

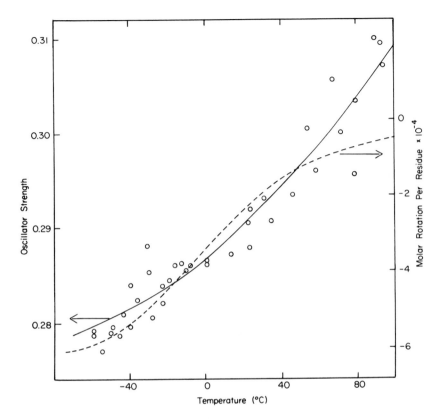

FIG. 4. A comparison of the temperature dependence of the absorption and rotation of ApA in 25.2% LiCl at pH 7.

for ApA have different temperature dependences as shown in Fig. 4. (ii) The ORD temperature dependence (from 0° to 90°C) of the fifteen dinucleoside phosphates studied can be fit by a model in which the two bases are assumed to be connected by a harmonic, torsional spring (13). (iii) The thermodynamic parameters for the two-state model calculated from different properties for ApA differ by a factor of two as shown in Table I.

Therefore, the logic and the evidence is consistent with the following picture. At −70°C and below the bases of all dinucleoside phosphates are rigidly stacked one above the other. As the temperature increases the structure becomes less rigid and the bases increase their relative oscillation. Both torsional and stretching oscillations will occur. When the amplitudes are large enough, solvent will be between the bases most of the time and the dinucleoside phosphate will be essentially unstacked. The potential energy describing this motion (which leads to all the thermodynamic variables) will

TABLE I

A COMPARISON OF $\Delta H°$ VALUES FOUND FOR UNSTACKING OF ApA

Optical property	$\Delta H°$ (kcal/mole)	Reference
Optical rotation	5.3	[a]
Optical rotation	6.5	[b]
Circular dichroism	8	[c]
Hypochromism	8.5	[d]
Hypochromicity	9.4	[e]
Hypochromicity	10	[f]

[a] This work.
[b] D. Poland, J. Vournakis, and H. Scheraga, *Biopolymers* **4**, 223 (1966).
[c] K. E. VanHolde, J. Brahms, and A. M. Michelson, *J. Mol. Biol.* **12**, 726 (1965).
[d] This work.
[e] J. Applequist and V. Damle, *J. Am. Chem. Soc.* **88**, 3895 (1966).
[f] M. Leng and G. Felsenfeld, *J. Mol. Biol.* **15**, 455 (1965).

be very complicated, but it can be tested by comparing calculated and measured optical properties for each dinucleoside phosphate. The optical rotatory dispersion shows that UpU is the least stacked, therefore uracil is the most solvated base. On the other hand, ApA is so strongly stacked that even at 90°C there is significant base-base interaction. This is shown in Fig. 5 where one observes that four nuclear magnetic resonance peaks are seen at all temperatures for the H-2 and H-8 protons on each of the two adenines. That this magnetic nonequivalence of the four protons is caused by interaction of the bases has been shown by appropriate controls (12).

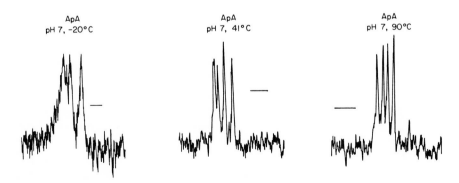

FIG. 5. The nuclear magnetic resonance of the aromatic protons of ApA in 25% LiCl in D_2O at three temperatures. The bar is 10 cps long.

II. Single-Strand Polynucleotides

Knowledge of the structures of the dinucleoside phosphates will only be useful to us if it can be applied to higher polymers. Earlier work has indicated that dimer geometry can indeed be applied to trimers (14) and long polymers (15). This work will not be reviewed here. The important fact is that the optical rotatory dispersion and absorption for polymers is an appropriate sum of the corresponding properties of dimers and monomers. This implies that the interaction between nearest neighbor bases (and between base and ribose) is similar in the polymer to that in the dimer.

III. Base Pairing

We have discussed base stacking in dinucleoside phosphates up to now, tacitly assuming concentrations were low enough to avoid intermolecular base pairing. We need to know however, the magnitude of base-pairing inter-actions and we also need to know the changes in ORD associated with base pairing. These changes had been roughly estimated from studies of homo-polynucleotides (15). The main conclusion was that base pairing led to a blue shift of the first cross-over wavelength in ORD.

To study intermolecular interactions systematically, we have measured the ORD of mixtures of oligonucleotides. A 0.01 M concentration of nucleotides was used which is the maximum practical for a 0.1 mm cell. Two solvents at pH 7 were used: one contained 0.1 M phosphate plus 0.5 M NaCl; the other contained 0.05 M phosphate plus 0.01 M MgCl$_2$. Studies were made at 1° and 26°C. The measure of base-base interaction was always a significant difference between the ORD of the 1:1 mixture of oligonucleotides and the sum of the individual components. We have only tried oligomers which could form Watson-Crick base pairs in an antiparallel arrangement. The dimer pairs of ApC and GpU were tried without success (16). Furthermore, ApApApA + UpUpUpU, ApGpU + ApCpU, and ApGpC + GpCpU also showed no definite sign of interaction. ApGpC, however, did self-aggregate at 0.01 M concentrations in the presence of Mg^{2+}. The first evidence of base-pair formation came with the mixture of GpGpC + GpCpC. This is shown in Fig. 6. There is a change in shape of the ORD and a 5 mμ shift in the cross-over wavelength. Similar behavior occurs in the absence of Mg^{2+}. The stoichiometry of the interaction was determined from the data (shown in Fig. 7) at 1°C in the presence of Mg^{2+}. It is clear that the complex contains two moles of GpGpC for every mole of GpCpC. This is also true in the absence of Mg^{2+}. From Fig. 7 it is possible to estimate that about 75% of the bases are involved in the complex. The thermodynamics of this complex have not been determined yet; the equilibria are complicated by the self-aggregation of GpGpC. A melting curve for the complex is given in Fig. 8.

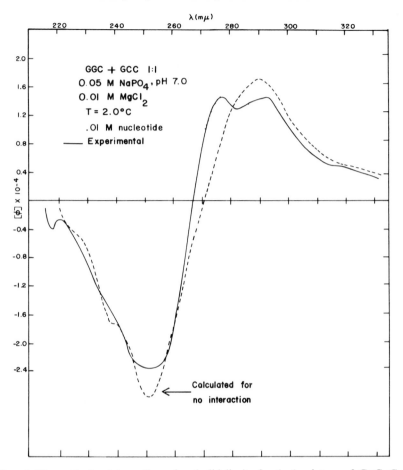

FIG. 6. The optical rotatory dispersion (solid line) of a 1 : 1 mixture of GpGpC and GpCpC at 2°C in 0.01 M MgCl$_2$ and 0.05 M phosphate buffer (pH 7). The dashed line is the average of the separate trimers under the same conditions.

The structure of the complex is not known. It is consistent with triple-strand complexes containing 2G : 1C and 2C : 1G found previously (17, 17a). We assume it involves hydrogen bonds among the bases, but, of course, it is possible that the complex is held together by intercalation of the bases and that it is not a simple ternary complex.

Our findings can be summarized and extrapolated as follows; solutions of trinucleoside diphosphates containing 0.01 M concentration of nucleotides at pH 7, 0.01 M Mg^{2+}, and 1°C will form a specific complex only if the molecules contain only guanine or cytosine. Furthermore, the complexes will contain $2G_3 : 1C_3$ or $2(G_2C) : 1(C_2G)$. These hypotheses are stated simply to encourage experimental test.

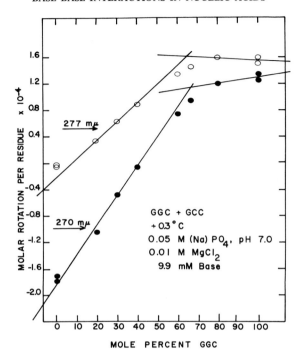

FIG. 7. Variation in the optical rotation as a function of mole fraction GpGpC at constant concentration of nucleotide. The solvent composition is the same as in Fig. 6.

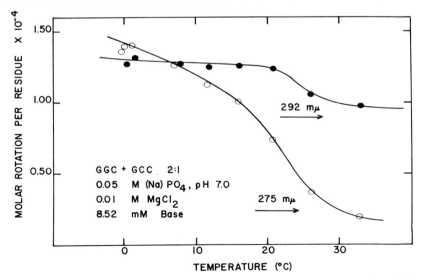

FIG. 8. The temperature dependence of the rotation of the 2GpGpC : 1GpCpC solution. The solvent composition is the same as in Fig. 6.

Miles *et al.* (18) have found from infrared studies that UpUpU +ApApA solutions in Mg^{2+} form a 2 : 1 complex at concentrations of about 0.1 *M*. This implies that any complementary trinucleoside diphosphates will form a complex at high concentrations. These results can obviously be applied to polynucleotide conformation and codon-anticodon recognition.

IV. Multistrand Regions in Polynucleotides

Although DNA seems to exist in a well-defined double-strand structure in solution, except for a few examples, RNA does not. The few exceptions

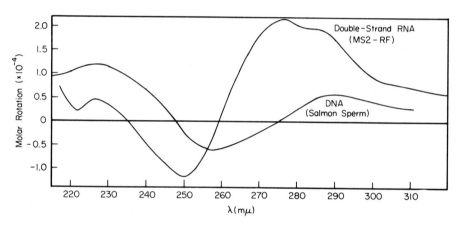

FIG. 9. The optical rotatory dispersion of salmon sperm DNA and the replicative form of MS2 RNA.

are complementary homopolynucleotides and the replicative form of viral RNA. These double-strand RNA's do not have the same geometry as DNA. This is shown clearly by X-ray studies (19) and by the large ORD differences between salmon sperm DNA and the replicative form of MS2 RNA illustrated in Fig. 9. Both double- and triple-strand helices have been found with homopolynucleotides (20).

For nominally single-strand RNA's the molecules presumably contain many small multistrand regions. These regions can be formed by the chain looping back to form antiparallel, complementary base-paired double strands (21). Triple strands can also be formed analogous to homopolynucleotide structures such as poly A : 2 poly U (22), 2 oligo G : poly C (17, 17a), oligo G : 2 poly C (17) and the triple-strand oligomer complex 2GpGpC : GpCpC. Multistrand regions involving other base-base interactions are, of course, also possible.

A. Loops

To assess the extent of multistrand regions in single-strand RNA molecules we must consider the probability of forming loops. Two factors are important: the interaction energies of the specific oligomers closing the loop and the entropy loss on loop formation. The equilibrium constants for interactions between oligomers can be measured directly. From these data it should be possible to obtain probabilities for the formation of the first base pair and for adding successive pairs; i.e., we should be able to calculate the equilibrium constant for complex formation between oligomers of any sequence and chain length.

1. Double-Strand Loops

We want to know the equilibrium constant (K_1) for a loop closed by the interaction of two oligomers. We are given the equilibrium constant (K_2) for the interaction of the free oligomers in solution. This is illustrated in Fig. 10.

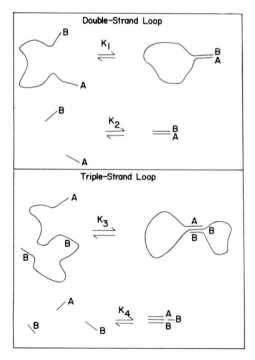

FIG. 10. The definition of the equilibrium constants for the double- and triple-strand loops.

The ratio of the equilibrium constants can be written in terms of the thermo-dynamics of the two reactions.

$$\frac{K_1}{K_2} = \left[e^{\frac{-(\Delta H°_1 - \Delta H°_2)}{RT}} \right] \left[e^{\frac{\Delta S°_1 - \Delta S°_2}{R}} \right]$$

As the bonding is assumed to be the same in the two reactions, it is customary (23, 24) to say that $\Delta H°_1 = \Delta H°_2$. The ratio of the two equilibrium constants then depends only on the entropies. This ratio can be equated to the ratio of the probabilities of finding the oligomers near each other when tied together in a loop and when free in solution (24).

$$K_1 = jK_2$$

For this equation the concentration units in K_2 must be molecules per cubic centimeter. Assuming that the distance between ends of the loop has a Gaussian distribution and that the contour length is long compared to the mean end-to-end distance, Jacobson and Stockmayer find (23)

$$j = \left(\frac{3}{2\pi n b^2} \right)^{3/2}$$

where n equals the number of statistical units of length b in the loop. The quantity nb^2 is equal to the mean square distance between the ends $\langle L^2 \rangle$. Better values of j (dependent on base sequence) can be obtained from the conformations of individual dinucleoside phosphates. At present we will assume that the above value of j is reasonable for n greater than or equal to 10.

Even without making calculations it is clear that having the oligomers tied together will favor their interaction. For quantitative conclusions it is most useful to have an expression involving the fraction f_1 of oligomer bound in the loop and the fraction f_2 bound in solution. The resulting expression is

$$\frac{f_1}{1 - f_1} = \frac{548}{n^{3/2} b^3} \left[\frac{f_2}{(1 - f_2)^2 m} \right]$$

Here b is given in Ångstroms and m is the initial concentration (moles of oligomer/liter) of each oligomer in solution which leads to fraction f_2 bound. We have no data for double-strand formation of trimers, but we will assume that they can form at an appropriate ionic strength at similar concentrations to triple-strand complexes. We found that about 75% triple-strand formation occurred for trimers containing three G's or C's at a concentration of 0.003 mole trimer/liter. Miles et al. (18) found about the same amount of triple-strand formation for trimers containing no G's or C's with a 10-fold increase in concentration. These experiments were done at 0°C in the presence of Mg^{2+}. This should represent the extreme possibilities for trimers; therefore,

TABLE II

FORMATION OF DOUBLE-STRAND AND TRIPLE-STRAND LOOPS
IN POLYNUCLEOTIDES[a]

| No. of monomers in loop (n) | Double-strand loops | |
| | Fraction (f_1) of loops closed for $b = 7$ Å | |
	For G-C base pairs ($m = 0.003$ mole/liter)[b]	For A-U base pairs ($m = 0.03$ mole/liter)[b]
10	0.99	0.95
20	0.99	0.88
100	0.86	0.39

| No. of monomers in each loop ($n = n'$) | Triple strand loops | |
| | Fraction (f_3) of double loops closed for $b = b' = 7$ Å | |
	For G-C base pairs ($m = 0.003$ mole/liter)[c]	For A-U base pairs ($m = 0.03$ mole/liter)[c]
10	1	0.97
20	1	0.81
100	0.77	0.033

[a] The loops are closed by complementary base pairs between trinucleotides.

[b] m = concentration of trimers in solution required to produce 75 % complex formation ($f_2 = 0.75$).

[c] m = concentration of trimers in solution required to produce 75 % complex formation ($f_4 = 0.75$).

we have used these figures in our calculations shown in Table II. The effective length of a mononucleotide was chosen as 7 Å; this number is approximately equal to the distance between phosphates in an extended dinucleotide. The critical assumption in the calculation is choosing the statistical unit as the mononucleotide. If there is strong stacking of bases this will not be true. Therefore, our conclusions will only apply to chains containing a sufficient number of unstacked (independent) bases. The calculations indicate that any complementary pair of trinucleotides is capable of closing a loop at low temperatures.

2. Triple-Strand Loops

For triple-strand regions we assume that there are three oligomers in the chain capable of interacting as illustrated in Fig. 10. The actual distances between oligomers have a Gaussian distribution and the probabilities of closing the two loops are independent. The appropriate equations are

$$K_3 = jj'K_4$$

Each loop has a value of j (or j') which depends on the number of links and value of b (or b') for that loop. For the fractions of ternary complex formed the corresponding equation is

$$\frac{f_3}{1 - f_3} = \frac{7.51 \times 10^4}{(nn')^{3/2}(bb')^3} \left[\frac{f_4}{(1 - f_4)^3 m^2} \right]$$

The fraction of double loops formed is f_3, and f_4 is the fraction of ternary complex formed in solution containing an initial concentration of m moles of A and 2 m moles of B oligomers per liter. Calculations are given in Table II for the same experimental conditions used for the double strand. Again the conclusion is that small triple-strand loops are capable of being closed by any complementary trimers.

V. Conformation of Ribonucleic Acids

At equilibrium the conformations of a ribonucleic acid are characterized by a minimum in the free energy of the solution containing the polymer. The distribution of single-strand, double-strand, and triple-strand regions will depend on competition among all the possible interactions. We have attempted to assess the likelihood of some of these interactions.

The qualitative factors to keep in mind when proposing an RNA structure are as follows.

(1) The pyrimidine nucleosides will be anti, but the purine nucleosides may be syn or anti.

(2) The bases A, G, and C will tend to stack, but U will be much more flexible.

(3) Loops of 10 or 20 bases will be common when three or more consecutive bases are capable of forming complementary base pairs.

(4) Triple-strand regions must be considered. Triple strands may be most important for transfer RNA. If there are long complementary sequences of bases, then Watson-Crick base-paired double strands will be formed. If only short complementary sequences exist, however, probably triple strands will be more stable. These can be either intermolecular or intramolecular.

(5) Non-Watson-Crick base pairs should be considered. The fact that ApGpC and GpGpC self-aggregate at low concentrations implies that similar bonds may be formed in RNA molecules.

To test a proposed structure [see, for example, the structures of transfer RNA's (25–28)] one can compare any measured property in solution with that calculated from the structure. By comparing properties as a function of temperature, ionic strength, and solvent composition one should be able to deduce a fairly accurate structure. This is not possible yet, but preliminary results are encouraging. The optical rotatory dispersion of tobacco mosaic virus has been fit by the sum of a single-strand and multistrand contribution as a function of temperature and ionic strength (29). The formation of one base pair in a transfer RNA should lead to a perceptible change in ORD (15).

VI. Codon-Anticodon Recognition

From the fact that most trimers will not interact in solution except at very high concentrations we can say very little quantitatively about the relative stability of codon-anticodon pairs. We can only conclude that the effective concentrations of the trimers must be increased in the neighborhood of their interaction. This is presumably done by binding the messenger RNA and the transfer RNA to the ribosome. Furthermore, if the interaction occurs in a groove in the ribosome (as found in enzymes), then the exclusion of water might favor complex formation.

ACKNOWLEDGMENTS

We wish to thank Professor Charles Cantor, Columbia University, for the preparation of many of the trinucleoside diphosphates and for permission to quote his unpublished work. We also thank Professor J. Wang, Berkeley, Professor P. O. P. T'so, Johns Hopkins, and Professor O. Jardetsky, Harvard, for many helpful discussions. This work was supported in part by grant GM 10840 of the National Institutes of Health and by the USAEC through the Lawrence Radiation Laboratory.

REFERENCES

1. J. Donohue and K. N. Trueblood, *J. Mol. Biol.* **2**, 363 (1960).
2. A. E. V. Haschemeyer and A. Rich, *J. Mol. Biol.* **27**, 369 (1967).
3. G. N. Ramachandran, D. M. Venkatachalam, and T. Krimm, *Biophys. J.* **6**, 849 (1966).
4. K. D. Gibson and H. A. Scheraga, *Proc. Natl. Acad. Sci. U.S.* **58**, 420 1967.
5. M. Sundaralingam and L. H. Jensen, *J. Mol. Biol.* **13**, 930 (1965).
6. M. Sundaralingam and L. H. Jensen, *J. Mol. Biol.* **13**, 914 (1965).
7. T. R. Emerson, R. J. Swan, and T. L. V. Ulbricht, *Biochemistry* **6**, 843 (1967).
8. E. Shefter, M. Barlow, R. Sparks, and K. Trueblood, *J. Am. Chem. Soc.* **86**, 1872 (1964).
9. M. P. Schweizer, S. I. Chan, and P. O. P. Ts'o, *J. Am. Chem. Soc.* **87**, 5241 (1965).
10. J. Brahms, A. M. Michelson, and K. E. Van Holde, *J. Mol. Biol.* **15**, 467 (1966).
11. M. M. Warshaw and I. Tinoco, Jr., *J. Mol. Biol.* **19**, 29 (1966).
12. R. C. Davis and I. Tinoco, Jr., *Biopolymers* (1968) (in press).
13. D. Glaubiger, Ph.D. Thesis, University of California, Berkeley, California (1965).
14. C. R. Cantor and I. Tinoco, Jr., *J. Mol. Biol.* **13**, 65 (1965).
15. C. R. Cantor, S. R. Jaskunas, and I. Tinoco, Jr., *J. Mol. Biol.* **20**, 39 (1966).
16. C. R. Cantor, Ph.D. Thesis, University of California, Berkeley, California (1966).
17. M. N. Lipsett, *J. Biol. Chem.* **239**, 1256 (1964).

17a. F. Pochon and A. M. Michelson, *Proc. Natl. Acad. Sci. U.S.* **53**, 1425 (1965).
18. H. T. Miles, J. Frazier, and F. M. Rottman, quoted in Felsenfeld and Miles (20).
19. S. Arnott, F. Hutchinson, M. Spencer, M. H. F. Wilkins, W. Fuller, and R. Langridge, *Nature* **211**, 227 (1966).
20. For a review, see G. Felsenfeld and H. T. Miles, *Ann. Rev. Biochem.* **36**, 407 (1967).
21. J. R. Fresco, B. M. Alberts, and P. Doty, *Nature* **188**, 98 (1960).
22. C. L. Stevens and G. Felsenfeld, *Biopolymers* **2**, 293 (1964).
23. H. Jacobson and W. H. Stockmayer, *J. Chem. Phys.* **18**, 1600 (1950).
24. J. C. Wang and N. Davidson, *J. Mol. Biol.* **19**, 469 (1966).
25. R. W. Holley, J. Apgar, G. A. Everett, J. T. Madison, M. Marquisee, S. H. Merrill, J. R. Penswick, and A. Zamir, *Science* **147**, 1462 (1965).
26. H. G. Zachau, D. Dütting, and H. Feldmann, *Angew. Chem.* **78**, 392 (1966).
27. J. T. Madison, G. A. Everett, and H. Kung, *Science* **153**, 531 (1966).
28. V. L. Raj Bhandary, S. H. Chang, A. Stuart, R. D. Faulkner, R. M. Hoskinson, and H. G. Khorana, *Proc. Natl. Acad. Sci. U.S.* **57**, 751 (1967).
29. D. W. McMullen, S. R. Jaskunas, and I. Tinoco, Jr., *Biopolymers* **5**, 589 (1967).

Oligonucleotide Interactions

A. M. MICHELSON

Service de Biochimie
Institut de Biologie Physico- chimique
Paris, France

In the past few years, the concept of single-strand helical structures has been developed for polynucleotides such as poly (A) (1, 2), poly (C) (3, 4), and poly (U) (5). This conformation, which arises from the stacking of successive bases in the polymer chain is not at all static compared with the hydrogen-bonded double helical structure of DNA, but it is a dynamic arrangement with fluctuating sequences of more-or-less ordered regions. As a result of this, little hydrodynamic rigidity is given to the molecule which behaves in some respects as a random coil. At what level of polymerization do the various physical parameters which arise from or which give rise to ordered structures appear? A variety of studies have now established that the essential interactions begin immediately at the dinucleotide level.

I. Hypochromicity

The earliest observations related to base stacking in oligonucleotides were concerned with hyperchromic-hypochromic effects (6, 7) which had previously been noted in nucleic acids and polynucleotides (8).

It was shown that similar (though smaller) effects on the ultraviolet absorption spectra could be observed in many dinucleotides as well as in higher oligomers. With a homologous series of oligonucleotides the size of the hypochromic effect at or near λ_{max} (and the actual value depends on the wavelength) increases with increase in chain length, but rapidly reaches a limit at chain lengths of 7–10 nucleotides (9). Over a range of chain length "n" good agreement for the relationship with ε_{max} is obtained with the function (10)

$$\varepsilon_n = \frac{1}{n}[2\varepsilon(A_2) + (n-2)\varepsilon A_\infty]$$

The hyperchromic effect depends on the nature of the bases in the oligomer, being much smaller, for example, for a given oligouridylate compared with an oligoadenylate; sequence can also have a marked effect. Thus, the ultra-violet absorption spectra of isomeric pairs such as ApC, CpA and ApU, UpA, are not identical. Again, the hypochromicity depends on pH and, in general, the effects are reduced if both bases lose a proton in alkali pH or accept a proton at acid pH. Ionization of one base only in a dinucleotide, however, does not necessarily remove the hypochromic effect. Finally, modification of the ultraviolet absorption spectrum is also a function of the nature of the internucleotide linkage whether $2' \rightarrow 5'$, $3' \rightarrow 5'$, $5' \rightarrow 5'$, or even $P^1\text{-}5'$ $P^2\text{-}5'$ pyrophosphate. Values of the hyperchromic effect at or near λ_{max} can be quite large even for dinucleotides, e.g., diadenosine pyrophosphate shows a 40% increase in absorption on degradation to the monomer. That hydrogen bonds directly between bases are not involved was readily shown by studies of dinucleotides containing 6-dimethylaminopurine which show a 35% effect (11). These early studies were based mainly on difference spectra of oligonucleotides before and after degradation at ambient (20°C) temperature. In fact, much of the hypochromicity of oligonucleotides is temperature sensitive, the absorption at λ_{max} decreasing at low temperature and increasing at high temperature. In addition, the integrated value of hypochromicity over a given range of wavelengths is probably more significant for comparative purposes than the value at an arbitrary wavelength.

II. Optical Rotatory Dispersion

More recently, this earlier work has been fully confirmed and amplified in the publications of Tinoco and co-workers, particularly by the application of optical rotatory dispersion techniques (12, 13). Comprehensive studies indicated a "degree of stacking" which varied with the dinucleotide sequence and the pH. All that I wish to add here is a demonstration that the π-electron system is essential. The three trinucleotides GAU, UGA, and GUA all possess a characteristic ORD spectrum. Irradiation in water at 260 mμ affects only the uracil residue to give a 4,5-dihydro-4-hydroxyuracil derivative, i.e., the major part of the uracil π-electron system is eliminated. In the case of the first two trimers the resultant ORD spectra closely resemble that of GpA, whereas with the third, in which the dihydrouridine separates the other two bases, no interaction is observed and the ORD spectrum is essentially a summation of those of the monomers A and G. This experiment also demonstrates that whereas optical effects may depend on direct interplanar interactions, it is not necessarily true that such interactions are primarily responsible for the ordered "stacked" conformation since, despite the considerable freedom possible with the sugar phosphate chain, the dihydrohydroxyuracil residue is not looped out to allow close contact of A and G, but is apparently maintained in a blocking position between the two (14).

III. Circular Dichroism

Recently, in collaboration with Dr. J. Brahms, we have studied the ultraviolet circular dichroism of homologous series of oligonucleotides and the various dinucleotide sequences (4, 15, 16). Such spectra have been measured as a function of temperature in 4.5 M KF at pH 7.0. This solvent was used since measurements can be made down to $-20°C$. Noncooperative changes indicating a process

$$\text{order} \rightleftharpoons \text{disorder}$$

(where order refers to a situation such that interaction between the bases occurs and disorder is a lack or marked decrease of this interaction) are observed. Since equilibrium constants can be readily obtained given the maximum and minimum values for a given measurable parameter, the curves obtained could be translated into van't Hoff plots (Figs. 1–3).

For a given series of homologous oligonucleotides such plots give parallel straight lines and, indeed, for various dinucleotide sequences the slopes of log K against $1/T$ do not differ enormously. From such plots the thermodynamic parameters can be readily determined. For the dinucleoside phosphate sequences, values of $\Delta H°$ vary from 6–8 kcal, $\Delta S°$ 21–28 eu and ΔF

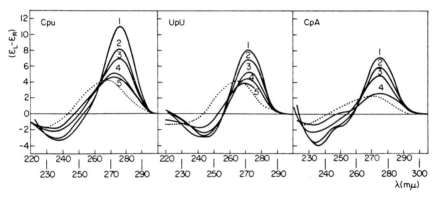

FIG. 1. Circular dichroism spectra of 3′ → 5′ dinucleoside phosphates at various temperatures in 4.7 M KF, 0.01 M Tris, pH 7.4. 3′ → 5′-CpU at: (1) $-20°C$; (2) 3°C; (3) 28°C; (4) 60°C; (5) 72°C. 3′ → 5′-UpU at: (1) $-18°C$; (2) 2°C; (3) 26°C; (4) 41°C; (5) 67°C. 3′ → 5′-CpA at: (1) $-20°C$; (2) 0°C; (3) 23°C; (4) 81°C. Dotted line: the circular dichroism spectra of the monomers.

at 0°C is of the order 0.2–0.7 kcal/mole as shown in Table I. Midpoints of transition vary from about 6° to 25°C depending on the dinucleotide sequence. The absence of major differences in thermodynamic parameters does not allow a division of dinucleotides into groups characterized by an ordered or disordered conformation, although clearly differences do occur. It may be noted that with oligonucleotides containing 2′ → 5′ internucleotide linkages instead of the natural 3′ → 5′ linkage the transition (as followed by circular dichroism) appears to occur at much lower temperatures, despite the fact that the hypochromicity of such derivatives tends to be even larger than in

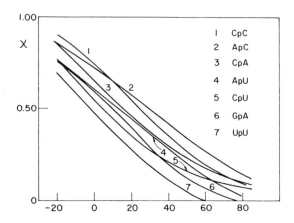

FIG. 2. The fraction of stacked bases (χ) as a function of temperature (conditions as described in Fig. 1).

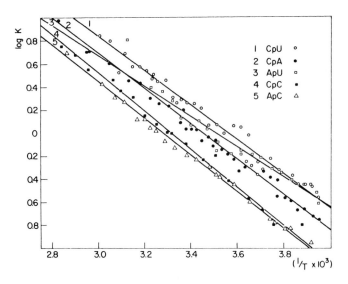

FIG. 3. A van't Hoff plot of the thermal denaturation of some $3' \to 5'$ dinucleoside phosphates.

TABLE I

THERMODYNAMIC PARAMETERS OF DINUCLEOSIDE PHOSPHATES
IN 4.7 M KF USING CIRCULAR DICHROISM

Oligomer	$\Delta H°$ (kcal/mole)	$\Delta S°$ (eu/mole)	$\Delta F°$ (kcal/mole at 0°C)	T_m (°C)
CpA	7.0	24	0.4	15
ApC	6.1	21	0.6	25
CpU	6.8	24	0.2	6
ApU	6.7	24	0.3	11
CpC	7.5	25	0.7	24
GpA	6.1	22	0.2	9
ApA	8.0	28	0.4	25

$3' \rightarrow 5'$ dinucleotides at 20°C. It thus appears, as has also been demonstrated by Ts'o and collaborators in studies of polyribonucleotides and polydeoxyribonucleotides (17), that the 2'-hydroxyl group may play an important role in stability by intramolecular hydrogen bonding.

IV. Energy Transfer

Various studies of the luminescence (fluorescence and phosphorescence) of oligonucleotides at very low temperatures (77° K) suggest that under these conditions considerable stacked overlap occurs. In particular, evidence indicating the formation of excimers (dimer stable only in the excited state) suggests a restricted interplanar separation (18, 19).

V. Effect of Chain Length

The effect of chain length in a homologous series of oligonucleotides on hypochromicity has been mentioned previously. A similar progression of effects with chain length reaching a limit at approximately the hepta- to decanucleotide has also been observed by means of optical rotatory dispersion [oligo(A)s] (20) and by circular dichroism [oligo(A)s and oligo(C)s] (Table II). Similarly, the chain length has a small but significant effect on the thermodynamic parameters, $\Delta S°$ decreasing somewhat while $\Delta F°$ increases slightly, i.e., the ordered form (or forms, since a static structure cannot be envisaged) of two successive bases is slightly more stable in a polymeric chain than in the corresponding free dinucleotide. The effects of substitution of the bases on the thermodynamic parameters of polynucleotides is shown in Table III.

TABLE II

EFFECT OF CHAIN LENGTH ON CERTAIN OPTICAL PROPERTIES
AND THERMODYNAMIC PARAMETERS

Oligo-mer[a]	Hyper-chromi-city (%)	ORD ampli-tude	Dichroic peak $R_0 \times 10^{40}$ at 0°C	$\Delta H°$ (kcal/mole)	$\Delta S°$ (eu/mole)	$\Delta F°$ (kcal/mole at 0°C)	T_m (°C)
A_2	12.9	272	16.0	8.0	28	0.4	25
A_3	21.7	319	18.7	8.0	28	0.4	—
A_5	30.5	330	22.3	8.1	28	0.5	—
A_7	42.2	423	26.3	8.1	27	0.6	—
A_{12}	55.0	494	27.5	7.8	26	0.8	—
Poly (A)	73.0	675	32.0	7.9	25	1.1	40
C_2	8.7	—	22	7.5	24.5	+0.7	24
C_3	16.2	—	31	6.3	20.5	+0.7	—
C_4	18.6	—	34	6.1	19.8	+0.7	—
C_5	21.7	—	34	6.2	20.2	+0.7	—
C_{10}	38.4	—	35	5.3	16.0	+0.9	59
Poly (C)	41.2	—	—	—	—	—	—
Poly (U)	—	—	—	6.0	21.0	+0.3	15

[a] For oligomers A through poly A: 0.1 M Na$^+$, pH 7.4. For C_2 through C_{10}: 4.7 M KF pH 7.5. Poly C: as for poly A.

TABLE III

THERMODYNAMIC PARAMETERS OF SINGLE-STRAND POLYNUCLEOTIDES
USING ULTRAVIOLET ABSORPTION[a]

Polymer	$\Delta H°$ (kcal/mole)	$\Delta S°$ (eu/mole)	$\Delta F°$ (kcal/mole at 0°C)	T_m (°C)
Poly (A)	13.5	40	1.9	45
Poly (HEA)	12.8	40	1.8	45
Poly (iso A)	9.4	28	1.8	65
Poly (C)	9.6	30	1.34	45
Poly (BrC)	10.8	32	2.05	65
Poly (Iodo C)	8.4	25	1.6	65
Poly (1 Me G)	12.0	38	1.65	42

[a] From Ref. 21.

VI. Nature of the Forces Involved

The word "stacked" with reference to single-strand polynucleotides is, of course, a geometrical term and has no implications concerning the nature of the forces involved. Clearly, a variety of factors may play a role. Among these are direct interactions such as van der Waals and London forces and dipole and monopole interactions. Many of these vary inversely as the fifth or sixth power of the distance and it must be remembered that interplanar distances even in hydrogen-bonded double helices can vary from 3.8 [acid form of poly (A)] to 3.11 Å [acid form of poly (C)]. In single-strand structures and particularly with certain blocked polynucleotide analogs intermolecular base-base hydrogen bonding does not occur. In the case of ribonucleotides, however, internal intramolecular hydrogen bonds between the 2'-hydroxyl group and a phosphate oxygen (4, 16) (or the 2-keto group of pyrimidines or N-3 of purines) (17) may add a small measure of stability. It appears, at present, that a major role is played by hydration and the local organization of solvent (so-called hydrophobic and hydrophilic forces) with respect to the total single-strand molecule. This, of course, implies not only relatively short-range ($\sim$3–4 Å) effects, but also long-range interactions via an organized "shell" of water molecules.

VII. Hydrogen-Bonded Complexes

The conformational properties of single-strand oligonucleotides and polynucleotides indicate that despite the presence of a defined helical structure there remains considerable vibrational freedom both along the chain (stretching vibration) and in a rotational sense (overlap of bases in an average of a dynamic situation). This freedom is considerably reduced in oligonucleotide complexes in which two or more strands are maintained in a secondary structure by hydrogen bonds between the bases. The formation of a double parallel-strand form of poly (A) at acid pH has been extensively studied (22). Such structures are also obtained with oligo (A) above a certain chain length. In 0.15 M Na$^+$ at pH 4.0 and 20°C hydrogen-bonded secondary structure commences at the heptanucleotide. This can be readily demonstrated by the appearance of abrupt changes in optical rotatory dispersion (20), circular dichroism (15), and ultraviolet absorption spectra of a series of oligo (A)s at pH 4.0. Perhaps the most striking demonstration lies in the spectrophotometric titration at 260 mμ. The apparent pK of protonation of the adenine residues reaches a minimum at heptaadenylic and coincident with this a new pK appears corresponding to formation of hydrogen-bonded secondary structures (Fig. 4).

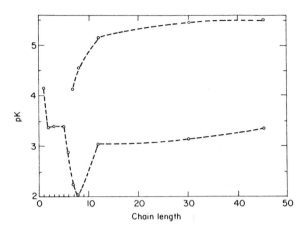

FIG. 4. Variation of apparent pK of oligoadenylics with chain length.

Similar studies with oligo (C)s at acid pH also show the appearance of organized secondary structure [corresponding to the acid form of poly (C)] at heptacytidylic acid (4).

Apart from such homologous hydrogen-bonded complexes, a wide range of heterologous oligonucleotide-polynucleotide interactions have been studied. These include the interaction of oligo (A)s with poly (U), poly (BrU), and poly (X); oligo (U)s with poly (A) and poly (iso A); oligo (C)s with poly (I); and oligo (I)s with poly (C), poly (BrC), and poly (iso A). In each case a series of thermal dissociations (followed by ultraviolet absorption changes) with distinct T_m values was obtained for the oligonucleotide series, both in the absence and presence of magnesium ions. Since the T_m depends to some extent on the concentration of oligonucleotide relative to poly-nucleotide, strict stoichiometry was observed. Typical ultraviolet absorption-temperature profiles are shown in Fig. 5. Now if such T_m values are plotted as various functions of chain length it is found empirically that linear relation-ships are obtained if log n is plotted against T_m (the T_m is, of course, equal to $-\Delta H/\Delta S$ since at $T_m \Delta F = 0$) or against $1/T_m$. In the latter case log n is equivalent to an apparent equilibrium constant log K which is a direct function of chain length, i.e.,

$$K = \frac{[\text{polymer}][\text{oligo}] \times n}{[\text{complex}]}$$

where n is the chain length of oligonucleotide. The general expression is, thus, that of a van't Hoff plot. Several characteristics are immediately apparent when all the results are plotted in this fashion (Fig. 6). First, for a given series of oligomers the points fall on straight lines which are parallel, e.g., oligo (I)

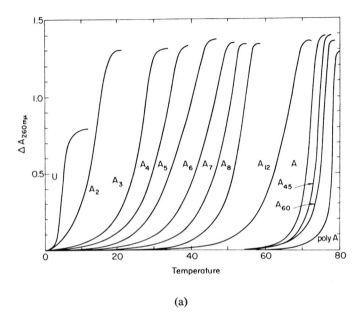

(a)

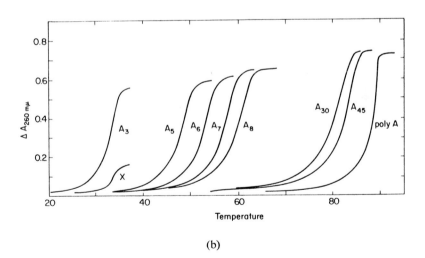

(b)

FIG. 5. (a) Ultraviolet absorption-temperature profiles at 260mμ of oligo (A) · 2 poly (U) in 0.1 M NaCl, 0.05 M sodium cacodylate, pH 7.0, 0.01 M MgCl$_2$, 0.1 μmole oligo (A), and 0.2 μmole poly (U) per ml. (b) Ultraviolet absorption-temperature profiles of oligo (A) · poly (X) in 0.1 M NaCl, 0.05 M sodium cacodylate, pH 7.0.

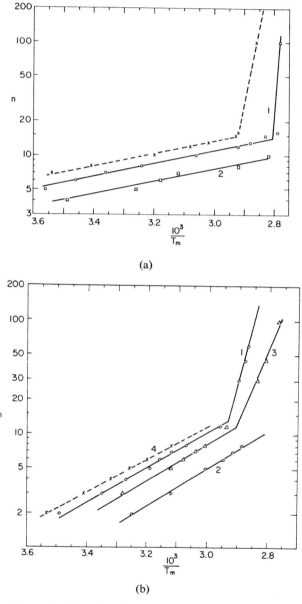

(a)

(b)

FIG. 6. Log (chain length) plotted against $1/T_m$ of complex. (a) Oligo (U) complexed with poly (A) (dashed line) and poly (iso A) (solid lines) in absence of Mg^{2+} (1) and in presence of Mg^{2+} (2). (b) Complexes of oligo (A) with 2 poly (U) (line 1), 2 poly (BrU) (line 2), and poly (X) (1:1) (line 3). Line 4 represents complexes between 2 poly (U) and a series of oligo (A) containing $2' \rightarrow 5'$ internucleotide linkages only.

complexed with poly (C), poly (BrC), or poly (iso A), in the presence or absence of magnesium ions. Second, the slope of these lines is characteristic of the oligonucleotide and not of the complementary polynucleotide which has an influence only on the T_m value. This slope, characteristic of the oligonucleotide, whether in homologous structures, e.g., acid form of oligo (A) or oligo (C) or in heterologous structures, e.g., oligo (U)·poly (A), can be given an arbitrary value from which an order for the series of natural bases is obtained (Table IV). It may be noted that whereas the nature of the complex does not modify this slope, other factors such as the presence or absence of terminal phosphate groups or $2' \rightarrow 5'$ internucleotide linkages instead of $3' \rightarrow 5'$ may produce a change.

TABLE IV

SLOPES OF LOG n/T AND LOG n/T^{-1}

Oligomer	Log n/T	Log n/T^{-1}
$(A)_n$	9.10	14.6
$(2' \rightarrow 5' A)_n$	9.70	14.6
$(G)_n$	5.30	8.3
$(G)_n p$	4.40	6.7
$(C)_n$	3.75	5.9
$(U)_n$	3.30	5.3
$(I)_n$	2.80	5.0
$(C)_n H^+$	3.60	6.4
$(A)_n H^+$	2.80	4.4

It may be noted that the T_m of oligo $(A)_n \cdot$ poly (U) is not the same as that of oligo $(U)_n \cdot$ poly (A), e.g., oligo $(A)_8 \cdot$ poly (U) (T_m 60°C) and oligo $(U)_8 \cdot$ poly (A) (T_m 20°C). Owing to the different slopes of the rate of increase of T_m with chain length, this difference is not constant.

Finally, it can be seen that a second function relating T_m and chain length appears quite abruptly (Fig. 6). In general, this occurs at chain lengths of about twelve nucleotides, i.e., one turn of a double helix.

The results appear to indicate conclusively that hydrogen-bonded base pairing itself does not contribute predominantly to the stability of double helices (23).

VIII. Thermodynamic Stability and T_m

It has been tacitly assumed that the T_m of double and triple helical complexes is truly a measure of the thermodynamic stability. In the case of polynucleotide-polynucleotide interactions at neutral pH this is probably true as shown by various displacement reactions which invariably lead to formation of the complex with the higher T_m (22, 24). This is true even when the T_m difference of two possible complexes is very small and of the order of one or two degrees. For example, the displacement reaction

Poly (I) · Poly (BrC) + Poly (Iodo C) → Poly (I) · Poly (Iodo C) + Poly (BrC)

$$T_m \, 89°C \qquad\qquad\qquad T_m \, 91°C$$

is unidirectional and is not reversible (25).

In the case of oligonucleotide-polynucleotide interactions, however, this assumption is no longer valid and, indeed, several examples can be given where the ultimate product is a complex with a T_m lower than that of the original structured form. Thus, oligo $(A)_3$ interacts readily with poly (X) to give a complex with a T_m 0.7° lower than that of the organized secondary structure present (26) in poly (X) itself. Similarly, hepta- and octacytidylic acids form complexes with poly (I) both of which posses a T_m significantly lower than that of poly (I). Hence, the thermal stability of small sequences hydrogen-bonded to a complementary sequence does not indicate the thermodynamic stability nor necessarily reflect the behavior of macromolecular interactions. This is of importance biologically since often enzymatic systems, such as those involving replication, transcription, and translation, are concerned with quite small sequences over a given time interval.

IX. Ribosome-Messenger tRNA Interactions

As an example of specific interactions in a biological system we have studied the thermal stability of ribosome-messenger oligo(poly) nucleotide-tRNA using a technique of filtration of samples preincubated at various temperatures, using either tRNA charged with a radioactive amino acid or radioactive messenger (27).* The binary associations mRNA-ribosome, tRNA-ribosome, and mRNA-tRNA possess low thermal stability, i.e., at say 20° to 30°C the respective equilibria are such that the phrases are dissociated rather than associated. If the complete three-body system is formed, however, then the T_m is very much higher and lies between 55° and 75°C. Like polymer-polymer interactions the process of dissociation is reversible. The precise value of the T_m is a function of the biological source of the ribosomes and the nature of the messenger polynucleotide and tRNA, i.e., the codon-anticodon interaction. Other factors include the actual chain length of the messenger used and the presence and position (3′ or 5′) of a terminal phosphate group.

---

* mRNA, messenger RNA; tRNA, transfer RNA.

If the anticodon present in the tRNA, however, is regarded as an oligonucleotide interacting with a polynucleotide supported by the ribosome, a direct correlation cannot be made with the thermal stabilities of the oligonucleotide-polynucleotide interactions previously described. Indeed the effects of chain length and the nature of the base are greatly diminished and even reversed as shown in Table V. The thermal dissociation curves are quite reproducible,

TABLE V
T_m OF RIBOSOME-mRNA-tRNA INTERACTION

mRNA	tRNA	T_m (°C)
Poly (A)	Lysine-RNA	60
$(pA)_3$	Lysine-RNA	58
Poly (U)	Phenylalanine-RNA	56
$(pU)_4$	Phenylalanine-RNA	41

and with a copolymer such as copoly (A, U) coding for both lysine- and tyrosine-RNA an apparently biphasic thermal dissociation curve is obtained due to the different T_m values involved (53°C for tyrosine and 60°C for lysine).

As shown by earlier work with the complete system for the *in vitro* synthesis of protein using polynucleotide analogs such as polyfluorouridylate and polybromocytidylate as messenger RNA, the association cannot be described in simplified physical terms (28). Thus, both the polymers mentioned are effective and, indeed, superior to the "natural" polymers in that poly (FU) does not stimulate amino acid incorporation other than phenylalanine, unlike poly (U), whereas poly (BrC) is much more effective than poly (C) for proline incorporation. The range of thermal stability for interactions of these two polymers with A and G, respectively, is of the order of 100° (relative difference of T_m values). Clearly, the ribosomal system, which must remain reversible in order to function, eliminates such major differences in a manner as yet unknown. The subtleties of the interactions involved in the ribosome-mRNA-tRNA complex, to say nothing of the mechanisms of action of the various enzymes which play a role in the actual synthesis of a peptide bond, have yet to be elucidated.

ACKNOWLEDGMENT

I should like to acknowledge the support given in much of this work by C. Monny and by F. Pochon, M. Leng, A. Favre, J. P. Henry, M. F. Isambert, and M. Dumas.

REFERENCES

1. D. N. Holcomb and I. Tinoco, *Biopolymers* **3**, 121 (1965).
2. K. E. van Holde, J. Brahms, and A. M. Michelson, *J. Mol. Biol.* **12**, 726 (1965).
3. G. D. Fasman, C. Lindblow, and L. Grossman, *Biochemistry* **3**, 1015 (1964).
4. J. Brahms, J. C. Maurizot, and A. M. Michelson, *J. Mol. Biol.* **25**, 465 (1967).
5. A. M. Michelson and C. Monny, *Proc. Natl. Acad. Sci. U.S.* **56**, 1528 (1966).
6. A. M. Michelson, *Nature* **182**, 1502 (1958).
7. A. M. Michelson, *J. Chem. Soc.* p. 1371, 3655 (1959).
8. M. Kunitz, *J. Biol. Chem.* **164**, 563 (1946).
9. A. M. Michelson, "The Chemistry of Nucleosides and Nucleotides," p. 539. Academic Press, New York, 1963.
10. J. F. Liebman, private communication (1967).
11. A. M. Michelson, *Biochim. Biophys, Acta* **55**, 841 (1962).
12. M. M. Warshaw and I. Tinoco, *J. Mol. Biol.* **13**, 54 (1965).
13. C. R. Cantor and I. Tinoco, *J. Mol. Biol.* **13**, 65 (1965).
14. A. Favre and A. M. Michelson, unpublished work (1966).
15. J. Brahms, A. M. Michelson, and K. E. van Holde, *J. Mol. Biol.* **15**, 467 (1966).
16. J. Brahms, J. C. Maurizot, and A. M. Michelson, *J. Mol. Biol.* **25**, 481 (1967).
17. P. O. P. Ts'o, S. A. Rapaport, and F. J. Bollum, *Biochemistry* **5**, 4153 (1966).
18. J. Eisinger, M. Gueron, R. Shulman, and T. Yamane, *Proc. Natl. Acad. Sci. U.S.* **55**, 1015 (1966).
19. C. Hélène and A. M. Michelson, *Biochim. Biophys. Acta* **142**, 12 (1967).
20. A. M. Michelson, T. L. V. Ulbricht, T. R. Emerson, and R. J. Swan, *Nature* **209**, 873 (1966).
21. M. Leng and A. M. Michelson, *Biochim. Biophys. Acta* (1967) (in press).
22. A. M. Michelson, J. Massoulié, and W. Guschlbauer, *Progr. Nucleic Acid Res. Mol. Biol.* **6**, 83 (1967).
23. A. M. Michelson and C. Monny, *Biochim. Biophys. Acta* **149**, 107 (1967.)
24. M. J. Chamberlin, *Federation Proc.* **24**, 1446 (1965).
25. J. Massoulié and A. M. Michelson, *Biochim. Biophys. Acta* **134**, 22 (1967).
26. A. M. Michelson and C. Monny, *Biochim. Biophys. Acta* **129**, 460 (1966).
27. C. S. Mclaughlin, J. Dondon, M. Grunberg-Manago, A. M. Michelson, and G. Saunders, *J. Mol. Biol.* (1967) (in press).
28. A. M. Michelson, *Bull. Soc. Chim. Biol.* **47**, 1553 (1965).

Some Effects on Noncomplementary Bases on the Stability of Helical Complexes of Polyribonucleotides*

OLKE UHLENBECK, RICHARD HARRISON, AND PAUL DOTY

Department of Chemistry
Harvard University
Cambridge, Massachusetts

After a decade of intensive investigation of the interactions of homopolynucleotides our knowledge of the helical complexes which can form is essentially complete and the rationalization of their relative stabilities is well advanced (1). This is a necessary preliminary to the more difficult and basic problem of establishing and understanding the conformations taken up by polynucleotides of specified sequence. One of the many directions of attack now lies in the use of synthetic polynucleotides whose sequences are predominantly those of one of the helical complexes, but which contain minor departures, or defects, inconsistent with complementary base pairing. The question can then be asked as to which alternative is followed by the odd base: Does it remain in the helical framework, perhaps with some distortion, or does it loop out and take up a position outside the helical framework?

This report tells of two studies, not yet complete, which indicate answers to such questions in two contexts. Of course, other contexts require exploration before we will have a body of knowledge of reliably predictive value that can be applied to naturally occurring nucleotide sequences such as those now available for several transfer RNA's.

One of the basic problems in the assessment of nucleotide interactions in polynucleotides is whether or not isolated noncomplementary bases remain in the helical framework of an otherwise complementary region. Since base interactions depend on stacking interactions, as well as hydrogen bonding and the extent of solvation interaction, the preferred conformation may depend not only on the specific nucleotides that are noncomplementary, but also on their neighboring nucleotides. Fresco and Alberts (2) examined one case—that of uracil bases within a predominantly adenine chain complexed with poly U. They found that the minimum of hypochromicity as a function of mole fraction of poly U occurred at the point where the moles of poly U equaled the moles of A in the copolymer (two-strand helix) or equaled twice

* This work was supported by the National Science Foundation grant (GB-4563).

that amount (three-strand helix). As a consequence they concluded that the U residues in the poly AU looped out of the helical framework allowing A residues to be opposite each U residue in the poly U.

To see if this conclusion was generally true we have undertaken an examination of the case which is most likely to be different, namely, poly AG complexed with poly U. Here the minor component, G, is expected to have as neighbors in the chain adenine bases with which interaction is strong and, moreover, by shifting about 2.5 Å, hydrogen bonds between guanine and uracil could form (3). Thus, there are two reasons for the G remaining in the helical framework.

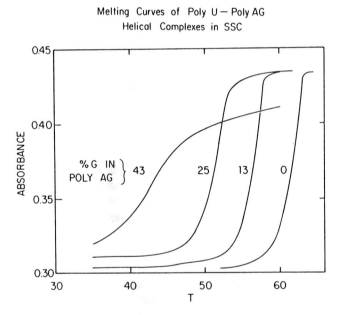

FIG. 1. Absorbance-temperature curves for complexes of poly U and poly AG in SSC.

Three AG copolymers containing 13, 25, and 43%, respectively, of guanylic acid were prepared and absorbance (260 mμ) temperature profiles determined for their equilibrated equimolar mixtures in SSC (0.15 M NaCl + 15 mM Na citrate, pH 7). The results are shown in Fig. 1 together with poly A complexed with poly U. The profiles for the 13 and 25% guanine polymers are quantitatively similar to the one for poly U + poly A except for translations to lower temperatures. Only at 43% guanine is the character of the profile changed; here the shift to lower temperatures continues, but, in addition, the transition is broadened and the total hypochromicity change is substantially diminished.

The most direct interpretation of these results is that in the first two copolymers the guanine bases remain in the helical framework since the hypochromic change is the same as for poly A + poly U; i.e., if the guanine bases were outside, a corresponding fraction of the hypochromic change for the transition would be absent. Since hypochromism comes mostly from stacking interaction between adjacent bases and is sensitive to the angular orientation of the transition dipoles, it is likely that the orientation of the guanine bases is not much different from that in the perfect helix since the hypochromicity in this case is close to the value for poly A + poly U. Thus, a shift to allow the "wobble" interaction (3) to occur seems unlikely. Moreover, the persistence of the sharpness of the transition indicates that the quanine residues do not interrupt the stability of the structure; i.e., the number of adjacent bases that must cooperate in the transition process seems unchanged and this must be on the average a number considerably larger than the stretches of uninterrupted adenine residues.

With the 43% guanine copolymer everything is different; this is most simply explained by assuming that most of the guanine residues loop out of the helical frame. Indeed, with about one-third of the hypochromicity lost it appears that all the guanine bases that are in clusters of two or more as well as single guanines that are not far distant from others do loop out producing local regions displaying quite a range of stability.

An attempt was made to discriminate between these looped-out regions and complementary helical regions by selective reaction with formaldehyde (4). However, at 25°C and concentrations of formaldehyde as low as 0.3% the extent of reaction appeared to be simply proportional to the extent of denaturation produced by the formaldehyde and prior reaction with non-helical regions could not be isolated.

Some confirmation of this interpretation of the absorbance-temperature profiles might be expected from determining the mixing curves, i.e., the absorbance at 260 mμ as a function of composition. Unfortunately, the results were not conclusive. The minima were well defined and occurred at 0.55 mole % uracil for poly AG$_{25}$ and at 0.46 for poly AG$_{43}$. These minima are much higher than those expected for complete looping out. If the behavior had been similar to that found for poly AU + poly U the corresponding mole fractions would have been 0.43 and 0.32, respectively. It seems likely that the higher uracil content of the complex is due to partial involvement of three-strand structures containing two poly U strands and thus introducing more uracil into the complex. Consequently, one can not deduce from these mixing curves any useful information other than that the behavior is very unlike the case where looping out of the noncomplementary base occured exclusively.

Although this study with copolymers can probably be carried to a more decisive state, it seemed clear that precise information could only come from the use of polynucleotides with sequences selected to answer specific problems. With this end in view we have prepared several related series of oligomers of known sequence using the synthetic techniques developed in this laboratory (5). In this phase we have elected to stay with the poly A + poly U framework and operate at a somewhat higher Na^+ concentration (0.25 M) where complexes are expected to be triple stranded and where the selective effect of chain length is very pronounced. We then replace poly A by an oligomer that is predominantly A. Thus far we have been able to determine what happens to noncomplementary base G, C, or U when it occurs at either end of pentamers and hexamers, and we have begun to answer the same question when the noncomplementary base is one removed from the end.

We can begin by examining the complexes formed between poly U and four oligomers: $(Ap)_4C$, $C(pA)_4$, $(Ap)_5C$, and $C(Ap)_5$. For mixing curves under conditions of triple-strandedness the minima should occur at 0.667 mole % poly U if the noncomplementary base remains in the helix and at 0.612 and 0.625 (for the pentamers and hexamers, respectively) if the odd base is outside. The mixing curves for the complexes with the two pentamers are shown in Fig. 2 where it is seen that the two minima are close to the value

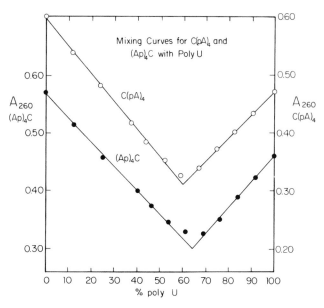

FIG. 2. Mixing curves for poly U and the pentanucleotides, $(Ap)_4C$ and $C(pA)_4$, in 0.25 M NaCl.

for looping out. Similar results were found for the two other series to be presented. Thus, we can conclude that a noncomplementary base at either end of a complementary oligomer will reside outside the helical frame, thereby allowing complete base pairing within the helix.

Within this general conclusion there is room for some interesting variations since the stability and hypochromicity of the complex can be expected to vary with the actual disposition of the odd base as well as its nature. Consignment outside the helix still permits different kinds of interactions with the solvent and the neighboring regions of the helix and the net effect of these with respect to the stability of the complex can be elucidated.

As an example, the absorbance-temperature curves for the poly U complexes with $(Ap)_4C$ and $C(pA)_4$ are shown in Fig. 3. Here it is seen that the precision employed reveals a quite significant difference in both the transition temperature (T_m) and the hypochromicity $(\% H)$, defined as the percentage decrease below the high-temperature limit of absorbance. The results for this pair and the hexamer pairs are shown in Table I. In this group as well as the others to be presented, a clear pattern can be discerned: the location

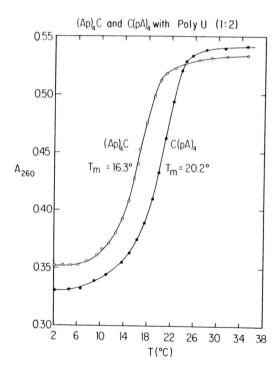

FIG. 3. Absorbance-temperature curves for complexes of poly U and the pentanucleotides, $(Ap)_4C$ and $C(pA)_4$.

of the noncomplementary nucleotide at the 5′ end of the oligomer chain leads to greater stability and greater hypochromicity.

The results for the corresponding oligomers containing guanine and uracil are shown in Tables II and III. The qualitative conclusion is seen to apply.

TABLE I

$(Ap)_n C$ AND $C(pA)_n +$ POLY U

Oligomer	T_m	$\%H$
$(Ap)_4 C$	16.3	33
$C(pA)_4$	20.2	38
$(Ap)_5 C$	22.7	32
$C(pA)_5$	28.6	40

TABLE II

$A_n G$ AND $GA_n +$ POLY U

Oligomer	T_m	$\%H$
$(Ap)_4 G$	13.2	31
$G(pA)_4$	14.4	36
$(Ap)_5 G$	21.2	35
$G(pA)_5$	23.0	37

TABLE III

$A_n U$ AND $UA_n +$ POLY U

Oligomer	T_m	$\%H$
$(Ap)_4 U$	12.1	32
$U(pA)_4$	15.2	35
$(Ap)_5 U$	20.1	36

Upon comparing the relative stabilities of the corresponding complexes containing the three different bases, it is evident that those involving cytosine are somewhat more stable. A similar conclusion was reached earlier from observations on the series $(Ap)_3 X$ (6). This situation probably reflects the greater solvation of cytosine arising from its considerably larger dipole moment.

To place the effect of the noncomplementary bases in a broader context it is useful to compare the foregoing results with those for oligomers of adenine of the same length. Such data are assembled in Table IV. Comparison shows that whereas the pentamers with an odd base show T_m values

TABLE IV

$(Ap)_n$ AND $(Ap)_n A + $ POLY U

Oligomer	T_m	$\%H$
$(Ap)_3 A$	19.0	36
$(Ap)_4$	11.5	34
$(Ap)_4 A$	28.0	39
$(Ap)_5$	20.1	37

from 12° to 20°C, the pentamer of adenine, $(Ap)_4 A$, displays a value of 28°C; that for $(Ap)_3 A$ is 19°C. Thus, the inclusion of an odd base in a terminal position is always destabilizing compared with having a complementary base instead. Less obvious, however, is the conclusion that the odd base is stabilizing relative to the absence of a base, that is, $X(pA)_4$ and $(Ap)_4 X$ form more stable complexes than $(Ap)_4$.

The other conclusion to be drawn from Table IV is that a 3'-terminal phosphate is destabilizing compared with its absence to the extent of about 8°C. This is a relatively small cost to pay for the inclusion of two additional negative charges in the helix for each oligomer.

Parallel to the variations in T_m on which these evaluations of stability were based, one sees a similar variation of hypochromicity. This suggests that the critical feature at play is the optimum positioning of the adenine residue that lies adjacent to the odd base. The less the adjacent odd base interferes with this, the better the stacking interaction, the higher the hypochromicity, and the greater the thermal stability.

It may be of interest to mention the preliminary results that have been obtained in the next more complex series. $ApU(pA)_4$ and $GpU(pA)_4$ have been prepared and complexed with poly U. The minima of the mixing curves indicate that all the adenines, but nothing else, are paired with the uridines. Thus, in $ApU(pA)_4$ the uracil appears to loop out as in the earlier copolymer studies (1), whereas the GpU of the other oligomer is disposed outside the helical frame. It is by progress in this kind of work that we hope to be able to learn enough about the pattern of noncomplementary base interaction to be useful in assessing the conformations taken up by naturally occurring sequences.

In conclusion, it remains to point out an interesting difference between dealing with thermally induced helix-coil transitions between polymers and those, such as employed here, that involve oligomers. This difference is the breakdown in the latter case of the independence of concentration usually displayed by the polymeric helix-coil transitions; i.e., with one "side" of the helix composed of short oligomers it is to be expected that these will enter and leave the helix in a manner controlled largely by a mass action equilibrium constant. With the equilibrium constant favoring occupation in the helix the effect of concentration of the oligomer in the solution may not be very great, but at sufficiently short oligomer length it should be noticeable. We have studied this with the pentamer, $(Ap)_4A$. The transition temperatures as a function of its concentration have been determined and are plotted in Fig. 4. It is seen that T_m varies from 22° to 36°C over a hundredfold variation

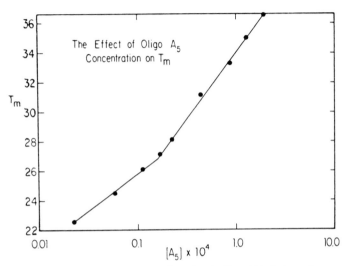

FIG. 4. The dependence of T_m of the complex of poly U and$(Ap)_4A$ on the concentration of $(Ap)_4A$.

of the concentration of the oligomer. An inflection occurs at the concentration corresponding to the composition of the three-strand helix. The elucidation of the interplay of this mass action effect on first-orderlike transition that is normally independent of the concentration of its components deserves careful examination. For the present, however, it is sufficient to point out that in obtaining T_m values for comparative purposes in systems involving oligomers it is important to work at the mole ratio corresponding to the composition of the complex being studies.

REFERENCES

1. For a review, see A. M. Michelson, J. Massouli, and W. Guschlbauer *Progr. Nucleic Acid Res. Mol. Biol.* **6**, 83 (1967).
2. J. R. Fresco and B. M. Alberts, *Proc. Natl. Acad. Sci. U.S.* **46**, 311 (1960).
3. F. H. C. Crick, *J. Mol. Biol.* **19**, 548 (1966).
4. H. Boedtker, *Biochemistry* **6**, 2718 (1967).
5. R. E. Thach and P. Doty, *Science* **147**, 1310 (1965); **148**, 632 (1965).
6. E. K. F. Bautz and F. A. Bautz, *Proc. Natl. Acad. Sci U.S.* **52**, 1476 (1964).

Some Practical Improvements in the Calculation of Intermolecular Energies*

PIERRE CLAVERIE

Service de Biochimie Théorique
Institut de Biologie Physico-chimique
Paris, France

I. Introduction

A. The Theoretical Works

Since the earlier work of London (see, e.g., London, 1937, or the review paper by Margenau, 1939), the concept of intermolecular forces based on the dipole approximation (namely, dipole-dipole, dipole–induced-dipole and London dispersion forces) has been largely widespread. Still, the inadequacy of these simple formulas, based on the dipole approximation, for short-range interactions was rapidly recognized. London (1942) already introduced two important ideas: the summation of bond contributions and the interpretation of the integrals involved as electrostatic interactions between charge distributions (deduced from the wave functions); such an interpretation allows for rational approximations of these integrals, corresponding to convenient approximations of the charge distributions themselves [London proposed to represent them by sets of point charges (monopoles)]. In the further theoretical works, the dispersion term was exclusively considered; Coulson and Davies (1952) used LCAO molecular orbitals, evaluating the integrals rather accurately, but they obtained the dispersion energy between the π systems only. Haugh and Hirschfelder (1955) evaluated the *total* dispersion energy, separating it into π-π, σ-π, and σ-σ contributions, each being evaluated separately in a special manner.

The purely theoretical work of Longuet-Higgins (1956) presented in a rigorous and systematic manner the interpretation of the integrals as electrostatic interactions between charge distributions.

More sophisticated methods (use of the polarizability for imaginary frequency) were developed, but they give new expressions for the dispersion energy only and these expressions do not seem to be useful for large systems (for a recent application, see Deal and Kestner, 1966). On the other hand, direct calculations of the complex treated as a single molecule were performed

* This work was supported by grant No. 67-00-532 of the Délégation Générale à la Recherche Scientifique et Technique (Comité de Biologie Moléculaire).

by the usual methods for molecular calculations, but sufficient accuracy for the evaluation of intermolecular energy could be obtained only for the smallest systems [two helium atoms, for instance, see Phillipson (1962) or Hirschfelder *et al.* (1964) p. 1210, note added to the page 1064]. Among the most recent attempts in this direction is the use of the many-electron theory of Sinanoğlu, adapted to the problem of molecular interactions (Kestner and Sinanoğlu, 1966) and applied to the interaction of two helium atoms (Kestner, 1966).

B. The Recent Practical Calculations on Large Molecules

For some years, a rather different kind of calculation has been developed, namely, effective calculations of the interaction energy between rather large molecules, so that various approximations of questionable validity had to be made. After the paper by De Voe and Tinoco (1962), who initiated this new direction of work, it is possible to mention the papers by Bradley *et al.* (1964), Nash and Bradley (1965, 1966), Pullman *et al.* (1966), Claverie *et al.* (1966), Rein and Pollak (1967), and Pollak and Rein (1967).

It appeared a general trend to improve the theoretical validity of the methods used; this question is explicitly considered by Rein and Pollak (1967), Pollak and Rein (1967) and also in more recent works by Rein *et al.* (1967a), Claverie and Rein (1967), and Claverie (1967). The purpose of this chapter is to indicate the theoretical status of the above-mentioned works and the most recent developments worked out in our laboratory, especially those concerning the short-range repulsion energy which was not considered in the previous works.

II. The Long-Range Interaction Energy

The qualification "long-range" simply applies to the parts of the energy which decrease slowly with intermolecular distance, e.g., $R^{-n}(n = 3, 4, 5, 6, \ldots)$, compared with parts decreasing exponentially as a function of R; since these last terms may be related to the electron overlap (exponentially decreasing when R increases) between the two molecules (when the L.C.A.O. approximation is used, a convenient order of magnitude of this overlap is given by the overlap integrals between atomic orbitals of the two molecules), it is considered that the long-range terms give a convenient approximation to the true interaction energy when this overlap is negligibly small. Such a condition is realized for distances larger than 5 Å, but it is *not* at distances corresponding to the region of the equilibrium position (3–4 Å). The examination of the supplementary "short-range"* terms will be the object of Section III. The present section will be devoted to the evaluation of the so-called long-range terms at any distance and, especially, at short distances.

---

* Some authors call it "medium-range" and use "short-range" for chemical bond distances (1–2 Å).

A. Theoretical Outline

1. General Perturbation Treatment

Only the main features, which are relevant for the subsequent discussion, will be presented here. The general theoretical basis has been known for a long time (London, 1937; Margenau, 1939; Longuet-Higgins, 1956) and the detailed discussion (concerning the development of the general theory in the case of large interacting molecules at short distances) will be given elsewhere (Claverie and Rein, 1967; Rein et al., 1967a). The treatment of Longuet-Higgins (1956) is used, with some changes in the notations—the Hamiltonian of the molecule i ($i = 1$ or 2) is noted H_i, the eigenfunctions are noted $|0^{(i)}\rangle$, $|1^{(i)}\rangle$, ... $|a^{(i)}\rangle$, ... $|b^{(i)}\rangle$ ..., the corresponding eigenvalues being $E_0^{(i)}$, $E_1^{(i)}$, ... $E_a^{(i)}$, ..., $E_b^{(i)}$, ... (labeled in order of increasing magnitude). Then, the eigenfunctions of $H_1 + H_2$ are all the simple products of an eigenfunction of 1 and an eigenfunction of 2:

$$|0^{(1)}0^{(2)}\rangle, \ldots, |0^{(1)}a^{(2)}\rangle, \ldots, |a^{(1)}0^{(2)}\rangle, \ldots, |a^{(1)}b^{(2)}\rangle, \ldots$$

with the corresponding eigenvalues:

$$E_0^{(1)} + E_0^{(2)}, \ldots, E_0^{(1)} + E_a^{(2)}, \ldots, E_a^{(1)} + E_0^{(2)}, \ldots, E_a^{(1)} + E_b^{(2)}, \ldots$$

The total Hamiltonian of the complex is $H = H_1 + H_2 + U$, where U is the interaction Hamiltonian:

$$U = e^2 \left[\sum_{\mu^{(1)}} \sum_{\mu^{(2)}} \frac{Z_{\mu^{(1)}} Z_{\mu^{(2)}}}{r_{\mu^{(1)}}} - \sum_{\mu^{(1)}} \sum_{i^{(2)}} \frac{Z_{\mu^{(1)}}}{r_{\mu^{(1)}i^{(2)}}} - \sum_{i^{(1)}} \sum_{\mu^{(2)}} \frac{Z_{\mu^{(2)}}}{r_{i^{(1)}\mu^{(2)}}} + \sum_{i^{(1)}} \sum_{i^{(2)}} \frac{1}{r_{i^{(1)}i^{(2)}}} \right]$$

$$(1)$$

where r equals the distance, μ refers to the nuclei, and i to the electrons; the superscript (1 or 2) refers to the molecule, Z to the nuclear charge.

This interaction U is considered as a perturbation (the unperturbed Hamiltonian being $H_1 + H_2$), so that the energy of interaction E^{int} may be expressed with Rayleigh-Schrödinger perturbation expansion:

$$E^{int} = \langle 0^{(1)}0^{(2)}|U|0^{(1)}0^{(2)}\rangle \text{ (electrostatic)} \qquad \text{1st order}$$

$$-\sum_{a^{(1)}}' \frac{|\langle 0^{(1)}0^{(2)}|U|a^{(1)}0^{(2)}\rangle|^2}{E_a^{(1)} - E_0^{(1)}} \text{ (polarization of 1 by 2)}$$

$$-\sum_{b^{(2)}}' \frac{|\langle 0^{(1)}0^{(2)}|U|0^{(1)}b^{(2)}\rangle|^2}{E_b^{(2)} - E_0^{(2)}} \text{ (polarization of 2 by 1)} \left.\right\} \text{2nd order}$$

$$-\sum_{a^{(1)}}' \sum_{b^{(2)}}' \frac{|\langle 0^{(1)}0^{(2)}|U|a^{(1)}b^{(2)}\rangle|^2}{(E_a^{(1)} + E_b^{(2)}) - (E_0^{(1)} + E_0^{(2)})} \text{ (dispersion)}$$

$$+ \cdots \quad \text{higher order terms} \qquad (2)$$

(in the symbol $\sum'$, the prime means that 0 is excluded from the summation).

As it may be seen from Eq. (2), the energy terms which are commonly considered (electrostatic, polarization, and dispersion) correspond to the first- and second-order perturbation terms only. The truncation of the series after the second-order terms is, still at the present time, a procedure whose validity is not mathematically proven, especially for the equilibrium region.

2. The Calculation of the "Long-Range" Terms at Short Distances

The question remains—to calculate these terms correctly for short distances, because, in the classic treatment, they are evaluated with the use of the dipole approximation; this procedure introduces a second level of approximation (the first one consists in truncating the perturbation expansion). Now, especially for large molecules, the second level approximation seems more drastic than the first one* and the purpose of this section is to examine how this second level approximation can be removed and to see how this goal has been approached in the practical calculations quoted in Section I,B. In the usual treatments (see, for instance, the textbook by Hirschfelder *et al.*, 1964), U is developed in a multipole expansion (i.e., a sum of multipole-multipole interactions), usually truncated after its first term (the dipole-dipole one for neutral molecules), and the integrals are then calculated exactly (in principle) using the approximated interaction operator. Such a procedure suffers of two major defects. (*a*) From the practical point of view it is very difficult to use terms beyond the dipole-dipole one, as it would be necessary to obtain better results at short distances. (*b*) From the theoretical point of view, the use of a multipole expansion is never rigorously legitimate, since it cannot represent exactly a molecular charge distribution.† As a result, the use of a complete multipole expansion would lead to a divergent series in (R^{-1}) as has been shown by Dalgarno and Lewis (1956); according to these authors, the series is asymptotic, so that, when truncated at some order, it gives an approximate

---

* As it will be quoted later, the dipole approximation is not valid when the intermolecular distance is of the same order of magnitude as the dimensions of the interacting molecules. Now, let us consider two linear or planar molecules, having a length of about 5 Å, lying parallel 5–6 Å apart. The overlap integrals are very small (order of magnitude about 10^{-4} at a distance of 5 Å between the centers for two $2p$ orbitals pointing one to each other), so that the neglect of the " short-range " terms is still possible, whereas the dipole approximation can by no means be justified.

† This is simply due to the fact that the charge density never vanishes at finite distance; now, the multipole expansion of a charge distribution converges only for points outside of a sphere centered at the origin and containing the whole charge distribution. Such a condition cannot be fulfilled for any molecular or atomic charge distribution, since they extend to infinity.

value which becomes better and better when R increases. But, at distances which are not significantly larger than the dimension of the molecules, the error becomes too large, so that it is not a convenient solution to introduce terms other than the dopole-dipole one. We may conclude that the multipole expansion of the interaction potential is a useful procedure as concerns the interaction between very small molecules or atoms (see, e.g., the works of Kołos, 1967, or Hirschfelder and Löwdin, 1966, 1959), but becomes not only difficult, but also not legitimate and, therefore, useless for large molecules at short distances.

Fortunately, there exist a much more convenient way of building approximations to the exact integrals, based upon the following fact: any integral $\langle a^{(1)}b^{(2)}|\mathsf{U}|c^{(1)}d^{(2)}\rangle$ may be interpreted (Longuet-Higgins, 1956) as the electrostatic interaction energy between two charge distributions with charge densities $f_{a^{(1)}c^{(1)}}(\mathbf{r})$ and $f_{b^{(2)}d^{(2)}}(\mathbf{r})$:

$$\langle a^{(1)}b^{(2)}|\mathsf{U}|c^{(1)}d^{(2)}\rangle = \iint \frac{f_{a^{(1)}c^{(1)}}(\mathbf{r}_1)f_{b^{(2)}d^{(2)}}(\mathbf{r}_2)}{|\mathbf{r}_1 - \mathbf{r}_2|} \, dv_1 \, dv_2 \qquad (3)$$

As indicated by the notation, the first charge distribution depends only on the wave functions $a^{(1)}$ and $c^{(1)}$ and is, therefore, related to molecule 1 only; a similar statement holds for the second charge distribution.

Any charge distribution $f_{ac}(\mathbf{r})$ corresponding to the states a and c of some molecule is given by the formula:

$$f_{ac}(\mathbf{r}) = \langle a|\wp(\mathbf{r})|c\rangle \qquad (4)$$

where $\wp(\mathbf{r})$ is the charge-density operator for the molecule under consideration:

$$\wp(\mathbf{r}) = e\left[\sum_v Z_v \, \delta(\mathbf{r} - \mathbf{r}_v) - \sum_i \delta(\mathbf{r} - \mathbf{r}_i)\right] \qquad (5)$$

where δ is the Dirac delta "function." $|a\rangle$ and $|c\rangle$ are functions of the nuclear coordinates $\mathbf{r}_\alpha$ and the electron coordinates $\mathbf{r}_i$. In $\langle a|\wp(\mathbf{r})|c\rangle$, the integration is performed over the $\mathbf{r}_\alpha$'s and $\mathbf{r}_i$'s, so that one is actually left with a function of the variable $\mathbf{r}$ only.

a may be identical to c; $f_{aa}(\mathbf{r})$ may then be called a "*state* charge distribution"; $f_{00}(\mathbf{r})$ is the *ground state* or *permanent* charge distribution. When a and c are different states, $f_{ac}(\mathbf{r})$ will be more specifically designated as a "*transition* charge distribution."

Equation (3) gives us the possibility of making approximations whose degree of accuracy may be easily known; we need only to replace the *exact* charge distributions involved by *approximate* ones, then the electrostatic interaction energy between the approximate charge distribution will be calculated exactly; in this way, we place ourselves in the domain of ordinary electrostatics, and the approximations which can be made in such problems have been extensively studied for a long time. The approximate charge distributions will be chosen, of course, in such a way that the calculation of the interaction energy will be as easy as possible. A tendency will therefore be to replace a continuous charge distribution by a set of charges (monopoles), dipoles, multipoles, in order to avoid the integrations.

3. Three Possible Approximations to the Charge Distributions

We shall now define three kinds of approximations for a charge distribution, which have been the only ones to be used in the practical calculations quoted in Section I,B. These three approximations are described with more detail elsewhere (Claverie and Rein, 1967).

a. The Dipole Approximation. A charge distribution is approximated by its dipole moment, located at some fixed point in the molecule. The values obtained for the matrix elements are exactly the same as those obtained by reducing U to its dipole-dipole part, so that the use of this approximation for all the charge distributions leads to the well-known formulas. The inadequacy of this approximation for charge distributions lying at distances of the order of magnitude of their own dimensions is self-evident and appears very clearly in the charge distribution scheme. A detailed discussion, including a comparison with the monopole approximation (see below) has been given by Pollak and Rein (1967) and Rein and Pollak (1967).

b. The Monopole Approximation. Practically, the nuclei will be assumed to be fixed at equilibrium positions r_v, so that the wave function will be reduced to its electronic part and the integration over the nuclear part of $\rho(r)$ will simply introduce the point nuclear charges multiplied by the scalar product of the electronic wave functions:

$$f_{ac}(\mathbf{r}) = e\left[\langle a|c\rangle \sum_v Z_v \delta(\mathbf{r} - \mathbf{r}_v) - \langle a| \sum_i \delta(\mathbf{r} - \mathbf{r}_i)|c\rangle\right] \tag{6}$$

The nuclear part will, therefore, be nonzero for the state-charge distributions only, since $\langle a|c\rangle = 0$ for $|a\rangle \neq |c\rangle$.

When, moreover, the electronic wave functions are expressed as Slater determinants Ψ (or more generally as sums of several ones), built of molecular *spin orbitals,*

$$\Psi = \frac{1}{\sqrt{n!}} |\varphi_1(\mathbf{r}_1)\sigma_1(1) \ldots \varphi_p(\mathbf{r}_n)\sigma_n(n)|^*$$

---

* p may be smaller than *n*, since *one* given orbital φ may be used for building *two* spin orbitals $\varphi\alpha$ and $\varphi\beta$. The $\sigma_1, \sigma_2, \ldots, \sigma_n$ stand for α or β.

The application of Eq. (4) leads to very simple results. For a state charge distribution

$$f_{\Psi\Psi}(\mathbf{r}) = e\left[-\sum_k n_k \varphi_k^2(\mathbf{r}) + \sum_v Z_v \delta(\mathbf{r} - \mathbf{r}_v)\right] \tag{7}$$

where n_k is the occupation number of the *orbital* φ_k.

For a transition-charge distribution, the result is nonzero when the two Slater determinants differ by *one* spin orbital only since $\wp(\mathbf{r})$ contains only monoelectronic operators; moreover, if the spin function were different, the result would still be zero [since $\wp(\mathbf{r})$ does not contain spin operators], so that the *orbitals* themselves must be different. Let us say φ_i in $|\Psi\rangle$ and φ_j in $|\Psi'\rangle$; then

$$f_{\Psi\Psi'}(\mathbf{r}) = -e[\varphi_i(\mathbf{r})\varphi_j(\mathbf{r})] \tag{8}$$

If now the LCAO approximation is used ($\varphi_k = \sum_\alpha c_{k\alpha}\chi_\alpha$, where the χ_α are atomic orbitals), the expressions $f_{\Psi\Psi}(\mathbf{r})$ and $f_{\Psi\Psi'}(\mathbf{r})$ become:

$$f_{\Psi\Psi}(\mathbf{r}) = e\left[-\left(\sum_\alpha \left(\sum_k n_k c_{k\alpha}^2\right)\chi_\alpha^2(\mathbf{r}) + \sum_{\alpha \neq \beta}\sum \left(\sum_k n_k c_{k\alpha} c_{k\beta}\right)\chi_\alpha(\mathbf{r})\chi_\beta(\mathbf{r})\right)\right.$$

$$\left. + \sum_v Z_v \delta(\mathbf{r} - \mathbf{r}_v)\right] \tag{9}$$

$$f_{\Psi\Psi'}(\mathbf{r}) = -e\left[\sum_\alpha c_{i\alpha} c_{j\alpha} \chi_\alpha^2(\mathbf{r}) + \sum_{\alpha \neq \beta}\sum c_{i\alpha} c_{j\beta} \chi_\alpha(\mathbf{r})\chi_\beta(\mathbf{r})\right] \tag{10}$$

$\sum\sum_{\alpha \neq \beta}$ means a double summation where the indices must be *different*.

We are left with the "elementary" or "basic" distributions $\rho_\alpha(\mathbf{r}) = \chi_\alpha^2(\mathbf{r})$ (total charge in electron unit: $\int \rho_\alpha(\mathbf{r})dv = 1$) and $\rho_{\alpha\beta}(\mathbf{r}) = \chi_\alpha(\mathbf{r})\chi_\beta(\mathbf{r})$ (total charge $\int \rho_{\alpha\beta}(\mathbf{r}) = S_{\alpha\beta}$, overlap integral between χ_α and χ_β). If we approximate the $\chi_\alpha^2(\mathbf{r})$ by their first multipole moment, namely, a point charge 1 (in electron unit) located at the nuclear center of orbital χ_α and the $\chi_\alpha(\mathbf{r})\chi_\beta(\mathbf{r})$ by two equal charges $\frac{1}{2}S_{\alpha\beta}$ located at the centers of the orbitals χ_α and χ_β, the total charge distribution reduces to a set of point charges, all located at the nuclei. In the case of a state-charge distribution, the total point charge on a given nucleus is given by the formula which defines the atomic net charge in quantum chemistry:

$$q_v = e\left\{Z_v - \sum_\alpha \left[\sum_k n_k \left(c_{k\alpha}^2 + \sum_{\beta \neq \alpha} S_{\alpha\beta} c_{k\alpha} c_{k\beta}\right)\right]\right\}^* \qquad (\chi_\alpha \text{ centered at } v) \tag{11}$$

---

* The formula is very often written for the special case where there is only *one* AO per atom; this special formula would be obtained from Eq. (11) by replacing α by v and suppressing the summation over α.

In the case of the transition-charge distribution, one obtains in a similar manner transition charges:

$$q_v(i, j) = -e \left\{ \sum_\alpha c_{i\alpha} c_{j\alpha} + \frac{1}{2} \sum_{\beta \neq \alpha} S_{\alpha\beta}(c_{i\alpha} c_{j\beta} + c_{i\beta} c_{j\alpha}) \right\} \qquad (\chi_\alpha \text{ centered at } v) \quad (12)$$

The approximation of a charge distribution by a set of point charges (called "monopoles") as just defined will be designated as the monopole approximation. It appears clearly that the above described choice of these monopoles is a quite natural consequence of the representation of the wave functions in the MO-LCAO scheme. But other choices would be, in principle, possible. In practice, the use of the monopoles makes it necessary to know the wave functions of the ground and excited states, including all the electrons (and not only the π system). This is not possible at the present time, especially for the excited states, and this practical impossibility explains that a third kind of approximation is used simultaneously with the monopole one. We shall now consider it.

c. The Bond Dipole Approximation. According to this approximation, a charge distribution is reduced to a set of bond dipoles, each of them being located usually at the middle of the bond. The advantage of this approximation is the introduction, in the further developments, of the theoretical expressions of quantities which may be obtained from experiment—bond dipoles and, especially, bond polarizabilities (whose theoretical expressions involve a summation over bond-transition moments). It is possible to derive, for the polarization and dispersion energy, formulas involving the bond polarizabilities (Claverie and Rein, 1967).

B. Practical Calculations

We would now make a survey of the practical calculations quoted in Section I,B, indicating for each of them the kind of approximation used for the evaluation of the various contributions to the "long-range" part of the interaction energy. In the cases where other parts of the energy are considered too, the fact will simply be mentioned; the study of these parts will be the purpose of the next section.

1. The Pure Dipole Approximation

De Voe and Tinoco (1962) calculated the three usual "long-range" contributions using the ordinary formulas, i.e., the dipole approximation for all charge distributions. Owing to the closeness of the interacting molecules (stacked base pairs in the DNA), this approximation is inadequate, and the numerical results obtained in this pioneering work may now have a qualitative interest only. The dipole approximation was also used by Gersh and Jordan (1965) (see Gilbert and Claverie, this symposium). The well-known formulas are given in the first line of Table I.

TABLE I. The "Long-Range" Terms in Various Approximations[a]

Interaction	Electrostatic	Polarization	Dispersion
Dipole-dipole	$-\boldsymbol{\mu}_1\cdot\dfrac{\overline{\overline{T}}_{12}}{(r_{12})^3}\cdot\boldsymbol{\mu}_2$	$-\dfrac{1}{2}\vec{\mathscr{E}}_d^{\,21}\cdot\overline{\overline{\mathscr{A}}}^{(1)}\cdot\vec{\mathscr{E}}_d^{\,21}$ $-\dfrac{1}{2}\vec{\mathscr{E}}_d^{\,12}\cdot\overline{\overline{\mathscr{A}}}^{(2)}\cdot\vec{\mathscr{E}}_d^{\,12}$	$-\dfrac{\Delta}{4}\dfrac{1}{(r_{12})^6}Tr[\overline{\overline{T}}_{12}\mathscr{A}^{(1)}\overline{\overline{T}}_{12}\overline{\overline{\mathscr{A}}}^{(2)}]$ (for isotropic polarizabilities α: $-\dfrac{3}{2}\Delta\{1/(r_{12})^6\}\tfrac{1}{3}\alpha^{(1)}\alpha^{(2)}$)
Monopoles-dipole	$\displaystyle\sum_{\mu(1)=1}^{N_1}\sum_{\nu(2)=1}^{N_2}\dfrac{Q_{\mu(1)}Q_{\nu(2)}}{r_{\mu(1)\nu(2)}}$	$-\dfrac{1}{2}\vec{\mathscr{E}}_m^{\,21}\cdot\overline{\overline{\mathscr{A}}}^{(1)}\cdot\vec{\mathscr{E}}_m^{\,21}$ $-\dfrac{1}{2}\vec{\mathscr{E}}_m^{\,12}\cdot\overline{\overline{\mathscr{A}}}^{(2)}\cdot\vec{\mathscr{E}}_m^{\,12}$	Identical
Monopoles-bond polarizabilities	Identical	$-\dfrac{1}{2}\displaystyle\sum_{u(1)=1}^{B_1}\vec{\mathscr{E}}_{u(1)}\cdot\overline{\overline{\mathscr{A}}}_{u(1)}\cdot\vec{\mathscr{E}}_{u(1)}$ $-\dfrac{1}{2}\displaystyle\sum_{v(2)=1}^{B_2}\vec{\mathscr{E}}_{v(2)}\cdot\overline{\overline{\mathscr{A}}}_{v(2)}\cdot\vec{\mathscr{E}}_{v(2)}$	$-\dfrac{1}{4}\displaystyle\sum_{u(1)=1}^{B_1}\sum_{v(2)=1}^{B_2}\left\{\dfrac{\Delta_{u(1)v(2)}}{(r_{u(1)v(2)})^6}\,Tr[\overline{\overline{T}}_{u(1)v(2)}\overline{\overline{\mathscr{A}}}_{u(1)}\overline{\overline{T}}_{u(1)v(2)}\overline{\overline{\mathscr{A}}}_{v(2)}]\right\}$

[a] r always designates a distance.

$$\overline{\overline{T}}=3\left(\frac{\mathbf{r}}{r}\otimes\frac{\mathbf{r}}{r}\right)-\overline{\overline{1}}$$

$$\vec{\mathscr{E}}_d(\mathbf{r}_0)=\frac{1}{(r_{\mu 0})^3}\overline{\overline{T}}_{\mu 0}\cdot\boldsymbol{\mu}$$

$$\vec{\mathscr{E}}_m(\mathbf{r}_0)=\sum_\rho\frac{Q_\rho}{(r_{\rho 0})^3}\mathbf{r}_{\rho 0}$$

(electric field created by a dipole $\boldsymbol{\mu}$—located at the origin—at a point $\mathbf{r}_0$)   (electric field created by a set of charges at a point $\mathbf{r}_0$).

$\vec{\mathscr{E}}_{u(1)}$ is the electric field created by the monopoles of molecule 2 at the middle of the bond $u^{(1)}$. $\overline{\overline{\mathscr{A}}}$ designates a polarizability tensor, $\overline{\Delta}$ a "mean value" of the excitation energies (usually approximated by the ionization potential). Tr designates the trace of a matrix (sum of the diagonal elements).

μ, ν label the atoms, u, v the bonds.

Further practical details are given by Pullman et al. (1966).

2. The Electrostatic Term in the Monopole Approximation

Bradley *et al.* (1964) and Nash and Bradley (1965) used the monopole approximation for the ground-state charge distributions leading to the electrostatic term and neglected the two others. Therefore, the results may be of some significance for problems where the electrostatic part plays a predominant role, e.g., in hydrogen-bonded configurations. When the relative position is changed, deep energetic minima (corresponding to the equilibrium positions) are obtained when the monopole approximation is used, whereas nothing similar appears in the dipole approximation [Nash and Bradley, 1966; in this paper, an evaluation of the polarization and dispersion terms (neglected in the previous works mentioned above) is attempted].

The formula for the electrostatic energy in the monopole approximation is quite evident and is given in Table I (second line, first column).

3. The Monopole and Dipole Approximation

Pullman *et al.* (1966) and Claverie *et al.* (1966) used the monopole approximation for the ground-state charge distributions and the dipole one for the transition-charge distributions; hence, the name " monopole-dipole " given to this procedure, which is in some sense " nonhomogeneous " (the formulas are given in the second line of Table I; for their derivation, see Claverie and Rein, 1967). The electrostatic term is the same as that used by Bradley *et al.* and Nash and Bradley. The dispersion term is the same as that in the pure dipole approximation and, therefore, not reliable from the quantitative point of view; the procedure was still able to give some interesting results.

4. The Pure Monopole Approximation

Rein and Pollak (1967) use the monopole approximation for both kinds of charge distributions. They also consider the " short-range " terms. Contrary to the previous treatments mentioned above, the calculations are purely quantum mechanical and do not use any experimental quantity (like the polarizability). But the wave functions involve the π system only; as a consequence, the calculated atomic charges cannot be the exact ones, the charge distributions associated with transitions involving a σ orbital are not taken into account, and this results in a strong underestimation of the polarizability, and consequently, of the polarization and dispersion energies. However, when total (σ and π) wave functions will become available, the Rein and Pollak formulas will be able to give very interesting results. On the other hand, they may be used immediately for practical purposes, if they are associated with some other procedure which represents the charges and the polarizability arising from the σ system (see below).

5. The Monopole and Bond Dipole Approximations; the Corresponding "Bond Polarizabilities" Procedures

The ground-state charge distribution is always approximated by total $(\sigma + \pi)$ monopoles.

a. As shown by Claverie and Rein (1967) the approximation of the transition-charge distribution by bond dipoles makes it possible to derive, for the polarization and dispersion energy, formulas in which the theoretical expression of the bond polarizability tensors appears, so that practical formulas involving the experimental bond polarizabilities are finally obtained (see Table I, third line). This "monopole and bond polarizabilities" procedure has been used by Rein *et al.* (1967a), Claverie (1967), Gilbert and Claverie (this symposium), Gilbert and Claverie (1967), Mantione (this symposium),* and Rein *et al.* (1967b).

b. The Mixed Transition Monopole and Bond Dipole Approximation. It is necessary to point out that the pure bond polarizability approximation is strictly valid only for localized bond excitations; otherwise, "cross-polarizabilities" between different bonds appear, and such quantities cannot be obtained from experiment. The hypothesis of additive bond polarizabilities may be valid for σ systems, but not for π systems. Hence, the idea of using simultaneously the bond dipole approximation just described for the σ-transition charge distributions and the monopole approximation for the π-transition charge distributions as explained above. This mixed method has been developed recently by Rein *et al.* (1967a), who derived all the practical formulas. The polarization energy may be divided in a σ term (monopoles-bond polarizabilities) and a π term (monopoles-transition monopoles) and the dispersion energy in a σ-σ term (bond polarizabilities-bond polarizabilities), a σ-π term (transition monopoles-bond polarizabilities), and a π-π term (transition monopoles-transition monopoles).

It is necessary to point out that, apart from the uncertainty due to the approximations, two "complementary" practical difficulties exist—when bond polarizabilities are introduced, some "mean transition energy" has automatically appeared in the derivation of the formula, due to the use of a closure approximation. It is rather difficult to fix an accurate value (in the calculations quoted above, the π-ionization potential has been used).

When transition monopoles are used, the summation of the second-order terms is performed directly, and the difficulty consists now in the unavoidable truncation; in the usual MO-LCAO method, the number of monoexcited

---

* In these four last papers, the short-range repulsion energy is evaluated by the means of a semiempirical formula (see Section III,B).

states is strongly limited. A possible check for the validity of this truncation is the theoretical calculation of the polarizability and a comparison with experimental values.

A detailed numerical comparison of the methods described above has been made by Rein *et al.* (1967a).

III. The Interaction Energy at Short Distances

Experimentally, it is observed that, when the distance between two molecules decreases, the energy of interaction reaches a minimum (the equilibrium distance has an order of magnitude of 3–4 Å for two planar conjugated molecules lying parallel to each other) and then increases very rapidly. These facts prove the existence of a short-range repulsion term, which varies very rapidly with distance, so that it becomes rapidly negligible with respect to the so-called "long-range" terms. This does not exclude the existence of short-range attraction terms (see Section III,C on charge transfer), it is simply necessary that the repulsion ones become the most important when the distance decreases below the equilibrium value. The "long-range" terms, as considered in Section II, do not exhibit the convenient property; at short distances, the second-order terms (polarization and dispersion) become predominant* and they always give rise to an attraction. It is necessary to point out that this argument is valid for the exact long-range terms as well as for approximate ones. It appears, therefore, necessary to explain theoretically the existence of these short-range supplementary terms and to derive practical formulas for calculating them (as well as possible).

A. The Theoretical Picture

The most accurate calculations existing at the present time have been performed on small systems, the complex of the two molecules being considered as a "super-molecule" treated by one of the numerous methods suitable for a molecular calculation (various calculations concerning the He–He system have been mentioned in Section I,A). Owing to the nonperturbative character of these treatments, there is no special problem concerning the "short distance" case. But the amount of calculation needed becomes rapidly prohibitive when the dimension of the interacting systems increases, and it may be difficult to isolate the interaction energy itself and to separate it into parts having a different behavior when the distance varies.

---

* At long distance, they vary like R^{-6}, whereas the first-order (electrostatic) term varies like R^{-3}. Although these simple laws are no longer valid at short distance, the second-order terms actually increase more rapidly than the first-order one when the distance decreases.

These slight defects would not exist in a perturbation treatment, and it is, therefore, interesting to build a perturbation expansion valid up to short distances. As it was just seen, a perturbation treatment limited to second order in a basis of simple products (of the wave functions of the separated molecules) is not sufficient. How can this treatment be improved?

As pointed out by Hirschfelder (1966), the answer is still not clear at the present time from the theoretical point of view. In papers devoted to this subject [Longuet-Higgins (1956), and chiefly Murrell *et al.* (1965), Salem (1966), Musher (1967), Musher and Salem (1966), Murrell and Shaw (1967), and Musher and Amos (1967)] it is stated that some a priori antisymmetrization must take place. Initially, Murrell *et al.* (1965) used the set of all antisymmetrized products, which set is actually a generating system for the totally antisymmetric subspace of the complex system (Claverie, 1966), but generally not a true basis [these antisymmetrized products will be dependent, in general (Claverie, 1966; Musher and Amos, 1967)].

It is rather difficult to perform a rigorous perturbation treatment in such an "overcomplete" set, so that in more recent works, Murrell and Shaw (1967) and Musher and Amos (1967) propose to antisymmetrize the product of the ground states only; the perturbation treatment may be performed much more easily in such a basis (it is now a true basis). Murrell and Shaw calculate the first- and second-order terms, each of them being developed according to the powers of the overlap integrals (which appear as a consequence of the exchange in the antisymmetrized product of the ground states). The terms which do not involve overlap* are identical with the "long-range" terms obtained in the basis of simple products.

This picture is physically satisfying: Since all terms involving overlap decrease exponentially when the distance increases, it is explained why the "long-range" terms are actually the only important ones for large intermolecular distances, and supplementary, rapidly varying, "short-range" terms are exhibited.

There still subsists a puzzling theoretical question: What is the behavior of the perturbation treatment built with the basis of simple products, when the intermolecular distance decreases? For not too small distances, the interaction is still rather small compared to the excitation energies of the molecules, so that convergence would be expected for the "simple products" treatment, as well as for Murrell and Shaw's treatment (of course, it would be necessary to really solve this question mathematically). If both treatments actually converge, are the resulting energies different, the resulting functions different? If they are, which of them is the right solution, and what represents the other one? Musher (1967) proves that the resulting functions cannot be the

---

* Explicitly (overlap integrals) or implicitly (hybrid and exchange integrals).

same,* but this does not necessarily imply that the energies are different, *owing to the exchange degeneracy*; if the energies were the same, it would be sufficient to antisymmetrize the function given by the "simple products" treatment in order to obtain the true physical totally antisymmetric function† so that there would be no essential difference between the two treatments.‡ It actually seems that no rigorous and clear answer to these questions exists at the present time.

B. An Attempt of Practical Evaluation

1. Method

The treatment given by Murrell and Shaw (1967) looks rather attractive since the familiar "long-range" terms are explicitly separated from the supplementary "short-range" terms: the first ones may be calculated by the methods described in Section II, and only the second ones present a new problem. Unfortunately, they involve new types of integrals (exchange and hybrid) which are much more difficult to calculate than the Coulomb integrals; on the other hand, it may be feared that these terms are rather sensitive to the quality of the wave function used (in opposition to the "long-range" terms). Much work will still be necessary before reliable results may be obtained in this way (for practical purposes, one would need a method which is not extremely time-consuming, and this practical condition excludes the direct numerical calculation of the integrals involved).

This is the reason why we tried an evaluation of the "short-range" repulsion term using one of the various semiempirical formulas proposed in the literature. We tried the formula proposed by Kitaygorodsky (1961) for C–C, C–H, H–H interaction and applied in a more extensive manner by Favini and

---

* His proof has been given again by Murrell and Shaw (1967).

† By the way, it seems convenient to point out that the argument given by Murrell and Shaw (1967) that their perturbed function Ψ is antisymmetric is not convincing. They state that, if $(E-H)T|\Phi\rangle$ is antisymmetric, $T|\Phi\rangle$ is then antisymmetric. This is not necessarily true because E is an eigenvalue of H, so that $T|\Phi\rangle$ may contain a nonantisymmetric eigenfunction of H relative to the eigenvalue E. This is not an argument against their treatment, since, if necessary, the antisymmetrization of their Ψ would give the correct function without changing the energy.

‡ According to Musher himself, the "simple products" treatment would be valid for two-electron systems (Musher, 1965). He asserts that this is no more true for more than two electrons (Musher, 1967), but, as stated above, his proof concerns the antisymmetric character of the perturbed wave function only and cannot, therefore, be considered as a complete proof of the inadequacy of the "simple products" treatment.

Simonetta (1963). The energy is a sum of atom-atom interactions (exactly as is the monopole electrostatic energy) given by the formula:

$$E = -C_1\left(\frac{r_0}{r}\right)^6 + C_2 e^{-\alpha(r/r_0)} \tag{13}$$

r_0 is the sum of the van der Waals radii (Pauling, 1939) of the two atoms, r is their distance.

Kitaygorodsky fitted the constants in order to reproduce the crystal structure and energy of various hydrocarbons, and the values are $\alpha = 13$, $C_1 = 0.14$ kcal/mole, $C_2 = 3 \times 10^4$ kcal/mole; the ratio C_1/C_2 is such that E reaches its minimum for $r = r_0$.

Since these hydrocarbons are nonpolar, the long-range terms reduce almost* exactly to the dispersion one. This corresponds to the attraction term $C_1(r_0/r)^6$. The repulsive term gives us the evaluation of the short-range energy that we needed, but it seems necessary, before using it, to check the "compatibility" of this semiempirical formula with the more theoretical formulas of Section II. This comparison is just possible for the second-order terms, and

TABLE II

COMPARISON OF THE SECOND-ORDER ENERGY IN THE BOND POLARIZABILITIES
APPROXIMATION AND THE KITAYGORODSKY ATTRACTION TERM[a]

Distance between the planes	E_{pol}[b]	E_{disp}[b]	E_{K1}[c]	$\dfrac{E_{K1}}{E_{disp}}$	$\dfrac{E_{K1}}{E_{pol} + E_{disp}}$
3.2	−0.083	−5.117	−5.274	1.031	1.014
3.4	−0.061	−3.757	−3.920	1.044	1.027
3.6	−0.045	−2.802	−2.957	1.056	1.039
3.8	−0.033	−2.119	−2.260	1.067	1.050
4.0	−0.025	−1.622	−1.748	1.077	1.061

[a] Example of two stacked benzene molecules. The distances are expressed in Å and the energies in kcal/mole.

[b] E_{pol} and E_{disp} are the polarization and dispersion energies calculated in the monopoles-bond polarizabilities approximation, using as "mean energy" the ionization potential (9.3 eV).

[c] E_{K1} is the "attractive" energy calculated as the first term of Kitaygorodsky's formula ($C_1(r_0/r)^6$).

* There is a very small electrostatic and polarization energy when the monopole approximation is used.

TABLE III

COMPARISON OF THE "LONG-RANGE" ENERGY WITH THE TOTAL ENERGY INCLUDING THE SHORT-RANGE REPULSION[a]

Interaction	Energy (monopoles and bond polarizabilities)	Polarization energy (sum of 4 bases interactions)	Polarization energy (interaction between the 2 pairs)	Short-range repulsion energy	Total energy (with repulsion)
G–C	−21.30	—	—	5.28	−16.02
A–T	−6.67	—	—	3.81	−2.86
1 ↑GC|↑CG | GC↓ | CG↓	−7.20	−3.64	−1.57	3.74	−1.26
2 ↑CG | GC↓	−10.96	−3.44	−1.48	3.77	−4.99
3 ↑GC | CG↓	−15.18	−4.24	−1.85	3.08	−9.80
4 ↑AT | GC↓ ↑CG | TA↓	−8.04	−1.69	−1.14	3.77	−3.50
5 ↑TA | GC↓ ↑CG | AT↓	−7.52	−2.11	−1.34	4.13	−2.41
6 ↑AT | CG↓ ↑GC | TA↓	−11.16	−2.40	−1.39	2.94	−7.25
7 ↑TA | CG↓ ↑GC | AT↓	−11.36	−2.79	−1.55	4.01	−6.04
8 ↑AT | AT↓ ↑TA | TA↓	−8.32	−0.90	−0.94	4.08	−4.08
9 ↑TA | AT↓	−7.28	−1.27	−1.25	4.58	−2.52
10 ↑AT | TA↓	−7.22	−0.71	−0.79	2.83	−4.48

Table II shows the results of such parallel calculations for two stacked benzene molecules. The agreement between the two evaluations is excellent, and we consider it as very encouraging as concerns the quality of the presently available theoretical approximations to the long-range terms.*

It could be thought, now, that it is possible to add directly the repulsion term

$$C_2 \sum_{\mu(1)} \sum_{\mu(2)} \exp\left(-\alpha \frac{r_{\mu(1)\,\mu(2)}}{r_{\mu(1)} + r_{\mu(2)}}\right)$$

(r_μ van der Waals radius of atom μ) arising from Kitaygorodsky's formula to the long-range terms arising from one of the elaborate methods described in Section II,B,5. But doing this leads to intermolecular equilibrium distances which are systematically too small in the case of two stacked planar conjugated molecules; the same phenomenon holds when the complete Kitaygorodsky's formula is used alone. This is not contradictory with the fitting mentioned above, because this fitting concerned cases where no such stacking occur, like methane crystal, or benzene crystal (in which the planes of consecutive molecules are *perpendicular*). There are some theoretical reasons to think that the repulsion energy is underestimated in this case and, more exactly, that C_2 is too small for such configurations.† It appears, therefore, convenient to perform a special fitting of C_2 for this case of stacked planar conjugated systems.

---

* It seems interesting to point out that the use of the dipole approximation would give a dispersion energy about two times larger than the bond polarizability approximation (see Rein *et al.*, 1967a), so that no agreement would then exist.

† In opposition to the sum of Coulomb integrals, the integrals involving overlap may exhibit a rather pronounced dependence on the relative orientation of the molecules, especially according to the existence of $2p$ orbitals directed one to each other (case of parallel conjugated planar systems), for which strongly predominant integral types appear. Since, on the other hand, the exponent α corresponds roughly to the screening parameters on the atomic orbitals, which are not concerned by the relative orientation, the directional dependence of the short-range energy should rather affect the coefficient C_2.

---

^{a}The energies are expressed in kcal/mole.

The difference between columns 2 and 3 comes out from the nonadditivity of the polarization energy for nonpolar interacting molecules (Pullman *et al.*, 1967). Column 1 has been calculated using the polarization energies of column 2, and column 5 using those of column 3. Therefore, (column 5) = (column 1) − (column 2) + (column 3) + (column 4). (Small discrepancies, not larger than 0.24, are simply due to small differences between the dispersion energies in the calculations corresponding to columns 2 and 3, owing to the use of slightly different $\triangle$'s for these two calculations.)

A multiplication of C_2 by a factor 2.7, giving a new constant $C_2' = 8.1 \times 10^4$ kcal/mole, has been found to be satisfying; this must not be considered, of course, as a very accurate and definitive value.

2. Applications

Qualitatively, the use of a short-range repulsion energy insures the existence of a minimum, i.e., of an equilibrium position; it was impossible to obtain this result with long-range energy terms only, so that it was necessary to compare different complexes by calculating their interaction energy at some fixed distance, assumed to be rather close to the equilibrium distance for all these complexes. This assumption is actually rather valid for complexes between planar conjugated systems lying parallel to each other, as proven by experimental data; this property appears also in the calculations including the short-range repulsion, and it explains the success of the correlations between experimental results and interaction energies including the long-range terms only. These values often varied in the right sense, although they were systematically too large. *Quantitatively*, indeed, the total interaction energies including the repulsion energy are noticeably smaller than the long-range energies alone and are actually more reasonable. An example is given in Table III. Figure 1 shows the variation of the interaction energy between two consecutive base pairs of the DNA when winding or unwinding takes place (see Claverie, 1967). This method has also been used by Gilbert and Claverie (1967), Gilbert and Claverie (this symposium); Mantione (this symposium).

C. The Charge-Transfer Terms

Since this question is treated in a rather detailed manner by Mantione (this symposium), some brief remarks only will be made here.

1. The Theoretical Point of View

If a complete basis could be used for each molecule, the interaction energy would be obtained exactly by perturbation theory as it has been described in the preceding section. To add to a complete basis antisymmetrized products of ionic states of the two molecules (charge-transfer state of the complex) would be completely superfluous and would make the set of basic states "overcomplete" (a very interesting discussion has been made by Musher and Amos, 1967). A so-called charge-transfer state may be developed as a linear combination of products of excited states (mainly the high excited ones and even the continuum), so that the mixing of such a state with the ground state (which gives rise to the so-called charge-transfer energy) would simply be a part of the polarization.

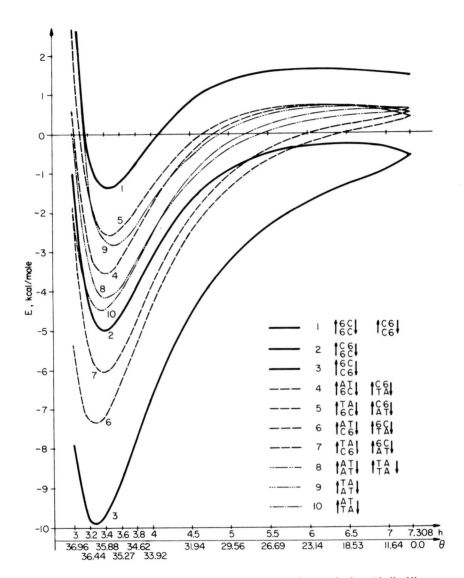

FIG. 1. The interaction energy between two consecutive base pairs in a "helical" movement.

The length L of the sugar phosphate chain is assumed to be constant. If R designates the radius of the cylinder on which the helix stands, θ (angle of the base pairs) and h (vertical distance) are related by $h^2 + R^2\theta^2 = L^2$. We took the values $R = 10.33$ Å $L = 7.308$ Å.

The interaction energy is the sum of the long-range terms (calculated in the monopoles-bond polarizabilities approximation) and the short-range repulsion term evaluated by the semiempirical treatment of Section III,B.

2. The Practical Point of View

In practice, the basis can never be complete and, especially for large molecules, they can represent only the lowest excited states, and these states are not localized far from the molecule, owing to the drastic limitation of the basis of atomic orbitals. As a consequence, excited states of a molecule which have a noticeable charge density in the region of the other molecule cannot be represented; then, the use of charge-transfer states allows us to compensate this gap and to complete the insufficient polarization (and also dispersion) energies given by the limited set of excited states.* But it is also rather unlikely that these charge-transfer states give a dominant contribution to the total interaction energy, in opposition with rather widespread concepts. The charge transfer occurs during the excitation, but probably plays a rather minor role in the ground state of most so-called "charge-transfer" complexes. [It is quite possible that, in some cases, the charge-transfer component becomes important (perhaps the dominant one), but in such cases, the whole spectrum of the complex should be different from the superposition of the spectra of the separated molecules and rather similar to the spectra of their ions. When this is not the case, we think that our statement (concerning the weakness of the charge-transfer contribution to the ground state of the complex) is likely to be true.]

IV. Conclusion

Our purpose was to show that, despite the number of works which are performed and the variety of methods and approximations used, there is not a complete disorder in the field of intermolecular forces. The practical calculations become less crude and there is a trend to use a theoretical background as rigorous as possible. The agreement between the monopoles-bond polarizabilities second-order terms and the attraction term of Kitaygorodsky's formula seems rather encouraging as concerns the closeness to the exact values. Some important problems still persist: to derive theoretical formulas for the short-range repulsion energy, to check the order of magnitude of the charge-transfer energy with respect to the other terms in practical calculations using a

---

* In a method like the bond dipole approximation, the summation is directly performed, leading to the bond polarizabilities, so that it could be thought that the charge-transfer energy is included. This is not true because, if all integrals are actually included, only the integrals corresponding to transition-charge distributions located each on its molecule are correctly approximated, and this condition is just not fulfilled for the transition-charge distributions associated with the high excited states under consideration, since the distribution then extends more or less on *both* molecules. The corresponding integrals are certainly strongly underestimated, so that it is probably not superfluous to add the charge-transfer energy when the bond polarizabilities approximation is used.

finite basis, and finally to check whether it is legitimate to truncate the perturbation expansion at second-order. Indeed, it seems to us that, at the present time, the practical available methods must be used rather carefully. A too crude calculation may make impossible the understanding of a phenomenon, but it may also lead to artifacts and erroneous interpretations. The certitudes are rather negative, at the present time; the dipole approximation is *not* reliable, but it is difficult to state to which extent the more elaborate approximations are, and the answer can be given both by improvement of the theory and comparison with experimental values. Unfortunately, most experimental results involve a solvent effect,* which itself is precisely a result of statistically averaged interactions involving the solute and the solvent together. Another possible check of intermolecular forces formulas consists in their application to molecular crystals. Only the experimental checking and the answer to the theoretical problems will possibly permit us to obtain a set of practical formulas suitable for the calculation of intermolecular energies in the equilibrium region with reasonable accuracy and for all kinds of molecules.

REFERENCES

Bradley, D. F., Lifson, S., and Honig, B. (1964). *In* "Electronic Aspects of Biochemistry" (B. Pullman, ed.), p. 77. Academic Press, New York.

Claverie, P. (1966). *Discussions Faraday Soc.* **40**, 174.

Claverie, P. (1967). *J. Chim. Phys.* (in press).

Claverie, P., and Rein, R. (1967). *Intern. J. Quantum Chem.* (submitted for publication).

Claverie, P., Pullman, B., and Caillet, J. (1966). *J. Theoret. Biol.* **12**, 419. There is an error in the Table 4 of this paper: in the line $\left|\begin{smallmatrix}\uparrow GC \\ CG\downarrow\end{smallmatrix}\right|$, $E_{\rho\rho}$ must be taken equal to -1.6 instead of 1.6. As a result, the column Sum should contain -8.1 instead of -4.9 and the column Total -27.3 instead of -24.1.

Coulson, C. A., and Davies, P. L. (1952). *Trans. Faraday Soc.* **48**, 777.

Dalgarno, A., and Lewis, J. T. (1956). *Proc. Phys. Soc.* (*London*), **A69**, 57.

Deal, W. J., and Kestner, N. R. (1966). *J. Chem. Phys.* **45**, 4014.

De Voe, H., and Tinoco, I., Jr. (1962). *J. Mol. Biol.* **4**, 500.

Favini, G., and Simonetta, M. (1963). *Theoret. Chim. Acta* **1**, 294.

Gersh, N. F., and Jordan, D. O. (1965). *J. Mol. Biol.* **13**, 138.

Gilbert, M., and Claverie, P. (1967). A theoretical study of the electrostatic interactions in the intercalation model of the DNA-dye complex, *J. Theoret. Biol.* (in press).

Haugh, E. F., and Hirschfelder, J. O. (1955). *J. Chem. Phys.* **23**, 1778.

Hirschfelder, J. O. (1966). *In* "Perturbation Theory and Its Applications in Quantum Mechanics" (C. H. Wilcox, ed.), p. 3. Wiley, New York.

---

* Only the experimental results concerning complexes in gaseous phase could be compared directly with theoretical calculations, and such results are not very numerous.

Hirschfelder, J. O., and Löwdin, P. O. (1959). *Mol. Phys.* **2**, 229.

Hirschfelder, J. O., and Löwdin, P. O. (1965). *Mol. Phys.* **9**, 491.

Hirschfelder, J. O., Curtiss, C. F., and Bird, R. B. (1964). "Molecular Theory of Gases and Liquids" (2nd printing, corrected, with notes added). Wiley, New York.

Kestner, N. R. (1966). *J. Chem. Phys.* **45**, 208(A) and 213(B).

Kestner, N. R., and Sinanoğlu, O. (1966). *J. Chem. Phys.* **45**, 194.

Kitaygorodsky, A. I. (1961). *Tetrahedron* **14**, 230.

Kołos, W. (1967). *Intern. J. Quantum Chem.* **1**, 169.

London, F. (1937). *Trans. Faraday Soc.* **33**, 8.

London, F. (1942). *J. Phys. Chem.* **46**, 305.

Longuet-Higgins, H. C. (1956). *Proc. Roy. Soc.* A **235**, 537.

Margenau, H. (1939). *Rev. Mod. Phys.* **11**, 1.

Murrell, J. N., and Shaw, G. (1967). *J. Chem. Phys.* **46**, 1768.

Murrell, J. N., Randié, M., and Williams, D. R. (1965). *Proc. Roy. Soc.* A**284**, 566.

Musher, J. I. (1965). *J. Chem. Phys.* **42**, 2633.

Musher, J. I. (1967). *Rev. Mod. Phys.* **39**, 203.

Musher, J. I., and Amos, A. T. (1967). "On the Theory of Atomic and Molecular Interactions" *J. Chem. Phys.* (in press).

Musher, J. I., and Salem, L. (1966). *J. Chem. Phys.* **44**, 2943.

Nash, A., and Bradley, D. F. (1965). *Biopolymers* **3**, 261.

Nash, A., and Bradley, D. F. (1966). *J. Chem. Phys.* **45**, 1360.

Pauling, L. (1939). "The Nature of the Chemical Bond." Cornell Univ. Press, Ithaca, New York (see p. 260 in the 3rd ed., 1960).

Phillipson, P. E. (1962). *Phys. Rev.* **125**, 1981.

Pollak, M., and Rein, R. (1967). *J. Chem. Phys.* **47**, 2045.

Pullman, B., Claverie, P., and Caillet, J. (1966). *Proc. Natl. Acad. Sci. U.S.*, **55**, 904.

Pullman, B., Claverie, P., and Caillet, J. (1967). *Proc. Natl. Acad. Sci. U.S.* (in press).

Rein, R., and Pollak, M. (1967). *J. Chem. Phys.* **47**, 2039.

Rein, R., Claverie, P., and Pollak, M. (1967a). *Intern. J. Quantum Chem.* (in press).

Rein, R., Goel, N. S., Fukuda, N., Pollak, M., and Claverie, P. (1967b). *Ann. N.Y. Acad. Sci.* (to be published).

Salem, L. (1966). *Discussions Faraday Soc.* **40**, 150.

Physics of Protein Synthesis

D. F. BRADLEY AND H. A. NASH

Laboratory of Neurochemistry
National Institute of Mental Health, Bethesda, Maryland

I. Introduction

Protein synthesis is a complex biochemical process. In the present context we use the term in the limited sense of the process by which genetic information stored in the nucleotide sequence of DNA is converted into the amino acid sequences of proteins. Recent biochemical work has defined a number of the steps involved. The time is now ripe to formulate molecular level models of these steps and some progress has been made in this direction, e.g., molecular models of DNA structure, condon-anticodon recognition, etc.

It is, therefore, perhaps not too early to begin thinking about the problem of the physics of protein synthesis. Thus, the molecular level models of protein synthesis describe the chemical reactions among the molecules involved. When these models have been fully developed, it will surely become of interest to know what causes these molecules to react as they do. This means that we shall have to know the physics of protein synthesis as well as its chemistry and biochemistry. Thus, when particular molecular structures are found to play a key role in translating, transforming, or transmitting genetic information, we shall have to know the physical forces which lead to the formation of these structures. When specific reactions between molecules play a key role, we shall have to know the detailed nature of the intermolecular forces involved.

There are two principal problems associated with developing the physics of any biochemical process: to obtain good sets of physical forces and to find tactics for applying them to specific situations. In this chapter we wish to present some results we have obtained while trying to work out the physics of two biochemical steps in protein synthesis: DNA replication and codon-anticodon recognition.

II. Interactions between Nucleotide Bases

At the heart of the molecular level models of DNA replication and codon-anticodon recognitions are complex formations between purine and pyrimidine bases. Numerous experimental studies on the types and strengths of such complexes and theoretical studies on the reasons for complex formation have been carried out in recent years. We await a comprehensive critical review, which interrelates experiment and theory on this subject.

Our theoretical work on the subject has concerned itself with the development of an electrostatic–hard-sphere representation of the interaction between bases with a view toward calculating equilibrium energies and geometries of base pairs in both unconstrained and externally constrained situations. As a longer range goal, we have attempted to develop a logical framework within which to relate these geometries and energies to chemical, biochemical, and/or genetic level observables.

The electrostatic–hard-sphere representation we employ has been described in detail previously (1, 2). Briefly, the permanent charge distribution in the bases are represented by point, nonintegral charges located at the atom centers. The electrostatic potential energy of a pair of bases with respect to infinite separation is obtained by summing over the contribution to the Coulomb potential of each of these charges on one base with every charge on the other base. The closest contacts of a pair of bases is defined by a preselected set of minimum atom-atom contact distances. No potential energy is assigned to such contacts, i.e., we employ a hard-sphere model.

In current studies (3) of complex formation between smaller molecules, the intermolecular potentials are represented by, in addition to the electrostatic R^{-1} term, an R^{-12} term for deformable spheres, and R^{-4} and R^{-6} terms for formal charge-polarizability and London polarizability-polarizability potentials. When the coefficients for these terms have been proved out to our own satisfaction, we plan to introduce them into any further calculations on the nucleotide problem.

III. Unconstrained Case

Honeywell 800 and IBM 360-50 computers were used to calculate the electrostatic–hard-sphere potential energies of pairs of bases in a high-density sampling of two-dimensional configuration space. Some typical results are shown in Figs. 1 and 2.

The energies at the deepest of the minima are shown as the upper figures in Table I for all possible pairs among adenine, uracil, cytosine, guanine, and inosine (2). The lower numbers are the enthalpies of pair formation in $CHCl_3$

---

[a] Upper values: V_{min} computed for methyl derivatives in kilocalories per mole of dimer. I- containing pairs were not considered in Ref. 2 but have been computed in this work using the same sources of input data as for the others. Lower values: $\Delta H^\circ_{298°K}$ determined for the cyclohexyl and ethyl derivatives of uracil (U) and adenine (A) respectively, and for the 2′, 3′-benzylidine-5′-trityl derivatives of cytosine (C), guanine (G), and inosine (I). Values are in kilocalories per mole of dimer. Values in parentheses are estimated from measured equilibrium constants and assumed entropies.

[b] Taken from Kyogoku et al. (4).

[c] Taken from Kyogoku et al. (5).

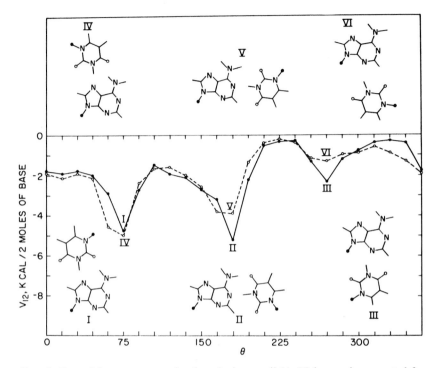

FIG. 1. Potential energy curves for the adenine-uracil (A–U) base pair computed from the electrostatic-hard sphere model. The configurations of the pair corresponding to each labeled minimum are shown. [Courtesy of Nash and Bradley (2).]

TABLE I

COMPARISON OF $V_{ij(min)}$ *in vacuo* WITH $\Delta H°$ IN $CHCl_3$ FOR VARIOUS COMBINATIONS OF NUCLEOTIDE BASES[a]

Base	A	G	C	U	I
A	−3.55	−6.61	−5.06	−6.07	−6.65
	−4.0 ± 0.8[b]	—	—	−6.2 ± 0.6[b]	(< −4)
G	−6.61	−14.38	−15.03	−9.81	−13.48
	—	(−8.5 to −10)[c]	(−10 to −11.5)[c]	—	—
C	−5.06	−15.03	−9.84	−4.77	−11.61
	—	(−10 to −11.5)[c]	−6.3 ± 0.6[c]	—	(−9)[c]
U	−6.07	−9.81	−4.77	−7.03	−9.41
	−6.2 ± 0.6[b]	—	—	−4.3 ± 0.4[b]	—
I	−6.65	−13.48	−11.61	−9.41	−12.64
	(< −4)	—	(−9)[c]	—	(−8)[c]

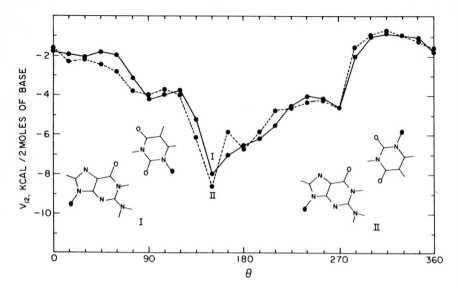

FIG. 2. Potential energy curves for the guanine–uracil (G–U) base pair computed from the electrostatic-hard sphere model. The input data (without curves) are given by Nash and Bradley (2).

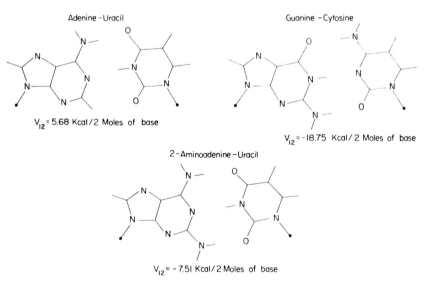

FIG. 3. Configurations corresponding to V_{min} for the A–U, G–C and 2-aminoadenine-uracil (AA–U) base pairs using Del Re σ and SCF π charge densities.

as measured by Kyogoku et al. (4, 5). There are, of course, many factors, both known and unknown, which could create a considerable difference between a minimum potential energy, which would be the enthalpy at $0°K$ in vacuo, and room temperature enthalpies in $CHCl_3$, e.g., zero-point energy, specific and nonspecific solvent-solute interactions, and the bulky 2',3'-benzylidine-5'-trityl side chains used to solubilize the bases in $CHCl_3$. Nevertheless, the fact that the two sets of values are comparable is encouraging.

Both sets of numbers reinforce the key concept that any base can form a reasonably stable complex with any other in this set, at least in vacuo and in $CHCl_3$. The absence of experimental values for some of the pairs can be explained thermodynamically (6) in terms of the requirement that for a mixed dimer to be observed, its enthalpy must be greater than the mean of the two corresponding self-associations.

As an illustration of the manner in which such computations can resolve certain experimental dilemmas, consider the aminoadenine-uracil problem (Fig. 3). The AA–U pair, as drawn, has three hydrogen bonds as does the G–C pair, while the A–U pair has only two. The usual explanation of the stability of such pairs in terms of number and types of hydrogen bonds would lead to the prediction that the G–C and AA–U pairs would have the same enthalpy of association which would be considerably larger (e.g., by 2 to 3 kcal) than that of the A–U pair.

There are three independent experimental results, however, which prove this prediction to be in error. First is the experiment of Shoup et al. (7) on NMR evidence of association of nucleotide bases in dimethyl sulfoxide, from which it is clear that the downfield proton shifts are in the order $GC \gg AA–T > AT$. Second is the demonstration by Howard et al. (8) that the melting temperature of a helical complex of poly AA + poly U is intermediate between that of poly A + poly U and that of G–C helices. Third is the demonstration of Kyogoku et al. (9) that the equilibrium constant for

TABLE II

V_{min} for Three Base Pairs[a]

Pair	$V_{min}{}^b$	$V_{min}{}^c$
A–U	-6.21	-5.68
G–C	-18.25	-18.75
AA–U	-8.46	-7.51

[a] Values in kilocalories per mole of dimer.
[b] Computed from Del Re σ and SCF π charge densities.
[c] Computed from Del Re σ and Hückel π densities.

association in chloroform between 9-ethylaminoadenine and 1-cyclohexy-luracil is only slightly greater than that of 9-ethyladenine and 1-cyclohexy-luracil, whereas that of guanine and cytosine derivatives is at least an order of magnitude greater. This dilemma, however, is resolved by computing the $V_{\text{min's}}$ for the three pairs (Table II). See also Pullman *et al.* (10).

IV. DNA Replication

Both theoretical and experimental data presented in Table I highlight the dilemma about DNA replication: If there is no particular specificity in the pairing of the bases, how is it that DNA replicates so faithfully? Thus, for example, if the base on the parental strand selects the entering base on the basis of the strength of their interaction, the filial strand would end up as primarily poly G with small amounts of C and T as shown in Table III.

TABLE III

EXPECTED BASE PAIRING IN DNA ASSUMING THAT THE ENERGY OF
BASE PAIRING DETERMINES THE BASE ADDED[a]

Base on parent strand	Base added to growing strand			
	A	G	C	T
A	1	100	8	42
G	<1	35	100	<1
C	<1	100	<1	<1
T	<1	100	<1	1

[a] Values are the relative (to 100) numbers of bases which would be added to the filial strand opposite particular bases on the parental strand provided that the four nucleoside triphosphates in the medium were at equal concentration. A simple method of calculation was employed (1), e.g., $(G–G)/(G–C) = 100 \exp(V_{GC}–V_{GG}/RT)$, which is equivalent to setting $V_{\text{min}} = \triangle F^{\circ}_{310^{\circ}K}$.

Whence, therefore, does the demonstrable specificity in replication arise. Perhaps it originates in complex formation between bases which although not specific in terms of exclusivity is specific in terms of geometric constraints which it imposes on another part of the system. Consider, for example, where the proximal phosphate of an entering nucleoside triphosphate would be (relative to the position it would occupy as part of the sugar phosphate backbone of a completed DNA helix) if the base were complexed with the base on the parental strand and the parental strand plus the previously created filial strand were already in a DNA geometry. Such distances, computed on the assumption of a rigid structure for the nucleoside triphosphate, are shown in Table IV.

TABLE IV
DISTANCES OF THE PROXIMAL PHOSPHATE OF THE ENTERING BASE[a]

Parental base	Entering base			
	A	G	C	U
A	12.67	4.59	14.39	0.39
G	3.32	4.52	0.51	2.34
C	8.90	0.51	11.57	2.53
U	0.51	3.87	2.06	3.48

[a] From where it would be if covalently linked to the growing filial DNA chain. Data given in Ångstroms. These values are the smallest distances that can be obtained by putting the pairs in any of their potential wells.

Therefore, to resolve the replication dilemma one needs only to postulate that the DNA polymerase enzyme which forms the phosphodiester linkage is fixed in position relative to the preexisting DNA and has a small (ca. 1Å) active site. Of course, once the entering base has been covalently linked to the growing filial chain its now-built-in-specificity (with respect to its parental complement) generates the other part of the geometric constraints which determine the specificity of the next entering base.

V. DNA Replication in the Presence of Base Analogs

This model of DNA replication predicts that a given base analog will successfully replace one of the normal nucleotides if it forms a stable base pair with a geometry which locates the phosphate group not too far from the position it would have in a B-form helix: For example, the I–C pair is strong and has a phosphate deviation of 0.21 Å. The xanthine-cytosine (X–C) pair is weaker and has a phosphate deviation of 1.74 Å. Therefore, the model explains the inability of X and the success of I in replacing G on both criteria (11).

As another example, consider the replacement of A by 2-aminopurine (Fig. 4). In this case a base pair of 2AP–U with $V_{(min)} = -4.90$ kcal/base pair and a phosphate deviation of 0.83 Å can be formed corresponding to the structure of Sobell (12). It should, therefore, be weakly able to replace A in DNA replication. In addition there is a base pair of 2AP–C with $V_{(min)} = -6.77$ kcal and a phosphate deviation of 1.97 Å. Because of the low energy with respect to G–C, 2AP in solution should not replace G. But if it has been incorporated into DNA (by replacing A in a previous cycle of replication) it could subsequently code for C instead of U a small percentage of the time; the greater stability of the 2AP–C pair in part compensating for the more favorable phosphate deviation of 2AP–U. Thus, our model is consistent with

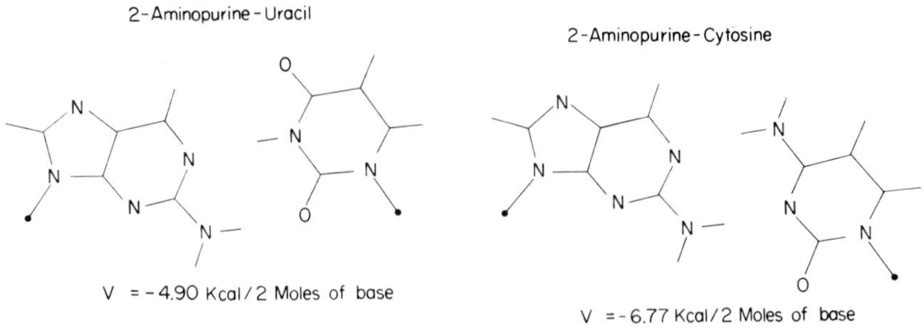

2-Aminopurine-Uracil

2-Aminopurine-Cytosine

V = -4.90 Kcal/2 Moles of base

V = -6.77 Kcal/2 Moles of base

FIG. 4. Configurations corresponding to V_{min} for the 2-aminopurine-uracil (2AP–U) and 2-aminopurine-cytosine (2AP–C) base pairs using Del Re σ and Hückel π charge densities.

the well-known phenomena of limited incorporation and high mutagenesis of 2AP (13).

It is interesting to note that in the codon-anticodon complex which, to be sure, involves significantly different external geometric constraints, 2AP does appear to form stable complexes with either U or C(14).

REFERENCES

1. Nash, H. A., and Bradley, D. F., *Biopolymers* **3**, 261 (1965).
2. Nash, H. A., and Bradley, D. F., *J. Chem. Phys.* **45**, 1380 (1966).
3. Minicozzi, W., Nash, H. A., and Bradley, D. F., in preparation.
4. Kyogoku, Y, Lord, R. C., and Rich, A., *J. Am. Chem. Soc.* **89**, 496 (1967).
5. Kyogoku, Y., Lord, R. C., and Rich, A., personal communication.
6. Pullman, B., Claverie, P., and Caillet, J., *J. Mol. Biol.* **22**, 373 (1966).
7. Shoup, R. R., Miles, H. T., and Becker, E. D., *Biochem. Biophys. Res. Commun.* **23**, 194 (1966).
8. Howard, F. B., Frazier, J., and Miles, H. T., *J. Biol. Chem.* **241**, 4293 (1966).
9. Kyogoku, Y., Lord, R. C., and Rich, A., *Proc. Natl. Acad. Sci. U.S.* **57**, 250 (1967).
10. Pullman, B., Claverie, P., and Caillet, J., *Compt. Rend.* **263**, 2006–2009 (1966).
11. Okazaki, T., and Kornberg, A., *J. Biol. Chem.* **239**, 259 (1964).
12. Sobell, H. M., *J. Mol. Biol.* **18**, 1 (1966).
13. Freese, E., in "Molecular Genetics" (J. H. Taylor, ed.), Part 1, p. 222. Academic Press, New York, 1963.
14. Wacker, A., Lodemann, E., Gauri, K., and Chandra, P., *J. Mol. Biol.* **18**, 382 (1966).

Note Added in Proof: Freese and Freese (*Proc. Natl. Acad. Sci. U.S.* **57**, 650 (1967)), from a study of mutation rates induced by base analogs in bacteria with either normal or mutant DNA polymerases, have arrived at a DNA replication paradigm similar to that described herein. That this model can be arrived at from two such different starting points reduces the likelihood of finding equally satisfying alternative paradigms.

Kinetics of Helix Formation and Slippage of the dAT Copolymer*

ROBERT L. BALDWIN

Department of Biochemistry
Stanford University School of Medicine
Palo Alto, California

I. Introduction

There are many examples known in which the secondary structure of a DNA plays a key role in determining its accessibility to enzymes or its effectiveness in a biological assay. For instance, in degrading natural DNA's to mononucleotides *Escherichia coli* exonuclease I displays an almost absolute preference for single-stranded as compared to double helical DNA—the preference is at least 40,000 : 1 (Lehman *et al.*, 1965). In the usual assays of transforming activity only double helical DNA is effective (Roger and Hotchkiss, 1961; Ginoza and Zimm, 1961).

A more subtle structural transformation of a DNA has recently been shown to have dramatic biological consequences: at the opposite ends of the DNA of bacteriophage λ there are short complementary stretches of single-stranded DNA, approximately 20 nucleotides in length (Wu and Kaiser, 1967). Allowing these ends to join noncovalently to form circular molecules destroys their infectivity in the helper assay used by Kaiser and Inman (1965).

We wish to discuss here the dynamic properties of a particular DNA helix and to consider a case in which these properties may determine the rate of replication of the template by DNA polymerase.† Short oligonucleotides $d(AT)_n$, $n = 4, 5, 6, \ldots$, will function as templates for synthesis of the alternating dAT copolymer (Kornberg *et al.*, 1964). If the usual rules for the action of *E. coli* DNA polymerase are obeyed in this unusual case, the nucleotide

---

* This work has been supported by research grants from the U.S. National Institutes of Health (AM 04763) and the National Science Foundation (GB 4061).

† Abbreviations: the notation for synthetic polynucleotides is that of Inman and Baldwin (1962a), in which dAT stands for an alternating DNA copolymer, secondary structure unspecified; dAT : dAT stands for the dAT double helix; and rA : rU stands for an RNA double helix whose complementary strands are homopolymers. Here A stands for adenine; T, thymine; $\overline{BU}$, 5-bromouracil; U, uracil; I, hypoxanthine (whose ribonucleoside is inosine); and $\overline{BC}$, 5-bromocytosine. Other abbreviations: T_m, melting temperature or the midpoint of a helix-random coil transition; T_{opt}, temperature at which the rate of dAT replication is optimal.

added to the growing end of the daughter strand (the polymer) is determined by base pairing with the template and the daughter strand remains hydrogen-bonded to the template. Thus, oligomer and polymer presumably form a double helix in which the growing end of the polymer is held opposite the oligomer, and to accomplish this the complementary strands must slip over each other during synthesis. The rate of synthesis may be determined by this slippage. We shall discuss the conformational properties of dAT in relation to its synthesis and describe progress with a system for measuring rates of the elementary processes of helix formation, the rates of helix initiation, and those of base pair closure and opening. These data are needed to evaluate possible mechanisms for slippage.

The dAT copolymer was first discovered in a test for DNA synthesis *in vitro* when one of the four deoxyribonucleoside triphosphates (dGTP) was omitted, but the other necessary ingredients (DNA polymerase, Mg^{2+}, and a template DNA) were present. After a long lag period without observable synthesis, a synthetic DNA was made *de novo* which proved to have the strictly alternating base sequence ... ATAT ... (Schachman *et al.*, 1960). A naturally occurring dAT copolymer has also been found in crabs (Sueoka, 1961); it contains 93% alternating AT sequences (Swartz, *et al.*, 1962). The enzymatically synthesized dAT usually has a high molecular weight (1–10 million) and it displays thermal melting curves characteristic of double helical DNA. X-ray diffraction studies indicate that its helical structure is like that of natural DNA's in the B form, lithium salt (Davies and Baldwin, 1963). Recent work in our own and other laboratories has furnished the following picture of the physicochemical behavior of this unusual DNA.

II. Formation of Hairpin Helices

In early studies of DNA melting curves, it was expected that those of dAT would be readily reversible because of the ease with which sequences can be matched to form a double helix, and this was confirmed experimentally (Marmur and Doty, 1959). It was also expected that a single dAT strand might form a double helix with a loop at one end (a " hairpin " helix), in which the complementary strands have the antiparallel orientation of natural DNA's (Schachman *et al.*, 1960). This hypothesis provided explanations for certain properties of dAT; for example, very high polymer concentrations are required to form a double helix from two separate strands, presumably because of the competing reaction in which hairpin helices are formed (Inman and Baldwin, 1962b). A clear demonstration that the dAT double helix can be formed from a single strand was provided by a study of the kinetics of helix formation. When a double helix is formed from a pair of

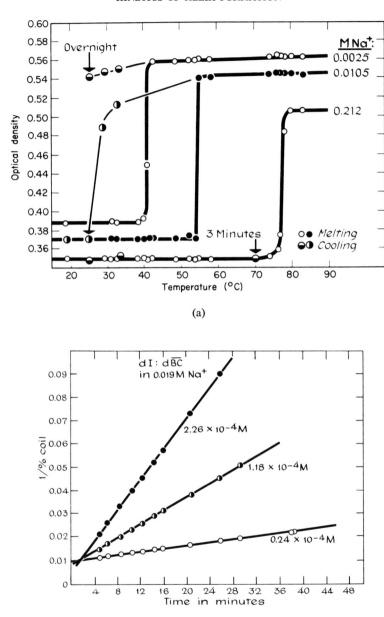

(a)

(b)

FIG. 1. Helix melting and reformation of a DNA homopolymer pair (dI : $\overline{BC}$). Figure 1a shows that helix formation is slow in low salt concentration, and Fig. 1b shows that it follows second-order kinetics. From Inman and Baldwin (1964).

DNA homopolymers such as dI and $\overline{dBC}$, helix formation is rapid in high salt concentrations, but becomes very slow in low salt concentrations, presumably because the large electrostatic repulsion between strands opposes contact. This is shown in Fig. 1a with the dI : $\overline{dBC}$ helix first being melted and then allowed to reform at various salt concentrations. Helix formation follows second-order kinetics (Fig. 1b), showing that initiation of the helix is rate-limiting, whereas winding up the helix is relatively fast. This type of

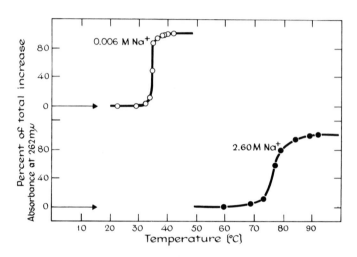

FIG. 2. Comparison of the melting curves shown by the dAT copolymer in low and high salt concentrations. On cooling, the double helical conformation reforms within the time of temperature equilibration (a few minutes) in both cases. From Inman and Baldwin (1962a).

behavior was first demonstrated by Ross and Sturtevant (1960) in their studies of the formation of the RNA double helix rA : rU. On the other hand formation of the dAT double helix is rapid both in high salt and low salt concentrations (Fig. 2), indicating that it is not necessary for separate strands to come together to initiate helix formation. Later studies with a temperature-jump apparatus have shown that the halftime of helix formation is actually in the millisecond range, even in low salt concentration (Spatz and Baldwin, 1965).

Recently hairpin helices have been seen directly in the electron microscope by Davidson *et al.* (1965) who photographed crab dAT before and after melting; after allowing the helices to reform on cooling they found branched molecules in which each branch presumably is a hairpin double helix but might contain G or C in the loop.

III. Branching

A second unusual property of dAT might also have been expected, but instead was discovered experimentally: the ability of the double helix to form a branched structure without undergoing melting. A plausible mechanism by which branching might occur is shown in Fig. 3. Segments of the DNA helix are assumed to open and close again, in a dynamic equilibrium. In the case of dAT each complementary strand can close on itself to form a hairpin helix, and this results in a branched structure. Measurements of deuterium exchange in DNA are compatible with the opening and closing of interior segments of the double helix (Printz and von Hippel, 1965).

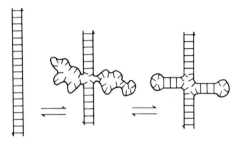

FIG. 3. A possible mechanism by which helical dAT can form a branched structure.

Branching was first suggested in a study of the intrinsic viscosity of the dAT helix at different temperatures. There is a slow drop in viscosity well below the melting zone, suggesting a gradual rearrangement to a more compact form (Inman and Baldwin, 1962a). The viscosity rises again on cooling, showing partial reversibility. Since the decrease in viscosity is more pronounced at high salt concentrations, suggesting that the helix forms more branches in high salt concentrations, this hypothesis could explain the remarkable broadening of the dAT melting curve with increasing salt concentration (Inman and Baldwin, 1962a; see Fig. 2). The slope of the melting curve at its midpoint should be inversely proportional to the average helix length (Crothers and Zimm, 1964), and so the broad melting curve indicates that the helical segments of partly melted dAT are relatively short in high salt concentrations as expected in the melting of a highly branched helix. On the other hand, the width of the melting curve of a DNA homopolymer pair, whose individual strands are not self-complementary, is essentially independent of salt concentration (Inman and Baldwin, 1964; see Fig. 1).

This interpretation was supported by a study of the kinetics of melting of the dAT copolymer (Spatz and Baldwin, 1965). The rate of melting of the dAT double helix is strongly dependent on its length (see Fig. 8). Rates of melting at the same temperature (T_m) were compared for temperature jumps from different initial temperatures below the melting zone. In low salt concentration there is little change in the rate of melting until the initial temperature is almost inside the melting zone. In high salt concentration a large increase in the rate of melting occurs as the initial temperature is raised, and the temperature zone in which this increase is found coincides with the zone in which the drop in viscosity occurs. A branched helix should melt more rapidly than an unbranched one of the same overall length, because the branches can unwind independently of each other.

A theoretical study of branching by Crothers and Zimm [unpublished; mentioned by Crothers and Zimm (1964)] indicates that branching should occur as the melting zone is approached whenever a single polynucleotide chain can loop out to form a hairpin helix.

The hypothesis that dAT double helices branch below the melting zone led Inman to make a striking prediction. Taking advantage of the fact that the melting zones of the dAT helix and its $d\overline{ABU}$ analog are separated by 9° in low salt concentration, he reasoned first that melting of a hybrid dAT : $d\overline{ABU}$ helix could be followed whether or not dAT : dAT and $d\overline{ABU}$: $d\overline{ABU}$ double helices are also present. Then, if these self-complementary double helices form branches below their melting zones, the dAT branches in a hybrid dAT : $d\overline{ABU}$ helix should melt as soon as they are formed, being above the dAT : dAT melting zone. Thus, his prediction was that the rate of melting of the dAT strand in a hybrid dAT : $d\overline{ABU}$ helix should equal the rate of branching of this helix and should be slow compared to the rates of melting dAT : dAT and $d\overline{ABU}$: $d\overline{ABU}$ helices. Also this slow melting should take place below the expected melting zone—less than halfway between the

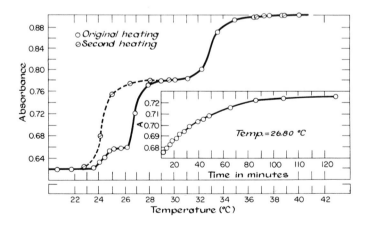

Fig. 4. Slow melting of a double helix containing complementary strands of dAT and $d\overline{ABU}$, leaving melted dAT and double-helical $d\overline{ABU}$. From Inman and Baldwin (1962b).

melting zones of dAT : dAT and d$A\overline{BU}$: d$A\overline{BU}$ helices. Both these predictions were verified for dAT : d$A\overline{BU}$ helices made by annealing under special conditions (Inman and Baldwin, 1962b), and also for dAT : d$A\overline{BU}$ helices made by using dAT as a template for synthesis of d$A\overline{BU}$, and vice versa (Wake and Baldwin, 1962). The slow melting of a hybrid dAT : d$A\overline{BU}$ helix is shown in Fig. 4.

IV. Mechanism Proposed for Reiterative Synthesis

The discovery that a dAT double helix can branch without prior melting suggested that the conformational mobility needed for branching might also be involved in the *de novo* synthesis of dAT. Crick had earlier proposed a model based on slippage (1959; quoted by Inman and Baldwin, 1962b). In his mechanism olignucleotides of different sequences are produced by an unspecified mechanism. An alternating dAT oligonucleotide is selected for more rapid chain lengthening by virtue of its ability to loop back and present a template strand at the site of synthesis. Slippage of the dAT chain would be needed to preserve this conformation. Once made, a dAT polymer serves as template in the usual way and rapid synthesis ensues.

The subject progressed from speculation to measurement when Kornberg *et al.* (1964) found that small dAT oligonucleotides function as templates for the enzymatic synthesis of dAT copolymer. Evidence that they are active as such, without first being elongated, came from a study of synthesis at different temperatures. When replication is measured with a d(AT)$_n$ oligomer ($n = 4$, 5, 6, ...) as template, the rate passes through a maximum with temperature and the optimal temperature is lower for smaller oligomers: for $n = 6$, $T_{opt} \simeq 37°C$, for $n = 5$, $T_{opt} \simeq 20°–25°$, and for $n = 4$, $T_{opt} \simeq 10°C$. The product is macromolecular dAT. Since d(AT)$_4$ is active at $10°C$ whereas d(AT)$_7$ is virtually inert, it seems unlikely that the oligomers must be lengthened before they will function (Kornberg *et al.*, 1964).

These authors proposed a general mechanism by which oligomers act as templates for reiterative synthesis: the first step is slippage, which leaves an overlapping end of the template strand, and the second step is replication, which again brings the ends in register (Fig. 5).

If their model is correct, either slippage or replication should be rate-limiting. The case in which replication is the slower process has been treated mathematically by Elson (1966, 1967). He finds that the concentration of overlapping ends predicted by helix-coil transition theory passes through a maximum with temperature, for a given oligomer-polymer double helix. This optimal temperature varies with the oligomer size in a manner which could explain the observations of Kornberg *et al.* (1964). In fact, Elson calculates

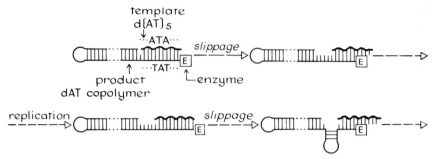

FIG. 5. A model for synthesis of the dAT copolymer from an oligomeric dAT template (Kornberg *et al.*, 1964).

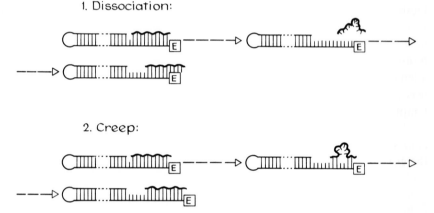

FIG. 6. Two possible models for slippage in the enzymatic synthesis of dAT copolymer from an oligomeric template.

rates of synthesis for different oligomeric templates which agree semi-quantitatively with the experimental ones, when he assumes that slippage is rapid compared to replication, so that the concentration of overlapping ends is always close to its equilibrium value.

V. Possible Mechanisms of Slippage

To study the possibility that slippage is rate-limiting, we need to know the mechanism of slippage. Two possible mechanisms are shown in Fig. 6. In the first, slippage occurs by complete dissociation of the oligomer from the product and recombination in a different conformation. Such a process undoubtedly occurs near, as well as inside, the melting zone when a short double helix must be formed from separate complementary strands. Whether this mechanism will also be operative when one or both complementary strands can form hairpin helices after dissociation is more questionable. One would expect formation of the hairpin helix to be faster than strand recombination. Of course, it is possible that the enzyme holds the growing polymer strand in a conformation favorable for recombination.

The second mechanism, termed "creep," could occur by the same processes involved to explain branching of the dAT polymer (Fig. 3). A segment of the double helix opens, and closure of each strand on itself results in the formation of hairpin branches. Then these hairpin helices move independently of each other to the right or left, in a random-walk fashion, by further opening and partial closing.

To decide whether either of these mechanisms can yield a rate of slippage fast enough to account for the rate of reiterative synthesis, we need to study slippage by direct measurement in a simpler system, uncomplicated by the presence of an enzyme. A system which may be suitable for this purpose is described at the end of this chapter.

The peculiar temperature dependence of oligomer replication could be explained also if slippage is rate-limiting and occurs by dissociation and recombination (Elson, 1966). Below the melting zone dissociation should be slow and strand recombination relatively fast, while as the temperature approaches T_m the rate of strand recombination should drop toward a low value (Ross and Sturtevant, 1960). Whether or not the rate of slippage via creep would pass through a maximum with increasing temperature is not clear.

VI. A System for Measuring the Rates of Elementary Processes in dAT Helix Formation

To evaluate properly a complex conformational reaction such as slippage we need to know first of all the rates of the basic steps in helix formation and melting: the rates of opening and closing a base pair next to an existing pair and the rate of helix initiation. In order to measure these parameters for the dAT helix we have begun a study of the kinetics of melting of $d(AT)_n$ oligomers (n between 5 and 25), which can form hairpin helices (Scheffler et al., 1967). The choice of hairpin helices appears especially suited to the problem of measuring the rate of base-pair closure; for dimer helices initiation of the double helix is likely to be rate-limiting, as illustrated in Fig. 1 for the DNA homopolymer pair dI : $d\overline{BC}$. These parameters can be found from the dependence of the kinetics of melting on helix length (as described below).

The dAT oligonucleotides were prepared by degrading macromolecular dAT to a controlled extent with bovine pancreatic deoxyribonuclease (DNase). Fractionation of the resulting oligonucleotides was achieved by a process of molecular sieving by electrophoresis in highly concentrated poly-acrylamide gels, which gives a large number of sharply resolved bands (Elson and Jovin, 1967; see Fig. 7). An unsuspected specificity on the part of pancreatic DNase helps to improve the separation; enzymatic tests show that almost all the oligonucleotides have T at their 5′ ends, so that successive oligomers should differ in size by two nucleotides (Scheffler, 1967). Also, these oligomers contain equimolar amounts of A and T, which simplifies such measurements as hypochromicity as a function of helix length.

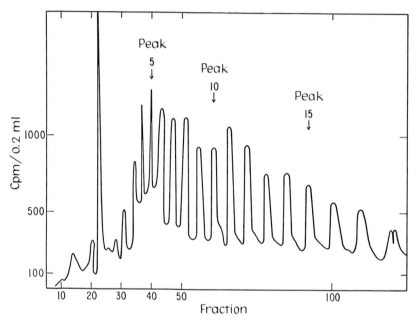

Fig. 7. Fractionation of dAT oligomers by electrophoresis in concentrated polyacrylamide gels (20% acrylamide, 3% bisacrylamide) (Elson and Jovin, 1967).

We have begun the measurement of the molecular weights of these oligomers in the ultracentrifuge, but for the present we refer to them by their order in acrylamide gels. The fraction numbers are approximately equal to the number of base pairs which can be formed in a hairpin helix with a loop size of four nucleotides. Molecular weight measurements indicate that the oligomers are chiefly in the form of hairpin helices inside their melting zones, although dimer helices may be formed below the melting zone.

When the kinetics of melting are measured in a temperature-jump apparatus (cf. Eigen and de Maeyer, 1963), we find that the rate of melting is in each case fairly homogeneous and that the terminal relaxation times are in the range 0.1–10 msec for the lower fractions. (For a simple unimolecular process $\tau^{-1} = k_{12} + k_{21}$ where τ is the relaxation time and k_{12} and k_{21} are rate constants for the forward and back reactions.) There is a strong dependence of τ on both chain length and temperature (Fig. 8). The relaxation times for fractions 10 and 16, which should differ by just six base pairs, are an order of magnitude apart.

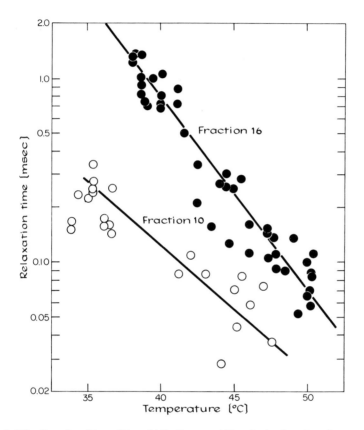

FIG. 8. Kinetics of melting of two dAT oligomers. Terminal relaxation times, measured in a temperature-jump apparatus, are shown for dAT oligomer fractions 10 and 16. Buffer: 0.05 M NaClO$_4$, 0.01 M Na cacodylate, pH 6.6. (Scheffler, 1967).

To illustrate how τ can be related to the elementary rate constants for helix formation, we will describe the kinetics for a simple case (too simple to be adequate for the dAT hairpins), in which melting and helix formation trace opposite paths along a single kinetic chain of events (Eq. 1), and melting occurs only by unwinding the hairpin helix at its open end. The rate of closing a base pair next to an already formed pair is k_f, the rate of opening it is k_b, and the rate of formation of the first base pair in the helix is σk_f (cf. Flory, 1961; Saunders and Ross, 1960).

$$A_0 \underset{k_b}{\overset{\sigma k_f}{\rightleftharpoons}} A_1 \underset{k_b}{\overset{k_f}{\rightleftharpoons}} A_2 \ldots A_{n-1} \underset{k_b}{\overset{k_f}{\rightleftharpoons}} A_n \qquad (1)$$

[In a more general treatment it is necessary to distinguish the rate of opening of the final base pair ($A_1 \rightarrow A_0$) from the others: cf. Kallenbach *et al.* (1963)].

If the change in concentration of each intermediate is small compared to the conversion of complete helix to random coil, one can make the steady-state approximation

$$\frac{dc}{dt} i = 0 \qquad (1 \leq i \leq n-1) \qquad (2)$$

where c_i is the concentration of any partly melted species. The melting process then is characterized by a single relaxation time, which shows a simple dependence on n, σ, and s (see Appendix).

$$\left[\frac{\tau^{-1}}{k_b} \right] = (s-1) \left[\sigma + \left(\frac{s-1}{s} \right)(1/s)^n \right] \qquad \text{for} \quad s^n \gg n, 1 \gg \sigma \qquad (3)$$

This equation shows that if σ, s, and n are known, one can calculate k_b from τ and k_f from s and k_b. Here s is the equilibrium constant for the formation of a base pair next to an already formed one, or

$$s = k_f/k_b \qquad (4)$$

and σs is the corresponding equilibrium constant for an isolated base pair. Thus, there is a direct relation between the relaxation time and the elementary rate constant for melting which can be used to compute k_b when the two equilibrium constants s and σs are known. The same type of result is usually obtained for more complicated cases, so long as the steady-state approximation can be applied.

A less approximate treatment of the problem yields n relaxation times for this reaction chain (cf. Eigen and de Maeyer, 1963). These can be found as the eigenvalues of a matrix by computer methods, which also can be used to find the changes in concentrations associated with each relaxation time, from the matrix of the eigenvectors. Preliminary computer calculations by Elson show that in the cases studied so far the largest eigenvalue gives a relaxation time of the same magnitude as Eq. (3), and the major eigenvector is associated with this relaxation time. However, the case of a single kinetic chain (Eq. 1) is too simple to describe the kinetics of melting dAT hairpin helices, since it neglects the numerous unsymmetrical species whose loops are not at the center of the strand, and which are able to form correspondingly fewer base pairs in the helical state. Inclusion of these species results in a kinetic chain with many branches, and the problem of relating theory to experiment is more complex, but we expect that it will be soluble.

In order to analyze the kinetic data, it is important to define the equilibrium properties of the melting process in order to obtain s and σ as functions of temperature and to be sure that the simple statistical model is adequate. The steps involved are (cf. Applequist and Damle, 1965): (a) measurement of molecular weights, both of the melted strands and the helices; (b) determination of the loop size at the end of a hairpin helix—now guessed to be about two base pairs; (c) measurement of the hypochromicity as a function of helix length; and (d) choice of a partition function which will accurately reproduce the melting curves for different fractions with a single set of parameters.

VII. A Possible Slippage Reaction Observed with dAT Oligomers

In the course of this work with dAT oligomers, Scheffler discovered a possible slippage reaction which can be measured directly by optical methods. It takes place at temperatures just below the melting zone (Fig. 9). The change is small, but easily measured, and follows apparent first-order kinetics (Fig. 10). At present, the evidence that this reaction may involve slippage is

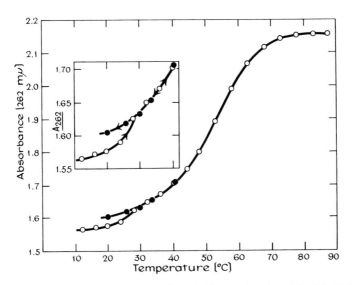

FIG. 9. The equilibrium melting curve of dAT oligomer fraction 9 in 0.5 M NaClO$_4$, 0.01 M Na cacodylate, pH 6.6. Note the shape of the curve at the beginning of the melting zone, and the hysteresis on cooling in this zone (the "slippage zone") (Scheffler, 1967).

simply that it can be observed by hypochromicity and is very slow compared to the rates of helix formation or melting—its half-time is about 10^6-fold larger than the half-times of melting.

What molecular processes are responsible for this reaction? A plausible explanation is that when hairpin helices are formed by quenching, only a fraction are symmetrical and in the remainder some base pairs cannot be formed. Then slippage converts the incomplete helices to more complete ones, by a mechanism to be determined. The formation of end-to-end dimers may be the first (fast) step (see below).

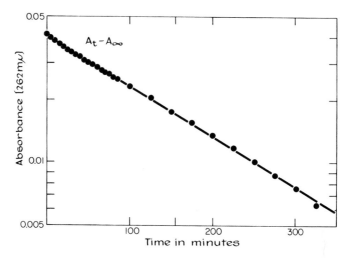

FIG. 10. Kinetics of slippage measured after cooling dAT oligomer fraction 9 just below the melting zone to 14.5° (see Fig. 9 for conditions) (Scheffler, 1967).

The melting curve of each oligomer seems to show a discrete transition just at the bottom of the curve (Fig. 9, inset). Since this is also the zone where the slow reaction is observed on cooling, the two phenomena are likely to be related. Preliminary measurements of molecular weight as a function of temperature suggest a conversion from dimer to hairpin helices with increasing temperature in this zone.

The significance of this finding lies in the possibility of studying slippage by direct optical methods in a reasonably simple system. Further study should elucidate the mechanism of this reaction. Some preliminary measurements of the rate of reaction as a function of chain length and temperature are given in Fig. 11. The rate at a given temperature decreases markedly with chain length, whereas the heat of activation (roughly −60 kcal) is not too different for these different oligomers.

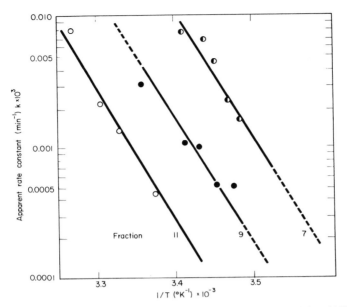

FIG. 11. Temperature dependence of the rate of slippage measured for dAT oligomer fractions 7, 9, and 11 in the buffer given under Fig. 9 (Scheffler, 1967).

VIII. Conclusion

The structure and conformational (or dynamic) properties of dAT are likely to be of importance in explaining the mechanism of its *de novo* enzymatic synthesis. Physical studies have shown that macromolecular dAT has a greater capacity for conformational variation than has natural DNA. At temperatures preceding thermal melting the dAT helix changes its conformation, apparently by the formation of self-complementary branches; these can account for properties such as the rapid kinetics of its melting. The mechanism of this rearrangement is likely to be related to the mechanism of slippage which occurs when dAT oligomers serve as templates for the replicative synthesis of dAT polymer. Current studies of the kinetics of melting of dAT oligomers should yield the rates of the basic steps in dAT helix formation and melting. A second, slow reaction has also been found during helix formation from dAT oligomers; it probably involves slippage. The results of these studies should make possible a critical examination of proposed mechanisms for the enzymatic synthesis of dAT.

IX. Appendix

In the steady-state approximation made here, the concentration of each intermediate is assumed constant so that the rate of disappearance of complete helix (species A_n) equals the rate of appearance of random coil (A_0) (cf. Saunders and Ross, 1960; Flory, 1961). Thus the rate of helix formation, v, is equal to the net forward rate at each step (see Eq. 1):

$$v = \sigma k_f c_0 - k_b c_1 \tag{5}$$

$$v = k_f c_i - k_b c_{i+1} \qquad (1 \le i \le n - 1) \tag{6}$$

where c_i is the concentration of A_i in moles per liter. For relaxation kinetics it is convenient to use the displacement from equilibrium, x_i (cf. Eigen and de Maeyer, 1963).

$$c_i = \bar{c}_i + x_i \tag{7}$$

where $\bar{c}_i$ is the final equilibrium concentration of i. Since $v = 0$ at equilibrium, substitution of Eq. (7) into Eqs. (5) and (6) gives, after dividing through by k_b and setting $s = k_f/k_b$,

$$v/k_b = \sigma s x_0 - x_1 \tag{8}$$

$$v/k_b = s x_i - x_{i+1} \qquad (1 \le i \le n - 1) \tag{9}$$

Summation of Eqs. (8) and (9) gives

$$n(v/k_b) = \sigma s x_0 + (s - 1) \sum_{i=1}^{n-1} x_i - x_n \tag{10}$$

Since the total number of moles of all species is constant,

$$x_0 + x_n = - \sum_{i=1}^{n-1} x_i \tag{11}$$

and we have

$$n(v/k_b) = x_0[1 - s(1 - \sigma)] - s x_n \tag{12}$$

A second equation containing x_0 and x_n is given by multiplying Eqs. (8) and (9) by s^{n-1-i}

$$(v/k_b)s^{n-1} = \sigma s^n x_0 - s^{n-1} x_1 \qquad (i = 0) \tag{13}$$

$$(v/k_b)s^{n-1-i} = s^{n-i} x_i - s^{n-1-i} x_{i+1} \qquad (1 \le i \le n - 1) \tag{14}$$

and then summing

$$(v/k_b)\left[\frac{s^n - 1}{s - 1}\right] = \sigma s^n x_0 - x_n \tag{15}$$

Defining the relaxation time, τ, as

$$\tau^{-1} \equiv - \frac{d \ln x_0}{dt} \tag{16}$$

and combining Eqs. (12), (15), and (16) gives the result

$$\left[\frac{\tau^{-1}}{k_b}\right] = (s-1)\left[\frac{\sigma s(s^n - 1) + (s-1)}{s(s^n - 1) - n(s-1)}\right] \tag{17}$$

Since the validity of this equation rests on the steady-state approximation, which in turn requires that $s^n \gg n$ and $\sigma \ll 1$, Eq. (17) may be simplified to:

$$\left[\frac{\tau^{-1}}{k_b}\right] = (s-1)\left[\sigma + \left(\frac{s-1}{s}\right)(1/s)^n\right] \tag{18}$$

To compute rate constants from our results on the kinetics of melting dAT oligomers, it will be necessary to treat a more complicated case, that of a kinetic chain with many branches. This problem is under study.

ACKNOWLEDGMENTS

The results and the ideas presented here are chiefly those of Ross Inman, Elliot Elson, and Immo Scheffler, who have also kindly allowed me to quote their unpublished work.

REFERENCES

Applequist, J., and Damle, V. (1965). *J. Am. Chem. Soc.* **87**, 1450.
Crick, F. H. C. (1959). Personal communication.
Crothers, D. M., and Zimm, B. H. (1964). *J. Mol. Biol.* **9**, 1.
Davidson, N., Widholm, J., Nandi, U. S., Jensen, R., Olivera, B. M., and Wang, J. C. (1965). *Proc. Natl. Acad. Sci. U. S.* **53**, 111.
Davies, D. R., and Baldwin, R. L. (1963). *J. Mol. Biol.* **6**, 251.
Eigen, M., and de Maeyer, L. (1963). *Tech. Org. Chem.* **8**, Part II, 895.
Elson, E. (1966). Ph.D. Thesis, Stanford University.
Elson, E. (1967). *Biopolymers* (1967) (submitted for publication).
Elson, E., and Jovin, T. (1967). To be published.
Flory, P. J. (1961). *J. Polymer Sci.* **49**, 105.
Ginoza, W., and Zimm, B. H. (1961). *Proc. Natl. Acad. Sci. U.S.* **47**, 639.
Inman, R. B., and Baldwin, R. L. (1962a). *J. Mol. Biol.* **5**, 172.
Inman, R. B., and Baldwin, R. L. (1962b). *J. Mol. Biol.* **5**, 185.
Inman, R. B., and Baldwin, R. L. (1964). *J. Mol. Biol.* **8**, 452.
Kaiser, A. D., and Inman, R. B. (1965). *J. Mol. Biol.* **13**, 78.
Kallenbach, N. R., Crothers, D. M., and Mortimer, R. G. (1963). *Biochem. Biophys. Res. Commun.* **11**, 213.
Kornberg, A., Bertsch, L. L., Jackson, J. F., and Khorana, H. G. (1964). *Proc. Natl. Acad. Sci. U.S.* **51**, 315.
Lehman, I. R., Linn, S., and Richardson, C. C. (1965). *Federation Proc.* **24**, 1466.
Marmur, J., and Doty. P. (1959). *Nature* **183**, 1427.
Printz, M. P., and von Hippel, P. H. (1965). *Proc. Natl. Acad. Sci. U.S.* **53**, 363.
Roger, M., and Hotchkiss, R. D. (1961). *Proc. Natl. Acad. Sci. U.S.* **47**, 653.
Ross, P. D., and Sturtevant, J. M. (1960). *Proc. Natl. Acad. Sci. U.S.* **46**, 1360.
Saunders, M., and Ross, P. D. (1960). *Biochem. Biophys. Res. Commun.* **3**, 314.

Schachman, H. K., Adler, J., Radding, C. M., Lehman, I. R., and Kornberg, A. (1960). *J. Biol. Chem.* **235**, 3242.

Scheffler, I. E. (1967). To be published.

Scheffler, I. E., Elson, E., and Baldwin, R. L. (1967). To be published.

Spatz, H. C., and Baldwin, R. L. (1965). *J. Mol. Biol.* **11**, 213.

Sueoka, N. (1961). *J. Mol. Biol.* **3**, 31.

Swartz, M. N., Trautner, T. A., and Kornberg, A. (1962). *J. Biol. Chem.* **237**, 1961.

Wake, R. G., and Baldwin, R. L. (1962). *J. Mol. Biol.* **5**, 201.

Wu, R., and Kaiser, A. D. (1967). To be published.

Some Aspects of RNA Transcription

E. PETER GEIDUSCHEK, EDWARD N. BRODY,
AND DAVID L. WILSON

Department of Biophysics
University of Chicago
Chicago, Illinois

I. Introduction — A Brief Review of RNA Transcription *in vitro*

Like polymerizations in general, RNA polymerization on its DNA template combines several distinguishable steps whose mechanisms must be understood both separately and in connection with each other: (1) binding of RNA polymerase to the DNA template, (2) initiation of synthesis, (3) propagation of synthesis, (4) termination of synthesis and detachment of enzyme from its template, and (5) detachment of the polymerized RNA chain from its template.*

A. RNA Polymerase-DNA Binding

In the binding of RNA polymerase to DNA templates, these important observations can be made. 1. Under some conditions of ionic strength, the binding of RNA polymerase to double-strand DNA is rapid, even at 0°C where the rate of subsequent RNA synthesis is extremely slow. The binding has been determined in a number of ways, including measurement of the amount of enzyme required to give maximal RNA synthesis, sedimentation measurements of DNA-enzyme complexes, ability of enzyme to bind DNA to nitrocellulose filters, and visualization of DNA-enzyme complexes in the electron microscope. The experiments have involved a number of DNA templates including T4, T7, λ, and doubly closed circular polyoma and papilloma DNA's. The quantities of enzyme bound to a number of different DNA templates correspond approximately to one enzyme molecule of molecular weight 800,000 to every 2000–2500 Å of DNA (Crawford *et al.*, 1965; Richardson, 1966b; Sternberger and Stevens, 1966; Jones and Berg, 1967). Since a 2500 Å stretch of DNA codes for a 240 residue polypeptide chain, it was thought that such average spacings of binding sites on DNA might correspond to the average spacing of cistronic transcription units.

---

* RNA, Ribonucleic acid; RNase, ribonuclease; CTP, cytidine triphosphate; UTP, uridine triphosphate; DMSO, dimethyl sulfoxide.

If this were so, then these binding sites could correspond to *in vivo* utilizable initiation sites for transcription. Although such speculations may not be valid (Naono and Gros, 1966; Thomas 1966), the observations do establish that binding sites on native DNA for *E. coli* RNA polymerase are, under certain ionic strength conditions, widely spaced and discrete.

2. On the other hand, denatured DNA binds much greater proportions of enzyme to nucleotide than does native DNA (Wood and Berg, 1964). The binding of enzyme to this much larger number of template sites in denatured DNA is reversible, as it is with native DNA. Thus, the (of course, reversible) dissociation of DNA secondary structure makes an affinity between many polynucleotide segments and RNA polymerase detectible by the methods which have been used thus far; when these segments are enclosed in ordered, double-strand DNA, their affinity for enzyme is so low as not to be detectible. These observations suggest that RNA polymerase, in binding to DNA, inter-acts with the secondary structure of the latter. On the other hand, it seems most unlikely that enzyme-binding sites are determined *only* by the secondary structure stability of sequences of nucleotide pairs. [Data bearing indirectly on this question come from studies of interaction of DNA with polyribonu-cleotides (Szybalski *et al.*, 1966; Kubinski *et al.*, 1966)]. Thus, it is not clear how the enzyme recognizes the (presumably specific) nucleotide sequences to which it binds. Models of this recognition have been proposed which require no distortion of the DNA secondary structure (Reich and Goldberg, 1964). It is presumably important, however, to take into account the dynamic mobility of DNA secondary structure which, for example, allows very rapid exchange of hydrogens enclosed in the purine-pyrimidine stacks of the DNA helix (Haggis, 1957; Printz and von Hippel, 1965; von Hippel and Printz, 1965; Crothers, 1964).

3. At high ionic strengths, binding of *E. coli* RNA polymerase to native or denatured DNA is prevented. At high salt concentrations, RNA poly-merase also dissociates into subunits (Zillig *et al.*, 1966; Richardson, 1966a; Stevens *et al.*, 1966; Pettijohn and Kamiya, 1967). The available experiments have not, however, clarified the connection between the state of polymeriza-tion of RNA polymerase and its function, particularly its binding to DNA.

4. Thus, at different salt concentrations, the capacity and presumably the affinity of DNA for RNA polymerase can be varied greatly. At low ionic strength the number of RNA polymerase molecules binding to circular polyoma DNA greatly exceeds one molecule per 2000 Å, but most of the enzyme is bound in a transcriptively inactive form (Pettijohn and Kamiya, 1967; see also Richardson, 1966b). A gradient of affinities between RNA polymerase molecules and λ and T7 DNA has recently been demonstrated (Stead and Jones, 1967). It seems probable that different templates will show individual and characteristic distributions of stronger and weaker binding

sites, but this remains to be investigated in detail. Until such data are available, it is difficult to state precisely the relation between enzyme-binding sites and sites for the initiation of RNA synthesis *in vivo* or *in vitro*.

B. Initiation of RNA Synthesis

The commencement of RNA synthesis on the ordered, double-strand DNA template considerably alters the enzyme-template interaction. The enzyme becomes much more tightly bound to its template, so that the complex is operationally irreversible (Bremer and Konrad, 1964) and stable toward salt concentrations that abolish attachment in the absence of RNA synthesis. The order-of-addition effects observed by Fox *et al.* (1965) and Berg (quoted in Berg *et al.*, 1965) are the result of this tight binding (together with the defective detachment of enzyme molecules from the template which will be further mentioned below). The reasons for the change of binding concomitant with RNA synthesis are not understood. It may be that there is a structural transition of the enzyme concomitant with the start of RNA synthesis. Alternatively, it may simply be that RNA synthesis engages a short segment of the coding strand of the DNA with its RNA complement and, therefore, frees the antisense strand of DNA to interact more strongly with the enzyme. (The latter alternative is equivalent to saying that the initiation of RNA synthesis further disrupts the secondary structure of the template beyond changes caused by the binding.)

As a result of this tighter binding, higher concentrations of salt (e.g., 0.4–0.5 M KCl), which completely prevent the binding of enzyme to DNA (and in the presence of which, RNA synthesis cannot be initiated), do not inhibit the continuation of RNA synthesis once it has been initiated (Richardson, 1966c; Anthony *et al.*, 1966; Fuchs *et al.*, 1967).

C. Propagation

The addition of ribonucleoside triphosphates to growing RNA chains has been thought to be the best understood aspect of *in vitro* transcription. Synthesis, was early shown to involve the complementarity of the Crick-Watson base pairs (Weiss and Nakamoto, 1961; Stevens and Henry, 1964; Hurtwitz *et al.*, 1962; Chamberlin and Berg, 1962) and the addition of nucleoside triphosphates to 3′ OH terminated polynucleotide chains (Shigeura and Boxer, 1964; Bremer *et al.*, 1965; Maitra and Hurwitz, 1965). The recent determination of the structure of a DNA-RNA hybrid synthesized *in vitro* on a template of f1 DNA (f1 is a filamentous bacteriophage of *E. coli* containing single-strand DNA) with *E. coli* RNA polymerase provides the strongest evidence yet available that the synthesized RNA chain is antiparallel to its template DNA strand (Milman *et al.*, 1967).

In fact, the basic assumption of the work described below is that once synthesis of RNA chains is properly underway in the presence of all four nucleoside triphosphates, propagation is straightforward and variations in the nature of the synthesized product can be studied as manifestations of processes controlling the initiation of transcription.

D. Termination Release

Once transcription is underway, RNA polymerase does not detach from native DNA templates *in vitro* under conditions thus far utilized (Bremer and Konrad, 1964; Kadoya *et al.*, 1964; Fox *et al.*, 1965). Accordingly, at least some of the RNA chains being synthesized in an enzymatic reaction continue to grow throughout the period of synthesis (Bremer and Konrad, 1964; Richardson, 1966c; Maitra and Hurwitz, 1965). With endonuclease-free RNA polymerase, the polynucleotide chains produced *in vitro* are very large indeed—in fact much longer than their *in vivo* counterparts (cf. e.g., Bremer and Konrad, 1964; Asano, 1965; Richardson, 1966c). Although the chain of argument is far from complete, such comparisons nevertheless lead to the supposition that termination signals have not been recognized in the hitherto reported *in vitro* RNA syntheses on native DNA templates.

On the other hand, release of enzyme molecules from single-strand DNA templates does appear to occur during *in vitro* transcription: the synthesized RNA consists of large numbers of small chains (Chamberlin and Berg, 1964) and the number of chains synthesized per RNA polymerase molecule is much greater with single-strand than with double-strand DNA templates (Maitra *et al.*, 1966). Thus, termination release appears to be dependent on secondary structure. It is as though the anticoding strand of the DNA template acted as a guiderail which holds or helps to hold RNA polymerase onto the nascent chain-template complex. One may conclude that the secondary structure of ordered, helical DNA restricts both the termination-release *and* binding-initiation steps of RNA polymerization.

E. Positive and Negative Control Elements of Transcription

The control of gene expression must involve a variety of mechanisms acting at the levels of translation and transcription. It is not our purpose to provide even a partial review of these mechanisms. The action of at least some of these control elements, however, can be related to *in vitro* transcription and it is useful to define two kinds of transcriptive controls: *positive* and *negative*. The operon theory (Jacob and Monod, 1961) specifies that a protein, which is the product of a regulator gene, acts in such a way as to block the initiation of transcription of three coordinately controlled genes of the *E. coli lac* operon.

The theory specifies that the *lac* repressor is a *negative* control element *restricting* transcription of a template that would otherwise be *open* to transcription. The opportunity for a direct experimental test of this hypothesis is perhaps provided by the isolation of the *lac* repressor protein by Gilbert and Muller-Hill (1967).

The existence of positive control elements for transcription has also been postulated for the control of a set of genes in *E. coli* connected with arabinose metabolism (Engelsberg *et al.*, 1965) and for the control of an early stage of the development of phage λ (Thomas, 1966) among others. A *positive* control element *permits* the transcription of a template which would otherwise be *closed* to transcription. It is conceivable that some positive control elements might be polymerases, whereas others might be modifiers of polymerase action. Alternatively, positive control elements could act on the template at the sites of polymerase binding and/or initiation. In general, it might be useful to distinguish control elements from more omnipotent participants in transcription on the basis of gene or chromosome specificity. The possible elaborations of definition are obvious. What is lacking is experimental evidence on the mode of action of positive control elements.

II. Selective Transcription of DNA Templates

In vivo transcription of RNA must proceed from defined initiation points to defined stopping points, yielding transcripts of DNA coding strands and not transcripts of their complements. *In vitro* transcription can manifest some or all of the same selectivity of initiation (Khesin *et al.*, 1963; Hayashi *et al.*, 1964; Geiduschek *et al.*, 1964; Luria, 1965; Naono and Gros, 1966). From what has been written in the introduction, it is clear that this selectivity originates in selective initiation of synthesis. The importance of selective *in vitro* transcription systems is that (*a*) they clearly possess at least some of the specific properties of *in vivo* RNA synthesis; (*b*) they are useful for classifying positive and negative control elements of transcription (see below); and (*c*) they are required for the study of the mechanism of action of control elements of transcription (Gilbert and Muller-Hill, 1967; Ptashne, 1967).

Two manifestations of selective transcription have been studied primarily: (1) asymmetry, i.e., the transcription of only one of the complementary template strands at any transcriptive locus and (2) selective transcription of only portions of certain DNA templates and nontranscription of others. Although the former of these criteria is the easier to describe, it is also fundamentally the less informative. Tests of asymmetry are simple and involve one of three choices: (*a*) tests of self-complementarity of synthesized RNA. As stated above, *in vivo* synthesized RNA is not self-complementary (e.g., Bautz, 1963), in contrast, for instance, to RNA-viral replicative forms.

Self-complementarity is manifested by formation of pancreatic or T1 ribo-nuclease-resistant duplexes. (b) When RNA transcription is asymmetric, the synthesized RNA is identical with, rather than complementary to, in vivo synthesized messenger. Thus, asymmetric in vitro synthesized RNA cannot form RNase-resistant RNA-RNA duplexes with homologous in vivo synthesized RNA, whereas symmetric RNA contains polynucleotide sequences complementary to in vivo RNA and can form RNA-RNA duplexes upon annealing with in vivo RNA. (c) When all the transcription units of a genome are arrayed on a chromosome with the same polarity, then all the sense (i.e., transcriptively active) sequences are located on the same DNA poly-nucleotide chain. This is the arrangement which occurs on ΦX 174 DNA (Hayashi et al., 1963), but not on λ DNA (Naono and Gros, 1966; Hogness et al., 1966; Gross et al., 1967; Taylor et al., 1967). The unipolar type of arrangement may also occur on the phage α chromosome; it certainly applies to messenger isolated at certain times after infection and (substantially) to the in vitro synthesized product (Geiduschek et al., 1964). Thus, under certain circumstances, fractionation of DNA chains permits them to be used to assay asymmetry of transcription.

A. Transcription Programs during Viral Development

Before discussing the selective transcription of viral DNA templates, we digress slightly to discuss transcription in viral development. Maturation of bacteriophage clearly involves the expression of different genes at different times (see, e.g., Stent, 1963). The analysis of viral development has progressed strikingly during the past few years as the result of the discoveries of Epstein and collaborators (1963). The transcriptive basis of viral development was first indicated by the experiments of Hall, Khesin, and their collaborators (Khesin et al., 1962; Hall et al., 1963) and Kano-Sueoka and Spiegelman (1962). Some detailed information is now available regarding at least three bacterial viruses: T4 (Hall et al., 1963; Bolle et al., 1968a, b, c; Baldi et al., 1967), λ (Joyner et al., 1966; Skalka, 1966; Naono and Gros, 1966; Taylor et al., 1967; Gros et al., 1967) and SPO1, which has been studied in our own laboratory (Gage et al., 1967; Gage and Geiduschek, 1967).

Both T4 and SPO1 developments are marked by the initial transcription of some viral genes during the first minutes of viral development. The tran-scription of these early genes is nonsynchronous as Bolle and collaborators (1968b) have shown in the case of T4 and some of these viral genes are transcribed even when bacteria are infected in the presence of chloram-phenicol and the absence of protein synthesis. Subsequent transcriptional events in the development of SPO1 (Fig. 1) do require protein synthesis, presumably because they have to be triggered by the action of viral gene products. A new class of RNA appears after a few minutes and as this RNA accumulates, RNA transcribed from the "earliest" genes disappears. After a brief period, this second set of genes is, in turn, shut off and a third class of messages starts to make its appearance at approximately the time that viral antigens are first synthesized; this last class of RNA probably corresponds to the typical late messenger which is also observed in T4 and λ development. Figure 1 presents a summary of these major transcriptional events during bacteriophage SPO1 development.

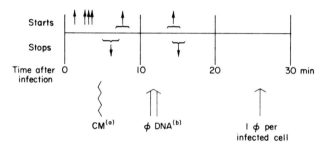

Fig. 1. Schematic representation of the major transcriptional events in the development of Bacteriophage SPO1. All times after infection refer to phage development at 37°C (Gage et al., 1967). (a) Chloramphenicol prevents the shutoff of early RNA synthesis and the turn-on of subsequent transcription. (b) DNA synthesis starts to be detected 9–11 minutes after infection.

Far less is known regarding SPO1 "late" transcription than about the analogous events in T4 and λ development. For example, the requirements for "late" messenger synthesis in T4-infected *E. coli* are: (*a*) DNA synthesis (and, accordingly, the action of all the viral genes that are required for DNA synthesis) and (*b*) the action of at least one and possibly two more gene products which control late transcription. One would be inclined to believe that the action of at least one of these control elements which turn on late T4 transcription is chromosome-specific and positive (Bolle *et al.*, 1968a, b).

B. *In vitro* RNA Synthesis on Viral DNA Templates

Experiments using φX 174 RF, T4, α, SPO1, and λ DNA with *E. coli* RNA polymerase have shown that one can preponderantly transcribe the *in vivo* coding strands of each of these DNA's (Hayashi *et al.*, 1964; Geiduschek *et al.*, 1964; Green, 1964; Naono and Gros, 1966). Under appropriate conditions, the synthesized RNA is neither self-complementary nor complementary to *in vivo* synthesized messenger RNA (TableI, part B, column A and Table II, part A). This has validly been regarded as evidence of selective initiation of

TABLE I

T4 RNA SYNTHESIS WITH *E. coli* RNA POLYMERASE

	Template for synthesis					
	A. Native DNA		B. Denatured DNA		C. Native DNA, 25% DMSO	
A. Hybridization-Competition[a]	Hybridized (cpm)	%	Hybridized (cpm)	%	Hybridized (cpm)	%
in vitro ^{3}H-RNA, denatured DNA	1324	(100)[b]	1325	(100)[c]	1679	(100)[d]
do + 2–2.5 mg/ml T4 early RNA	78	5.9	287	21.7	225	13.5
do + 2–2.1 mg/ml T4 late RNA	48	3.6	109	8.2	56	3.4

	Template for synthesis					
	A. Native DNA		B. Denatured DNA		C. Native DNA, 25% DMSO	
B. Antimessenger content	*in vitro* ^{3}H-RNA (µg/ml)	RNase resistant (%)	*in vitro* ^{3}H-RNA (µg/ml)	RNase resistant (%)	*in vitro* ^{3}H-RNA (µg/ml)	RNase resistant (%)
in vitro ^{3}H-RNA	0.27	2	0.2	6	0.26	8
do + T4 early RNA						
850 µg/ml	0.27	4.5	—	—	—	—
800 µg/ml	—	—	—	—	0.26	25
do + late RNA, 550 µg/ml	0.27	7	0.2	38	0.26	28

Footnotes on facing page.

TABLE II
SYNTHESIS OF SPO1 RNA WITH *E. coli*
RNA POLYMERASE

Template	Tests of self-complementarity and antimessenger content RNase resistant (%)
A. Native DNA	
2.9 μg/ml *in vitro* ^{3}H-RNA	3.6
0.72 μg/ml *in vitro* ^{3}H-RNA +	
1000 μg/ml SPO1 early RNA	5.5
B. Denatured DNA	
0.1 μg/ml ^{3}H *in vitro* RNA	
+210 μg/ml SPO1 "early" RNA	17
+225 μg/ml SPO1 "intermediate" RNA	30
+250 μg/ml SPO1 "late" RNA	28

Template	Hybridization-competition tests Hybridized (cpm)	%
C. Native DNA		
0.35 μg/ml *in vitro* ^{3}H-RNA 6.3 μg/ml		
denatured DNA	1250	(100)
do + 180 μg/ml SPO1 chloramphenicol		
early RNA	213	17
do + 1130 μg/ml SPO1 late RNA	665	53

[a] These competition experiments identify both messenger *and* antimessenger collectively as early or late. Presumably two kinds of competition are observed in this instance—the unlabeled RNA competes with labeled messenger sequences of *in vitro* synthesized RNA by forming DNA-RNA hybrids and with labeled antimessenger sequences by forming RNA-RNA duplexes. Although we have not measured the relative rates of these processes, this kind of result has been obtained repeatedly in our work. It is in contrast with the report of Green (1964) that one could identify asymmetry of *in vitro* transcription on the basis of hybridization-competition experiments alone and that messenger did not compete antimessenger in DNA-RNA hydridization-competition experiments.

[b] 5.4 μg/ml *in vitro* T4 ^{3}H-RNA and 1.6 μg/ml denatured T4 DNA.

[c] 3μg/ml *in vitro* T4 ^{3}H-RNA and 1.6 μg/ml denatured T4 DNA.

[d] 2.4 μg/ml *in vitro* ^{3}H-RNA and 1.6 μg/ml denatured T4 DNA.

transcription. (On the other hand, the converse assumption that symmetric synthesis implies indiscriminate initiation of RNA synthesis is not valid.) A more significant analysis of *in vitro* synthesized RNA comes from the comparison between the species synthesized *in vitro* and *in vivo*. The method of analysis that we have found convenient for comparison of *in vivo* and *in vitro* synthesized RNA is a hybridization-competition, isotope-dilution test, first used by Hall and co-workers (1963) and Khesin and co-workers (1962), which is now in common use (e.g., Skalka, 1966; Green, 1964; Luria, 1965; Denis, 1966; Naono and Gros, 1966). The principle of the method is sketched in Fig. 2. When the transcription products of templates with many transcription units are analyzed by this method, quantitative information about the rates of synthesis at different transcription units is relatively difficult to extract. Nevertheless, it provides considerable detailed information (Bolle *et al.*, 1968a). Hybridization-competition comparisons of *in vitro* synthesized T4 and T2 RNA with unlabeled *in vivo* messenger show that the asymetric *in vitro* product contains only early species (Fig. 3 and Table I, part A, column A). Unlabeled *in vitro* synthesized RNA effectively competes label from hybrids formed with early *in vivo* RNA. Moreover, it competes almost as much label from hybrids formed with late *in vivo* RNA as does unlabeled early *in vivo* RNA (Fig. 3b). Therefore, the *in vitro* product must contain almost *all* the early species.

In many important respects, this analysis of T-even early RNA transcription *in vitro* is incomplete. There are many unanswered questions which are the subject of current experiments. How many transcription units are active *in vitro*? What is their distribution on the viral DNA? What is the temporal sequence of *in vitro* transcription? How do the relative abundances of the products of different transcription units compare *in vivo* and *in vitro*, and how can the relative template activities of different transcription units be manipulated?

Regarding the last of these questions some preliminary indications are available from comparative hybridization-competition studies of *in vitro* and *in vivo* labeled T4 messenger RNA with unlabeled *in vivo* messenger. (1) The relative competing power of unlabeled early and late T4 *in vivo* RNA for labeled T4 early (5 minutes, 30°C) RNA and for labeled *in vitro* synthesized RNA is almost identical (Fig. 3c). (2) There is evidence that T4 early messenger (RNA extracted from cells 5 minutes after infection at 30°C) contains a number of different species whose abundances vary widely (Bolle *et al.*, 1968a). (3) It is also clear that during the first few minutes after infection, radioactivity is predominantly incorporated into those messenger species that become less abundant later on in infection (Bolle *et al.*, 1968a, c; Baldi *et al.*, 1967; Friesen *et al.*, 1967).

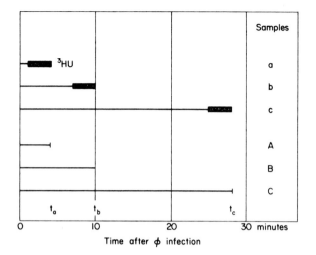

FIG. 2a–e. Hybridization-competition analysis of transcription.

FIG. 2a. Schedule for preparation of (nonuniformly) labeled and unlabeled RNA at various times, t_a, t_b, t_c, of bacteriophage development.

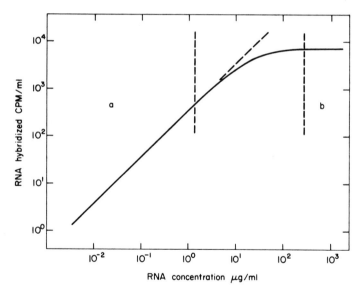

FIG. 2b. A hypothetical titration or saturation curve for labeled RNA hybridizing to denatured viral DNA, which identifies conditions of (a) excess of DNA over all hybridizable, labeled RNA and (b) excess of all hybridizable labeled RNA over homologous sites on DNA. Such binding curves can, in principle, be used to determine abundance distributions of messenger (McCarthy and Bolton, 1963; Bolle et al., 1968a).

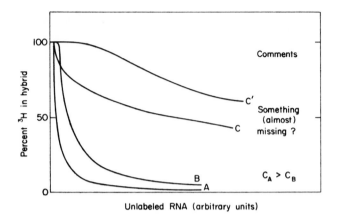

Unlabeled RNA (arbitrary units)

FIG. 2c. Hypothetical hybridization-competition curves for excess of DNA over labeled RNA.

Method. Samples to be hybridized contain constant amounts of denatured DNA (excess) and labeled RNA and varying quantities of unlabeled RNA. They are analyzed by any of several widely used methods (e.g., Nygaard and Hall, 1963; Gillespie and Spiegelman, 1965). Data are normalized to the radioactivity hybridizing to DNA when no unlabeled RNA has been added.

Analysis of competition curves. Consider the hypothetical case (Fig. 2a) of labeled RNA (a), and unlabeled RNA's A (which is identical with a), B, and C.

(1) *Gross features.* The control competition curve of A for a is shown in curve A. Sufficiently high concentrations of the unlabeled RNA entirely dilute the radioactivity hybridizing to DNA.

Curve B is a test for whether the species labeled just before t_a are still present at t_b. Unlabeled competitor B dilutes the isotopic label of those species in a that hydridize to DNA; therefore the species synthesized at time t_a, are still present at t_b. The competing power of B is less than that of A. Therefore, on the average, the concentrations of the species labeled just before t_a are lower at t_b than at t_a. If the shapes of competition curves B and A are the same, then the abundance distributions of the species (i.e., the species labeled in sample a) being tested may be the same at t_a and t_b. In that case the concentration ratio for samples A and B is simply the scale factor that brings curve A into curve B.

Curves C (or C′) are tests for whether the species labeled at t_a are still present at t_c. The conclusion to be drawn from such curves is that a large fraction of the label is incorporated into RNA species at t_a that are rare or absent at t_c.

(2) *More detailed analyses.* Bolle, Epstein, and Salser have pointed out that the competition curve A can, in principle, be analyzed to yield the abundance distribution of the labeled messenger species in sample a. Further details are given elsewhere (Bolle *et al.*, 1967). The shapes of competition curves C and C′ can, in principle, provide information on whether the messenger species common to RNA samples a and C are abundant or rare in either RNA. In comparison with curve A, curves C and C′ permit an upper limit to be placed on the relative abundance of the "missing" species (at t_c relative to t_a). A more sensitive way of determining this ratio has been described (Bolle *et al.*, 1968a).

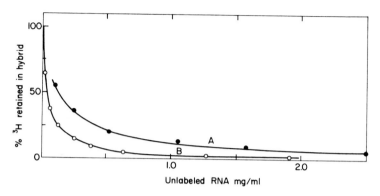

FIG. 2d. Competition of ³H-uridine labeled early (0–5 minutes after infection) T4 RNA by unlabeled RNA extracted 20 (late, curve A) and 5 minutes (early, curve B) after infection (at 30°C). Denatured DNA 5 μg/ml; labeled RNA (total RNA from infected cells) 1.4 μg/ml. Hybridization by the nitrocellulose filter method of Nygaard and Hall (1964; in this and all subsequent figures of this paper). All data corrected for a small background of label bound to filters in the absence of DNA. Details in Bolle et al. (1968a).

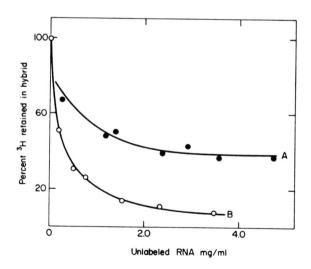

FIG. 2e. Competition of labeled late (20 minutes after infection) T4 RNA by unlabeled RNA extracted 5 (curve A) and 20 minutes (curve B) after infection (at 30°C). Denatured DNA 40 μg/ml; labeled RNA 45 μg/ml. Data corrected as above.

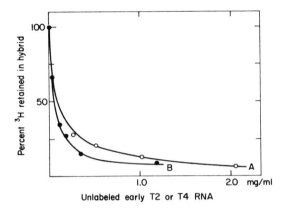

Unlabeled early T2 or T4 RNA

FIG. 3a–c. Comparison of *in vitro* synthesized T2 and T4 RNA with early and late messenger.

FIG. 3a. Competition of labeled *in vitro* synthesized T2 and T4 RNA by the respective early (5 minute, 30°C) unlabeled *in vivo* RNA's. *Curve A* (O): Each sample to be hybridized contained 0.9 μg/ml ^{3}H-labeled (CTP) RNA synthesized on a template of native T4 DNA with *E. coli* RNA polymerase, 3 μg/ml denatured T4 DNA, and varying concentrations of unlabeled RNA extracted 5 minutes after T4 infection at 30°C. Data are normalized to the amount of label hydridizing, when competitor RNA is omitted, and are corrected for the background mentioned in the legend for Fig. 2d (for details of analysis and synthesis, see Geiduschek *et al.*, 1966). *Curve B* (●): Each sample contained 5.8 μg/ml ^{3}H-labeled (CTP) RNA synthesized on a template of native T2 DNA with *E. coli* RNA polymerase, 10 μg/ml denatured T2 DNA, and varying concentrations of unlabeled RNA extracted 5 minutes after T2 infection at 30°C. Treatment of data as above. Experimental details in Geiduschek *et al.* (1966). In this experiment, DNA is saturated with respect to abundant *in vitro* ^{3}H-RNA species. Hybridization-competition under these conditions provides a somewhat more sensitive test of the presence, at low relative abundance, of *in vitro* synthesized late species.

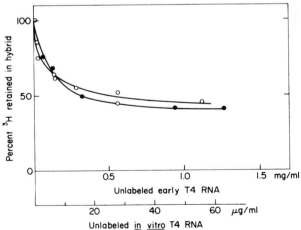

Unlabeled early T4 RNA

Unlabeled in vitro T4 RNA

FIG. 3b. Competition of ^{3}H-labeled late T4 messenger (17–20 minutes, 30°C) with un-labeled early (5 minutes, 30°C) *in vivo* RNA and *in vitro* synthesized RNA. ^{3}H-RNA 47 μg/ml; denatured DNA 3.1 μg/ml. (O), unlabeled *in vitro* RNA (lower scale); (●), unlabeled T4 + 5 minute RNA (upper scale).

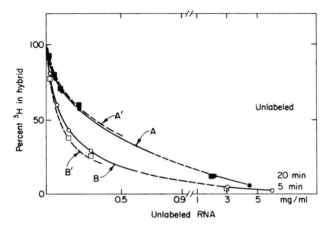

Fig. 3c. Comparison of hybridization-competition of *in vitro* and early T4 RNA. The *in vitro* ³H-labeled (UTP) RNA was synthesized on native T4 DNA with an excess of RNA polymerase (i.e., synthesis was limited by the amount of DNA present). The hybridization-competition was done with denatured DNA in excess over labeled RNA: 18.8 μg/ml denatured DNA and 0.16 μg/ml *in vitro* ³H-RNA or 3.16 μg/ml ³H-labeled (0–5 minutes, 30°C) early T4 RNA. A single pair of unlabeled early (5 minutes) and late (20 minutes) RNA preparations was used for the competition. Curve A, labeled *in vitro* RNA, unlabeled late RNA. Curve A′, labeled early RNA, unlabeled late RNA. Curve B, labeled *in vitro* RNA, unlabeled early RNA. Curve B′ labeled early RNA, unlabeled early RNA.

These three sets of observations and the previously cited evidence suggest the following conclusions.

1. The *in vitro* and early *in vivo* transcriptions are mainly confined to the same set of transcription units.

2. The *in vitro* transcription also produces predominantly those RNA species that become less abundant later on in viral development.

3. If the relative abundance distributions of *in vitro* and *in vivo* T4 RNA were totally different, then complete isotope dilution of each by the other would not be observed.

4. The information thus far available suggests, but does not prove, that similar abundance distributions exist for early *in vivo* and *in vitro* synthesized T4 RNA. This would be the result if the level of activity of different early T4 transcription units relative to each other were to be determined by the intrinsic* template activity of the corresponding DNA sequences with the host RNA polymerase.†

---

* Intrinsic means determined by interaction of DNA with RNA polymerase. The absolute, propagation rates of RNA synthesis *in vitro* and in rapidly growing *E. Coli* are different, and there is abundant evidence of coupling between protein and RNA synthesis (for discussion, see e.g., Stent, 1964; Maaløe and Kjelgaard, 1966). We postulate that, in this instance and perhaps others, the *relative* level of transcriptive activity is determined by the *intrinsic* DNA-polymerases interaction.

† During the first few minutes of *in vivo* transcription of the viral template, the abundance distribution of messenger species is likely to be dominantly determined by synthesis rather than stability.

In view of the incomplete nature of the evidence this discussion of messenger abundance distributions must be regarded as quite preliminary and speculative. The analysis thus far is not as quantitative as it should be, and there is no information available on the degree to which the relative rates of transcription of different species can be manipulated *in vitro*. Thus, we omit, as premature, discussion of the effects that nontermination of *in vitro* RNA synthesis or clustering of transcription units of comparable activity and other factors would have on the outcome of such comparisons as these. On the other hand, the question of the basic "setting" of the transcriptional level of different genes is an obvious one which has, as far as we are aware, received little or no attention.

Some effort has been devoted to finding out whether conditions can be found for the transcription of late T4 messages on native T4 DNA templates using *E. coli* RNA polymerase. The following tabulation lists some conditions under which only T4 early messenger RNA synthesis is observed.

SELECTIVE *in vitro* SYNTHESIS OF T4 EARLY RNA

Native DNA plus:
    Long synthesis time
    DNA excess
    Enzyme excess, sequential addition
    $\pm$ Spermidine
    Mg^{2+} or Mn^{2+}
    DNA treated with pronase
    DNA treated with $NaClO_4$
    DNA sheared
    B. subtilis or *E. coli* enzyme

Clearly, a wide variety of conditions—enzyme or DNA excess, Mn^{2+} or Mg^{2+}, spermidine, duration of synthesis, and various methods of phage DNA purification—leads only to early messenger synthesis. One might ask whether it would be possible to synthesize late messenger asymmetrically under conditions in which the secondary structure of DNA might be more readily dissociable or more mobile. It is tempting to talk, in an impressionistic way, about the "opening up" of DNA templates by RNA polymerase and about the possibility that DNA replication might "open up" late cistrons for transcription. The genetic evidence and other evidence are against this being the sole requirement for T4 late messenger synthesis (e.g., Bolle *et al.*, 1968b). Nevertheless, it seemed worthwhile to inquire whether such an *in vitro* synthesis might not be a model for related *in vivo* mechanisms. What has been attempted is a synthesis of RNA in water-DMSO mixtures which are known to destabilize the secondary structure of DNA, although DMSO is relatively innocuous in its effects on proteins. RNA synthesis can, in fact, be carried out in 25% DMSO, albeit at a rather lower rate than in water. The analysis of one such RNA is shown in Table I. Late messenger transcription is concomitant with antimessenger synthesis. In fact, experiments in which denatured DNA templates are used for RNA synthesis (e.g., Table I; other results not presented) have shown that even with a disordered single-strand template, the relative abundance of late messenger synthesis is relatively low. Hybridization-competition experiments with varying ratios of labeled RNA to DNA input show that many late regions of the DNA template are transcribed symmetrically, but at a relatively low rate. Even with denatured T4 DNA, one evidently need not have, and perhaps cannot achieve, uniform transcription of the template. Thus, the function of the gene products which are responsible for turning on T4 late messenger cannot be merely to dissociate the secondary structure of DNA at initiation sites of late messenger synthesis, but must evidently have a more positive directing effect on late messenger transcription.

Experiments on the *in vitro* transcription of SPO1 DNA show that with this template also, conditions can be found for the asymmetric transcription of RNA; the synthesized RNA is early messenger and contains very few polynucleotide sequences complementary to early messenger (Table II). It contains those polynucleotide sequences which are abundant at the beginning of infection and become less abundant late in infection.

Native DNA is required for asymmetric transcription on the T4 and SPO1 viral DNA templates; denaturation of the DNA template yields substantial synthesis of antimessenger and some transcription of those late species whose synthesis is excluded on the native DNA template (Tables I and II). T4 DNA can be fragmented, however, by shearing into small pieces (of the order of one-fortieth of the intact molecule) in such a way that the synthesis is still restricted to asymmetric, early RNA (see also Green, 1964).

C. *In vitro* Transcription Experiments and the Designation of Control Mechanisms

In closing, we consider the implications of these experiments on restricted *in vitro* synthesis for control mechanisms operating on the transcriptive programs of viral development.

The transcription of native T4 and SPO1 DNA *in vitro* predominantly yields the respective early viral messengers, as we have pointed out above. These selective syntheses do not require homology of template and enzyme, i.e., they do not require the use of enzymes from the corresponding viral hosts (Snyder *et al.*, 1967). The simplest interpretation of these results is that mature viral DNA is normally open to the transcription of these RNA species and normally closed to the transcription of other segments of the viral genome. One must also consider the alternate possibility that such selective transcription is not intrinsic to the DNA, but is due to special conditions of synthesis, contamination of DNA or enzyme by substances that regulate the transcription, presence (or absence) or nucleases, etc. In the case of T4, a number of such possibilities have been eliminated (Table III). The appearance of late messenger as a transcription product of mature viral DNA is also found to be concomitant with antimessenger synthesis (Table I; other data not shown). Thus, the currently available evidence suggests that the above simple interpretation is also likely to be the correct one and that, both in T4- and SPOl-infected cells, the initial stage of the transcriptive program is dominated by the transcription of the available phage genes by the host polymerase. It may well be that any regulation involved at this stage of the development is inherent in the properties of the DNA, the location and size of transcription units, and their affinity for the host polymerase. The aspects of these processes that are concerned with the sequence of transcription of different parts of the early genome and the relative abundances of products are the subject of current experiments.

During SPOl development, the initial "early" transcription is terminated after a few minutes (Fig. 1). The *in vitro* experiments help to identify this step as resulting from the action of *negative* control(s). Similarly, the late transcription in T4 and the intermediate transcription in SPOl are the result of the action of *positive* control elements, since the mature viral templates are evidently *closed* to the transcription of these products by enzyme preexisting in the host cell.

ACKNOWLEDGMENTS

Our research is supported by grants from the Institute of Child Health and Human Development, National Institutes of Health (HD 01257), and the National Science Foundation (GB 2120). We also gratefully acknowledge a postdoctoral fellowship of the National Science Foundation (to ENB), a predoctoral fellowship of the National Institutes of Health (to DLW), and a USPHS Research Career Development Award (to EPG).

REFERENCES

Anthony, D. D., Zeszotek, E., and Goldthwait, D. A. (1966). *Proc. Natl. Acad. Sci. U.S.* **56**, 1026.
Asano, K. (1965). *J. Mol. Biol.* **14**, 71.
Baldi, M. I., Doskocil, J., and Haselkorn, R. (1967). Manuscript in preparation.
Baldwin, R. L. (1967). This symposium.
Bautz, E. K. F. (1963). *Cold Spring Harbor Symp. Quant. Biol.* **28**, 205.
Berg, P., Kornberg, R. D., Fancher, H., and Dieckmann, M. (1965). *Biochem. Biophys. Res. Commun.* **18**, 932.

Bolle, A., Epstein, R. H., Salser, W., and Geiduschek, E. P. (1968a). *J. Mol. Biol.* **31**. In press.
Bolle, A., Epstein, R. H., Salser, W., and Geiduschek, E. P. (1968b). *J. Mol. Biol.* In press.
Bolle, A., Epstein, R. H., and Salser, W. (1968c). To be published.
Bremer, H., and Konrad, M. W. (1964). *Proc. Natl. Acad. Sci. U.S.* **51**, 807.
Bremer, H., Konrad, M. W., Gaines, K., and Stent, G. S. (1965). *J. Mol. Biol.* **13**, 540.
Chamberlin, M., and Berg, P. (1962). *Proc. Natl. Acad. Sci. U.S.* **48**, 81.
Chamberlin, M., and Berg, P. (1964). *J. Mol. Biol.* **8**, 297.
Crawford, L. V., Crawford, E. M., Richardson, J. P., and Slayter, H. S. (1965). *J. Mol. Biol.* **14**, 593 and 597.
Crothers, D. M. (1964). *J. Mol. Biol.* **9**, 712.
Denis, H. (1966). *J. Mol. Biol.* **23**, 269 and 285.
Engelsberg, E., Irr, J., Power, J., and Lee, N. (1965). *J. Bacteriol.* **90**, 946.
Epstein, R. H., Bolle, A., Steinberg, C. M., Kellenberger, E., Boy de la Tour, E., Chevallier, R., Edgar, R. S., Susman, M., Denhardt, G. H., and Lielausis, A. (1963). *Cold Spring Harbor Symp. Quant. Biol.* **28**, 375.
Fox, C. F., Gumport, R. I., and Weiss, S. B. (1965). *J. Biol. Chem.* **240**, 2101.
Friesen, J. D., Dale, B., and Bode, W. (1967). *J. Mol. Biol.* **28**, 413.
Fuchs, E., Millette, R. L., Zillig, W., and Walter, G. (1967). *F.E.B.S. J.* In press.
Gage, L. P., and Geiduschek, E. P. (1967). *J. Mol. Biol.* In press.
Gage, L. P., Wilson, D., and Geiduschek, E. P. (1967). *Federation Proceedings* **26**, 873.
Geiduschek, E. P., Tocchini-Valentini, G. P., and Sarnat, M. T. (1964). *Proc. Natl. Acad. Sci. U.S.* **52**, 486.
Geiduschek, E. P., Snyder, L., Colvill, A. J. E., and Sarnat, M. (1966). *J. Mol. Biol.* **19**, 541.
Gilbert, W., and Muller-Hill, B. (1967). *Proc. Natl. Acad. Sci. U.S.* **56**, 1891.
Gillespie, D., and Spiegelman, S. (1965). *J. Mol. Biol.* **12**, 829.
Green, M. (1964). *Proc. Natl. Acad. Sci. U.S.* **52**, 1388.
Gros, F., Naono, S., Cukier, R., and Sheldrick, P. (1967). Private communication.
Haggis, G. H. (1957). *Biochim. Biophys. Acta* **23**, 494.
Hall, B. D., Green, M., Nygaard, A. P., and Boezi, J. (1963). *Cold Spring Harbor Symp. Quant. Biol.* **28**, 201.
Hayashi, M., Hayashi, M. N., and Spiegelman, S. (1963). *Proc. Natl. Acad. Sci. U.S.* **50**, 664.
Hayashi, M., Hayashi. M. N., and Spiegelman, S. (1964). *Proc. Natl. Acad. Sci. U.S.* **51**, 351.
Hogness, D., Doerfler, W., Egan, J., and Black, L. (1966). *Cold Spring Harbor Symp Quant. Biol.* **31**, 129.
Hurwitz, J., Furth, J. J., Anders, M., and Evans, A. H. (1962). *J. Biol. Chem.* **237**, 3752.
Jacob, F., and Monod, J. J. (1961). *J. Mol. Biol.* **3**, 318.
Jones, O. W., and Berg, P. (1967). *J. Mol. Biol.* **22**, 199.
Joyner, A., Isaacs, L. N., Echols, H., and Sly, S. W. (1966). *J. Mol. Biol.* **19**, 174.
Kadoya, M., Mitsui, H., Takagi, Y., Otaka, E., Suzuki, H., and Osawa, S. (1964). *Biochim. Biophys. Acta* **91**, 36.
Kano-Sueoka, T., and Spiegelman, S. (1962). *Proc. Natl. Acad. Sci. U.S.* **48**, 1942.
Khesin, R. B., Gorlenko, Zh. M., Shemyakin, M. F., Bass, I. A., and Prozorov, A. A. (1963). *Biokhimiya* **28**, 1070.
Khesin, R. B., Shemyakin, M. F., Gorlenko, Zh.M., Bogdanova, S. L., and Afanaseva, T. P. (1962). *Biokhimiya* **27**, 1092.
Kubinski, H., Opara-Kubinska, Z., and Szybalski, W. (1966). *J. Mol. Biol.* **20**, 313.
Luria, S. (1965). *Biochem. Biophys. Res. Commun.* **18**, 735.

Maaløe, O., and Kjelgaard, N. O. (1966). "Control of Macromolecular Synthesis." Benjamin, New York.

McCarthy, B. J., and Bolton, E. (1964). *J. Mol. Biol.* **8**, 184.

Maitra, U., and Hurwitz, J. (1965). *Proc. Natl. Acad. Sci. U.S.* **54**, 815.

Maitra, U., Cohen, S. N., and Hurwitz, J. (1966). *Cold Spring Harbor Symp. Quant. Biol.* **31**, 113.

Milman, R., Langridge, R., and Chamberlin, M. (1967). *Proc. Natl. Acad. Sci U.S.* **57**, 1804.

Naono, S. and Gros, F. (1966). *Cold Spring Harbor Symp. Quant. Biol.* **31**, 363.

Nygaard, A. P., and Hall, B. D. (1963). *Biochem. Biophys. Res. Commun.* **12**, 98.

Nygaard, A. P., and Hall, B. D. (1964). *J. Mol. Biol.* **9**, 125.

Pettijohn, D. E., and Kamiya, T. (1967). *J. Mol. Biol.* In press.

Printz, M. P., and von Hippel, R. H. (1965). *Proc. Natl. Acad. Sci. U.S.* **53**, 363.

Ptashne, M. (1967). *Proc. Natl. Acad. Sci. U.S.* **57**, 306.

Reich, E., and Goldberg, I. H. (1964). *Prog. Nucleic Acid Res. Mol. Biol.* **3**, 183.

Richardson, J. P. (1966a). *Proc. Natl. Acad. Sci. U.S.* **55**, 1616.

Richardson, J. P. (1966b). *J. Mol. Biol.* **21**, 83.

Richardson, J. P. (1966c). *J. Mol. Biol.* **21**, 115.

Shiguera, H. T., and Boxer, G. E. (1964). *Biochem. Biophys. Res. Commun.* **17**, 758.

Skalka, A. (1966). *Proc. Natl. Acad. Sci. U.S.* **55**, 1190.

Snyder, L., Wilson, D. L., and Brody, E. N. (1967). To be published.

Stead, N. W., and Jones, O. W. (1967). *J. Mol. Biol.* **26**, 131.

Stent, G. S. (1963). "Molecular Biology of Bacterial Viruses." Freeman, San Francisco, California.

Stent, G. S. (1964). *Science* **144**, 816.

Sternberger, N., and Stevens, A. (1966). *Biochem. Biophys. Res. Commun.* **24**, 937.

Stevens, A., and Henry, J. (1964). *J. Biol. Chem.* **239**, 196.

Stevens, A., Emery, A. J., Jr., and Sternberger, N. (1966). *Biochem. Biophys. Res. Commun.* **24**, 929.

Szybalski, W., Kubinski, H., and Sheldrick, P. (1966). *Cold Spring Harbor Symp. Quant. Biol.* **31**, 123.

Taylor, K., Hradecna, Z., and Szybalski, W. (1967). *Proc. Natl. Acad. Sci. U.S.* **57**, 1618.

Thomas, R., (1966). *J. Mol. Biol.* **22**, 79.

von Hippel, P. H., and Printz, M. P. (1965). *Federation Proc.* **24**, 1458.

Weiss, S. B., and Nakamoto, T. (1961). *Proc. Natl. Acad. Sci. U.S.* **47**, 1400.

Wood, W. B., and Berg, P. (1964). *J. Mol. Biol.* **9**, 452.

Zillig, W., Fuches, E., and Millette, R. (1966). *In* " Procedures in Nucleic Acid Research " (G. L. Cantoni and D. Davies, eds.), Vol. 1, p. 323. Academic Press, New York.

Influence of the Structure of Transfer RNA on Its Interaction with Enzymes and Divalent Cations

M. GRUNBERG-MANAGO, M. COHN,* M. N. THANG,
B. BELTCHEV,† A. DANCHIN, AND L. DIMITRIJEVIC

Service de Biochimie
Institut de Biologie Physico-chimique
Paris, France

I. Introduction

The primary structures of five transfer RNA's (tRNA) have now been determined, namely, alanine (Holley *et al.*, 1965), serine (Zachau *et al.*, 1966), tyrosine (Madison *et al.*, 1966), valine (Bayev *et al.*, 1967), and phenylalanine (Raj Bhandari *et al.*, 1967). No unique secondary structure follows, however, as a necessary consequence of these primary structures. The most favored one has been the "clover leaf" (Fig. 1), although some recent experimental findings do not seem to be consistent with this model. If one estimates the maximum base pairing possible in RNA's, the percentage in tRNA is not very different from that for ribosomal and viral RNA (Cox, 1966). Nevertheless, evidence is accumulating that the three-dimensional structure for tRNA is strikingly different from the other RNA's.

The subject which will be discussed in this chapter concerns the types of experimental evidence which we have found to support the uniqueness of tRNA structure as compared to other RNA structures. The structural features of tRNA as revealed by the specificity of its interaction with enzymes and divalent cations may well be related to its biological function.

The properties which we have studied to better understand interactions dependent on secondary and tertiary structures are the following: (1) reactivity and specificity as a substrate in various enzymatic reactions and (2) the effect of the macromolecular structure on the molecular motion of water at the binding site of divalent metal ions, i.e., on the hydration sphere of Mn-tRNA as measured by the longitudinal proton-relaxation rates of water in NMR spectroscopy. The binding constants and number of binding sites of manganese can also be determined.

---

* Permanent address: Department of Biophysics and Physical Biochemistry, School of Medicine, University of Pennsylvania, Philadelphia, Pennsylvania.

† Present address: Institute of Biochemistry, Bulgaria Academy of Sciences, Sofia, Bulgaria.

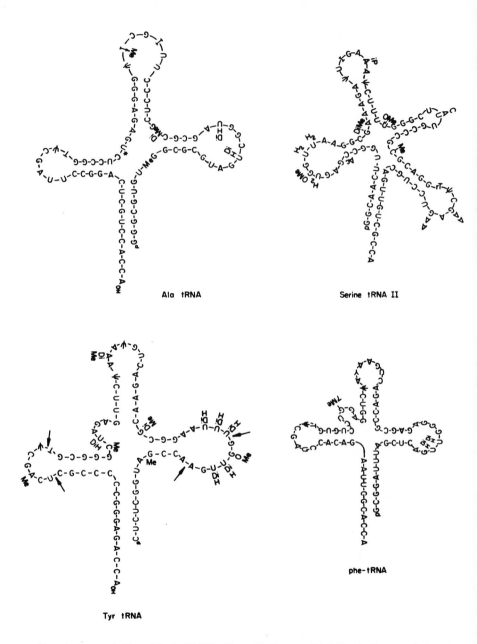

FIG. 1. Clover leaf model of tRNA's. From Holley *et al.* (1965), Zachau *et al.* (1966), Madison *et al.* (1966), and Raj Bhandari *et al.* (1967).

II. Reactivity and Specificity in Various Enzymatic Reactions

Two enzymatic degradations of tRNA have been studied in some detail: phosphorolysis catalyzed by polynucleotide phosphorylase and hydrolysis catalyzed by sheep kidney nuclease. The interaction with polynucleotide phosphorylase which always results in a total degradation of a given molecule yields information concerning secondary and tertiary structure from the rate and extent of phosphorolysis; the sheep kidney endonuclease hydrolysis yields similar information from the specificity of the bond cleavage.

Polynucleotide phosphorylase, in the presence of inorganic phosphate, phosphorolyzes ribopolynucleotides with a liberation of nucleoside diphosphates by the reaction shown in the following equation.

$$n\text{NDP} \xrightleftharpoons{\text{Mg}^{2+} \text{ or } \text{Mn}^{2+}} (\text{NMP})_n + n\text{P}_i$$

where N equals uracil, cytosine, guanine, adenine, or analogs of bases. The enzyme has a broad specificity for bases; polymers composed of base analogs (such as pseudo U or methylated bases) are substrates for the enzyme. It should be noted that the enzyme is devoid of endonuclease activity. The phosphorolysis of polynucleotides proceeds from the 3'-OH end in a stepwise fashion; mononucleotides are liberated one by one, but the degradation is not synchronous. Once the enzyme attacks a molecule it degrades it completely before leaving it and attacking another one (Thang et al., 1967a). The nonsynchronous attack of the enzyme is particularly clear in the case of tRNA which will be discussed later. Factors such as molecular configuration of different RNA's and biosynthetic polymers affect the rate of phosphorolysis, as measured by the rate of formation of diphosphates. The enzyme readily phosphorolyzes polymer having a single-strand structure, but acts more slowly on multistrand configurations (Grunberg-Manago, 1959). There are various examples—in dilute salt solution the rate of phosphorolysis of poly I and poly A are the same, but in high salt concentrations, poly I forms a triple-strand helix and is attacked very slowly; the rate of phosphorolysis of poly A which does not form a triple-strand helix is unaffected by salt concentration (Fig. 2). In general, the phosphorolysis of multistrand polymers is slower than for single-strand ones.

Although all polyribonucleotide homopolymers, copolymers, or RNA's, such as ribosomal or viral RNA's, are degraded completely (Grunberg-Manago, 1959), albeit with different rates, by polynucleotide phosphorylase, tRNA from various sources is virtually resistant to phosphorolysis at low temperature (Singer et al., 1960; Monier and Grunberg-Manago, 1962) (Table 1). For instance, under normal conditions (30°C and catalytic amount

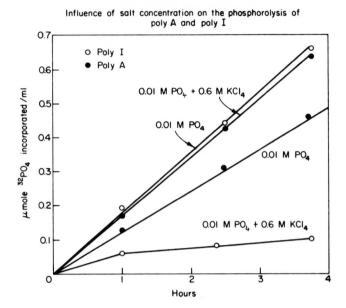

FIG. 2. Influence of salt on the phosphorolysis of poly A and poly I. From Grunberg-Manago (1959). The incubation mixture contained: $^{32}PO_4$ buffer, pH 7.4, 7.8 mM; MgCl₂, 3 mM; poly A (or poly I), 1 mg/ml; polynucleotide phosphorylase, 0.8 units/ml; temperature, 37°C. For experiments in the presence of salt, the polymers were preincubated in KCl 0.6 M for 30 minutes at 20°C before addition of the enzyme. The phosphorolysis was followed by the standard $^{32}PO_4$ incorporation assay.

TABLE I

EXTENT OF PHOSPHOROLYSIS OF DIFFERENT RNA's OR POLYRIBONUCLEOTIDES[a]

Ribopolynucleotides	Phosphorolysis (%)	
	8 hours	23 hours
tRNA from rat liver	15	18
tRNA from *E. coli*	10	12
tRNA from *E. coli*		
(5-FU) 25%	19	19
tRNA from yeast	14	14
Homopolynucleotides A, U, C	100	—
Copolymers of A, G, U, C in		
various combinations	100	—
Ribosomal RNA from *E. coli*,		
yeast	100	—

[a] From Monier and Grunberg-Manago (1962).

of enzyme) a mixture of *Escherichia coli* tRNA is 70–80% resistant to phosphorolysis.

This resistance of tRNA to phosphorolysis could be ascribed to several causes which were tested experimentally. That the phenomenon is due to the inhibition of the enzyme by the products of the reaction is eliminated by the following observations. (1) Addition of more enzyme only slightly stimulates the reaction, whereas a second addition of tRNA brings about a liberation of nucleoside diphosphates at the same rate and to the same extent as the initial

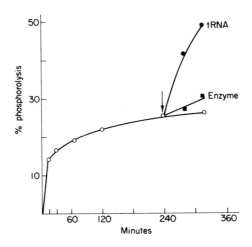

Fig. 3. Phosphorolysis of unfractionated tRNA. From Thang *et al.* (1967a). The incubation mixture contained: Tris, pH 8, 50 mM; MgCl$_2$, 0.5 mM; ^{32}PO$_4$, 10 mM; tRNA, A_{260} = 3.6; polynucleotide phosphorylase, 25 units/ml; temperature, 30°C. At the indicated time, the same amount of tRNA or enzyme was added to duplicate tubes and phosphorolysis followed by comparison to the control without addition.

addition (Fig. 3). (2) The phosphorolysis products of tRNA initially added to the reaction mixture do not inhibit the phosphorolysis of tRNA (Fig. 4a), nor does the dialysis of the products during the course of the reaction increase the extent of the reaction (Fig. 5). (3) In the presence of these products, poly A which has a Michaelis constant similar to that of tRNA reacts only at a slightly lower rate than it does in the absence of the phosphorolysis products of tRNA (Fig. 4b). (4) In the presence of arsenate, where the final products are monophosphates and not diphosphates, the reaction stops at the same extent of arsenolysis as for phosphorolysis. All these experiments lead to the conclusion that the reaction ceases before complete degradation, not because the enzyme is inactive, but because the limiting factor resides in the substrate, tRNA. There are several possible explanations for this behavior of the substrate.

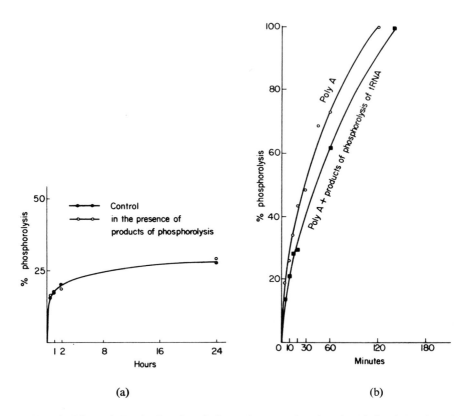

(a)   (b)

FIG. 4. Effect of tRNA phosphorolysis products on the phosphorolysis of tRNA and poly A. From Thang *et al.* (1967b). (a) The phosphorolysis products of tRNA were prepared by complete phosphorolysis at 60°C in the presence of $^{32}PO_4$. The enzyme was inactivated by heating at 100°C. The incubation mixture contained: Tris, pH 8, 50 mM; MgCl$_2$, 0.5 mM; $^{32}PO_4$, 10 mM (with the same specific activity as the ^{32}P incorporated into the ND ^{32}P released by complete phosphorolysis of tRNA and used in this experiment); tRNA, $A_{260} = 5.0$; polynucleotide phosphorylase, 23 units/ml. (b) For the poly A experiments, tRNA was replaced by poly A at a concentration of 0.40 mM (nucleotides), and the concentration of polynucleotide phosphorylase was 1.4 units/ml. The ratio between the products of phosphorolysis and the nucleotides of RNA (or poly A) was 1 to 2.

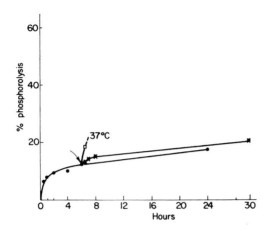

FIG. 5. Effect of dialysis on the phosphorolysis of tRNA. From Thang *et al.* (1967b). The incubation mixture contained: Tris, pH 8, 50 mM; MgCl$_2$, 0.5 mM; ^{32}PO$_4$, 10 mM; tRNA, $A_{260} = 4.8$; polynucleotide phosphorylase, 22 units/ml. The phosphorolysis was carried out at 25°C for 6 hours. Two samples were withdrawn from the incubation mixture and treated as follows. One was kept at 4°C for 15 hours, while the other was dialyzed at 4°C for 15 hours against a buffer containing Tris, pH 8, 50 mM; MgCl$_2$, 0.5 mM; and cold PO$_4$, 10 mM. Then ^{32}P was added back to give the same specific activity as that of the initial reaction mixture. Both dialyzed and undialyzed samples were reincubated at 25°C. One aliquot from the dialyzed sample was reincubated at 37°C to make sure that the tRNA was not damaged at 4°C.

1. The 30% phosphorolysis corresponds to a contaminant RNA in the tRNA preparation. This possibility was eliminated as a result of the following experiments (Thang *et al.*, 1967a). (*a*) The base ratio of the nucleoside diphosphates liberated after 30% phosphorolysis is the same as the base ratio of the total tRNA. (*b*) tRNA enzymatically labeled with ^{32}P in the terminal adenine nucleotide (tRNA pCpC^{32}pA) was phosphorolyzed by polynucleotide phosphorylase; there is a good correlation between the percentage of terminal adenine liberated and the extent of phosphorolysis. This also indicates that the 30% phosphorolysis corresponds to 30% of the ribopolynucleotide molecules being completely degraded while the remaining 70% is totally resistant, their terminal adenine not even being cleaved, which is consistent with the mechanism of phosphorolysis described above.

2. The 30% is due to preferential phosphorolysis of a particular tRNA. This possibility has also been eliminated since the loss of amino acid acceptor activity parallels the percentage of phosphorolysis, i.e., the specific activity of the remaining tRNA is the same after phosphorolysis, at least for the amino acids tested (valine, leucine, serine, and phenylalanine) (Table II).

TABLE II

CORRELATION OF LOSS OF ACCEPTOR ACTIVITY WITH EXTENT OF PHOSPHOROLYSIS[a]

Extent of phosphorolysis (%)	Loss of acceptor activity[b] (%)		
	Valine	Leucine	Phenylalanine
33	30	43	44
55	50	58	53

[a] From Thang et al. (1967a).

[b] The initial specific acceptor activities (mμmoles/mg RNA) were: 0.33 for valine; 1.14 for leucine; and 1.58 for phenylalanine. The values for valine acceptor is abnormally low in this experiment; generally the specific activity for this amino acid is about 1.0–1.5.

The incubation mixture (per ml) for phosphorolysis contained: Tris, pH 8, 50 mM; $MgCl_2$, 0.5 mM; $^{32}PO_4$ (or (PO_4), 10 mM, $A_{260} = 2.5$; polynucleotide phosphorylase, 5 units. Phosphorolysis was carried out at 60°C. The extent of degradation was followed by ^{32}P incorporation into NDP released and the acceptor activity was assayed after phosphorolysis, on parallel experiments with nonlabeled PO_4.

The most cogent evidence that the resistance to phosphorolysis is an inherent property of tRNA comes from the phosphorolysis of pure serine tRNA which is also phosphorolyzed only to an extent of 30% at 30°C. Since 30% is completely degraded, there are no features of the primary structure of tRNA which are responsible for the resistance of this nucleic acid to polynucleotide phosphorylase.

It is tempting to postulate that tRNA exists in two forms—one S, susceptible to polynucleotide phosphorylase, and the other R, resistant to polynucleotide phosphorylase (originally designated A and B, respectively; Thang et al., 1967a)—and that the two forms are not in dynamic equilibrium under the conditions of phosphorolysis at low temperatures, i.e., at temperatures below 45°C. As shown in Fig. 6 at these temperatures the phosphorolysis is not complete, the percentage degraded increasing with increasing temperatures. The effect of temperature (cf. Fig. 6) is most striking between 37° and 45°C; at 37°C, 60% of the molecules remain undegraded, at 42°C there still is 30% undegraded, but at 45°C, only 3° higher, or above, the phosphorolysis proceeds to completion. At low temperatures, the phosphorolysis after a fast initial rate, approaches a plateau asymptotically at a definite percentage of substrate phosphorolyzed consistent with complete degradation of the S form and an immeasurably slow rate of conversion of S into R. At higher temperatures the rate of a second slow phase of the reaction becomes observable, consistent with a slow conversion of the resistant form into the sensitive form. The dependence of the ratio of S and R forms on temperature can be seen most readily in Fig. 7, where the extent of phosphorolysis reaction at

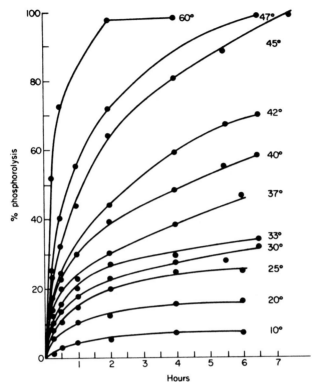

Fig. 6. Phosphorolysis of tRNA as a function of temperature. From Thang *et al.* (1967a). The incubation mixture contained: Tris, pH 8, 50 mM; MgCl$_2$, 0.5 mM; ^{32}PO$_4$, 10 mM; tRNA, A_{260} = 3.6; polynucleotide phosphorylase, 24 units/ml.

constant time (taken arbitrarily at 22 hours when the rate has reached a plateau at low temperature) is plotted as a function of temperature. A sharp rise is observed at 40°C, which could correspond to the temperature of the onset of a rapid conversion of the R form into the S form.

As would be expected, on the basis of a nonequilibrium between the two forms at low temperature, if a second addition of tRNA is made after a plateau has been reached in the phosphorolysis, the reaction begins again immediately and proceeds to the same limited extent, i.e., only the S form is degraded, as in the first reaction. The equilibrium between S and R may, however, be "unfrozen" after a plateau has been reached, e.g., at 25°C, by heating at 100°C in the phosphorolysis buffer (in the presence of 0.5 mM Mg^{2+}). After cooling the mixture may again be incubated at 25°C with addition of fresh enzyme and the phosphorolysis proceeds once more at the same rate and to the same extent as the first incubation (Fig. 8). This cycle can be repeated until total degradation is obtained.

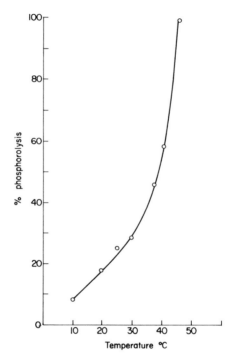

Fig. 7. Extent of phosphorolysis of tRNA as function of temperature. From Thang *et al.* (1967a). Experimental conditions identical to those described in Fig. 6. The extent corresponds to the percentages obtained in 22 hours at indicated temperatures.

The reestablishment of equilibrium between S and R forms by heating above 45°C (100°C in Fig. 8) may be similar to the hysteresis effect that can be observed with pH and temperature in the association of polynucleotides (see, e.g., Michelson *et al.*, 1967). Such a phenomenon occurs (Massoulié, 1967) when one of the structures is more stable thermodynamically under given conditions, but there is a kinetic barrier which prevents rapid attainment of the equilibrium state.

There are some aspects of the behavior of tRNA phosphorolysis, however, which are difficult to explain on the basis of the postulated simple "frozen" equilibrium between two forms. The chief difficulty is that in spite of the "frozen" equilibrium, in the temperature region below 45°C, there are different ratios of S and R as defined by a different extent of phosphorolysis (see Fig. 6). If the equilibrium is "frozen" at the cessation of phosphorolysis, how was it established initially? In particular, when phosphorolysis has reached a plateau, at low temperature, 10°C as shown in Fig. 9, and the temperature is subsequently raised to 25°C—a temperature at which the equilibrium is still "frozen" by the criterion of the cessation of degradation at 25%—the reaction nevertheless begins again without any lag phase. This

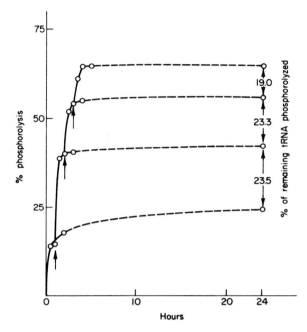

FIG. 8. Effect of heating and cooling on the extent of phosphorolysis. From Thang *et al.* (1967b). The incubation mixture contained: Tris, pH 8, 50 mM; MgCl$_2$, 0.5 mM; ^{32}PO$_4$, 10 mM; tRNA, $A_{260} = 4.0$; polynucleotide phosphorylase, 20 units/ml; temperature, 25°C. The phosphorolysis was followed as a function of time. When the reaction showed a sharp diminution in rate, an aliquot was taken and heated immediately at 100°C for 3 minutes; the solution was cooled and an amount of fresh enzyme equivalent to that initially present in the aliquot was added. The mixture was then reincubated at 25°C and the same procedure repeated.

phenomenon cannot be understood in terms of a simple "frozen" equilibrium between two forms at 25°C, since obviously more of the S form had *immediately* become available upon raising the temperature, in spite of the fact that the R form does not continue to be converted to the S form as the phosphorolysis proceeds; to wit, the phosphorolysis stops after 25% degradation.

The two configurations of tRNA may differ either in secondary or tertiary structure. Intermolecular association is unlikely, since there is no evidence that it occurs under the phosphorolysis conditions. The tRNA was incubated in the phosphorolysis mixture in the absence of enzyme and separated on a Sephadex column; it had a sedimentation constant of 4 S. The extent of phosphorolysis is not changed whether only traces of Mg are present or if the metal ion concentration is increased to 5×10^{-3} M (Table III). Also, no evidence of aggregation is found under the conditions of phosphorolysis by independent measurement of divalent cation binding, as will be discussed later.

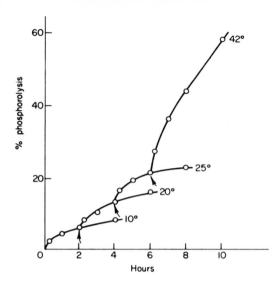

FIG. 9. "Stepwise" phosphorolysis of tRNA in function of temperature. From Thang *et al.* (1967b). The incubation mixture contained: Tris, pH 8, 50 mM; MgCl$_2$, 0.5 mM; ^{32}PO$_4$, 10 mM; tRNA, $A_{260} = 3.84$; polynucleotide phosphorylase, 22 units/ml. The phosphorolysis was first carried out at 10°C. When the reaction showed a sharp diminution in rate, an aliquot was brought up to 20°C. The phosphorolysis was followed again in the same way. The same procedure was then repeated at 25°C.

TABLE III

NONVARIANCE OF EXTENT OF PHOSPHOROLYSIS WITH Mg CONCENTRATION[a]

Concentration of Mg^{2+} (M)	Phosphorolysis (%)
10^{-4}	31.5
5.10^{-4}	31.2
10^{-3}	30.0
5.10^{-3}	28.4

[a] From Thang *et al.* (1967b).

Two clearly differentiated forms of tRNA structure which polynucleotide phosphorylase can distinguish are the "denatured" and "renatured" forms of leucine III tRNA purified from yeast and described by Lindhal *et al.* (1966). They have shown that the leucine acceptor activity of a specific leucyl-tRNA, separated by countercurrent distribution, could be increased by heating for 2 minutes at 60°C in the presence of Mg. On the contrary, heating at 60°C for 2 minutes in the absence of Mg and in the presence of ethylenediaminetetraacetic acid (EDTA), results in tRNA losing practically all its ability to accept leucine. After treatment with Mg, the form of tRNA which resulted was designated renatured, which is equivalent to a native form of tRNA if denaturation is avoided during isolation; the form resulting after treatment in the presence of EDTA was designated the denatured form. These forms are slowly interconvertible under certain conditions. After dialysis against Tris buffer, in the presence of 0.05 M salt containing 0.001 M

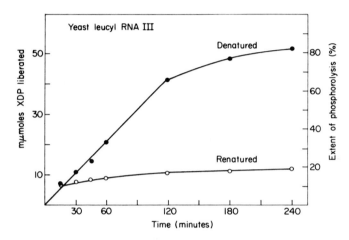

FIG. 10. Phosphorolysis of renatured and denatured yeast leucyl-III tRNA. From Thang *et al.* (1967c). The purified yeast leucyl-III tRNA was treated according to Lindhal *et al.* (1966) to obtain the renatured and denatured forms: tRNA was first denatured in Tris pH 8, 10 mM containing EDTA, 1 mM at 60°C for 3 minutes; to one-half of the solution MgCl$_2$ was added to a final concentration of 10 mM, and the sample reheated at 60°C for 3 minutes. Both denatured and renatured samples were dialyzed against buffer containing Tris, pH 8, 10 mM; NaCl, 50 mM; MgCl$_2$, 1 mM; dialysis overnight at 3°C. The incubation mixture for phosphorolysis contained: Tris, pH 8, 10 mM; MgCl$_2$, 1 mM; NaCl, 50 mM; ^{32}PO$_4$, 10 mM; tRNA, $A_{260} = 5.0$; polynucleotide phosphorylase, 20 units/ml.

Mg, the acceptor activity for leucine remains the same for both the denatured form (very low activity) and the renatured form (20-fold higher activity) (Table IV). This medium is suitable for the action of polynucleotide phosphorylase. The native form is essentially resistant (Fig. 10); by contrast, the denatured leucine RNA is almost completely phosphorolyzed and, furthermore, the phosphorolysis reaction slows down as renaturation of the RNA begins (Thang *et al.*, 1967c) (Table IV).

The renatured and denatured forms of leucine III RNA differ from each other with respect to their amino acid acceptor activity and can be separated from each other by countercurrent distribution. On the other hand, the S and R forms gave no evidence of a difference with respect to their amino acid

TABLE IV

RENATURATION OF LEUCINE III tRNA UNDER PHOSPHOROLYSIS CONDITIONS[a]

Incubation time (hour)	^{14}C-Leucine incorporated (cpm)	
	Renatured RNA	Denatured RNA
0	4389	206
2	4436	1416
4	4663	5154

[a] Samples are incubated in phosphorolysis buffer without enzyme at 30°C. At indicated time intervals charging of leucine was assayed directly without further treatment.

acceptor activity. Furthermore, when mixtures of tRNA have been subjected to the same treatment as the one used for the interconversion of denatured and renatured leucine tRNA, no significant difference in the amino acid acceptor activity of the resulting product was observed (Thang et al., 1967a). A significant difference appeared, however, between the denatured and renatured total tRNA in their interaction with sheep kidney nuclease.

The sheep kidney nuclease (Kasai and Grunberg-Manago, 1967) shows no apparent base specificity, nor any specificity for the sugar moiety of polynucleotides. The unique feature of this endonuclease is the fact that its digestion products are oligonucleotides, most of which are tetramers or higher and, furthermore, $5'-PO_4$ ended; no mononucleotides are produced. This pattern of products is in sharp contrast with that of nucleases such as ribonuclease (RNase). The sheep kidney enzyme, however, displays extremely rigorous specificity with regard to the secondary structure of its substrate, as illustrated by the fact that native DNA (Fig. 11), complexes between poly A and poly U (Fig. 12), and polyinosinic acid in high salt concentration are completely resistant to the enzyme. In this connection, the term "structure" is used to cover a variety of structures differing from the random coil. The stacked structure of neutral poly A and poly C are attacked very much more slowly than the less structured poly U; ribosomal RNA and especially tRNA are only slowly attacked. The effect of temperature on the digestion of poly A clearly demonstrates this structural inhibition; the increase in the rate of hydrolysis with temperature is the same as that observed for the highly structured tRNA and nearly four times the increase observed for poly U.

Total mixed tRNA was treated in the same way as described for leucine III RNA (Lindhal et al., 1966) to yield the denatured form (60°C in the presence of EDTA) and the renatured form (60°C in the presence of Mg). The two forms were then hydrolyzed by the sheep kidney nuclease. Total loss of amino acid acceptor activity was observed in both cases which suggests that all the molecules were attacked.

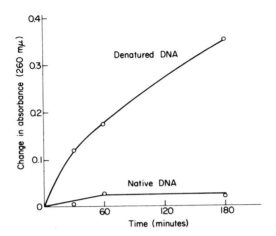

FIG. 11. Hydrolysis of DNA. From Kasai and Grunberg-Manago (1967). Denatured DNA was obtained by heat treatment at 100°C for 10 minutes, followed by rapid cooling. The incubation mixture contained: Tris-HCl, pH 7.5, 50 mM; β-mercaptoethanol, 25 mM; MgCl₂, 5 mM; E. coli DNA, 50 μg/ml; bovine serum albumin, 1 mg/ml; and sheep kidney nuclease, 1.6 μg/ml. Incubation at 37°C.

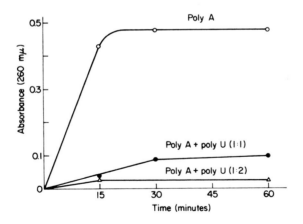

FIG. 12. Hydrolysis of poly A + poly U complex. From Kasai and Grunberg-Manago. (1967). The incubation mixture contained: Tris-HCl, pH 7.5, 50 mM; MgCl₂, 1.5 mM; β-mercaptoethanol, 25 mM; polyribonucleotide, 2 mM; sheep kidney nuclease, 15μg/ml. Before addition of enzyme the mixture was preincubated at 37°C during 1 hour for formation of the complex.

After hydrolysis, the products were isolated on a Sephadex G-50 column. Figure 13 shows the elution pattern of the hydrolysis product of denatured and renatured tRNA. One can see that in the case of denatured RNA, 17% of the product was in the form of large fragments, 30% intermediary frag-

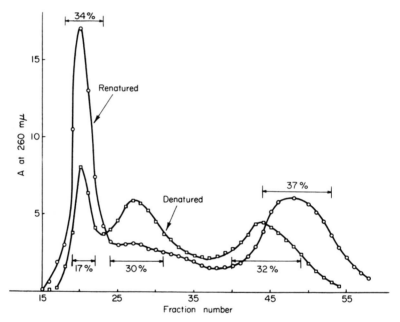

FIG. 13. Degradation of denatured and renatured *E. coli* total tRNA by sheep kidney nuclease. From Dimitrijevic and Grunberg-Manago. (1967). The denatured tRNA was obtained by treatment at 60°C for 2 minutes in the presence of EDTA, 1 mM in 10 mM Tris buffer, pH 8. The renatured tRNA was prepared as described in Fig. 10. The incubation mixture contained: Tris-HCl, pH 8, 50 mM; MgCl$_2$, 0.5 mM; β-mercaptoethanol, 25 mM; *E. coli* tRNA, 1.85 mg/ml; bovine serum albumin 1% and enzyme, 0.115 mg/ml. The reaction was carried out at 45°C for 140 minutes. The hydrolysis products were separated on a column of Sephadex G-50 (50 × 1 cm). Elution with 0.2 M ammonium formiate. Fractions of 1 ml were collected.

ments, and 32% small oligonucleotides; with the renatured RNA there is a twofold increase in the percentage of the large fragments, practically no intermediary size fragments, and 37% very small oligonucleotides. The observed differences between the denatured and renatured form are not maximal since partial renaturation undoubtedly occurred during the course of the enzymatic hydrolysis. Nevertheless, the results clearly indicate that there are more sites of attack by the enzyme with denatured RNA as substrate than with the renatured form.

The analysis of the hydrolysis products of the renatured RNA shows that the high molecular weight peak (see Fig. 13) contained mainly fragments of polynucleotides (average chain length 40 nucleotide units), whereas the average chain length of the oligonucleotides was 4.5 nucleotide units. The polynucleotides contain pG terminal units in 5′ and TψGC sequences. This result indicates that the TψGC sequence is located in the RNA structure in a position that it is not accessible to the enzyme. If one accepts that the enzyme preferentially hydrolyzes the loops, as suggested by other data, this means that this sequence is not located in a loop (as is usually assumed), or that the tertiary structure of tRNA renders this region inaccessible to the enzyme.

These results are in very good agreement with those of Wagner and Ingram (1966), and Armstrong *et al.* (1966) who used pancreatic ribonuclease and RNase T_1; from experiments with purified alanine tRNA, they concluded that the central loop containing the anticodon is the region most sensitive to hydrolysis, and that the half of the molecule containing TψGC is the most stable one.

Thus, there is a form of tRNA—designated as renatured by Lindhal *et al.* (1966)—in which only bonds in specific regions are hydrolyzed by sheep kidney enzyme. This form, which is resistant to polynucleotide phosphorylase, at least in the case of yeast leucyl III tRNA, is also probably the only one which can serve as substrate for amino acid activating enzymes.

III. Measurement of Proton Relaxation Rates of Water at the Binding Sites of Manganese

Many investigations have been made of the effect of divalent metal ions on the structure of nucleic acids, but we have investigated the inverse, namely, the effect of the macromolecule on the rotational motion of water in the first hydration sphere of the bound metal ion. The spectroscopic property which is measured is the longitudinal relaxation rate of water protons ($1/T_1$) in manganese complexes, i.e., the rate of attainment of equilibrium over nuclear spin states of water. Our justification for using such an apparently indirect parameter is the experimental finding that this parameter is far more sensitive to changes in structure, e.g., in various modifications of tRNA, than any spectroscopic property which can be measured of the nucleic acid itself.

The physical principles underlying the enhancement of the proton relaxation rate (PRR) due to Mn complexes will be described briefly. The principal mechanism of relaxation in pure water resides in the magnetic dipolar interaction between protons, but in the presence of a paramagnetic ion the much larger proton-electron dipolar magnetic interaction dominates the relaxation rate. Since the relaxation rate is inversely proportional to the sixth power of the distance between the water protons and the manganese ion, the dominant contribution to the effect is due to the interaction of Mn^{2+}

with water in its first hydration sphere. Consequently, the magnitude of the effect is very sensitive to the immediate environment of the Mn^{2+}. The displacement of H_2O by other ligands in the Mn^{2+} aquocation would be expected to decrease the effectiveness of Mn^{2+} on the PRR and, indeed, such an effect has been observed upon chelation of Mn^{2+} with ethylenediamine-tetraacetic acid (EDTA) (King and Davidson, 1958). On the other hand, an unexpected enhancement of the effectiveness of Mn^{2+} in increasing $1/T_1$ was observed upon complexing Mn^{2+} with nucleic acids and other polynucleotides by Eisinger et al. (1962, 1965). Cohn and associates (Cohn, 1963; Mildvan and Cohn, 1963) have observed a similar phenomenon upon binding Mn to proteins and have shown that the magnitude of the enhancement is monotonically related to the degree of activity of different conformations of the protein at the site of binding (O'Sullivan and Cohn, 1966), provided that the number of water ligands remains constant and the rate of chemical exchange between water in the hydration sphere and the bulk water is not rate limiting.

Eisinger et al. (1962) defined an enchancement factor ε equal to the ratio of relaxation rates in the presence and absence of complexing agent, and with a few approximations of the fundamental Bloembergen-Solomon equations pointed out the following relationship.

$$\varepsilon = \frac{1/T_1^* - 1/T_1(0)}{1/T_1 - 1/T_1(0)} \approx \frac{p^* \tau_c^*}{p \tau_c}$$

where ε is the enhancement factor; T_1, observed relaxation time; $T_1(0)$, relaxation time in absence of Mn^{2+}; p, the ratio of the number of water protons in hydration shell to the total water protons; τ_c, dipolar correlation time; and the asterisk indicates the presence of complexing agent.

Since p can only decrease when H_2O is replaced in the Mn aquocation by ligands group from the complexing agent, one would expect the enhancement to be less than one as found in Mn-EDTA; when the enhancement is greater than 1, it becomes necessary to invoke an increase in τ_c in the Mn complex. If it is assumed that τ_c in the complex is determined primarily by the rotational correlation time, τ_r, as demonstrated in the Mn aquocation (Bloembergen and Morgan, 1961), then it may be concluded that the relative rotational motion of the Mn and water is hindered by the binding of the macromolecule and the relaxation rate thereby increased. It must be remembered that the observed relaxation rate is the weighted average of the H_2O in the hydration sphere and in the bulk of the solution which is normally exchanging rapidly with the H_2O in the hydration sphere; an increase in relaxation rate upon complexing Mn^{2+} ($\varepsilon > 1$) implies that Mn^{2+} is at least partially accessible to the water in the bulk of the solution and is, therefore, located at a site not completely buried within the macromolecule.

The investigation of this spectroscopic property of tRNA is particularly pertinent since all biological functions of tRNA are manifested in the presence of divalent cations. Although the binding of Mn may not be identical with that of Mg, the Mn complex may substitute for the Mg complex in all the biological activities of tRNA. The methods of analyzing the data to obtain ε_b, the PRR enhancement of the binary complex; K_d, the dissociation constant of the binary complex; and n, the number of binding sites per

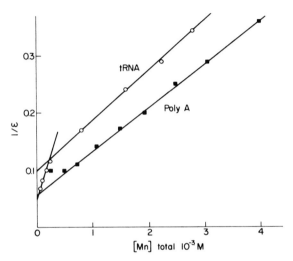

FIG. 14. Comparison of titrations of tRNA and poly A with Mn^{2+}. Plot $1/\varepsilon$ vs. Mn_{total} for tRNA at a concentration of 2.6 mM in nucleotide units in 0.05 M KCl and 0.05 M Tris-HCl buffer, pH 7.5; for poly A at a concentration of 0.25 mM in 0.01 M Tris-HCl buffer pH 8.

molecule, have been described previously (Eisinger *et al.*, 1965; Mildvan and Cohn, 1963). The measurements reported in this paper were obtained at 22°–25°C with a Bruker pulsed NMR spectrometer operating at 15 Mc.

In a titration of tRNA with increasing concentration of Mn^{2+}, using ε as the parameter, tRNA, unlike other polynucleotides, exhibits two types of binding sites for Mn as is evident from Fig. 14 which shows a comparison of titration of poly A and tRNA. Only a single type of site is observed for poly A, but the tRNA curve is complex as previously observed by Eisinger *et al.*, (1965). The data may be interpreted as representing multiple binding sites, roughly divided into two classes. The enhancement of strong binding sites, is 19.5 and indicates that the bound manganese is highly immobilized; the number of strong binding sites is approximately 10% of the total phosphate groups. The value of ε is invariant with ionic strength in the region investigated, 0.01 to 0.1, but the slope of the titration curves changes, indicating

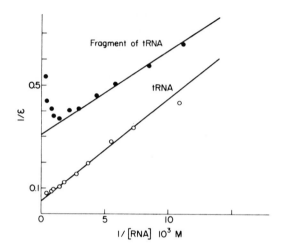

FIG. 15. Comparison of tRNA and fragment of tRNA (approximately 40 nucleotide chain length) obtained from action of sheep kidney nuclease. Plot of $1/\varepsilon$ vs. $1/RNA$, RNA variable, $MnCl_2$ concentration, 0.1 mM; 0.05 M KCl and 0.05 M Tris-HCl, pH 7.5; ε_b (tRNA) is 19.1; ε_b (fragment of tRNA) is 3.8.

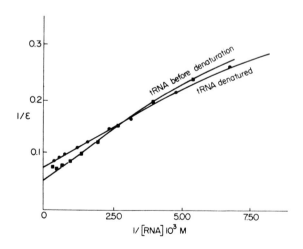

FIG. 16. Comparison of tRNA and "denatured" tRNA. Plot of $1/\varepsilon$ vs. $1/RNA$. Conditions same as Fig. 15 ε_b(tRNA) is 19.5, ε_b("denatured" tRNA) is 13.0.

that K_d or n or both change with the ionic strength. If the tRNA is degraded, a dramatic decrease in enhancement results. For example, as shown in Fig. 15, a fragment of tRNA, 40 nucleotides long, obtained as a product of sheep kidney nuclease action, has an enhancement at the strong site about one-fifth that of intact tRNA. In Fig. 15, only the strong sites are titrated when Mn is held constant, and the concentration of complexing agent is varied. It will be noted that ε_b for the tRNA fragment has a maximum value at 0.7 mM nucleotide units and then decreases, undoubtedly due to aggregation where some of the bound Mn becomes inaccessible to the water; this phenomenon occurs with intact tRNA or poly A only at much higher concentrations.

Two forms of intact tRNA were then compared, the usual preparation and the one designated as denatured by Lindhal *et al.* (1966). As has already been mentioned, these two forms of leucine III tRNA differ in countercurrent distribution, amino acid acceptor activity, and susceptibility to polynucleotide phosphorylase. The "denatured" mixture of tRNA's differs in the reaction with the sheep kidney nuclease. As shown in Fig. 16, the enhancement of the tight binding site after denaturation is reduced to 13 from 19.5. This difference would be expected to be much larger for the denatured and renatured forms of leucine III tRNA since in the mixture of tRNA's used in the experiment only a few of tRNA's exist in the two forms under the conditions used.

Last, we have investigated whether a difference could be detected between tRNA and its aminoacyl derivative. As shown in Fig. 17, the value of the

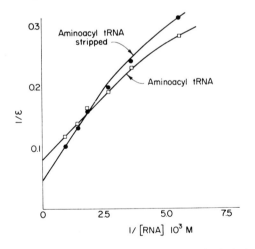

FIG. 17. Comparison of amionacyl tRNA before and after hydrolysis of the aminoacy group. The concentration of $MnCl_2$ is 0.1 mM, KCl 0.05 M, Tris-HCl buffer, 0.05 M, pH 7.5. The value of ε_b of tRNA before aminoacylation was 19.5; ε_b of aminoacyl tRNA is 12.8 and after hydrolysis of the aminoacyl group is 19.1.

strong binding site is reduced from 19.5 to 12.8 upon attachment of the aminoacyl group. The true value for the aminoacyl form is probably even lower since the sample measured was only about 70% charged. To make certain that the lowering of the enhancement was not due to partial degradation of tRNA during the charging reaction, the aminoacyl derivative was discharged and measured again, and it was found to return to the value for intact tRNA as shown in Fig. 17.

The preliminary results which have been described with the PRR method are encouraging in that we have a physical parameter which reflects subtle changes in the structure of tRNA, as illustrated by the significant change observed by the attachment of an aminoacyl group to a single nucleotide residue. More careful examination may also reveal differences between individual tRNA's. Although the theory of relaxation rates of paramagnetic ion complexes with simple molecules is well understood, the theoretical interpretation of the magnitude of the enhancement factor in terms of correlation with detailed structure is not yet at hand. Hopefully further studies with model structures and studies of temperature dependence of PRR as well as auxiliary measurement of ^{31}P NMR of the metal complexes will make such an interpretation possible.

It is obvious that the ideal method for determining three-dimensional structures is X-ray crystallography which has been so successful for proteins and DNA. Unfortunately, nobody has succeeded so far in crystallizing tRNA. Therefore, one has to be satisfied with more indirect methods. We have outlined a few of these methods which appear promising for probing the rather subtle aspects of the structure of this molecule in solution which are undoubtedly intimately involved with its biological activity; these methods should be even more useful for the study of the interactions of tRNA with components it encounters in biological systems in solution and for elucidating the relation between structure and function of tRNA in its complexes with cellular components.

It may be seen from the results that have been described that certain groups in the tRNA molecule are accessible to specific enzymes and also to binding by divalent cations. Changes in the structure, e.g., aminoacylation, denaturation, or fragmentation, are revealed in interactions with enzymes and are also reflected in changes in the environment of the binding sites of divalent cations.

ACKNOWLEDGMENTS

We are most grateful to Dr. Sadron and Dr. Ptak, Biophysics Department of the Museum d'Histoire Naturelle in Paris for kindly letting us use their NMR spectrometer. We wish to thank Mme. L. Dondon for her skillful technical assistance and Miss M. Graffe for her exacting preparation of polynucleotide phosphorylase.

This work was supported by the following grants: No. C-04580 of the United States Institutes of Health; Convention 6600 020 of Délégation Générale à la Recherche Scientifique et Technique (Comité de Biologie Moléculaire); French National Research Council (RCP No. 24); L.N.F.C.C. (Comité de la Seine); F.R.M.F.; and a participation from the French Atomic Energy Commission.

REFERENCES

Armstrong, A., Hagopian, H., Ingram, V. M., and Wagner, E. K. (1966). *Biochemistry* **5**, 3027.

Bayev, A., Venkstern, T., Mirzabekov, A., Krutilina, A., Axelrod, V., Li, L., and Engelhardt, V. (1967). *In* "Properties and Functions of Genetic Material" (in press).

Bloembergen, N., and Morgan, L. O. (1961). *J. Chem. Phys.* **34**, 842.

Cohn, M. (1963). *Biochemistry* **2**, 623.

Cox, R. A. (1966). *Biochem. J.* **98**, 841.

Dimitrijevic, L., and Grunberg-Manago, M. (1967). Unpublished data.

Eisinger, J., Shulman, R. G., and Szymanski, B. M. (1962). *J. Chem. Phys.* **36**, 1721.

Eisinger, J., Fawaz-Estrup, F., and Shulman, R. G. (1965). *J. Chem. Phys.* **42**, 43.

Grunberg-Manago, M. (1959). *J. Mol. Biol.* **1**, 240.

Holley, R. W., Apgar, J., Everett, G. A., Madison, J. T., Marquisee, M., Merril, S. H., Penswick, J. R., and Zamir, A. (1965). *Science* **147**, 1462.

Kasai, K., and Grunberg-Manago, M. (1967). *European J. Biochem.* **1**, 152.

King, J., and Davidson, N. (1958). *J. Chem. Phys.* **29**, 787.

Lindhal, T., Adams, A., and Fresco, J. R. (1966). *Proc. Natl. Acad. Sci. U.S.* **55**, 941.

Madison, J. T., Everett, G. A., and Kung, H. (1966). *Science* **153**, 531.

Massoulié, J. (1967). *European J. Biochem.* (in press).

Michelson, A. M., Massoulié, J., and Guschlbauer, W. (1967). *Progr. Nucleic Acid Res.* **6**, 83.

Mildvan, A. S. and Cohn, M. (1963). *Biochemistry* **2**, 910.

Monier, R., and Grunberg-Manago, M. (1962). *In* "Acides nucléiques et polyphosphates," Vol. 106, p. 163. C.N.R.S., Paris.

O'Sullivan, W. J., and Cohn, M. (1966). *J. Biol. Chem.* **241**, 3116.

Raj Bhandari, U. L., Chang, S. H., Stuart, A., Faulkner, R. D., Hoskinson, R. M., and Khorana, H. G. (1967). *Proc. Natl. Acad. Sci. U.S.* **57**, 751.

Singer, M. F., Luborsky, S., Morrison, R. A., and Cantoni, G. L. (1960). *Biochim. Biophys. Acta.* **38**, 568.

Thang, M. N., Guschlbauer, W., Zachau, H. G., and Grunberg-Manago, M. (1967a). *J. Mol. Biol.* **26** 403.

Thang, M. N., Beltchev, B., and Grunberg-Manago, M. (1967b). In preparation.

Thang, M. N., Adams, A., Lindhal, T. Fresco, J. R., and Grunberg-Manago, M. (1967c). Unpublished data.

Wagner, E. K., and Ingram, V. M. (1966). *Biochemistry* **5**, 3019.

Zachau, H. G., Dütting, D., and Feldmann, H. (1966). *Z. Physiol. Chem.* **347**, 212.

The Interaction of Aromatic Hydrocarbons with Nucleic Acids and Their Constituents

E. D. BERGMANN

Department of Organic Chemistry
Hebrew University
Jerusalem, Israel

The discovery that certain aromatic polycyclic hydrocarbons can induce cancerous growth was perhaps one of the most unexpected observations made in the field of the correlation of molecular structure and biological activity. All our previous experience had suggested that a biologically active substance must possess at least one polar group, or to use the terminology of Paul Ehrlich, a haptophoric group. That we deal here, indeed, with an unusual phenomenon is emphasized by the fact that the introduction of polar groups such as hydroxyl or amino groups into these hydrocarbons, abolishes their carcinogenic activity.

These hydrocarbons have two distinguishing features. One is that they are completely or almost completely flat (1); any deviation from planarity, such as, e.g., the hydrogenation of one of the benzene rings to a nonplanar cyclohexane ring, destroys the biological activity. Second, these hydrocarbons have a large number of delocalized π electrons. Both these features have to be taken into consideration if one wishes to correlate structure and activity.

In fact, there are *two* phenomena which one has to explain: not only are aromatic polycyclic hydrocarbons capable of some biologically significant interaction with cell constituents, but also the active hydrocarbons must have a certain size (mostly five aromatic rings) and shape (that of the phenanthrene system). The biologically significant interaction with the cell has, therefore, also an aspect of specificity—and that is what is expressed in the theory of Pullman (2) as the insistence on the importance of the electron density in the so-called k region of the molecule.

It is obvious that the first question to be asked is: Is there a cell constituent which combines with this kind of hydrocarbon and, of course, has a special affinity to the carcinogenic hydrocarbons? An answer to this question would make it possible to obtain a deeper insight into the mode of action of these hydrocarbons. Two theories have been proposed: that of Heidelberger (3) referring to proteins and that of Boyland (4) referring to nucleic acids as the point of attack. The theory of Boyland is, we believe, intuitively more attractive to the chemist because the determinant role of the nucleic acids in biological events would make it seem more reasonable that *they* are connected with so fundamental an event as the transformation of a normal into a cancerous cell. To this should be added that at least some of the carcinogenic hydrocarbons appear to have mutagenic effects on microorganisms (5).

Without wanting to make a decision between the two alternative theories—they may in the end turn out to be complementary—we would like to discuss the evidence existing for an interaction of polycyclic hydrocarbons with the constituents of nucleic acids, i.e., the purines and pyrimidines, and also the questions arising from this evidence.

Brookes and Lawley (6) found in experiments with DNA, RNA, and protein (the skin protein of the mouse), that they all bind polycyclic hydrocarbons; but only in the case of DNA does a parallel exist between the affinity for the high molecular component and the carcinogenic power of the hydrocarbons; it was shown that 0.6 mole of 9,10-dimethyl-1,2-benzanthracene combines with one mole of DNA (20,000 molecules).

De Maeyer (7) studied in detail the interesting observation that methylcholanthrene, 3,4-benzopyrene, and 9,10-dimethyl-1,2-benzanthracene, the three most important carcinogens, inhibit the biosynthesis of interferon in tissue cultures of rat cells infected with sindbis virus. This biosynthesis is DNA-dependent, and the analogous inhibition by actinomycin D is ascribed to its complex formation with DNA. Equally, benzopyrene and dimethylbenzanthracene inhibit plaque formation by viruses (herpes simplex, vaccinia) that are DNA-dependent, but not by RNA-dependent viruses. It is important that noncarcinogenic hydrocarbons do not have this effect, e.g., the isomeric 1,2-benzopyrene, or 1,2-benzanthracene, the inactive parent substance of the carcinogenic 9,10-dimethyl-1,2-benzanthracene.

Experiments of Hsu and others (8) have somewhat modified this picture; they worked with *Escherichia coli* spheroplasts, preincubated with the carcinogens, and observed certain differences between them and actinomycin—the latter does not inhibit the replication of RNA viruses, the carcinogens do. On the other hand, contrary to the antibiotics, the carcinogens have no effect on RNA *synthesis* in *E. coli* spheroplasts. There is one exception from the parallelism between carcinogenic action and the depressing effect on viral replication: 1,10-dimethyl-1,2-benzanthracene, which is not a carcinogen, acts on the *E. coli* spheroplasts as if it were a carcinogen. The total of these data, however, appears to support the view that the carcinogenic hydrocarbons have an affinity for DNA.

What chemical reactions can be expected? It is interesting to recall (9) that benzopyrene, mixed with the natural pyrimidines, e.g., uracil, thymine, cytosine, and 5-methylcytosine, as well as guanine, gives under the influence of ultraviolet light, stable addition products; the unsubstituted bases purine and pyrimidine do not react analogously. A (potential) conjugated carbonyl group in the pyrimidine ring appears to be necessary for the formation of the reaction product. Based on spectroscopic evidence and assuming that the k region of benzopyrene will participate in the reaction, a formula such as

has been derived for the product (the synthesis of such a compound should not present unsurmountable difficulties). This reaction of individual pyrimidine bases would be a good model for the observation of Ts'o (10) that benzopyrene combines under the influence of light with DNA in its heat-denatured form, i.e., in the uncoiled or uncoiling state. Each thousand molecular units of DNA combine with one molecule of benzopyrene, although one cannot, of course, be sure that all active sides are occupied. The fact that the combination with DNA has a hypochromic effect points to a similar reduction of the number of conjugated double bonds in the hydrocarbons as in the above case. It is of importance to note that the affinity of benzopyrene for DNA is—under the conditions of the experiment—a thousand times higher than for adenine and 10^4 to 10^5 times higher than for thymine; also it is higher than the affinity for RNA.

We have thus some idea of the possible final step in the interaction of such carcinogenic hydrocarbons with the constituents of nucleic acids, although we have to assume tentatively that in the cell a mechanism is operative similar to the photochemical one we have discussed and which represents a *chemical* reaction of the hydrocarbons. However, the question remains: What physical forces initially bring the two components of the system together?

Much attention has been paid to the possibility that charge-transfer complexes are formed between polycyclic aromatic hydrocarbons and purine and pyrimidine bases (11). Generally speaking, the formation of a charge-transfer complex expresses itself in the appearance of a new band in the electronic spectrum, as it appears, for instance, in the system purine-isalloxazine (12) or purine-actinomycin (13). No such band has been observed in any combination of these donors and acceptors. We have investigated the electronic spectra of nine polycyclic hydrocarbons and several purines and pyrimidines—a study greatly hampered by problems of solubility in general and solubility in common solvents in particular (14). With tetramethyluric acid, which we selected in parallel to a study to which I will refer presently, only one hydrocarbon showed a (very slight) change in sprectrum in chloroform solution, namely 3,4-benzopyrene. Incidentally, solutions of all these hydrocarbons in chloroform showed a gradual change in the spectrum on standing in daylight, so that all work with such solutions should be done in the dark (at present we are studying this unexpected photochemical effect). The fact that no charge-transfer band has been observed, seems *a priori* to rule out the possibility of the formation of a charge-transfer *complex*. One can argue, however, that the complex might be so weak that the band will not appear in the visible, where it has been sought, but in the far ultraviolet, where it has not yet been looked for. Indeed, the electron-acceptor properties of these hydrocarbons are very much smaller than (not more than 25%) of those of the usual electron acceptors.

In looking for other data which might be interpreted as indications of an interaction between polycyclic hydrocarbons and purine or pyrimidine bases, one may refer to the well-known phenomenon that uric acid and tetramethyluric acid make these hydrocarbons soluble in water (15). Also, other purines and pyrimidines show this effect, but to a lesser degree; in fact, there is an impressive parallelism between the solubilizing power and the electron-donor properties to which we will refer in a moment. If we disregard the present discussion (16) of whether this effect is a real solubilization or a colloid-chemical phenomenon, one finds that the distribution of various hydrocarbons between a nonpolar solvent and an aqueous phase containing tetramethyluric acid is dependent on the structure of the hydrocarbons (17). The more benzene rings the structure has, the more it is inclined to complexation with the purine derivative, and the more compact the molecule, the greater its complex-forming ability. If we remember that all carcinogens contain phenanthrene systems, it is interesting to note that phenanthrenes are better solubilized than anthracenes; but unfortunately, there is no apparent specificity for carcinogenic hydrocarbons; e.g., the fact that the complex formation is parallel to the number of rings does not reflect the observation mentioned before that a number of about five benzene rings appears optimal for carcinogenic activity.

The same disappointing conclusion (18) has been drawn by Pullman and co-workers from theoretical studies. Pullman (19) has calculated the energies of the highest filled and lowest empty energy levels of the various purines and pyrimidines; these are the basic data for our considerations, as the charge-transfer complex is formed by transition of an electron from the highest filled energy level of the donor to the lowest empty energy level of the acceptor. These two energies are, as Mulliken (20) first pointed out, related to the ionization potential of the donor I_D and to the electron affinity of the acceptor E_A, so that the energy E, comprising the binding energy of the complex and the transition energy to the excited state, is

$$E = I_D - E_A + \Delta$$

where Δ is a stabilization term which can be assumed to be constant in a series of similar chemical compounds. Table I shows that the purines can be expected to be good donors, but bad acceptors for electrons. The best of the purines is uric acid; methylation improves the donor properties, so that it could be expected that tetramethyluric acid, which we have used before, is, indeed, a very good donor. The pyrimidines, on the other hand, are only moderate donors and acceptors.

TABLE I

ELECTRON-DONOR AND ELECTRON ACCEPTOR PROPERTIES OF
PURINES and PYRIMIDINES[a]

Compound	K_{HFMO}	K_{LEMO}
Adenine	0.49	−0.87
Guanine	0.31	−1.05
Hypoxanthine	0.40	−0.88
Uric acid	0.17	−1.19
Uracil	0.60	−0.96
Thymine	0.51	−0.96
Cytosine	0.60	−0.80
5-Methylcytosine	0.53	−0.80
Barbituric acid	1.03	−1.30

[a] K, energy coefficients of the highest filled and lowest empty molecular orbitals; energy of the orbitals $E = \alpha + K\beta$.

Unfortunately, very few data are available on the ionization potentials (21) and even less on the electron affinities of the compounds concerned. We have undertaken, together with Dr. Pullman, to fill this gap and have recently employed a simple method for the determination of ionization potentials, using a mass spectrograph and measuring, in principle, the threshold energy required to extract the electron from the compound studied (22). It is interesting to note that the *sequence* of the compounds that emerges from these measurements is the same as that calculated by Pullman and co-workers (19), although there seems to be a constant difference of about 0.8 eV between the calculated and the experimental values for the ionization potentials. We are now trying to develop a good experimental method for the determination of the electron affinities.

The energies of the lowest empty molecular orbitals of the polycyclic hydrocarbons, that is their acceptor abilities, which express their electron affinities have been calculated by Streitwieser (23) (Table II). In accordance

TABLE II

ELECTRON-ACCEPTOR PROPERTIES OF POLYCYCLIC
AROMATIC HYDROCARBONS

Hydrocarbon	K_{LEMO}
Anthanthrene	0.291
Naphthacene	0.295
Perylene	0.307
3, 4-Benzopyrene	0.365
1, 2-Benzopyrene	0.497
Pyrene	0.445
Anthracene	0.414
1, 2-Benzanthracene	0.452
2, 3, 6, 7-Dibenzanthracene	0.437
1, 2, 5, 6-Dibenzanthracene	0.474
1, 2, 3, 4-Dibenzanthracene	0.492
Phenanthrene	0.605
Chrysene	0.520
3, 4-Benzophenanthrene	0.528

with the solubilization experiments we find that the acceptor ability of the hydrocarbons is a function of the number of rings and the compactness of the molecule. Again, there is no difference between carcinogenic and non-carcinogenic hydrocarbons; this is most clear for the dibenzanthracenes, of which only the 1,2,5,6-compound is carcinogenic, and for 3,4-benzopyrene which appears as a worse electron acceptor than the noncarcinogenic 1,2-isomer.

Apart from charge-transfer, there is another type of force that may bring about the first interaction between polycyclic hydrocarbons and the heterocyclic bases, namely, dispersion forces (24). These forces might explain in the solid state and perhaps also in solution the stacking of layers of the one component above and below the layer of the other (or the intercalation of the hydrocarbons in suitable free spaces in the structure of DNA). These dispersion forces can be calculated from the polarizabilities of the molecules concerned. Thus, Pullman has calculated the "dispersion interaction" E_L between benzopyrene and various purines and pyrimidines assuming a distance between the base layer and the hydrocarbon layer of 4 Å. The data obtained (Table III) show conclusively that there are purines and pyrimidines which will permit stacking more easily than others (again the purines are better than the pyrimidines, and the best is tetramethyluric acid), but it is obvious that this method too gives no explanation for the specificity of carcinogens: 1,2- and 3,4-benzopyrene and all the dibenzanthracenes have the same calculated polarizability, but do not have the same carcinogeneity.

TABLE III
POLARIZABILITY α OF PURINES AND PYRIMIDINES AND
DISPERSION INTERACTION E_L WITH 3, 4-BENZOPYRENE

Compound	α	E_L
Tetramethyluric acid	21.4	−14.8
Guanine	14.6	−11.8
Adenine	14.2	−11.2
Hypoxanthine	13.1	−10.3
Cytosine	11.1	− 8.9
Thymine	12.0	− 9.4
Uracil	10.0	− 8.0

However, we would like to add that the last word may not have been said in this respect either. The polarizability which is calculated from the atomic or bond refractions, is not always identical with the experimental value. There are cases of "exalted distortion polarization," which appear to be specific for certain hydrocarbons (25). This has been observed for dibiphenyleneethene and a number of fulvenes, and an attempt at a theoretical explanation has been made (26). In order to determine the polarizability of the molecules accurately, it will be necessary to measure the molecular refraction in the far infrared (27); until then, the possibility should not be completely excluded that dispersion forces can bind specific aromatic hydrocarbons and purine and pyrimidines bases together.

We would like to submit, however, that there is a possible explanation of the failure of the two assumptions we have discussed so far. Individual purine and pyrimidine bases may not constitute good model substances, and the interaction that we study may be related to the DNA molecule as a whole or at least to part of it—an oligonucleoside or oligonucleotide. It is interesting to note that according to the calculations of Pullman the pairs adenine-thymine and guanine-cytosine are better electron donors and electron acceptors than the individual purines and pyrimidines studied (K_{HFMO} 0.43 and 0.41; K_{LEMO} − 0.87 and − 0.78, respectively), the second pair being even better than the former. This new hypothesis makes it reasonable to contemplate that the polycyclic hydrocarbons attach themselves to a certain "active site" on the helix of the DNA which has the same geometrical size and shape as the hydrocarbons. This would make possible a specificity in size (and shape) for certain hydrocarbons, namely, the carcinogenic hydrocarbons.

Let us consider that point a little further. Huggins and Yang (28) have already drawn attention to the fact that the size and shape of the pair guanine-cytosine is at least very similar to that of the carcinogens 9,10-dimethyl-1,2-benzanthracene and 3,4-benzopyrene and that their thickness [3.6 Å] is also identical with that of the pair, so that these hydrocarbons can intercalate in the double-strand DNA without disrupting the sugar-phosphate side chain. The importance of the spatial structure of the hydrocarbons is also emphasized by the recent observation that there is only one substitution in the carcinogens which does not destroy their biological activity, namely, that made by fluorine. Without going into a detailed discussion of these facts, I would like to state that certain fluorinated carcinogens, prepared partly in M. S. Newman's and partly in our own laboratory, have been found to be strongly carcinogenic (29). Now, it is well known that fluorine is almost identical in size to the hydrogen atom (van der Waals radius 1.35 Å, as compared with 1.2 Å for hydrogen); replacement of the hydrogen by a fluorine atom will thus not affect the spatial structure of the molecule.

It is interesting to note in this context that the activity of thalidomide, which would reasonably be expected to affect the nucleic acids as carriers of the genetic properties of the cell, has also been ascribed (30) to the striking similarity of this molecule to the nucleoside deoxycytidine—the glutarimide moiety forms an angle with the phthalimide residue exactly as the deoxyribosyl radical does with the pyrimidinic part of the nucleoside molecule.

Another case of interest to this discussion is that of the interaction of DNA (or part of it) with steroids (31). Huggins and Yang (28) have also shown that testosterone, progesterone, and 17β-estradiol have a shape and size very similar to the pair guanine-cytosine, but their thickness (5–6 Å) is greater than that of the pair. Thus, intercalation is impossible without a distortion of the polymer chain. Nevertheless, these data appear to permit a molecular understanding of the following observations: when breast tumors, produced in 50- to 60-day-old albino rats by a single feeding of carcinogenic hydrocarbons, are treated with the steroid hormones mentioned, the number of fatal cases decreases and about 30% of the cancers disappear completely. It is not difficult to make a mechanical picture of the interaction of the components of this system.

If we compare the model which we have suggested with the interaction of such compounds as acridine orange, proflavine (32) or chloroquine (33) with DNA, we will find that, for instance, chloroquine (which is known to be active in its di-cationic form) can also be pictured as forming a sheath around the active site of the DNA molecule; thus the protective action of the anti-malarial drugs against denaturation can be explained mechanistically. The interaction of these drugs with DNA is an electrostatic one, whereas in the case of the hydrocarbons, forces arising from the *matching of the molecular surfaces* are responsible for the interaction.

The importance of the shape of the molecules for their activity in biological systems is not a new concept. In fact, the theory of Ehrlich concerning anti-metabolite action is based on this concept: sulfanilamide is toxic for cells requiring aminobenzoic acid because the shape of the molecules is so similar that the cell cannot discriminate between them, and the former enters biosynthetic pathways instead of the latter. It is interesting that 3-fluoro-4-aminobenzoic acid is equally an antimetabolite to aminobenzoic acid (34). We have already pointed out that the fluorine atom is so similar to hydrogen that its introduction does not change the shape of the molecule and therefore the response of the cell to it. The attempts to elucidate the active site of an enzyme by the study of specific inhibitors is another aspect of this principle of matching shapes and perhaps one should mention in conclusion another field which seems to lead to an analogous conclusion. Among the psychopharmacologically active substances we find a large number of tricyclic compounds containing such systems as phenothiazine, xanthene, thioxanthene, fluorene, dibenzocycloheptadiene, dibenzocycloheptatriene, and dibenzazepine. What could be more reasonable than to assume that this obvious similarity in structure reflects a similarity to the shape and size of the chemoreceptors on which these compounds act? Thus, the final answer to our problem may be found in studies of the behavior of the polycyclic hydrocarbons in the presence of synthetic oligonucleosides or oligonucleotides.

REFERENCES

1. F. Bergmann, *Cancer Res.* **2**, 660 (1942).
2. A. Pullman and B. Pullman, "La cancérisation par les substances chimiques et la structure des molecules." Masson, Paris, 1955.
3. C. W. Abell and C. Heidelberger, *Cancer Res.* **22**, 931 (1962). P. M. Bhargava, H. I. Hadler, and C. Heidelberger, *J. Am. Chem. Soc.* **77**, 2877 (1955); P. M. Bhargava and C. Heidelberger, *ibid.* **78**, 3671 (1956).
4. E. Boyland and B. Green, *Brit. J. Cancer* **16**, 347 and 507 (1962); *Biochem. J.* **83**, 12P (1962); **84**, 54P (1962); **87**, 14P (1963).
5. E. L. Tatum, *Ann. N. Y. Acad. Sci.* **49**, 87 (1947); L. C. Strong, *Chem. Abstr.* **45**, 6286 (1951); A. Graffi and D. Fritz, Naturwissenschaften **45**, 320 (1958).
6. P. Brookes and P. D. Lawley, *Nature* **202**, 781 (1964); *J. Cellular Comp. Physiol.* **64**, Suppl. **1**, 11 (1964); cf. C. Heidelberger, quoted by E. C. Miller and J. A. Miller, *Pharmacol, Rev.* **18**, 805 (1966).
7. E. de Maeyer and J. de Maeyer-Guignard, *Virology* **20**, 536 (1963); Science **146**, 650 (1964).
8. W.-T. Hsu, J. W. Moohr, and S. B. Weiss, *Proc. Natl. Acad. Sci. U.S.* **53**, 517 (1965).
9. J. M. Rice, *J. Am. Chem. Soc.* **86**, 1444 (1964).
10. P. O. P. Ts'o and P. Lu, *Proc. Natl. Acad. Sci. U.S.* **51**, 272 (1964).
11. R. Mason, *Nature* **181**, 820 (1958); *Brit. J. Cancer* **12**, 469 (1958); M.-J. Mantione and B. Pullman, *Compt. Rend.* **262**, 1492 (1966); A. Pullman and B. Pullman, *in* "Quantum Theory of Atoms, Molecules, and the Solid State " (P. O. Löwdin, ed.). Academic Press, New York, 1966.
12. G. Weber, *Biochem. J.*, **47**, 114 (1950); H. A. Harbury, K. F. La Noue, P. A. Loach, and R. M. Amick, *Proc. Natl. Acad. Sci. U.S.*, **45**, 1708 (1959); cf. D. A. Wadke and D. E. Guttman, *J. Pharm. Sci.* **54**, 1293 (1965).

13. B. Pullman, *Biochim. Biophys. Acta* **88**, 440 (1964); M. T. A. Behme and E. H. Cordes, *ibid.* **108**, 312 (1965); cf. W. Kersten, H. Kersten, and W. Szybalski, *Biochemistry* **5**, 236 (1966).

14. E. D. Bergmann and H. Weiler-Feilchenfeld, unpublished results (1967).

15. A. M. Liquori, B. deLerma, F. Ascoli, C. Botre, and M. Trasciatti, *J. Mol. Biol.* **5**, 521 (1962); E. Boyland and B. Green, *ibid.* **9**, 589 (1964); J. K. Ball, J. A. McCarter, and M. F. Smith, *Biochim. Biophys. Acta* **103**, 275 (1965).

16. B. C. Giovanella, L. E. McKinney, and C. Heidelberger, *J. Mol. Biol.* **8**, 20 (1964).

17. J. D. Mold, T. B. Walker, and L. G. Veasey, *Anal. Chem.* **35**, 2071 (1963); cf. the analogous study by G. Cilento and D. L. Sanioto, *Arch. Biochem. Biophys.* **110**, 133 (1965).

18. S. S. Epstein, I. Bulon, J. Koplan, M. Small, and N. Mantel, *Nature*, **204**, 750 (1964).

19. B. Pullman and A. Pullman, "Quantum Biochemistry," Wiley (Interscience), New York, 1963.

20. R. S. Mulliken, *J. Phys. Chem.* **56**, 801 (1952); *J. Am. Chem. Soc.* **74**, 811 (1952); *Rec. Trav. Chim.* **75**, 845 (1956); *J. Chim. Phys.* **61**, 20 (1964).

21. Ionization potentials have been determined in the solid state: D. R. Kearns and M. Calvin, *J. Chem. Phys.* **34**, 2026 (1961).

22. Ch. Lifschitz, E. D. Bergmann, and B. Pullman, *Tetrahedron Letters*, in press (1967).

23. A. Streitwieser, "Molecular Orbital Theory for Organic Chemists," Wiley, New York 1961.

24. B. Pullman, P. Claverie, and J. Caillet, *Science* **147**, 1305 (1965); *Compt. Rend.* **260**, 5915 (1965).

25. E. D. Bergmann, E. Fischer, and J. H. Jaffe, *J. Am. Chem. Soc.* **75**, 3230 (1953).

26. B. Pullman, quoted by E. D. Bergmann, *Progr. Org. Chem.* **3**, 81 (1955).

27. J. H. Jaffe, *J. Opt. Soc. Am.* **41**, 166 (1951).

28. C. Huggins and N. C. Yang, *Science* **137**, 257 (1962).

29. E. D. Bergmann, J. Blum, and A. Haddow, *Nature* **200**, 480 (1963); J. A. Miller and E. C. Miller, *Cancer Res.* **23**, 229 (1963); *Lab. Invest* **15**, 217 (1966); *Pharmacol. Rev.* **18**, 805 (1966).

30. S. Furberg, *Acta Chem. Scand.* **19**, 1266 (1965).

31. G. Molinari and G. F. Lata, *Arch. Biochem. Biophys.* **96**, 486 (1962).

32. G. G. Hammes and C. D. Hubbard, *J. Phys. Chem.* **70**, 2889 (1966).

33. S. N. Cohen and K. L. Yielding, *J. Biol. Chem.* **240**, 3123 (1965); R. L. O'Brien, J. L. Allison, and F. E. Hahn, *Federation Proc.* **25**, 558 (1966); R. Ladda and J. Arnold, *ibid.* **25**, 558 (1966).

34. O. Wyss, M. Rubin, and F. B. Strandskov, *Proc. Soc. Exptl. Biol. Med.* **52**, 855 (1943).

On the Solubilization of Aromatic Carcinogens by Purines and Pyrimidines[*]

JACQUELINE CAILLET AND BERNARD PULLMAN

Service de Biochimie Théorique
Institut de Biologie Physico- chimique
Paris, France

We have recently studied (B. Pullman *et al.*, 1965a,b) the nature of the intermolecular forces involved in the association of purines and pyrimidines with polybenzenoid hydrocarbons as manifested by the solubilization of the latter by the former, adopting the hypothesis of a "stacking" model of interaction strongly suggested by the work of Ts'o and collaborators (Ts'o and Lu, 1964) and exemplified also by the crystal structure of complexes formed between 1,3,7,9-tetramethyluric acid (TMU) and, for example, pyrene or benzpyrene (Damiani *et al.*, 1965, 1966).

We have shown that the calculation of the van der Waals-London inter-ation energies between a given carcinogen (3,4-benzpyrene) and a series of purines and pyrimidines accounts for the relative order of the solubilizing power of the heterocyclic bases.

In this chapter we are extending these calculations to the complementary evaluation of the interaction energies between a given purine, in fact, TMU, and a series of aromatic hydrocarbons. The calculations are being carried out in the same general scheme as the previous ones, although some refinements recently developed in our laboratory have been introduced. The most significant developments are the following.

(1) The calculations are carried out in the monopole approximation, account is being taken of the small net σ charges on the C and H atoms of the hydrocarbons. This refinement introduces a small electrostatic component ($E_{\rho\rho}$).

(2) The polarization ($E_{\rho z}$) and dispersion (E_L) energies are evaluated by considering bond polarizabilities instead of the total molecular polarizability. This refinement is important in the present case (huge molecules at small distances from each other) as it takes into account the appropriate bond distances of the atoms instead of a mean distance (see Claverie, this symposium).

(3) Repulsive (exchange forces) (E_R) operating particularly at short distance

* This research was supported by grant CR-66-236 of the Institut National de la Recherche Médicale (Intergroupe Cancer et Leucémie).

have been included by using Kitaygorodsky's semiempirical formula (see Claverie, this symposium). This refinement enables the determination of the equilibrium position for the interaction, which is found invariably for a separation of about 3.4 Å between the planes of the interacting molecules, provided that the van der Waals radii for the methyl groups of TMU are considered as those of their four constituent atoms and not those of the "group" (with a van der Waals radius of 2Å for the "methyl group" the predicted equilibrium distance would be about 3.7 Å). This refinement also enables the determination of the probable geometry of the association, which will be discussed separately.

(4) Finally, the interaction energies thus obtained are "corrected" by making a correction on the components involving the polarizabilities. Thus, the $E_{p\alpha}$ and E_L components are evaluated with reference to benzene and such a calculation does not take into account the increase of the π-polarizabilities with the decrease of the excitation energy in the series of the hydrocarbons. As there exists in this series of molecules a nearly linear relation between the excitation energy and the ionization potential, it seems reasonable to apply the sum $(E_{p\alpha} + E_L)$ a correction $K = I_{benzene}/I_{hydrocarbon}$ always greater than 1.

The results of the calculations are presented in Table I and compared with

TABLE I

INTERACTION ENERGIES OF TMU WITH AROMATIC HYDROCARBONS

Hydrocarbon	E_{pp}	$E_{p\alpha}$	E_L	E_R	E_{total}	E_{total} corrected	K_B/K_A
1, 2, 3, 4-Dibenzo-pyrene	−1.93	−1.11	−10.37	4.95	−8.46	−11.33	4.80
Anthanthrene	−1.34	−0.93	−8.72	4.79	−6.20	−9.67	4.80
Perylene	−1.35	−0.86	−7.80	4.25	−5.76	−8.54	3.32
1, 2-Benzopyrene	−1.77	−0.96	−9.23	4.51	−7.45	−9.74	3.25
3, 4-Benzopyrene	−1.38	−0.86	−8.64	4.48	−6.40	−8.61	2.63
Pyrene	−1.46	−0.78	−7.76	4.12	−5.88	−7.73	2.05
1, 2, 5, 6-Dibenzan-thracene	−1.09	−0.79	−8.23	3.88	−6.23	−8.16	1.98
Chrysene	−1.43	−0.77	−8.11	4.16	−6.15	−7.54	1.88
1, 2-Benzanthracene	−1.18	−0.80	−7.92	4.15	−5.75	−7.71	1.85
Phenanthrene	−1.57	−0.65	−7.13	3.77	−5.58	−6.76	1.76
Anthracene	−1.27	−0.78	−6.84	3.67	−5.22	−6.93	1.60

the experimental enhancement of solubility as determined by Mold et al. (1963) in their studies on the selective separation of polycyclic aromatic compounds by countercurrent distribution with a solvent system containing tetramethyluric acid. The extent of this increased solubility is given by the ratio of the distribution coefficients K_B/K_A, where K_A is the distribution coefficient in aqueous methanol containing TMU and K_B is that of the same system without TMU. A higher value of this ratio indicates greater solubilization by the purine.

The overall parallelism observed between the theoretical and the experimental results confirms the predominant role of the Van der Waals-London forces in this type of association. The calculations seem to underestimate somewhat, however, the ability for complexation of hydrocarbons with highly fused, compact structures like pyrene and perylene. It could be that the reason for such discrepancies resides in "exalted distortion polarizations" observed in this type of molecules (Bergmann et al., 1953).

The results also confirm the nonspecificity of this type of interaction with respect to carcinogenic activity and, thus, indicate that this interaction probably has no direct significance for the mechanism of carcinogenesis by this type of molecule. This viewpoint is in agreement with the general lines of K–L region theory of carcinogenesis (A. Pullman and B. Pullman, 1955; B. Pullman, 1964, 1965), which implies that carcinogenesis is produced by a strong specific chemical interaction of the hydrocarbons with an appropriate cellular receiver. Such strong specific chemical interactions have been observed both with proteins (e.g., Oliverio and Heidelberger, 1958) and nucleic acids (Brookes and Lawley, 1964; Brookes, 1966) and they correlate much better with the carcinogenic activity of the molecules. Whether it is the interaction with the proteins or the nucleic acids that is essential for carcinogenicity is, however, an unresolved question.

Note. Mold et al. (1963) have also investigated the solubilization of some heterocyclics by TMU. In relation to the problem of the role of van der Waals-London forces in charge-transfer complexes, as discussed in this symposium by Mantione, it may be useful to remark that the computed interaction energies decrease in the series carbazole > dibenzothiophene > dibenzofurane (-5.05, -4.95, and -4.45 kcal/mole, respectively) in agreement with the order of decreasing values of K_B/K_A.

REFERENCES

Bergmann, E. D., Fischer, E., and Jaffé, J. H. (1963). *J. Am. Chem. Soc.* **75**, 3230.

Brookes, P. (1966). *Cancer Res.* **26**, 1994.

Brookes, P., and Lawley, P. D. (1964). *Nature* **202**, 781.

Damiani, A., de Santis, P., Giglio, E., Liquori, A. M., Puliti, R., and Ripamonti, A. (1965). *Acta Cryst.* **19**, 340.

Damiani, A., Giglio, E., Liquori, A. M., Puliti, R., and Ripamonti, A. (1966). *J. Mol. Biol.* **20**, 211.

Mold, J. D., Walker, T. B., and Veasey, L. G. (1963), *Anal. Chem.* **35**, 2071.

Oliverio. V. T., and Heidelberger, C. (1958). *Cancer Res.* **18**, 1094.

Pullman, A., and Pullman, B. (1955). *Advan. Cancer Res.* **3**, 117.
Pullman, B. (1964). *J. Cellular Comp. Physiol.* **64**, Suppl. 1, 91.
Pullman, B. (1965). *In* "Molecular Biophysics" (B. Pullman and M. Weissbluth, eds.), p. 117. Academic Press, New York.
Pullman, B., Claverie, P., and Caillet, J. (1965a). *Science* **147**, 1305.
Pullman, B., Claverie, P., and Caillet, J. (1965b). *Compt. Rend.* **260**, 5925.
Ts'o, P.O.P., and Lu, P. (1964). *Proc. Natl. Acad. Sci. U.S.* **51**, 17.

The Interaction of Heterocyclic Compounds with DNA

D. O. JORDAN

Department of Physical and
Inorganic Chemistry
University of Adelaide
Adelaide, South Australia

I. Introduction

Studies of the interaction of nucleic acids with organic molecules and ions are not only of considerable biological interest in that they provide an opportunity for elucidating the molecular mechanism of the action of mutagens, but they also have added to our knowledge of the detailed secondary and tertiary structure of DNA. The aminoacridines act mutagenically on bacteriophage possibly by causing the insertion or deletion of a base pair in its DNA (Brenner *et al.*, 1961; Orgel and Brenner, 1961). Ethidium bromide (2,7-diamino-9-phenyl-10-ethylphenanthridinium bromide), [Fig. 1(V)] inhibits the synthesis of nucleic acids in a variety of organisms (Newton, 1957, 1963; Kerridge, 1958; Kandaswamy and Henderson, 1962; Tomchick and Mandel, 1964), while in cell-free systems it inhibits the DNA-dependent DNA polymerase and the RNA polymerase of *Escherichia coli* (Elliott, 1963; Waring, 1964). The addition of aminoacridines or ethidium bromide to circular polyoma DNA changes the structure from that of a superhelix to a circle (Vinograd, 1966; Crawford and Waring, 1967) and these compounds also markedly increase the thermal stability of long-chain DNA (Gersch and Jordan, 1965; Waring, 1966a).

II. The Experimental Study of the Interaction of DNA and Heterocyclic Compounds

The change in the spectra of heterocyclic compounds and dyes when nucleic acids or other biological or synthetic polyions are added to aqueous solutions is well known and has been used to study the nature of the binding process. These compounds in solution do not usually obey Beer's law and this behavior is attributed to the aggregation of the heterocyclic or dye ion at high concentrations. This aggregation is generally accompanied by a blue spectral shift (Michaelis, 1947). The spectral changes which occur in the presence of macroions are dependent upon the nature of the macroion and in the case of DNA and other double helical polyions also upon the ratio of the

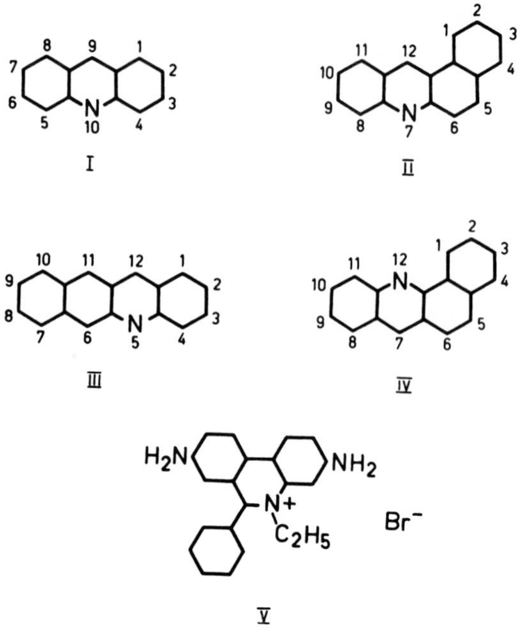

FIG. 1. The structure of acridine (I); benz[a]acridine (II); benz[b]acridine (III); benz[c]-
acridine (IV); and ethidium bromide (V). Proflavine is 3,6-diaminoacridine and acridine
orange is tetramethyl-3,6-diaminoacridine.

concentrations of DNA to heterocyclic or dye ion. With single chain polyanions,
e.g., heparin (Bradley and Wolf, 1959; Gersch, 1966) and isotactic or atactic
polystyrene sulfonate (Jordan et al., 1967), the aminoacridines show a blue
spectral shift. At low ratios of DNA to aminoacridine (Bradley and Wolf,
1959) or to rosaniline (Lawley, 1956) a blue spectral shift is also observed,
whereas at high ratios of DNA to aminoacridine a red spectral shift is
observed (Peacocke and Skerrett, 1956; Gersch, 1966). The blue spectral
shift has been interpreted as being caused by the interaction between adjacent
dye ions in the aggregate (Michaelis, 1947) or stacked on the surface of the
polyion (Bradley and Wolf, 1959). The red spectral shift, however, is usually
attributed to the interaction of the heterocyclic ring system of the bound
aminoacridine with the purine and pyrimidine bases of the DNA (Bradley
and Wolf, 1959; Drummond et al., 1965), a view which has been substantiated
by a study of the fluorescence spectra of the complex formed between proflavine and DNA (Weill and Calvin, 1963) and confirmed by the absence of
the red spectral shift when DNA interacts with 9-amino-1,2,3,4-tetrahydro-
acridine (Drummond et al., 1965).

The binding of heterocyclic compounds to DNA may be determined by a variety of techniques, the most usually employed being spectrophotometric titration and equilibrium dialysis. Typical binding curves for aminoacridines and aminobenzacridines (Fig. 1) to native and denatured DNA are shown in Fig. 2. From these data (particularly curves 1, 4, and 5) it has been inferred that there are two distinct stages in the binding process (Peacocke and Skerrett, 1956; Drummond *et al.*, 1965; Gersch, 1966; Jordan and Ellerton, 1967). In the primary stage, at low concentrations of aminoacridine (i.e., a high DNA/aminoacridine ratio), there is a strong binding of aminoacridine ions leading to the formation of a plateau in the curve for the variation of r, the number of ligand molecules bound per atom of DNA phosphorus, with c, the concentration of free, unbound ligand (Fig. 2). The value of r in this

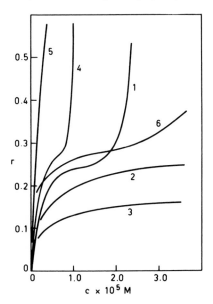

FIG. 2. Binding isotherms for 9-aminoacridine and 7-aminobenz[*c*]acridine in DNA. (1) 9-Aminoacridine, $\mu = 0.001$, pH 6.0, 20°C (Jordan and Ellerton, 1967). (2) 9-Aminoacridine, $\mu = 0.001$, pH 6.9, 25°C (Drummond *et al.*, (1965). (3) 9-Aminoacridine, $\mu = 0.1$, pH 6.9, 25°C (Drummond *et al.*, 1965). (4) 7-Aminobenz[*c*]acridine, $\mu = 0.1$, pH 6.2, 20°C (Jordan and Ellerton, 1967). (5) 7-Aminobenz[*c*]acridine, $\mu = 0.001$, pH 6.2, 20°C (Jordan and Ellerton, 1967). (6) 9-Aminoacridine and denatured DNA, $\mu = 0.001$, pH 6.9, 20°C (Drummond *et al.*, 1965).

plateau region is dependent upon the nature of the aminoacridine, the ionic strength, and the temperature. In the secondary stage, at higher concentrations of aminoacridine (i.e., low DNA/aminoacridine ratio), the value of r increases rapidly with increase of c and approaches or may exceed $r = 1.0$.

Not all systems show this second stage of weaker binding due to experimental difficulties in carrying out spectrophotometric titrations at high ligand concentrations. The secondary stage of the binding process is also markedly dependent upon ionic strength, the nature of the ligand, and the temperature. From the spectra of the bound heterocyclic molecules, it is concluded (Bradley and Wolf, 1959; Bradley and Felsenfeld, 1959; Stone and Bradley, 1961) that in the primary stage the ligand molecules are sufficiently separated from each other when bound to the DNA molecule not to interact with each other, whereas in the secondary stage the characteristic spectral shifts associated with the interaction between ligand molecules occur.

The general binding equation for p different types of binding sites,

$$r = \sum_{j=1}^{j=p} \frac{n_j k_j c}{1 + k_j c} \tag{1}$$

where r and c have the meaning given above, n_j is the number of binding sites of type j per macromolecule, and k_j is the corresponding binding constant, reduces for two types of binding sites to

$$r = \frac{n_1 k_1 c}{1 + k_1 c} + \frac{n_2 k_2 c}{1 + k_2 c} \tag{2}$$

where the subscripts 1 and 2 refer to the two binding stages. However in view of the probable nature of the binding process, particularly that of the secondary stage where the presence of the bound ligand molecule may lead to an increased affinity for further binding, the binding constants, k_1 and k_2, are unlikely to be independent of r and the binding isotherm cannot be used in the simple form represented by Eq. (2). An increase in the value of k_2 with r is the likely explanation of the rapid increase of r with c in the secondary stage of the binding process (Fig. 2).

The concept of two distinct binding mechanisms is further confirmed by studies of the binding of ethidium bromide to polyribonucleotides and mixtures of polyribonucleotides (Waring, 1966b). Ethidium bromide interacts with DNA to give spectral changes and binding curves very similar to those obtained with aminoacridines (Waring, 1965); comparison between the behavior of the two types of heterocyclic compounds is thus possible. Ethidium bromide binds to mixtures of polyribonucleotides to yield curves for the variation of r with c comparable with those obtained for the primary stage of the binding of aminoacridines (Fig. 3). The secondary stage of binding was not clearly observed although the results for poly(A + U) and poly(A + I) show that at the values of c attained, this binding stage was just beginning. The binding of ethidium bromide to polyadenylic acid, polyuridylic acid, and polyinosinic acid shows a behavior very different from that observed with nucleic acids and the polyribonucleotide mixtures (Fig. 3). The binding curves indicate little interaction at low concentrations of ligand showing that the strong primary binding process is absent. The increasing slope of the curves when binding occurs at high ligand concentrations is characteristic of the secondary binding stage in which the presence of the bound ligand molecules leads to an increased affinity for further binding. The secondary structure of the homopolymers in solution, however, is a little uncertain and they may consist of mixtures of single- and double- or triple-strand helices.

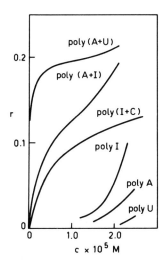

FIG. 3. The binding of ethidium bromide to polynucleotides and polynucleotide mixtures in 0.004 M Tris-HCl buffer, pH 7.9 (Waring, 1966b).

The binding of aminoacridines to DNA increases the thermal stability of DNA, the melting temperature, T_m, of the aminoacridine-DNA complex being appreciably higher than that of native DNA (Gersch and Jordan, 1965). The shape of the thermal denaturation curve for the complex is also different from that of native DNA, there being a gradual increase in optical density as typified by the region AB in Fig. 4 which is greater than that associated with the heating of native DNA in the same temperature range. This effect has been shown to be due to a decrease in dye binding with increase of temperature (Gersch, 1966), but the proportion of dye so released compared to the total amount bound is small. Whether the dye released was originally bound by the primary or secondary process has not been determined. Because of this effect, the melting temperature, T_m', has been defined as the temperature at which 50% hyperchromicity is attained from the onset of the rapid increase of the optical density above about 70°C until no further increase in optical density is recorded, i.e., in the region BC in Fig. 4. The value of T'_m increases with r (Fig. 5) reaching a maximum value at the value of r corresponding to the plateau in the binding isotherms (Kleinwächter and Koudelka, 1964; Gersch and Jordan, 1965). This suggests that the increased stability is caused mainly by the primary binding process and that the secondary binding has little effect on the thermal stability of the DNA-aminoacridine complex.

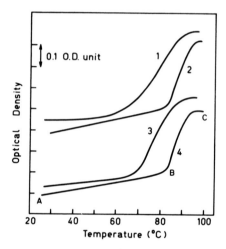

FIG. 4. Thermal denaturation curves for DNA-aminoacridine complexes. The values of r are close to 0.13 in all cases. (1) DNA-proflavine in 0.001 M NaCl. (2) DNA-proflavine in 0.1 M NaCl. (3) DNA-3-aminoacridine in 0.001 M NaCl. (4) DNA-3-aminoacridine in 0.1 M NaCl. T_m for native DNA, 82°C in 0.1 M NaCl, 55°C in 0.001 M NaCl (Gersch and Jordan, 1965).

Direct evidence that the aminoacridine molecules bound by the primary binding process are intimately associated with the double helical structure of DNA arises from the binding studies of Chambron et al. (1966a,b), who have determined the binding of proflavine to DNA at a range of temperatures above and below the thermal denaturation temperature, T'_m. There is a small but steady decrease in r as the temperature is increased (Fig. 6), the release of aminoacridine over this temperature range confirming the interpretation of the thermal denaturation curve (Fig. 4). The main release of proflavine occurs, however, over a short range of temperature corresponding to the melting of the DNA-proflavine complex. The release of the proflavine is clearly a cooperative phenomenon, and the double helical structure of DNA evidently is essential for the primary binding process to occur.

The binding of heterocyclic ions to DNA in concentration ranges corresponding to the primary binding process brings about an increase in the

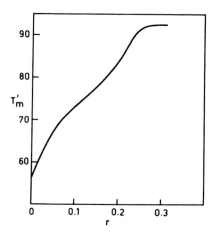

FIG. 5. Variation of T'_m with r for proflavine in 0.001 M NaCl. The secondary binding stage starts at about $r = 0.25$ (Gersch and Jordan, 1965).

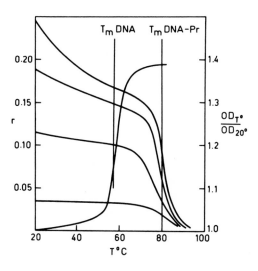

FIG. 6. The variation of r with temperature for the DNA-proflavine complex $\mu = 0.01$ (Chambron et al., 1966b).

specific viscosity (Lerman, 1961; Drummond *et al.*, 1966), which is dependent on the particular aminoacridine ion present (Gersch, 1966). The sedimentation coefficient of DNA shows a marked decrease on complexing the DNA with aminoacridines (Lerman, 1961; Gersch, 1966), which would correspond to a decrease in the mass per unit length of DNA on complexing with the amino-acridine. From these two observations Lerman (1961) concluded that there was an extension of the molecule on binding of the heterocyclic ion to DNA. More direct evidence for this extension of the DNA molecule has been obtained from X-ray studies (Luzzati *et al.*, 1961; Neville and Davies, 1966), autoradiography (Cairns, 1962), and light scattering studies (Mauss *et al.*, 1967).

III. Models for the Complex Formed by the Binding of Heterocyclic Molecules and Ions to DNA

Michaelis (1947) had hoped to elucidate the structure of nucleic acids by dye-binding studies and although this attempt was not successful his comments on the structure of the DNA-dye complex are worth repeating. "Nucleic acid, whether of high or low molecular weight, may be imagined to consist of strings or bundles of nucleotides arranged in such a way that the pyrimidine, or purine rings lie parallel to each other, connected by phosphate groups; the dye molecules attached to the negatively charged end of the phosphate group. Each dye cation combined with one phosphate group must lie in the space between the planes of the pyrimidine or purine rings, and so they are prevented from approaching each other in such a way as to interfere optically with each other and from exhibiting the spectrum of a higher dyestuff aggregate." This structure was reiterated by Oster (1951) and again by Heilweil and Van Winkle (1955). Such interaction between the planar heterocyclic ions and the planar bases of the DNA molecule is not compatible with the Watson-Crick model, however, since in native DNA the bases are in close van der Waals contact. Some distortion of the native double helix is therefore necessary and Lerman (1961, 1963) suggested that if the DNA molecule was extended, the planar heterocyclic ions could intercalate between successive base pairs. This has the result that the native DNA helix [with orientations of base pairs with respect to helix and dyad axes as pro-posted by Langridge *et al.* (1960) in their model 3] is extended so that the distance between two adjacent pairs is increased from 3.36 to 6.72 Å. This expansion of the helix according to Lerman involves an uncoiling process through 45° and the original right-handed helix, in which the angle between adjacent base pairs is 36°, becomes left-handed with an angle of 9°. We shall call this model Ia. Fuller and Waring (1964), while intercalating the hetero-cyclic ion in an identical way to Lerman, have found it necessary, on the basis of more accurate models, only to make an uncoiling of 12°. The DNA helix thus remains right-handed with an angle between adjacent base pairs (6.72 Å apart) of 24°. This model we shall call Ib.

In both models Ia and Ib the arrangement of the acridine molecule is as originally proposed by Lerman (1961, 1963) and as shown in Fig. 7. Pritchard *et al.* (1966) have recently proposed a modified intercalation model, which we shall call Ic, in which an aminoacridine does not interact with two hydrogen-bonded base pairs, but with two adjacent bases on the same polynucleotide chain (Fig. 7). The increased spacing between the two adjacent base pairs is

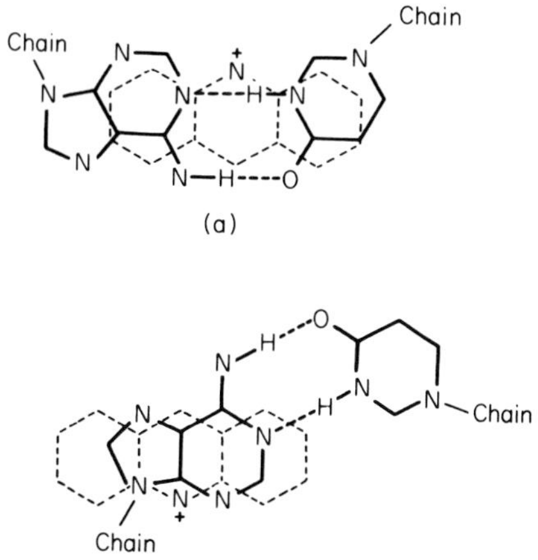

(a)

(b)

FIG. 7. (a) The relative position of an acridine nucleus and a base pair in the intercalation model according to Lerman (1961). (b) The relative position of the acridine nucleus and the purine of a base pair in the intercalation model of Pritchard *et al.* (1966).

approximately the same as in models Ia and Ib and the degree of uncoiling presumably 12° as in model Ib. An important feature of this model according to the authors is that, unlike models Ia and Ib, a negatively charged oxygen atom on the phosphate group between the two bases can swing in toward the inside of the chain and take up a position adjacent to the positively charged ring nitrogen of the aminoacridinium ion.

In an alternate model for the DNA-aminoacridine complex, the aminoacridine is associated with the phosphate group on the outside of the double helix, the dimensions of the helix remaining essentially unchanged. As originally described by Bradley and Wolf (1959), the dye molecules were stacked perpendicularly to the helix axis. This model we shall call model II. This model has been refined by Mason and McCaffery (1964) who, on the basis of conclusions drawn from optical rotation measurements of streaming solutions, consider the aminoacridine to be attached through the ring NH group to a phosphate group, the heterocyclic ring being oriented at an angle between 45° and 90° to the helix axis. In the intercalated models (Ia, Ib, and Ic) external, edgewise binding of the heterocyclic molecules, as in model II, is clearly also possible.

IV. Free Energy Calculations for Models Based on Intercalation and External, Edgewise Binding of Aminoacridines

A. General Principles

The total free energy of interaction between two molecules, charged or uncharged, having permanent and induced dipole moments, is given by

$$F_{total} = F_{ES} + F_L \tag{3}$$

where F_{ES} is the electrostatic free energy from interactions between charges, permanent, and induced dipoles, and F_L is the London dispersion energy. F_{ES} may be considered as the total of five separate interactions:

$$F_{ES} = F_{\rho\rho} + F_{\rho\mu} + F_{\rho\alpha} + F_{\mu\mu} + F_{\mu\alpha} \tag{4}$$

where ρ, μ, and α represent charge, dipole moment, and polarizability, respectively. The various interaction energies of Eq. (3) and (4) are given by:

$$F_{\rho\rho} = \sum_i \sum_{j>1} \frac{\rho_i \rho_j}{\varepsilon_{ij} R_{ij}} \tag{5}$$

$$F_{\rho\mu} = \sum_i \sum_{j \neq i} \frac{1}{\varepsilon_{ij}} \rho_i \mu_j C_{i,j} \tag{6}$$

$$F_{\rho\alpha} = -\tfrac{1}{2} \sum_i \sum_{j \neq i} \sum_{l=1}^{3} \frac{1}{\varepsilon_{ij}} \rho_i^2 \alpha_{jl} C_{i,jl} \tag{7}$$

$$F_{\mu\mu} = \sum_i \sum_{j>1} \frac{1}{\varepsilon_{ij}} \mu_i \mu_j G_{i,j} \tag{8}$$

$$F_{\mu\alpha} = -\tfrac{1}{2} \sum_i \sum_{j \neq i} \sum_{l=1}^{3} \frac{1}{\varepsilon_{ij}} \mu_i^2 \alpha_{jl} G_{i,jl}^2 \tag{9}$$

$$F_L = -\frac{h}{4} \sum_i \sum_{j>i} \sum_{l=1}^{3} \sum_{m=1}^{3} \frac{v_i v_j}{v_i + v_j} \frac{\alpha_{il} \alpha_{jm} G_{il,jm}^2}{\varepsilon_{ij}} \tag{10}$$

where $C_{i,j}$ and $G_{i,j}$ are geometric factors involving the unit vectors $\mathbf{e}_i$ and $\mathbf{e}_j$ of the group dipoles μ_i and μ_j and are given for two interacting molecules or ions by:

$$C_{i,j} = \frac{1}{R_{ij}^3}\,\mathbf{e}_j\,\mathbf{R}_{ij} \tag{11}$$

$$G_{i,j} = \frac{1}{R_{ij}^3}\left[\mathbf{e}_i \cdot \mathbf{e}_j - \frac{3}{R_{ij}^2}(\mathbf{e}_i \cdot \mathbf{R}_{ij})(\mathbf{e}_j \cdot \mathbf{R}_{ij})\right] \tag{12}$$

where the distance vector between i and j is denoted by $\mathbf{R}_{ij}$; ε_{ij} is the dielectric constant, α_{jl} the component of the polarizability along the principal polarization axis l, and v_i is the characteristic frequency.

Equations (6)–(10) were used by DeVoe and Tinoco (1962) to determine the free energy of the native DNA helix which was found to be dependent both on the base composition and the base sequence. The calculations of the interactions for the different models of the complex formed between DNA and aminoacridines have been made using similar methods and approximations (Gersch and Jordan, 1965; Jordan and Ellerton, 1967). The unit vectors representing the dipole moments and polarizability components were considered to be at the geometric centre of each molecule. Quadrupole moments were not included in the calculations. For the bases of DNA, the values of μ_i, α_i, and hv_i used were those given by DeVoe and Tinoco (1962) and are given in Table I. For the aminoacridines, hv_i was found to be 200 kcal

TABLE I

GROUP PROPERTIES OF DNA BASES AND THE PROTONATED FORMS OF
THE AMINOACRIDINES USED FOR ENERGY CALCULATIONS

Molecule	μ_i (D)	θ_i (°)[a]	α_i (Å^3)	hv_i (kcal mole^{-1})
Adenine	2.8	88	14	200
Thymine	3.5	33	11	240
Guanine	6.9	324	14	200
Cytosine	8.0	108	11	240
Proflavine	4.5	180	20	200
2-Aminoacridine	0.8	235	18	200
3-Aminoacridine	2.6	244	18	200
9-Aminoacridine	3.0	0	18	200
12-Aminobenz(a)acridine	3.0	94	21	200
2-Aminobenz(b)acridine	6.6	276	21	200
7-Aminobenz(c)acridine	6.1	40	21	200

[a] Values of θ for the aminoacridines refer to the rotation from the OX axis in an anti-clockwise direction, where the positive direction of the OX axis bisects the —CNC— bond angle directed away from the ring.

mole^{-1} from experimental values of the dispersion of benzene and pyridine. The average group polarizability, α_i, was estimated to be 18 Å^3 for the monoaminoacridines, 20 Å^3 for proflavine (3,6-diaminoacridine), and 21 Å^3 for the aminobenzacridines from values of atomic refractions (Fajans, 1959). It is not possible to measure the dipole moments of a charged molecule, and the dipole moments of the aminoacridines in the protonated form (which is the form in which the aminoacridines studied exist in solution at the pH used) were calculated in the following way (Jordan and Ellerton, 1967). The bond and hybrid moments used by DeVoe and Tinoco (1962) for the various atoms were used to give the σ moments. The π moments were obtained using calculated electronic distributions and the method described by Daudel *et al.* (1959). The center of mass of the molecule was taken as the origin. The σ and π moments so obtained were added vectorially to give the values recorded in Table I. In the calculations no charge-charge interactions have been considered and $F_{\rho\rho}$ has, therefore, been neglected. It has been assumed that the association of gegenions with DNA and aminoacridine is such as to neutralize these charges. This would certainly be true at high ionic strengths, but at low ionic strengths the value of $F_{\rho\rho}$ may not be negligible. This assumption was, in effect, also made by DeVoe and Tinoco (1962).

B. The Intercalation Model

The models adopted for these calculations were those suggested by Lerman (1961, 1963), model Ia, and by Fuller and Waring (1964), model Ib. In both these models the helix is extended so that the distance between two adjacent base pairs is 6.72 Å, but in model Ia the angle between the adjacent base pairs is 9° with a left-handed rotation, whereas for model Ib the angle is 24° with a right-handed rotation. The aminoacridine is then inserted between the base pairs in such a way that the helix axis passes through the center of the acridine molecule. The angle between an aminoacridine molecule and the base pair immediately below it will be 4.5° with a left-handed rotation in model Ia and 12° with a right-handed rotation in model Ib. The case considered is that in which an aminoacridine molecule is intercalated between each base pair, i.e., the case for maximum binding by the intercalation mechanism, when $r = 0.5$.

The geometric factors $C_{i,j}$ and $G_{i,j}$ necessary for the calculation of $F_{\rho\mu}$, $F_{\rho\alpha}$, $F_{\mu\mu}$, $F_{\mu\alpha}$, and F_L were calculated at a number of angles of orientation of the dipole-moment unit vector of the intercalated molecule (the angle θ in Fig. 8). Thus, for proflavine, where the dipole moment vector is oriented at 180° with respect to the pyridine dipole moment vector, the possible values of θ are 175.5° and 355.5° for model Ia and 192° and 12° for model Ib. In addition to calculating the free energy of interaction of the aminoacridine and the adjacent base pairs, calculations have also been made for the free energy of interaction between a given aminoacridine and the next intercalated aminoacridine 6.72 Å above and the base pair 10.08 Å above. The free energy arising from interactions between base pairs 6.72 Å apart has not been included since DeVoe and Tinoco (1962) found that the contribution to the free energy from interactions between base pairs more than 3.36 Å apart was negligible.

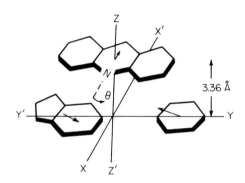

FIG. 8. Relative positions of an adenine-thymine base pair and an aminoacridine molecule, as used in the free energy calculations for the intercalated models Ia and Ib. The three dipole moment vectors are represented by arrows and are not drawn to scale (Gersch and Jordan, 1965).

For the intercalated model, the aminoacridine and adjacent base pairs are in van der Waals contact. The value of the dielectric constant, ε_{ij}, has therefore been taken as 1.0, as did DeVoe and Tinoco (1962) for DNA. Solvation of the helix would reduce the contribution of the electrostatic interactions by increasing the value of ε_{ij} and, owing to the presence of the charged dye ions within the helix, this effect is likely to be greater for the intercalated models than for DNA.

The repeating unit for DNA is one base pair. The energy calculations for a repeating unit may be considered in either of two ways. Either one-half the sum of the eight interactions between the given base pair and that above and that below, together with the interaction energy between the bases of the given base pair, is obtained; or the sum of the four interactions between the given base pair and that above, together with the average base-pair energies of the two base pairs involved, is taken. The latter definition of the repeating unit involves less complications when base-pair sequences are considered and is the one used here.

The repeating unit for DNA is shown in Fig. 9a. For the purpose of the comparison with the results obtained here, it has been necessary to modify the method of giving the values of the total free energies for the base pairs of DNA used by DeVoe and Tinoco (1962) so that the free energies are quoted as kilocalories per repeating unit. The values quoted here for DNA are different from those given by DeVoe and Tinoco (1962), but only because of the need to change the definition of the repeating unit. The repeating unit for the intercalation model (I) is shown in Fig. 9b.

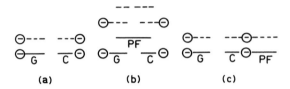

(a)   (b)   (c)

FIG. 9. Diagrammatic representation of the repeating units for (a) DNA; (b) intercalated models Ia and Ib; (c) external binding model IIa. All interactions involving the molecules represented by a solid line with the molecules represented by a broken line are included in the total free energy calculations for one repeating unit. PF represents proflavine; G and C guanine and cytosine, respectively (Gersch and Jordan, 1965).

The values of $F_{\mu\mu}$ and $F_{\mu\alpha}$ are dependent upon the orientation of the dipole moment vector of the intercalated molecule. In Fig. 10 is given the variation of $F_{\mu\mu}$, $F_{\mu\alpha}$, and $(F_{\mu\mu} + F_{\mu\alpha})$ with orientation for intercalation with AT and GC base pairs. For convenience, the dipole moment of the intercalated molecule is taken as 1.0 D and α_i as 20 Å^3. The values presented in the tables were obtained by combining the value of the dipole moment of the inter-calated molecule with the values of $F_{\mu\mu}$ and $F_{\mu\alpha}$ obtained from Fig. 10 at the required angle. F_L, $F_{\rho\mu}$, and $F_{\rho\alpha}$ are invariant with θ. The values for the inter-action free energies between the bases of the AT and GC base pairs were taken as 0.2 and -3.9 kcal per base pair (DeVoe and Tinoco, 1962).

The most extensive calculations have been made for the Lerman (1961, 1963) intercalated model, Ia, with proflavine as the intercalated molecule. Less extensive data have been obtained for the monoaminoacridines and the monoaminobenzacridines for both model Ia and Ib. In Table II are given the free energy values for the interaction of proflavine with nearest-neighbor base pairs according to the model Ia and a significant dependence on the nature of the base pair is observed. The energies of interaction of a proflavine molecule with a neighboring intercalated proflavine molecule and with the base pair 10.08 Å above or below the given intercalated molecule are quite small (Tables III and IV) in comparison to the interaction with adjacent base pairs (Table II).

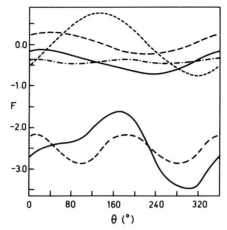

Fig. 10. The free energy (F) of dipole-dipole and dipole-induced dipole interactions (in kilocalories per base pair) for the intercalated molecule as a function of θ, the angle of orientation of the dipole moment of the intercalated molecule. The values obtained are for the intercalated molecule above *or* below the base pair indicated in parentheses. Upper curves: $F_{\mu\mu}(AT)$, $---$; $F_{\mu\alpha}(AT)$, $-\cdot-$; $F_{\mu\mu} + F_{\mu\alpha}(AT)$, $—$; $F_{\mu\mu}(GC)$, $----$. Lower curves: $F_{\mu\alpha}(GC)$, $---$; $F_{\mu\mu} + F_{\mu\alpha}(GC)$, $—$. Dipole moment of intercalated molecule taken as 1.0 D and α_i as 20 Å^3. (Gersch and Jordan, 1965).

TABLE II

CALCULATED FREE ENERGY VALUES FOR THE INTERACTION OF PROFLAVINE WITH NEAREST-NEIGHBOR BASE PAIRS FOR THE INTERCALATED MODEL Ia[a]

Base pair[b]	$F_{\mu\mu}$	$F_{\mu\alpha}$	F_L	$F_{\rho\mu}$	$F_{\rho\alpha}$	F_{total}
G .... C proflavine†	5.4	−5.6	−13.8	−11.8	−16.6	−46.3
G C	(−5.4)	(−5.6)	(−13.8)	(−11.8)	(−16.6)	(−57.1)
A .... T proflavine†	−1.8	−2.0	−13.8	−26.6	−16.6	−60.6
A .... T	(1.8)	(−2.0)	(−13.8)	(−26.6)	(−16.6)	(−57.0)
C .... G proflavine†	4.5	−5.2	−13.8	−11.8	−16.6	−46.8
C .... G	(−4.5)	(−5.2)	(−13.8)	(−11.8)	(−16.6)	(−55.8)
T .... A proflavine†	−1.8	−2.0	−13.8	−26.6	−16.6	−60.6
T .... A	(1.8)	(−2.0)	(−13.8)	(−26.6)	(−16.6)	(−57.0)

[a] Data given in kilocalories per base pair. $\varepsilon_{ij} = 1.0$.

[b] Dagger indicates that the proflavine molecule is considered to interact with all other molecules in the unit shown. The values within and without parentheses refer to values of the angle θ of 355.5° and 175.5°, respectively.

TABLE III

CALCULATED FREE ENERGY VALUES FOR THE INTERACTION OF ADJACENT
PROFLAVINE MOLECULE FOR THE INTERCALATED MODEL Ia[a]

Molecule[b]	$F_{\mu\mu}$	$F_{\mu\alpha}$	F_L	$F_{\rho\mu}$	$F_{\rho\alpha}$	F_{total}
Proflavine† base pair	0.86	−0.17	−0.47	0.0	−0.98	−0.8
Proflavine†	(−0.86)	(−0.17)	(−0.47)	(0.0)	(−0.98)	(−2.5)

[a] Data given in kilocalories per repeating unit. $\varepsilon_{ij} = 1.0$.
[b] Dagger indicates that only interactions between these two molecules are considered. The values within and without parentheses refer to angles of 171° and 351°, respectively, between proflavine dipole moment vectors.

TABLE IV

CALCULATED FREE ENERGY VALUES FOR THE INTERACTION OF PROFLAVINE
WITH THE BASE PAIR 10.08 Å ABOVE OR BELOW[a]

Molecule[b]	$F_{\mu\mu}$	$F_{\mu\alpha}$	F_L	$F_{\rho\mu}$	$F_{\rho\alpha}$	F_{total}
Proflavine† base pair dye G .... C†	0.01	−0.01	−0.03	−0.65	−0.17	−0.9
Proflavine† base pair dye A .... T†	−0.07	0.0	−0.03	−1.01	−0.17	−1.3
Proflavine† base pair dye C .... G†	0.06	−0.01	−0.03	−0.65	−0.17	−0.8
Proflavine† base pair dye T .... A†	0.0	0.0	−0.03	−1.01	−0.17	−1.2

[a] Data given in kilocalories per repeating unit. $\varepsilon_{ij} = 1.0$.
[b] Dagger indicates that only interactions between this proflavine and this base pair are considered. The values are for $\theta = 166.5°$ only.

C. The Model for External, Edgewise Binding

The model used for the calculation of the geometric factors for amino-acridines bound to the outside of the helix is shown as a projection on the XY plane in Fig. 11. If the aminoacridine molecules are oriented at the same angle to the helix axis, but not perpendicular to it, as shown in Fig. 11, there will be no effect on the values of the free energy obtained provided that the angle between the dipole moment vectors remains at 36°. The free energy calculations indicate that a variation in the angle of orientation of the two molecules with respect to each other within the limits imposed by the model brings about a change in F_{total} of no more than ± 0.5 kcal per repeating unit. The repeating unit corresponding to those chosen for DNA and the inter-calated model is shown in Fig. 9c.

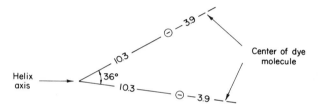

FIG. 11. Projection on the XY plane showing atomic distances used in the free energy calculations for model IIa, involving external, edgewise binding. The arrows represent the dipole moment vectors, not drawn to scale, of the charge aminoacridine molecules (Gersch and Jordan, 1965).

Since only the situation for which $r = 0.5$ has been considered for the inter-calated model (i.e., the case of maximum binding by intercalation), the case where $r = 0.5$ has also been considered for the external edgewise-binding model. This does not represent the situation for maximum binding which would correspond to $r = 1.0$ and leads to the possibility of two models, that where the aminoacridine is bound to all the phosphate groups on one strand of the double helix (model IIa) or that in which the aminoacridines are bound, on the average, to every alternate group on both helices. The model, IIa, has been that considered in these calculations, since interaction between adjacent ligand molecules will be greater for this model.

The charge on the phosphate group was considered as a point charge 10.3 Å from the helix axis, giving rise to charge-dipole and charge-polariz-ability interactions with the bound molecules. It has been assumed that no marked change of the charge distribution in the ligand molecule occurs by induction through the bonds and, hence, the dipole moment of the protonated ligand molecule is taken to be the same as that used for the intercalated model.

For this model (IIa or IIb) association of solvent molecules with bound aminoacridines will occur and will thereby increase the effective dielectric constant. The charge separation between the negative phosphate group and the positive bound aminoacridinium ion in the complex is 3.9 Å (Fig. 11). The values of 35 and 10 for ε_{ij} were calculated using the two models described by Hasted *et al.* (1948). These are taken as limiting values for the effective dielectric constant, now denoted ε'_{ij}, used in the calculations of the interaction between the charge on the phosphate group and the dipole moment and polarizability of the bound molecule. The charge separation between bound adjacent ligand positive charges is about 9.0 Å (model IIa) and a value of ε_{ij} of 70 was taken; for model IIb the corresponding charge separation is 11.7 Å and ε_{ij} will have a value close to 78.

The free energy of interaction for model IIa given in Table V is invariant

TABLE V

CALCULATED FREE ENERGY VALUES FOR EXTERNAL, EDGEWISE BINDING
OF PROFLAVINE (MODEL IIa)[a]

Molecule[b]	$F_{\mu\mu}$	$F_{\mu\alpha}$	F_L	$F_{\rho\mu}$	$F_{\rho\alpha}$	F_{total}
—Proflavine† —Proflavine†	0.06	−0.02	−0.10	−3.13	−0.48	−0.05 ($\varepsilon_{ij} = 70$)
—† — Proflavine†				−2.43	−0.54	−0.04 ($\varepsilon_{ij} = 70$)
— —† Proflavine†				−20.4	−17.2	−3.8 ($\varepsilon_{ij} = 10$) −1.1 ($\varepsilon_{ij} = 35$)

[a] Data given in kilocalories per repeating unit.

[b] — represents the negative phosphate group; a dagger indicates that only interactions between these molecules are considered. All values are for $\varepsilon_{ij} = 1.0$ unless otherwise stated.

with base composition, and the major contribution to F_{total} arises from the $F_{\rho\mu}$ and $F_{\rho\alpha}$ terms. The contribution of $F_{\mu\mu}$, $F_{\mu\alpha}$, and F_L is almost negligible, a result which is due to the considerable distance between the bound molecules (9.0 Å) and the high value of ε_{ij} which has to be used (70). Should additional stacking of the dyes occur with intercalation of aminoacridine molecules between molecules bound to adjacent phosphate groups, and this is geometrically possible, the contribution of $F_{\mu\mu}$, $F_{\mu\alpha}$, and F_L would increase markedly. This would, however, indicate that the maximum value of r for this type of binding is 2.0, a value which has not been approached in binding studies even at high ligand concentrations.

V. Discussion

The decrease in the free energy, ΔF, for the reaction

DNA repeating unit + aminoacridine → [DNA repeating unit–aminoacridine]

for different base sequences for the intercalated model, Ia, and for the external, edgewise-binding model, IIa, is given in Table VI for the interaction between DNA and proflavine. Similar calculations have been made for the intercalated model, Ib, but the change in angle between base pair and intercalated molecule from $-4.5°$ to $12°$ does not significantly affect the

TABLE VI

COMPARISON OF CALCULATED FREE ENERGY VALUES FOR NATIVE DNA, PROFLAVINE INTERCALATED IN DNA (MODEL Ia), AND EXTERNAL, EDGEWISE BINDING OF PROFLAVINE (MODEL IIa)[a]

Repeating unit	Native DNA (DeVoe and Tinoco, 1962); $\varepsilon_{ij} = 1.0$	Intercalated model Ia; $\varepsilon_{ij} = 1.0$		External attachment model IIa			
				$\varepsilon'_{ij} = 10.0$		$\varepsilon'_{ij} = 35$	
	F_{total}	F_{total}	ΔF	F_{total}	ΔF	F_{total}	ΔF
CG GC	-35.7	-48.3	-12.6	-39.6	-3.9	-36.9	-1.2
GC GC	-19.9	-48.0	-28.1	-23.8	-3.9	-21.1	-1.2
TA CG	-20.5	-55.3	-34.8	-24.4	-3.9	-21.7	-1.2
AT CG	-13.7	-55.3	-41.6	-17.6	-3.9	-14.9	-1.2
AT GC	-12.9	-55.2	-42.3	-16.8	-3.9	-14.1	-1.2
TA GC	-12.5	-55.2	-42.7	-16.4	-3.9	-13.7	-1.2
GC CG	-7.5	-48.2	-40.7	-11.4	-3.9	-8.7	-1.2
TA AT	-10.6	-62.7	-52.1	-14.5	-3.9	-11.8	-1.2
AT AT	-10.2	-62.7	-52.5	-14.1	-3.9	-11.4	-1.2
AT TA	-3.4	-62.6	-59.2	-7.3	-3.9	-4.6	-1.2

[a] Data given in kilocalories per repeating unit.

relative sequence of free energy change for the different base pairs (Jordan and Ellerton, 1967). The values given in Table VI show that for the intercalated model, the order of increased stabilization is AT : TA > AT : AT > TA : AT > TA : GC > AT : GC > AT : CG > GC : CG > TA : CG > GC : GC > CG : GC. Thus, the base sequence with the highest free energy in native DNA is stabilized the greatest in the DNA-proflavine complex. For model IIa, however, with $\varepsilon'_{ij} = 10$, ΔF is only -3.9 kcal per repeating unit and this value is independent of the base sequence. If the bound proflavine molecule is tilted inward toward a base, ΔF becomes -4.5 or -4.9 kcal per repeating unit depending on whether the proflavine molecule is bound to an A or T base or to a G or C base, respectively. Tilting of the bound molecules thus implies that GC sites bind slightly more strongly than AT sites. The experimental evidence obtained by Tubbs et al. (1964) from measurements of fluorescence quenching of DNA-acriflavine solutions shows that the relative affinity of acriflavine for binding sites in DNA is AT : AT > AT : GC > GC : GC, in agreement with the sequence predicted by the calculations for the intercalated model, Ia. The dependence on base content of the difference, ΔT_m, between the T_m of the DNA-acridine orange complex and that for the DNA alone for samples of DNA with different base composition (Kleinwächter and Koudelka, 1964) is also in agreement with the results given in Table VI, ΔT_m increasing with rise in AT content.

TABLE VII

FREE ENERGY CHANGE ACCOMPANYING INTERCALATION OF AMINOACRIDINES IN DNA ACCORDING TO MODEL Ia[a]

Molecule	GC : GC	θ (degrees)	AT : AT	θ (degrees)
Proflavine	-38.1	355.5	-51.7	177.5
2-Aminoacridine	-30.3	230.9	-47.7	230.9
3-Aminoacridine	-32.0	239.6	-49.0	239.6
9-Aminoacridine	-33.6	355.5	-48.8	175.5
12-Aminobenz(a)acridine	-37.6	269.5	-51.1	269.5
2-Aminobenz(b)acridine	-43.6	271.3	-54.5	271.3
7-Aminobenz(c)acridine	-36.5	35.3	-54.9	215.3

[a] The angle between the aminoacridine and the adjacent base pairs is 4.5°. (The angle giving the most favorable value of ΔF has been taken in each case.) Data given in kilocalories per repeating unit.

In Tables VII and VIII are given the calculated values of the free energy change for the intercalation of different aminoacridines with GC : GC and AT : AT base pairs (Jordan and Ellerton, 1967). The orientation of the intercalated molecule giving the maximum free energy change has been taken in

TABLE VIII

FREE ENERGY CHANGE ACCOMPANYING INTERCALATION OF AMINOACRIDINES
IN DNA ACCORDING TO MODEL Ib[a]

Molecule	GC : GC	θ (degrees)	AT : AT	θ (degrees)
Proflavine	−36.3	12.0	−52.0	192.0
2-Aminoacridine	−31.0	247.4	−47.7	247.4
3-Aminoacridine	−32.9	256.1	−48.8	256.1
9-Aminoacridine	−32.5	12.0	−49.1	192.0
12-Aminobenz(a)acridine	−38.0	286.0	−50.7	286.0
2-Aminobenz(b)acridine	−44.6	287.8	−53.9	287.8
7-Aminobenz(c)acridine	−36.5	231.8	−54.9	231.8

[a] The angle between the aminoacridine and the adjacent base pairs is 12°. (The angle giving the most favorable value of ΔF has been taken in each case.) Data given in kilocalories per repeating unit.

each case. It is evident that the change in rotation angle between models Ia and Ib does not cause a great change in the values of ΔF and produces only minor changes in the sequence of interaction energies.

For model Ib, the sequence of free energy change for the AT : AT base pair for those aminoacridines which we have also studied experimentally is: 7-aminobenz[c]acridine > 12-aminobenz[a]acridine > proflavine > 9-aminoacridine > 3-aminoacridine > 2-aminoacridine. The sequence for the approximate strength of binding by the primary process as measured by the value of r at the plateau in the binding isotherm is: 7-aminobenz[c]acridine > 12-aminobenz[a]acridine > 9-aminoacridine > proflavine > 3-aminoacridine > 2-aminoacridine, which is in good agreement with the predicted sequence. The sequence obtained from values of $\Delta T'_m$, the increase of T'_m of the complex compared to the original DNA, is not, however, in such good agreement with the calculated sequence. The sequence is: 7-aminobenz[c]acridine $\approx$ 9-aminoacridine > 12-aminobenz[a]acridine > proflavine > 3-aminoacridine. It is, however, to be noted that the main interchange in the sequences is between 12-aminobenz[a]acridine, 9-aminoacridine, and proflavine, and in every case the appropriate values for these different ligands are close (e.g., Table VIII).

For the intercalation models, the maximum value of r is 0.5, where all the sites suitable for intercalation are occupied. For $r > 0.5$, binding to the outside of the helix must also occur. For model Ia, the centers of two ligand molecules bound to adjacent phosphate groups on the same helix are found to be about 7 Å apart and for model Ib the value is 8.7 Å; in each case the plane of one ligand is 6.72 Å above the plane of the other. For native DNA, however, owing to the larger angle of 36° between the ligand molecules the distance between centers of these molecules is 9.4 Å, even though the distance between the planes of the ligand molecules is 3.36 Å. The distance between the centers of two externally bound dye molecules is greater for native DNA than for DNA with the maximum number of intercalated molecules. Thus, the interaction forces will be smaller for dye bound to native DNA than for dye bound externally to DNA extended to the maximum amount by intercalation. Furthermore, because of the greater distance in native DNA, the amount of associated water will be increased and the value of ε_{ij} will be larger; this will also have the effect of decreasing the interactions. These effects will be small, however, as the contribution of dye-dye interactions to the stability of model IIa is very small.

It is evident that the intercalation model offers a satisfactory explanation of the two stages of binding as shown by the binding isotherms (Fig. 2). The primary stage of strong binding corresponds to an intercalation mechanism and the secondary stage of weak binding to external, edgewise binding. Although the marked difference in the values of the free energy change for these two binding processes is in qualitative agreement with the experimental designation of strong and weak binding, respectively, refinement of the calculations is necessary before comparison of experimental and calculated values of binding constants can be contemplated.

The intercalation model, whether of type Ia, Ib, or Ic, appears to be supported strongly by both the experimental data and the theoretical calculations. Experimental studies concerned with the dimensions of the DNA molecule whether in solution or in fiber form indicate an increase in the length of the molecule on binding heterocyclic molecules or ions which can only be explained on the basis of the intercalation model. Likewise, the binding isotherms and the specific binding to AT sites is also capable of explanation by the intercalation model as the calculations of free energy change show. Finally, the data of Chambron et al. (1966b) which demonstrate that ligands bound strongly by the primary binding process are released cooperatively during the disruption of the helix on thermal denaturation are best interpreted on the basis of the intercalation model.

ACKNOWLEDGMENT

I wish to record my thanks to Mrs. Nerida Ellerton for considerable assistance in the preparation of this paper.

REFERENCES

Bradley, D. F., and Felsenfeld, G. (1959). *Nature* **184**, 1920.
Bradley, D. F., and Wolf, M. K. (1959). *Proc. Natl. Acad. Sci. U.S.* **45**, 944.
Brenner, S., Barnett, L., Crick, F. H. C., and Orgel, A. (1961). *J. Mol. Biol.* **3**, 121.
Cairns, J. (1962). *Cold Spring Harbor Symp. Quant. Biol.* **27**, 311.
Chambron, J., Daune, M., and Sadron, C. (1966a). *Biochim. Biophys. Acta* **123**, 306.

Chambron, J., Daune, M., and Sadron, C. (1966b). *Biochim. Biophys. Acta* **123**, 319.
Crawford, L. V., and Waring, M. J. (1967). *J. Mol. Biol.* **25**, 23.
Daudel, R., Lefebvre, R., and Moser, C. (1959). "Quantum Chemistry." Wiley (Interscience), New York.
DeVoe, H., and Tinoco, I. (1962). *J. Mol. Biol.* **4**, 500.
Drummond, D. S., Simpson-Gildemeister, V. F. W., and Peacocke, A. R. (1965). *Biopolymers* **3**, 135.
Drummond, D. S., Pritchard, N. J., Simpson-Gildemeister, V. F. W., and Peacocke, A. R. (1966). *Biopolymers* **4**, 971.
Elliott, W. H. (1963). *Biochem. J.* **86**, 562.
Fajans, K. (1959), *Techn. Org. Chem.* Part 2, 1169.
Fuller, W., and Waring, M. J. (1964). *Ber. Bunsenges. Physik. Chem.* **68**, 805.
Gersch, N. F. (1966). Ph. D. thesis, University of Adelaide.
Gersch, N. F., and Jordan, D. O. (1965). *J. Mol. Biol.* **13**, 138.
Hasted, J. B., Ritson, D. M., and Collie, C. H. (1948). *J. Chem. Phys.* **16**, 1.
Heilweil, H. G., and Van Winkle, Q. (1955). *J. Phys. Chem.* **59**, 939.
Jordan, D. O., and Ellerton, N. F. (1967). Unpublished results.
Jordan, D. O., Kurucsev, T., and Martin, M. L. (1967). Unpublished results.
Kandaswamy, T. S., and Henderson, J. F. (1962). *Biochim. Biophys. Acta* **61**, 86.
Kerridge, D. (1958). *J. Gen. Microbiol.* **19**, 497.
Kleinwächter, V., and Koudelka, J. (1964). *Biochim. Biophys. Acta* **91**, 539.
Langridge, R., Marvin, D. A., Seeds, W. E., Wilson, H. R., Hooper, C. W., and Wilkins, M. F. H. (1960). *J. Mol. Biol.* **2**, 38.
Lawley, P. D. (1956). *Biochim. Biophys. Acta* **22**, 451.
Lerman, L. S. (1961). *J. Mol. Biol.* **3**, 18.
Lerman, L. S. (1963). *Proc. Natl. Acad. Sci. U.S.* **49**, 94.
Luzzati, V., Masson, F., and Lerman, L. S. (1961). *J. Mol. Biol.* **3**, 634.
Mason, S. F., and McCaffery, A. J. (1964). *Nature* **204**, 468.
Mauss, Y., Chambron, J., Daune, M., and Benoit, H. (1967). Personal communication.
Michaelis, L. (1947). *Cold Spring Harbor Symp. Quant. Biol.* **12**, 131.
Newton, B. A. (1957). *J. Gen. Microbiol.* **17**, 718.
Newton, B. A. (1963). *In* "Metabolic Inhibitors" (R. M. Hochster and J. H. Quastel, eds.), Vol. 2, p. 285. Academic Press, New York.
Neville, D. M., and Davies, D. R. (1966). *J. Mol. Biol.* **17**, 57.
Orgel, A., and Brenner, S. (1961). *J. Mol. Biol.* **3**, 762.
Oster, G. (1951). *Trans. Faraday Soc.* **47**, 660.
Peacocke, A. R., and Skerrett, J. N. H. (1956). *Trans. Faraday Soc.* **52**, 261.
Pritchard, N. J., Blake, A., and Peacocke, A. R. (1966). *Nature* **212**, 1360.
Stone, A. L., and Bradley, D. F. (1961). *J. Am. Chem. Soc.* **83**, 3627.
Tomchick, R., and Mandel, H. G. (1964). *J. Gen. Microbiol.* **36**, 225.
Tubbs, R. K., Ditmars, W. E., and Van Winkle, Q. (1964). *J. Mol. Biol.* **9**, 545.
Vinograd, J. (1966). Personal communication.
Waring, M. J. (1964). *Biochim. Biophys. Acta* **87**, 358.
Waring, M. J. (1965). *J. Mol. Biol.* **13**, 269.
Waring, M. J. (1966a). Personal communication.
Waring, M. J. (1966b). *Biochim. Biophys. Acta* **114**, 234.
Weill, G., and Calvin, M. (1963). *Biopolymers* **1**, 401.

Recherches Théoriques sur l'Intercalement des Aminoacridines dans l'ADN*

M. GILBERT et P. CLAVERIE

Service de Biochimie Théorique
Institut de Biologie Physico-chimique
Paris, France

I. Introduction

De nombreux colorants, dérivés de l'acridine, forment des complexes avec les acides nucléiques. Ces complexes sont généralement classés en deux grandes catégories. Les complexes de la première catégorie (complexe I) sont caractérisés par un déplacement du maximum d'absorption du colorant vers des longueurs d'onde plus longues (Peacocke et Skerrett, 1956; Semmel et Daune, 1967). Le complexe II, qui se forme en présence d'un excès de colorant, est caractérisé par une énergie de liaison relativement faible et par l'apparition d'une bande d'absorption à des longueurs d'onde plus courtes. Beaucoup d'auteurs admettent que le complexe I avec l'acide déoxyribonucléique (ADN) natif, se forme par intercalement des colorants entre les plateaux des bases avec détorsion locale de la double hélice (Lerman, 1961; Gersch et Jordan, 1965; Chambron et coll., 1966). Ce modèle d'intercalement suggéré par Lerman en 1961, permet, en effet, d'interpréter plusieurs propriétés importantes du complexe I. Les résultats expérimentaux liés à la viscosité des solutions d'ADN en présence de colorants et à la réactivité chimique des aminoacridines en présence d'ADN sont en bon accord avec un modèle de ce type (Lerman, 1961; 1964). Les propriétés optiques du complexe I paraissent également compatibles avec l'intercalement (Weil et Calvin, 1963; Lerman, 1963; Tubbs et coll., 1964).

Quelques incertitudes demeurent, cependant, quant à la possibilité d'interpréter certaines autres caractéristiques du complexe I, sur la base d'un processus d'intercalement. On ne comprend pas encore parfaitement, par exemple, comment ce modèle pourrait permettre d'expliquer la forte dépendance de la force ionique manifestée par ces interactions. L'existence d'états de saturation maximum apparents r_M très inférieurs à 0.5 et fonctions de la force ionique du milieu demeure également inexpliquée. Drummond et coll. (1965), puis Chambron et coll. (1966) ont lié ces maxima et leurs variations,

---

* Ce travail a bénéficié de la subvention CR-66-236 de l'Institut National de la Santé et de la Recherche Médicale (Intergroupe Cancer).

soit avec la force ionique, soit avec la nature du ligand, à des variations du nombre total de sites disponibles. Cette hypothèse nous paraît, a priori, peu raisonnable et nullement requise par l'invariance apparente des rapports r_A/r_B pour deux colorants.* Nous nous proposons d'analyser ici la compatibilité du modèle proposé par Lerman, avec quelques résultats expérimentaux concernant le complexe I. Nous considérons plus précisément les quatre questions suivantes :

(*i*) Est-ce que le modèle d'intercalement permet d'expliquer l'existence de degrés d'association (états de saturation) maxima inférieurs à 0.5 ?

(*ii*) Comment l'énergie de liaison d'un colorant est-elle susceptible de varier avec la force ionique ?

(*iii*) Comment l'énergie de liaison varie-t-elle dans une série de dérivés d'acridine ?

(*iv*) Quel est l'effet de la composition de l'ADN en nucléotides, sur l'énergie de liaison d'un colorant ?

II. Théorie

Considérons une molécule d'ADN constituée de N paires de bases. Chaque surface de contact entre deux paires de bases superposées, peut être considérée comme une fonction acceptrice potentielle vis-à-vis d'un colorant A. Si l'on suppose que le ligand n'est pas autoassocié (faibles concentrations) et que chaque site est monoaccepteur, il intervient $N - 1$ équilibres successifs :

$$ADN \qquad + A \rightleftharpoons ADN—A$$

$$ADN—A \qquad + A \rightleftharpoons ADN—A_2$$

$$\overline{\qquad\qquad\qquad\qquad\qquad}$$

$$ADN—A_{n-1} + A \rightleftharpoons ADN—A_n$$

$$\overline{\qquad\qquad\qquad\qquad\qquad}$$

$$ADN—A_{N-2} + A \rightleftharpoons ADN—A_{N-1} \qquad\qquad (1)$$

L'état de saturation r de l'ADN est défini par le rapport $n/2N$, n et $2N$ étant respectivement le nombre de molécules de "colorant lié" et le nombre d'ions phosphate de l'ADN. En principe, une molécule d'ADN typique donne lieu à dix types différents de sites accepteurs s_j. Si n_j désigne la fraction des sites du type j on a : $\sum_j n_j = 0.5$. On doit en outre prévoir une variation progressive du pouvoir accepteur des sites liée à l'état de saturation du substrat. Si l'on suppose que cette variation ne dépend pas de la nature des sites, l'état de

---

* On peut montrer que, dans les petits intervalles de concentration étudiés, $\Delta(r_A/r_B)/(r_A/r_B)$ peut facilement être inférieur à 5% même si les énergies de fixation des colorants A et B sont assez différentes.

saturation r peut être exprimé en fonction de la concentration c du colorant libre, par l'équation (2).

$$r = \sum_j \frac{n_j \exp[-(\Delta \bar{F}_j + \Delta \bar{F}(r))/RT]c}{1 + \exp[-(\Delta \bar{F}_j + \Delta \bar{F}(r))/RT]c} \tag{2}$$

En absence d'effets coopératifs liés à la détorsion locale de la double hélice,* $\Delta \bar{F}(r)$ est la différence d'énergie libre molaire partielle qui résulte de la répulsion des cations intercalés, de l'attraction des ions phosphate pour ces cations, de la répulsion des ions phosphate et de la réaction de la solution à cette distribution de charges fixes.

$$\Delta \bar{F}(r) = \frac{\partial}{\partial n} (\Delta F^E - T\delta(\Delta S^L)) \tag{3}$$

ΔF^E représente la différence d'énergie libre purement ionique entre l'ADN et le complexe, en solution de même force ionique; $\delta(\Delta S^L)$ est la correction à l'entropie de localisation associée à ces interactions ioniques. On peut négliger la variation avec r de l'entropie de mélange des solutés au solvant puisque la concentration du colorant est toujours très faible (10^{-5} M/litre). Les $\Delta \bar{F}_j$ regroupent tous les autres termes d'énergie libre. Chacun d'eux contient un terme caractéristique du site j considéré et un terme commun à tous les sites.

$$\Delta \bar{F}_j = \Delta F_j^{\text{det}} + \Delta F_j^{\text{int}} + \Delta F^{\text{com}} \tag{4}$$

ΔF_j^{det} est l'énergie de détorsion locale de la double hélice impliquée dans la création du site j. [Le changement d'énergie libre des ions phosphate accompagnant la détorsion de l'hélice est inclu dans $\Delta \bar{F}(r)$.] ΔF_j^{int} désigne l'énergie d'interaction verticale entre un cation intercalé et les quatre bases de l'ADN formant le site de type j. Les interactions verticales entre un cation intercalé et les bases plus éloignées sont, soit négligeables (énergie de dispersion), soit inclues sous forme paramétrique dans le calcul de ΔF^E (énergies électrostatique et de polarisation). Les autres contributions qui dépendent peu ou pas de la nature du site d'intercalement sont regroupées dans ΔF^{com}. Le changement d'énergie de cavitation (Sinanoğlu et Abdulnur, 1965) qui accompagne la fixation d'une molécule de colorant, l'énergie d'interaction d'un colorant libre avec la solution et les changements d'énergies thermiques (rotation et translation) des molécules de colorant sont les plus importantes de ces contributions.

Même si le calcul de tous ces termes d'énergie libre est concevable, il ne saurait être effectué dans le cadre d'une approximation suffisamment homogène

---

* Bien que de tels effets ne puissent être totalement exclus, a priori, nous allons considérer qu'ils n'interviennent pas dans le processus d'intercalement.

pour conduire à des résultats significatifs. Nous utilisons donc une méthode semi-empirique qui nous permet de recueillir un maximum d'information du calcul détaillé de quelques unes seulement des contributions énumérées plus haut. Les problèmes concernant l'existence de r_M apparents peuvent, en effet, être abordés à partir de deux points extrêmes:

1. On peut supposer que la contribution purement ionique $\Delta \bar{F}(r)$ n'est pas fonction de l'état de saturation (la répulsion entre les cations intercalés est négligeable) et que l'hétérogénéité énergétique des sites d'intercalement suffit à expliquer l'aspect des courbes de liaison (r en fonction de c). L'état de saturation du substrat est alors donné par l'équation (5).

$$r = \sum_{j} \frac{n_j \exp(-\Delta \bar{F}_j{}^a/RT)c}{1 + \exp(-\Delta \bar{F}_j{}^a/RT)c} \tag{5}$$

$\Delta \bar{F}_j{}^a$ inclut les interactions électrostatiques entre les colorants et les ions phosphate. En ce cas, à condition de pouvoir choisir l'origine des énergies à une force ionique donnée, on devrait pouvoir reproduire théoriquement les courbes de liaison des aminoacridines en ne considérant explicitement que les énergies de détorsion et d'interaction verticale.

2. On peut, au contraire, supposer que l'hétérogénéité énergétique des sites est négligeable et que les interactions ioniques répulsives sont seules responsables de l'établissement d'états de saturation maximum, fonctions de la force ionique. On a alors:

$$r = \frac{0.5 \exp[-(\Delta \bar{F}(r) + C)/RT]c}{1 + \exp[-(\Delta \bar{F}(r) + C)/RT]c} \tag{6}$$

Le terme constant C est traité comme un paramètre ajustable caractéristique d'un colorant et il suffit d'analyser les interactions purement ioniques. Nous nous plaçons d'abord dans le cadre de la première hypothèse et ne considérons explicitement que les énergies d'interactions verticales et de détorsion. Nous nous plaçons ensuite dans le cadre de la seconde hypothèse et tirons les conséquences sur le rôle des interactions ioniques, de ce qui aura été appris dans la première partie.

III. Les Interactions Verticales

A. Méthode de Calcul

L'énergie d'interaction verticale entre les cations intercalés et les paires de bases (termes électrostatique $E_{\rho\rho}$, de polarisation $E_{\rho\alpha}$ et de dispersion E_L) est calculée dans le cadre de l'approximation «monopôles et polarisabilités de liaison» (Claverie et Rein, 1967; Claverie, 1967).

Les avantages de cette approximation par rapport aux approximations utilisées précédemment ont été souvent signalés (Rein et coll., 1967; Claverie, 1967). La répulsion liée à la possibilité d'échange des électrons entre les molécules est évaluée à l'aide d'une formule semi-empirique (Kitaygorodsky, 1961; Claverie, 1967).

Nous avons adopté, pour ces calculs, le modèle proposé par Lerman (1961). La molécule de colorant est insérée entre deux plateaux de bases distants de 6.72 Å et faisant entre eux un angle de 9 degrés à gauche. Comme Gersch et

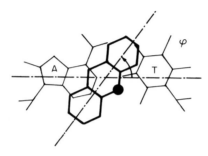

FIG. 1. Position d'un colorant intercalé (configuration axiale). Dans les configurations axiales, l'axe de la double hélice passe par le centre géométrique de la molécule d'acridine. L'angle φ définit l'orientation de la molécule de colorant par rapport à la paire de bases adjacente.

Jordan (1965), nous supposons d'abord que l'axe de l'hélice passe par le centre géométrique de la molécule intercalée (Fig. 1). Les dix types de couples de plateaux formant des sites énergétiquement différents sont considérés. Pour déterminer les configurations produisant les minima d'énergie, les cations sont placés successivement dans plus de quarante orientations. L'effet d'orientation que pourraient avoir les ions phosphate est négligé. Rien ne garantit, cependant, que les minima obtenus ainsi soient des minima absolus. Nous considérons donc aussi deux autres configurations (extra-axiales) qui paraissent, a priori, relativement favorables (Fig. 2). Gersch et Jordan (1965), puis Jordan (1967) ont fait des calculs analogues. Toutefois l'essentiel de leurs conclusions concernant l'hétérogénéité énergétique des sites accepteurs est fondé sur des artefacts liés à l'utilisation de l'approximation dipolaire d'une part et à l'identification arbitraire du centre de charge et du centre géométrique des colorants d'autre part.

B. L'Hétérogénéité Énergétique des Sites

1. Les calculs ont été effectués pour l'acridine et deux de ses dérivés (la proflavine et la 9-aminoacridine). Le Tableau I rassemble les énergies d'interaction verticale correspondant aux orientations les plus favorables de la

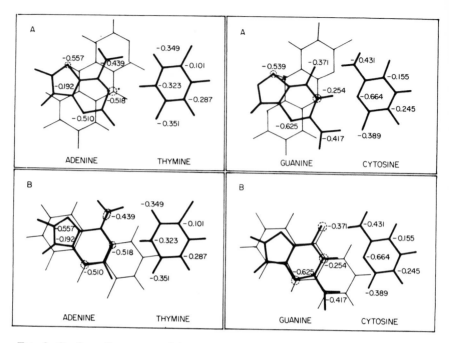

FIG. 2. Configurations extra-axiales. Les configurations A et B on été choisies parce qu'elles permettent le recouvrement des charges atomiques les plus négatives de la paire de bases voisine par les charges atomiques les plus positives de la molécule de colorant intercalée.

configuration axiale. Les résultats correspondant aux deux configurations extra-axiales sont donnés dans le Tableau II. A l'exception de sites

$$\left|\begin{matrix}\uparrow G \smallfrown C \\ G \smallsmile C\end{matrix}\right|_\downarrow$$

tous les sites accepteurs paraissent favoriser une fixation avec recouvrement maximum des paires de bases par la molécule de colorant. Ce résultat semble exclure tout intercalement partiel quand l'intercalement complet est stériquement possible. Le petit nombre de configurations étudiées impose cependant d'accueillir cette conclusion avec beaucoup de prudence. Dans la suite de cette communication, nous ne considérons que les configurations axiales.

On remarque que, dans le vide, les paires ↑G—C↓ et ↑C—G↓ stabilisent légèrement plus les molécules intercalées que les paires ↑A—T↓ et ↑T—A↓. Etant données les approximations faites, ce résultat n'est cependant pas significatif. La formule semi-empirique utilisée dans l'évaluation des énergies de répulsion surestime les répulsions impliquant la thymine (à cause de son groupe méthyl). D'autre part les énergies électrostatique et de polarisation, qui sont en grande partie responsables du plus grand pouvoir stabilisateur

TABLEAU I
INTERACTIONS VERTICALES: COLORANT—PAIRE DE BASES

	Electrostatique	Polarisation	Dispersion	Repulsion	Totale	Angle φ
Acridine:						
↑A—T↓	−6.84	−4.81	−8.529	5.03	−15.15	171°
↑T—A↓	−6.84	−4.81	−8.529	4.60	−15.58	189°
↑G—C↓	−7.45	−5,27	−8.359	4.64	−16.44	−18°
↑C—G↓	−7.82	−5.25	−7.600	3.84	−16.83	36°
Proflavine:						
↑A—T↓	−6.46	−4.58	−9.586	5.45	−15.18	0°
↑T—A↓	−6.46	−4.58	−9.572	4.95	−15.66	0°
↑G—C↓	−6.84	−4.96	−9.707	5.44	−16.07	0°
↑C—G↓	−6.84	−4.95	−9.707	5.44	−16.06	0°
9-Aminoacridine:						
↑A—T↓	−6.61	−4.61	−8.657	4.20	−15.68	−18°
↑T—A↓	−6.72	−4.66	−9.079	4.50	−15.96	−18°
↑G—C↓	−7.10	−5.05	−8.829	4.21	−16.77	−18°
↑C—G↓	−7.28	−5.03	−9.272	4.98	−16.60	−18°

TABLEAU II
INTERACTIONS VERTICALES: ACRIDINE—PAIRE DE BASES
(MODÈLES EXTRA-AXIAUX)[a]

		Configuration	
1	2	A	B
A—T	A—T	−8.14	−12.50
	T—A	−6.29	−11.94
	G—C	−11.80	−14.38
	C—G	−7.01	−12.79
T—A	T—A	−8.24	−12.42
	G—C	−9.06	−12.87
	C—G	−11.72	−14.07
G—C	G—C	−15.48	−16.26
	C—G	−9.77	−13.73
C—G	C—G	−15.20	−15.73

[a] Chaque couple formé en associant une paire de bases de la colonne 1 à une paire de bases de la colonne 2 constitue un site d'intercalement différent quant à l'énergie d'interaction verticale. Les énergies d'interaction des configurations A et B sont exprimées en kilocalories par paire de bases; elles doivent être comparées aux énergies du Tableau I. Chaque terme est la moyenne des énergies obtenues en plaçant, tour à tour, les paires de bases de la colonne 1 et celles de la colonne 2 dans les positions decrites dans la Fig. 2.

des paires G—C, sont sensiblement réduites par la solution. Il n'est donc pas impossible que l'ordre des énergies de stabilisation puisse être renversé par un calcul amélioré. Dans l'état actuel des calculs théoriques, il faut admettre qu'il n'existe pas de différences appréciables dans le pouvoir stabilisateur des paires de bases.

La comparaison des résultats obtenus pour les trois colorants étudiés est intéressante. Deux traits ressortent de cette comparaison: (a) Les énergies de stabilisation des trois colorants sont à peu près égales. (b) L'ordre des énergies de stabilisation n'est pas parallèle à l'affinité de ces molécules pour l'ADN (acridine < proflavine ⩽ 9-aminoacridine). Les faibles écarts observés rendent particulièrement évident le rôle joué par les conditions de réaction et les effets de solvant dans l'établissement des affinités apparentes des colorants. Ainsi, les énergies d'interaction verticale laissent supposer que l'acridine s'intercale mieux que la proflavine, ce qui n'est pas le cas (Lerman, 1961). Le pK_a de l'acridine étant relativement faible, la concentration d'acridine protonée, à pH 6–7, est plus petite que celle de son dérivé. Par ailleurs, la plus grande stabilisation de la 9-aminoacridine ne traduit l'ordre des affinités observées que dans la mesure où la proflavine est mieux stabilisée que la 9-aminoacridine en solution aqueuse (les coefficients de partition huile d'olive/eau de la proflavine et de la 9-aminoacridine à pH 7.1 sont respectivement 0.7 et 1.2; Albert et coll., 1943). Pourvu que l'on compare des molécules également protonées et qu'il n'y aie pas de restrictions stériques, un colorant cationique devrait s'intercaler d'autant mieux qu'il est plus grand et qu'il porte moins de substituants polaires. Dans l'état actuel de nos connaissances des effets de solvant et de la géométrie précise de l'intercalement, une règle qualitative comme celle-là est plus éclairante que le calcul détaillé des seules énergies d'interaction verticale, surtout si ces calculs sont effectués dans l'approximation dipolaire.

2. Le Tableau III résume les énergies de fixation intrinsèques des différents sites de l'ADN. On voit que le modèle d'intercalement (sans restriction directionnelle due aux ions phosphate) donne lieu à une hétérogénéité énergétique des sites accepteurs étalée sur un intervalle de 3.7 kcal/mole. La distribution est entièrement déterminée par les seules énergies de détorsion. On remarque que 55% des sites accepteurs ont une énergie de liaison intrinsèque comprise dans un intervalle de 0.85 kcal/mole. Ces résultats sont très différents de ceux obtenus par Gersch et Jordan (1965) dans le cadre de l'approximation dipolaire. Ils peuvent être analysés du point de vue de l'existence de r_M apparents inférieurs à 0.5 et de celui de l'effet de la composition de l'ADN sur la formation du complexe I. Ces deux problèmes peuvent être étudiés simplement en traçant des courbes de liaison théoriques (r en fonction de c), basées sur les résultats du Tableau III, à l'aide de l'équation (5). L'origine des énergies est fixée, en ajustant un point de la courbe théo-

rique au point correspondant de la courbe de fixation de la 9-aminoacridine à l'ADN de thymus de veau en force ionique 0.1 (Fig. 3). Toutes les courbes théoriques sont situées au-dessus de la courbe expérimentale. Une molécule d'ADN (A—T/G—C = 1) donnerait lieu à un r_M voisin de 0.4 alors que la valeur de r_M observée est voisine de 0.2. L'hétérogénéité des sites accepteurs ne contribue donc que faiblement à la limitation du processus d'intercalement. Le fait que la fixation de la proflavine sur le poly A—U, plafonne à environ 0.32 ($\mu = 0.1$; Semmel et Daune, 1967) suggère d'ailleurs qu'un mécanisme limitatif joue, même en absence d'hétérogénéité. On remarque,

TABLEAU III
ENERGIES DE LIAISON INTRINSÈQUES DES SITES

Sites	Energies de détorsion	Energies de liaison		
		Acridine	Proflavine	9-Amino-acridine
↑ T⤬A / G⤬C ↓	3.34	−28.66	−28.14	−29.16
↑ A—T / T—A ↓	4.21	−27.79	−27.27	−28.29
↑ T—A / A—T ↓	4.24	−27.76	−27.24	−28.26
↑ A⤬T / G⤬C ↓	4.55	−27.45	−26.93	−27.95
↑ A⤬T / A⤬T ↓	4.79	−27.21	−26.69	−27.71
↑ C—G / G—C ↓	4.91	−27.09	−26.57	−27.59
↑ G⤬C / G⤬C ↓	5.08	−26.92	−26.40	−27.42
↑ A⤬T / C⤬G ↓	6.00	−26.00	−25.48	−26.50
↑ T⤬A / C⤬G ↓	6.21	−25.79	−25.27	−26.29
↑ G—C / C—G ↓	7.02	−24.98	−24.46	−25.48

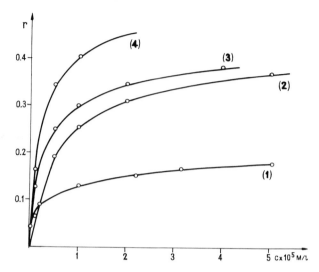

Fig. 3. Courbes de liaison. (1) Courbe expérimentale correspondant à la formation du complexe I entre la 9-aminoacridine et l'ADN de thymus de veau à force ionique 0.1 (Drummond et coll., 1965). (2) Courbe théorique pour une molécule d'ADN ne contenant que

$$G\text{—}C\left(\begin{vmatrix}\uparrow G\text{—}C \\ \times \\ G\text{—}C\downarrow\end{vmatrix}, \quad \begin{vmatrix}\uparrow C\text{—}G \\ G\text{—}C\downarrow\end{vmatrix}, \quad \begin{vmatrix}\uparrow G\text{—}C \\ C\text{—}G\downarrow\end{vmatrix}\right)$$

(3) Courbe théorique pour une molécule d'ADN dont le rapport A—T/G—C = 1. (4) Courbe théorique pour une molécule d'ADN ne contenant que

$$A\text{—}T\left(\begin{vmatrix}\uparrow A\text{—}T \\ \times \\ A\text{—}T\downarrow\end{vmatrix}, \quad \begin{vmatrix}\uparrow A\text{—}T \\ T\text{—}A\downarrow\end{vmatrix}, \quad \begin{vmatrix}\uparrow T\text{—}A \\ A\text{—}T\downarrow\end{vmatrix}\right)$$

Les sites différents sont présents en proportions égales et repartis au hasard.

également, qu'une molécule d'ADN riche en A—T devrait favoriser davan-tage le processus d'intercalement qu'une molécule riche en G—C. Ce résultat théorique est en accord avec les observations de plusieurs chercheurs (Tubbs et coll., 1964; Semmel et Daune, 1967). Il doit cependant être enregistré avec prudence car il est très sensible aux méthodes de calcul et rien ne permet d'affirmer qu'il ne sera pas changé par un calcul moins approximatif. Les énergies de polarisation ne sont pas additives comme nous l'avons supposé. L'introduction de la non-additivité de ces énergies n'est pas susceptible de modifier sensiblement l'étalement des énergies de liaison intrinsèques des sites mais pourrait modifier le classement des sites en fonction de leur pouvoir accepteur.

IV. Les Interactions Ioniques

A. Méthode de Calcul

On représente une molécule d'ADN ou une molécule du complexe par un cylindre infini de rayon $b = 10$ Å. Les ions phosphate sont décrits par des bandes de largeur $\delta = 1$ Å à la surface de ce cylindre. Chacune de ces bandes porte une charge négative $2fe$, correspondant à deux ions phosphate partiellement compensés. Les charges des aminoacridines intercalés sont représentées d'une façon analogue, par des bandes situées à la surface d'un cylindre coaxial du premier et de rayon $a = 1$ Å (Fig. 4). Ces bandes décrivent les charges des acridines et non les molécules d'acridine elles-mêmes.

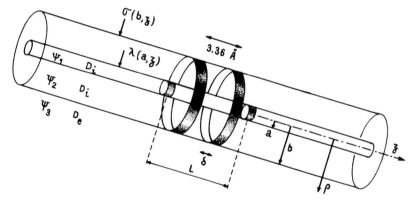

Fig. 4. Le modèle. (Voir les détails dans le texte.)

La Fig. 4 décrit un segment du modèle représentant le complexe à $r = 0.25$. Le nombre m de bandes d'ions phosphate entre deux cations consécutifs est égal à 2. D_i et D_E désignent les constantes diélectriques intérieures et extérieures. Les densités superficielles σ et λ sont données par les séries de Fourier suivantes :

$$\sigma(b, z) = \frac{m\sigma_0}{L} + 2\sigma_0 \sum_{n=1} \left[\sum_{k=1,2}^{m/2} \cos\left(\frac{2\pi nk}{n+1}\right) + \alpha \right] \frac{\sin(\pi n\delta/L)}{\pi n\delta} \cos(h_n z)$$

$$\lambda(a, z) = \frac{\lambda_0}{L} + 2\lambda_0 \sum_{n=1}^{\infty} \frac{\sin(\pi n\delta/L)}{\pi n\delta} \cos(h_n z)$$

$h_n \equiv 2\pi n/L$; $\alpha = 0$ si m est pair; $\alpha = \cos(\pi n)$ si m est impair.

MacGillivray et McMullen (1966) ont récemment utilisé un modèle analogue dans leur étude des propriétés électrostatiques de l'ADN. Nous ne considérons que des distributions régulières de cations intercalés. On peut donc toujours développer en séries de Fourier les densités de charges λ et σ (σ' dans le cas de l'ADN) à la surface des cylindres *a* et *b* respectivement. Les approximations inhérentes à cette description des charges ont fait l'objet d'une discussion détaillée dans un article récent (Gilbert et Claverie, 1967). On trouvera également dans cet article une description détaillée de la méthode résumée ici.

L'énergie électrostatique du complexe et celle de l'ADN sont données par les deux intégrales suivantes:

$$W_c = n\left[\frac{1}{2}\int_0^{2\pi}\int_{-L/2}^{L/2}(\Psi_1(a, z)\lambda(a, z)a + \Psi_2(b, z)\sigma(b, z)b)\,d\theta\,dz - W_s\right] \quad (7a)$$

$$W_{\text{ADN}} = \frac{nm}{2}\int_0^{2\pi}\int_{-1.68}^{+1.68}\Psi_i(b, z)\sigma'(b, z)b\,d\theta\,dz \quad (7b)$$

où W_s désigne la self-énergie d'une bande cylindrique représentative d'un cation intercalé. Les potentiels Ψ_1, Ψ_2, Ψ_3 et Ψ_i, à l'intérieur du cylindre *a*, entre les cylindres *a* et *b* et dans la solution environnant le complexe et l'ADN, sont déterminés par les méthodes habituelles d'application de la théorie de Debye et Hückel (Westheimer et Kirkwood, 1938; MacGillivray et McMullen, 1966). La différence d'énergie libre ΔF^E est donnée par la différence des intégrales (7a, b). L'entropie de localisation est calculée sur la base d'un modèle de Ising incluant les interactions électrostatiques ω_c entre les sites adjacents occupés (Harris et Rice, 1954; Everett, 1950).

$$\Delta\bar{S}^L = -R\log\frac{2r}{1-2r} - 2R\log\frac{2(1-2r)}{\beta+1-4r} + \left(1+\frac{4r-1}{\beta}\right)\frac{\omega_c}{T} \quad (8)$$

où

$$\beta^2 \equiv 1 - 8r(1-2r)\left(1 - \exp\left(-\frac{\omega_c}{RT}\right)\right)$$

à la différence de Harris et Rice, nous utilisons une énergie d'interaction ω_c, déduite de l'énergie électrostatique W_c du complexe en solution et corrigée, dans une certaine mesure pour tenir compte de l'interaction de sites plus distants.

B. Les États de Saturation Maximum

La différence d'énergie libre $\Delta\bar{F}(r)$ a été calculée pour différentes valeurs de *r*, à cinq forces ioniques. Les résultats sont rassemblés dans le Tableau IV et résumés dans les différentes courbes représentatives des variations de $\Delta\bar{F}(r)$ en fonction de *r* (Fig. 5A). Nous pouvons faire les remarques suivantes à propos de ces courbes: (a) la différence d'énergie libre croît avec l'état de saturation, quelque soit la force ionique de la solution; (b) la différence

TABLEAU IV
LA DIFFÉRENCE D'ÉNERGIE[a] LIBRE $\Delta\bar{F}$

μ r	0.05	0.10	0.15	0.20	0.25	0.30	0.35
∞	−1.925	−1.243	−0.566	0.077	0.618	1.110	1.673
1	−2.276	−1.581	−0.819	−0.150	0.382	0.904	1.506
0.1	−3.070	−2.303	−1.395	−0.459	0.080	0.507	1.153
0.01	−4.584	−3.525	−2.405	−1.334	−0.680	−0.130	0.631
0.001	−6.608	−5.185	−3.756	−2.425	−1.662	−1.075	−0.116

[a] Énergies exprimées en kilocalories/mole. La constante diélectrique intérieure D_i a été prise égale à 5.5. Le paramètre compensateur f est égal à 0.5. Ces énergies correspondent à l'intercalement d'un colorant cationique avec une partie de sa couche d'hydratation et de son nuage d'ions compensateurs.

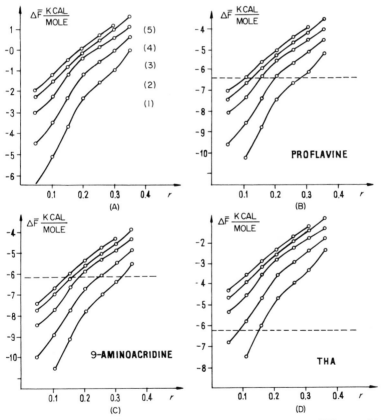

FIG. 5. La différence d'énergie libre du complexe ADN colorant. Cette différence d'énergie libre inclut le terme de localisation ($\Delta\bar{F} = \Delta\bar{F}^E - 300\Delta\bar{S}^L$). (1) $\mu = 0.001$; (2) $\mu = 0.01$; (3) $\mu = 0.1$; (4) $\mu = 1$; (5) $\mu = \infty$. L'axe de r des parties B, C et D est fixé d'une façon semi-empirique.

d'énergie libre augmente avec la force ionique. En accord avec les observations de Drummond et coll. (1965), cette augmentation est plus marquée pour $10^{-3} \leqslant \mu \leqslant 10^{-1}$ que pour $10^{-1} \leqslant \mu \leqslant \infty$.

On peut tirer de ces résultats des renseignements plus précis, à condition de savoir fixer le paramètre C de l'équation (6), c'est-à-dire l'axe des r de la Fig. 5B, C, D. La différence d'énergie libre qui accompagne la fixation d'une molécule de proflavine supplémentaire sur l'ADN de thymus de veau quand $r = 0.2$ et $\mu = 0.01$ est de -6.45 kcal/mole (Drummond et coll., 1965). Comme la valeur de $\Delta \bar{F}$ à ce point est de -1.33 kcal/mole, le paramètre C_{pro} pour la proflavine doit être fixé à -5.11 kcal/mole. On fixe d'une manière analogue les paramètres de la 9-aminoacridine et de la 9-amino-, 1,2,3,4-tetrahydroacridine (THA). On obtient $C_{9\text{-}A} = -5.47$ kcal/mole et $C_{THA} = -2.43$ kcal/mole. L'origine des énergies ainsi fixée, les points de rencontre entre les courbes représentatives de $\Delta \bar{F}(r)$ et l'axe des r, correspondent aux états de saturation maximum apparents. Leurs valeurs pour les trois colorants

TABLEAU V
ETATS DE SATURATION MAXIMUM

	∞	1	0.1	0.01	0.005	0.001
Proflavine						
r_M^{exp}	0.07^a		0.15	0.20	0.25	
$r_M^{\ T}$	0.09	0.10	0.15	0.20^b		0.28
9-Aminoacridine						
r_M^{exp}			0.17	0.26	0.30	
$r_M^{\ T}$	0.15	0.16	0.19	0.26^b		0.36
Tetrahydroacridine						
r_M^{exp}			0.07	0.088	0.125	
$r_M^{\ T}$			0.02	0.09^b		0.150

[a] Valeur extraite de Peacocke et Skerrett (1956) pour $\mu = 2$. Les autres valeurs expérimentales sont tirées de Drummond et coll. (1965).
[b] Les valeurs ainsi marquées ont été ajustées aux valeurs expérimentales.

étudiés sont comparées aux valeurs expérimentales dans le Tableau V. On remarque que l'accord avec l'expérience est excellent, sauf pour la THA. Ce désaccord lui-même n'est pas sans intérêt puisqu'il concerne une molécule qui ne saurait s'intercaler complètement. Enfin, dans la Fig. 6, nous reproduisons les courbes de liaison théoriques de la proflavine et de la 9-aminoacridine à force ionique 0.1. On voit que ces courbes coïncident presque parfaitement avec les courbes expérimentales.

Le paramètre compensateur f que nous avons utilisé pour les ions phosphate est plus grand (0.5) que celui qu'ont utilisé Schildkraut et Lifson (1965). Il y a deux raisons qui expliquent cette différence. La proximité des ions phosphate et des aminoacridines diminue probablement l'approche d'ions compensateurs de l'un et de l'autre. D'autre part, il faut rappeler que nous

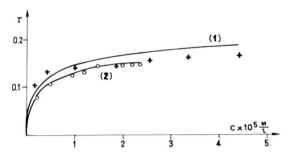

FIG. 6. Courbes de liaison. (1) Courbe théorique pour la 9-aminoacridine à $\mu = 0.1$. (2) Courbe théorique pour la proflavine à $\mu = 0.1$. O: points expérimentaux pour la proflavine; +: points expérimentaux pour la 9-aminoacridine.

nous sommes placés dans le cadre de la seconde hypothèse. L'effet de l'hétérogénéité des sites négligé explicitement est par conséquent inclu implicitement dans le paramètre compensateur. Une étude plus rigoureuse devrait inclure simultanément les effets d'hétérogénéité et les effets purement ioniques. Il ne nous est pas apparu souhaitable, cependant, de lier intimement le calcul microscopique des énergies d'interaction verticale au calcul macroscopique des interactions purement ioniques.

V. Conclusion

L'ensemble des résultats de notre analyse est largement favorable à un modèle d'intercalement avec recouvrement maximum des paires de bases par les colorants. Nous n'avons pas décelé d'incompatibilité entre les implications de ce modèle et les propriétés du complexe I. Nous avons au contraire montré que:

(a) L'existence de degrés d'association maxima apparents inférieurs à 0.5 peut être interprétée, principalement par la répulsion des colorants intercalés et secondairement, par la présence de sites d'intercalement légèrement moins favorables que les autres

$$\left(\left| \begin{matrix} \uparrow G—C \downarrow \\ C—G \end{matrix} \right| , \left| \begin{matrix} \uparrow T—A \downarrow \\ C—G \end{matrix} \right| , \left| \begin{matrix} \uparrow A—T \downarrow \\ C—G \end{matrix} \right| \right)$$

(b) La disparition progressive des interactions stabilisatrices entre les colorants et les ions phosphate fournit une interprétation simple de la diminution de la fixation qui accompagne l'augmentation de la force ionique.

(c) L'énergie de solvatation des molécules de colorant (cavitation et interaction avec le solvant) est un facteur essentiel de leur aptitude à s'intercaler. Dans la mesure où les empêchements stériques ne sont pas déterminants, une molécule devrait s'intercaler d'autant mieux qu'elle est plus grande et qu'elle est moins polaire.

(d) Le modèle d'intercalement permet de rendre compte du plus grand pouvoir d'adsorption des molécules d'ADN riches en adénine et en thymine.

Ces conclusions, la dernière en particulier, ne doivent cependant pas être dissociées des nombreuses réserves que nous avons faites au cours de l'exposé.

BIBLIOGRAPHIE

Albert, A., Goldacre, R., et Heymann, E. (1943). *J. Chem. Soc.* p. 651.

Chambron, J., Daune, M., et Sadron, Ch. (1966). *Biochim. Biophys. Acta* **123**, 306.

Claverie, P., et Rein, R. (1967). *Intern. J. Quantum Chem.* (*en cours de publication*).

Drummond, D. S., Simpson-Gildemeister V. F. W., et Peacocke, A. R. (1965). *Biopolymers* **3**, 135.

Everett, D. H. (1950). *Trans. Faraday Soc.* **46**, 942.

Gersch, N. F., et Jordan, D. O. (1965). *J. Mol. Biol.* **13**, 138.

Claverie, P. (1967). Ce volume.

Gilbert, M., et Claverie, P. (1967). (*en cours de publication*). *J. Theoret. Biol.*

Harris, F. E., et Rice, S. A. (1954,) *J. Phys. Chem.* **58**, 733.

Jordan, D. O. (1967). Ce volume.

Kitaygorodsky, A. T. (1961). *Tetrahedron* **14**, 230.

Lerman, L. S. (1961). *J. Mol. Biol.* **3**, 18.

Lerman, L. S. (1963). *Proc. Natl. Acad. Sci. U.S.A.* **49**, 94.

Lerman, L. S. (1964). *J. Mol. Biol.* **10**, 367.

MacGillivray, A. D., et McMullen, A. I. (1966). *J. Theoret. Biol.* **12**, 75.

Peacocke, A. R., et Skerrett, J. N. H. (1956). *Trans. Faraday Soc.* **52**, 261.

Rein, R., Claverie, P., et Pollak, M. (1967). *Intern. J. Quantum Chem.* (en cours de publication).

Schildkraut, C., et Lifson, S. (1965). *Biopolymers* **3**, 195.

Semmel, M., et Daune, M. (1967). (en cours de publication).

Sinanoğlu, O., et Abdulnur, S. (1965). *Federation Proc.* (24 Suppl.) **15**, S-12.

Tubbs, R. K., Ditmars, W. E., et Van Winckle, Q. (1964). *J. Mol. Biol.* **9**, 545.

Weil, G., et Calvin, M. (1963). *Biopolymers* **1**, 401.

Westheimer, F. H., et Kirkwood, J. G. (1938). *J. Chem. Phys.* **6**, 513.

Statistical Mechanical Analysis of Binding of Acridines to DNA

D. F. BRADLEY AND S. LIFSON

National Institute of Mental Health
Bethesda, Maryland and The Weizmann Institute of Science
Rehovoth, Israel

I. Introduction

In recent years much interest has been evidenced in the binding of small molecules to biopolymers. Particular attention has been given by many authors to the binding of acridines to DNA and polynucleotides. This is partly due to the remarkable mutagenicity of acridines, which interfere in the replication of DNA by causing deletions or insertions of base pairs (1). Apart from biological importance, however, the binding process brings to fore a number of fascinating problems for the physical chemist, worthy of experimental and theoretical study on their own merit.

A large number of physicochemical properties related to the binding of acridines on DNA have been studied in detail. These include equilibrium dialysis (2), shifts and intensity changes of spectral absorption bands (2–7), induced optical rotation (8–10) and circular dichroism (11, 12), effect of DNA fiber orientation on linear and circular dichroism (7, 11, 13), fluorescence (13, 14), changes in X-ray fiber diagrams (15, 7), viscosity (15–16), low-angle X-ray scattering (17), autoradiography (18), and amino group reactivity of bound acridines (19).

In spite of the accumulation of so much experimental knowledge, or rather because of the complexity of the binding process, as indicated by the collected data, there does not emerge as yet a unified picture or model of the binding process which will account for all observed phenomena. In order to understand all these phenomena using a unified model it is necessary to know what the binding sites are, how many sites each dye occupies when bound, whether dyes bind in more than one way, and what are the changes incurred on the DNA and dye molecule upon binding. For a quantitative description of the binding process by statistical mechanical methods it is necessary, furthermore, to know the energies or the equilibrium constants of the various possible modes of binding, as well as the nature and values of intermolecular interactions between the bound dyes. To include the equilibrium distribution of the dye between the dissolved and the bound states, it is also necessary to know the activity of the dye in the solution. To the extent that the molecular parameters are not known, it is possible to introduce them as unknown parameters in one or more plausible models and to derive the functional relations between these parameters and the various experimentally observable quantities. The examination of the models and their theoretical predictions versus experimental data will then determine their acceptability, give us a theoretical understanding of the binding phenomenon, help to discard wrong concepts, and may lead to a better design of experiments.

Statistical mechanical analysis has been employed hitherto by few authors in the analysis of the DNA-dye binding and several models have been suggested. Thus, Peacocke and Skerrett (2) assumed the existence of two modes of binding of dye molecules, a strong binding and a weak binding, characterized by intrinsic association constants and the absence of mutual interactions.

Bradley and co-workers (3, 20, 21) derived an equation relating the effects of interactions between neighboring bound dyes to the relative distribution of isolated and stacked dyes along polymer chains. Following Michaelis (22) they assumed that the shift of the absorption spectrum with the degree of binding is due to the gradual shift from the state of isolated dye binding at low degrees of binding to the full stacking of the molecules at maximum binding. They were able to represent with their equation the spectral changes of a large number of different polymers by varying the stacking coefficient from polymer to polymer.

Lerman (13, 15, 19) studied conformational changes of DNA due to dye binding and concluded that the dye molecule intercalates between the purine-pyrimidine pairs of DNA. This model explains the observed changes in viscosity, sedimentation, and X-ray fiber diagrams, as well as a number of other experimental observations, and suggests a rationale for the mutagenic effect of acridines.

No attempt has been made as yet, however, to examine the above mentioned models with respect to all experimental observations simultaneously and to put them to a comprehensive critical test. It seems to us that none of the above models will withstand such a test without major modifications.

It would be useful, therefore, to have at hand a general statistical mechanical method by which the analysis of experimental results could be worked out for various models in a uniform, convenient, and systematic manner. The sequence-generating function (SGF) method (23) appears to be such a method and we would like to show herein how it can be applied to the present class of problems. We will consider examples of models relating to the binding of acridines to DNA, including rather complex ones, so that the power of the SGF method will be demonstrated. We do not intend, however, to analyze at present all the data on the acridine-DNA systems in sufficient detail or to scan systematically all conceivable models pertaining to the data. We hope rather to show that the SGF method makes it possible to obtain the quantitative relations between observable quantities for any conceivable model once the statistical weights of the various possible microscopic states pertaining to the model have been chosen.

II. General Procedure

We give here a concise resume of the sequence-generating function (SGF) method (23) adapted to the class of models to be discussed presently. One begins by defining accessible states for the binding sites of the DNA polymer. For example, one may assume that the binding site may have two states, filled and empty. In more complex models one may distinguish more than two states for each site. Next, one defines sequences in which all the sites are in the same state. Each sequence has a length, which is the number of sites in the sequence, and is assigned a statistical weight proportional to the probability of occurrence of such a sequence. The statistical weight is determined by the considerations and concepts which underlie the model chosen to represent the physical system. A partition function of the system is obtained from the statistical weights of the sequences and the various average properties calculated, in a way which is easy to see from the following examples of the two-state models.

III. Two-State Models

We begin with a simple model in which a site may either be occupied by a dye in one particular way (filled or bound site) or unoccupied (empty site). We assign statistical weights u_i and v_j to sequences of empty and filled sites of length i and j, respectively. The sequence-generating functions [$U(x)$, $V(x)$] are defined as decreasing power series in a parameter, x, with the statistical weights as coefficients.

$$U(x) = \sum_{i=1}^{\infty} u_i x^{-i}; \qquad V(x) = \sum_{j=1}^{\infty} v_j x^{-j}$$

The SGF method gives a relation between the SGF's and the partition function of the system. This relation is based on the fact that the partition function $Z^{(P)}$—where P is the total number of sites on the polymer—can be written for sufficiently large P in the form

$$Z^{(P)} = x^P \tag{1}$$

It has been shown (23) that formally

$$\sum_{P=0}^{\infty} Z^{(P)} x^{-P} = \sum_{s=0}^{\infty} (UV)^s = (1 - UV)^{-1} \tag{2}$$

and since this sum has to diverge, because of Eq. (1), x has to be the largest root of the equation

$$f(x) \equiv U(x)V(x) - 1 = 0 \qquad (3)$$

The relations between observable properties of the system can be derived from $Z^{(P)}$.

Before going on to consider cases which have not been treated previously, we would like to illustrate the SGF method with a simple example. Consider an equilibrium binding of noninteracting dye molecules to a polymer in a solution where the dye has an activity a and q_0 and q_1 are the intrinsic, concentration-independent, statistical weights determined by the free energies of the empty and bound sites, respectively. The statistical weights of the corresponding sequences are given for this case by (23, 34)

$$u_i = q_0{}^i, \qquad v_j = (q_1 a)^j$$

The sequence-generating functions, to be denoted by U_0 and U_1 respectively, are

$$U_0(x, q_0) = q_0/(x - q_0) \qquad (4)$$

$$U_1(x, q_1, a) = q_1 a/(x - q_1 a) \qquad (5)$$

so that Eq. (3) for x may be written as

$$f(x) = (x - q_0)(x - q_1 a) - q_0 q_1 a = 0 \qquad (6)$$

Its largest root is $x = q_0 + q_1 a$ and its partition function is

$$Z^{(P)} = (q_0 + q_1 a)^P \qquad (7)$$

This simple result could, of course, have been derived easily without sequence-generating functions. The partition function for a single site having two states, empty and filled with statistical weights q_0 and $q_1 a$, is $q_0 + q_1 a$. If the sites are assumed to be noninteracting and, therefore, independent, the partition function of P sites is evidently $(q_0 + q_1 a)^P$. The powerful features of the SGF method are recognized, however, when it is applied to more realistic and thus more complex models.

We consider next the effect of neighbor interactions between the bound dyes. Let k represent the interaction coefficient, i.e., $-RT \ln k$ is the molar free energy of interaction of first-neighbor bound dyes. In a sequence of j bound dyes there are $j-1$ such neighbors, and its statistical weight is therefor $(q_1 a)^j k^{j-1}$. The sequence-generating function of bound dyes is now

$$U_1(x, q_1, k, a) = q_1 a/(x - q_1 k a) \qquad (8)$$

while $U_0(x)$ remains unchanged, so that Eq. (3) for x may be written as

$$f(x) = (x - q_0)(x - q_1 k a) - q_0 q_1 a = 0 \qquad (9)$$

The calculation of various average properties is the ultimate purpose of any statistical mechanical model, as it yields the required relations between observables. Such calculations are particularly simple with the SGF method and can be obtained by suitable partial differentiations of the function f, so there is no need to solve $f(x) = 0$ and obtain the partition function $Z^{(P)}$ explicitly. This is based on a general rule stating that the partial derivative of f with respect to any parameter which represents any particular state is proportional to the average number of occurrences of that state, i.e., to its relative abundance. For example, q_1 represents a bound state, whereas x^{-1} represents any state, bound or empty. Accordingly, the ratio of the number of bound dye molecules to the number of binding sites is

$$\frac{D}{P} = \frac{\partial f / \partial \ln q_1}{\partial f / \partial \ln x^{-1}} \tag{10}$$

Similarly, the ratio of first-neighbor bound molecules to the total number of bound molecules is $(\partial f / \partial \ln k)/(\partial f / \partial \ln q_1)$.

One can also calculate various other averages which are not readily derivable by other methods of statistical mechanics. An important class of such averages is F_m, the fraction of bound sequences of length m, i.e., the average number of such sequences divided by the average total number of bound dyes. (The corresponding relative number of dye molecules in such sequences is mF_m.) Such averages are of interest in the analysis of the change of spectral shifts with the degree of binding D/P. It is known that the spectral shift of isolated bound dyes is different from that of stacked dyes, and the question may be asked, in general, how are the spectral shifts and other spectral properties dependent on the length of the sequence. To obtain F_m we note that a sequence of length m is represented in the sequence-generating function $U_1(x)$ by its statistical weight $k^{m-1}(q_1 a)^m$. In order to differentiate with respect to this term, leaving all other terms constant one uses the following device. Define

$$U_1(x, q_1 a, k, \theta) = U_1(x, q_1 a, k) + (\theta_m - 1)k^{m-1}(q_1 a/x)^m \tag{11}$$

thus replacing $k^{m-1}(q_1 a/x)^m$ by $\theta_m k^{m-1}(q_1 a/x)^m$ in U_1; now differentiate f with respect to $\ln \theta_m$ and then put $\theta_m = 1$.

Thus, we may write

$$F_m = \frac{\partial f / \partial \ln \theta_m}{\partial f / \partial \ln (q_1 a)}\bigg|_{\theta_m = 1} \tag{12}$$

An evaluation of this equation and elimination of $q_0 x^{-1}$ and $q_1 a x^{-1}$ using Eq. (9) and (10) gives F_m as a function of the degree of binding D/P. The result for F_1, the relative number of isolated bound dyes as a function of D/P, is

$$P/D = [1 + (k - 1)F_1]/(1 - F_1^{1/2}) \tag{13}$$

Other F_m are derived by the same procedure, and it is easy to verify that F_m is related to F_1 in the present model by

$$F_m = F_1(1 - F_1^{1/2})^{m-1} \tag{14}$$

It will be noted that the ratio of singlets to doublets and higher multiplets is a function of F_1 only. Still, F_m depends on D/P and k according to Eq. (13).

Equations (13) and (14) have previously been derived (21) by a combinatorial method and applied to the analysis of spectra of a number of dye-polymer systems (3–5). Generally, good fit has been obtained with k values ranging from 1.2 for the acridine orange-DNA (AO–DNA) complex to about 4×10^4 for acridine orange-acid polysaccharide complexes.

An advantage to the SGF method is that it relates various observables to the same set of parameters. For example, from (9) and (10) we can obtain

$$k^2 a\left(\frac{q_1}{q_0}\right) + a^{-1}\left(\frac{q_1}{q_0}\right)^{-1} = \left(\frac{D}{P-D}\right) + \left(\frac{D}{P-D}\right)^{-1} + 2(k-1) \tag{15}$$

Thus, if one carries out binding studies (e.g., D/P vs. a), it is possible to obtain q_1/q_0 and $k^2(q_1/q_0)$ from the low and high occupancy values of $D/a(P-D)$, respectively. Such measurements would, therefore, not only provide an important parameter ratio appearing in the model but an independent measure of k, which as we have seen, has previously been determined to be 1.2 on the basis of the fit of F_1. P/D data on the AO–DNA system. Other experimental approaches may be found which can provide further independent measures of k and (q_1/q_0). If the values of either or both show considerable variation depending on the experiment by which they were determined, it would be advisable to begin to search for another model. Since Eq. (15) together with the spectrally determined value of $k = 1.2$ jointly predict a higher apparent binding constant $\equiv D/a(P-D))$ for AO on DNA at the higher accupancies, whereas most acridines show lower apparent binding constants at higher D/P's it would not be surprising to find that the k values determined independently from spectral shift and binding data were mutually inconsistent. We need not, however, wait for such detailed testing of the one-site-with-neighbor interaction model before examining models for AO–DNA with more than one binding mode, as will be seen in the next section. It must be emphasized that the relevance of a particular model to a particular physical system must be examined independently of its relevance to other systems. A one-site-with-neighbor interaction model which may not apply to acridine-DNA may apply to methylene blue-RNA, acridine-polyadenylate, Biebrich scarlet-histone, or oligonucleotide-polynucleotide complexes.

IV. Models of Competitive Modes of Binding

We wish to consider now models where several modes of binding compete with each other for the available binding sites on the DNA. A generalized SGF method can handle such models adequately. Let there be m different modes of binding. The considerations and conclusions drawn from experimental data with respect to these modes can be expressed by defining sequence-generating functions $U_0(x)$, $U_1(x)$, ..., $U_m(x)$ for the sequences of empty sites and the m modes of binding, respectively. The determinantal equation, considered as an equation for x,

$$f(x) = \begin{vmatrix} -1 & U_0 & \cdots & & U_0 \\ U_1 & -1 & U_1 & \cdots & U_1 \\ \vdots & & & & \\ U_m & \cdots & & U_m & -1 \end{vmatrix} = 0 \qquad (16)$$

has the required property, that its largest root x is related to the partition function $Z^{(P)}$ of the system by $Z^{(P)} = x^P$, i.e., by Eq. (1), as in the two-state models. All average properties of interest may be calculated from $f(x)$ by the same methods as those used for the two-state models, namely, by Eqs. (10)–(12) and other similar relations.

We shall now examine some of the experimental evidence for the assumption that there are several modes of binding and analyze the relations between the different possible modes. We shall then present for each mode its appropriate SGF.

A. Phosphate Binding

The following seems to be sufficient experimental evidence that one single mode of binding is predominant at high degrees of binding, i.e., the binding of the dye molecules on the phosphate groups.

1. The binding of acridines to DNA reaches saturation at a stoichiometric 1 : 1 ratio of phosphate groups to dye molecules (2, 4).

2. This 1 : 1 saturation ratio is not changed by denaturation (4).

3. The degree of binding decreases when the ionic strength of the solution increases (2, 6), indicating electrostatic interactions between the negative phosphate groups and the positive acridine ions.

B. Base-Pair Binding

There is also sufficient evidence that interaction between the dye molecule and the base pairs of the DNA participates in the binding process, particularly, at low degrees of binding.

1. X-Ray fiber diagrams (7, 15) show a partial disruption of the double helical structure at low D/P of acridines. They indicate either intercalation of the acridine dye between adjacent base pairs (7, 15) or possibly a rotation of adjacent base pairs relative to each other, effected by binding the dye molecule parallel to the base pairs (7). Whatever the exact nature of this binding may be, it is a mode essentially different from the external binding to the phosphate groups.

2. The specific viscosity of DNA increases with binding (15, 16) at low D/P, indicating elongation or stiffening of the DNA molecule. Autoradiographic measurements (18) of the length of DNA molecules show a statistical increase of the average length upon binding. Low-angle X-ray scattering shows a decrease of mass per unit length which may be interpreted in favor of intercalation.

C. Competition between the Two Modes

If base-pair binding would be the sole mode of binding, then the saturation ratio would be 1 : 2, dyes to nucleotides. If base-pair binding would exist parallel to phosphate binding, each mode being independent of the other, the saturation ratio would be 3 : 2. As the actual ratio is neither of the two, but rather is a 1 : 1 ratio, we are led to consider competition between phosphate binding and base-pair binding as an explanation for the observed behavior. For competition to take place, the two modes cannot coexist at the same time on the same nucleotide pair. This implies that the base-pair binding blocks in some way the two phosphate binding sites of the nucleotide pair. The choice is, then, between either one base-pair binding site or two phosphate binding sites. Naturally, the first one is favored at low degrees of binding, whereas the second is preferred at high degrees of binding (unless the binding constant of one mode is much larger than the other).

Further insight into the competitive nature of the two modes of binding is obtained from the change of the absorption spectrum with binding. At high degrees of binding the maximum absorption of AO shifts completely toward the 4650 Å region, which is typical of the stacked form of the dye. We might, therefore, assume that stacking stabilizes the phosphate-binding mode and helps to put this mode in a better competitive position at high D/P, where such stacking is more probable.

If two dye molecules are necessary to convert one base-pair binding site into a pair of phosphate-binding and if the dye-dye stacking strongly supports this mode, then we may neglect the binding of an isolated dye molecule to a phosphate group as insignificant. Then, phosphate binding occurs in pairs only, with each pair having a weight $k_1 q_1^2 a^2$, where k_1 is the stacking parameter of the pair of dye molecules. The SGF of such a sequence is

$$U_1(x, k_1, q_1, a) = \sum_{j=1}^{\infty} (k_1 q_1^2 a^2 x^{-2})^j = k_1 q_1^2 a^2 / (x^2 - k_1 q_1^2 a^2). \qquad (17)$$

The SGF of base-pair binding is similarly given by

$$U_2(x, k_2 q_2, a) = \sum_{j=1}^{\infty} k_2^{j-1}(q_2 ax^{-2})^j = q_2 a/(x^2 - k_2 q_2 a) \qquad (18)$$

Here the statistical weight of a sequence of j dye molecules bound to j base pairs is $k_2^{j-1}(q_2 a)^j$; if intercalation is assumed, then adjacent dye molecules are separated by a base pair; therefore, neighbor interactions should be neglected, i.e., $k_2 = 1$.

D. The Possibility of a Third Mode of Binding

Such a mode is tentatively indicated from some of the data related to the dependence of the Cotton effect on D/P and ionic strength for different adsorption bands. The Cotton effect behavior of acridines bound on DNA is, in general, a rich source of experimental data (8–12), but one which is, however, difficult to interpret quantitatively and unambiguously. We shall not attempt to discuss it in detail here, except for the particular behavior of the 4880 Å band, which led to the suggestion that it represents a special mode of binding (12).

The three circular dichroism bands of AO bound on DNA, at 5050, 4880, and 4650 Å, all tend to zero as $D/P \to 0$; they increase with D/P to a maximum and then reduce back to zero or to a small fraction of the maximum, as $D/P \to 1$. The maximum of the 4880 Å band, however, is at a rather low degree of binding, i.e., at $D/P \sim 0.1$, whereas the other bands have their maxima at D/P about 0.25 and 0.33, respectively. Moreover, the 4880 Å circular dichroism band increases, though mildly, with ionic strength, whereas the other bands decrease when ionic strength increases.

A simple explanation of these facts is that this band represents a particular mode of binding. Such a mode may be envisaged as an association to a relatively large number of sites, say, by binding along the external groove, thus covering a number of phosphate groups (the length of the AO molecule, including the van der Waals radii of the methyl groups, is almost 20 Å). Such a flat external adsorption may be due to the simultaneous effect of electrostatic, hydrophobic, and other forces. This mode would be limited to low D/P because each bound dye molecule apparently blocks a relatively large number of binding sites of the other modes.

The SGF of this mode is, therefore,

$$U_3(x, k_3, q_3 a) = \sum_{j=1}^{\infty} k_3^{j-1}(q_3 ax^{-s})^j = q_3 a/(x^s - k_3 q_3 a) \qquad (19)$$

where s is the number of sites occupied by one dye molecules in this mode.

The explicit evaluation of Eq. (16) and the derivation of various observable quantities, as well as their comparison with experiments, will not be attempted here.

As stated in the introduction, it is not the purpose of this chapter to decide upon various models of binding of acridines to DNA, but to provide a general method of establishing the relationships between observables which must follow from any proposed physical model. It is clear that qualitative relationships derived by intuition for models will not suffice to prove the consistency or inconsistency of models which have been proposed. The advantage of the SGF method lies in its ability to handle more complex physical models than other methods. As an introduction to its use we have described several models which have in whole or in part been treated by other statistical mechanical methods. Subsequently, a model which has heretofore not been presented and appears from a qualitative point of view to explain a number of heretofore puzzling features of binding of acridine dyes to DNA has been presented in some detail. We hope that the examples presented will serve as a guide for others in using the SGF method to test models of their own choosing.

REFERENCES

1. Brenner, S., Barnett, L., Crick, F. H. C., and Orgel, A. (1961). *J. Mol. Biol.* **3**, 121.
2. Peacocke, A. R., and Skerrett, J. N. H. (1956). *Trans. Faraday Soc.* **52**, 261.
3. Bradley, D. F., and Wolf, M. K. (1959). *Proc. Natl. Acad. Sci. U.S.* **45**, 944.
4. Bradley, D. F., and Felsenfeld, G. (1959). *Nature* **184**, 1920.
5. Stone, A. L., and Bradley, D. F. (1961). *J. Am. Chem. Soc.* **83**, 3627.
6. Drummond, D. S., Simpson-Gildemeister, V. F. W., and Peacocke, A. R. (1965). *Biopolymers* **3**, 135.
7. Neville, D. M., Jr., and Davies, D. R. (1966). *J. Mol. Biol.* **17**, 57.
8. Neville, D. M., Jr., and Bradley, D. F. (1961). *Biochim. Biophys. Acta* **50**, 397.
9. Blake, A., and Peacocke, A. R. (1965). *Nature* **206**, 1009.
10. Blake A., and Peacocke, A. R. (1966). *Biopolymers* **4**, 1091.
11. Mason, S. F., and McCaffery, A. J. (1964). *Nature* **204**, 468.
12. Gardner, B. J., and Mason, S. F. (1967). *Biopolymers* **5**, 79.
13. Lerman, L. S. (1963). *Proc. Natl. Acad. Sci. U.S.* **49**, 94.
14. Weill, G., and Calvin, M. (1963). *Biopolymers* **1**, 401.
15. Lerman, L. S. (1961). *J. Mol. Biol.* **3**, 18.
16. Drummond, D. S., Pritchard, N. J., Simpson-Gildemeister, V. F. W., and Peacocke, A. R. (1966). *Biopolymers* **4**, 971.
17. Luzzati, V., Masson, F., and Lerman, L. S. (1961). *J. Mol. Biol.* **3**, 634.
18. Cairns, J. (1962). *Cold Spring Harbor Symp. Quant. Biol.* **27**, 311.
19. Lerman, L. S. (1964). *J. Mol. Biol.* **10**, 367.
20. Bradley, D. F. (1961). *Trans. N.Y. Acad. Sci.* [2] **24**, 64.
21. Geisser, S., and Bradley, D. F., (1962). *Bull. Inst. Int. Statist.* **39**, 269.
22. Michaelis, L. (1947). *Cold Spring Harbor Symp. Quant. Biol.* **12**, 131.
23. Lifson, S. (1964). *J. Chem. Phys.* **40**, 3705.
24. Lifson, S. (1957). *J. Chem. Phys.* **26**, 727.

Alterations of T4 DNA Synthesis in the Presence of 9-Aminoacridine*

L. S. LERMAN AND S. ALTMAN†

*Department of Molecular Biology, Vanderbilt University
Nashville, Tennessee and
Department of Biophysics, University of Colorado
School of Medicine, Denver, Colorado*

I. Intercalation in Mutagenesis

A. The Requirements for Intercalability

The mechanism of the mutagenicity of the acridines differs from that of the base analog mutagens; instead of coding alterations that can be understood as the substitution of one base by another, the acridines elicit the appearance of the sequences which differ from the original by the deletion of one or more nucleotides or the insertion of one or more extra nucleotides. Although this analysis was first proposed on the basis of purely genetic arguments, it is now strongly supported by the comparison of amino acid sequences in normal and altered lysozymes produced by wild-type T4 and strains carrying a mutually suppressing pair of acridine mutations (Streisinger *et al.*, 1966). Mutagenic properties similar to these, which were established with the simple amino-acridines in T-even phages (and meiotic yeast), have recently been found for acridines with alkylating side chains and for quinacrine (which would not be expected to exhibit alkylating activity) in *Escherichia coli* (Whitfield *et al.*, 1966). It seems clear that all insertion-deletion generating compounds currently recognized can bind to DNA by intercalating into the helix, but despite the attractive correspondence between the intercalated structure and the character of the sequence alterations, it has not yet been demonstrated either that the insertions and deletions correspond to sites of intercalation or that these alterations depend on intercalation in a significant way.

* This work was supported by a research grant from the United States Public Health Service (GM-13767) and by a grant from the National Science Foundation (GB-4119).

† University Fellow, University of Colorado; present address: Biological Laboratories, Harvard University, Cambridge, Massachusetts.

B. Correlation of the Concentration Dependence of Mutation and Intercalation

1. Photodynamic Inactivation Studies

Suggestive evidence for the role of intercalation in the mutagenic mechanism is provided by measurement of the fractional yield of induced phage mutants as a function of acridine concentration in the medium. In the example shown in Fig. 1 it can be seen that there is an optimal concentration above which the fractional yield of mutants declines; it lies between 10^{-5} and 10^{-4} M.

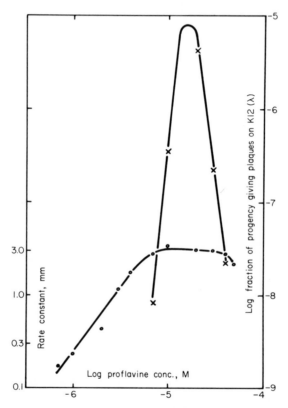

FIG. 1. The fractional yield of mutants after one-step growth in the concentration of proflavine specified on the abscissa is shown by the upper curve (crosses). The same lysate was exposed without dilution to white light from a tungsten projection lamp; the intensity at the solution was about 600 ergs mm^{-2} sec^{-1}. The rate constant for the exponential part of the phage inactivation kinetics, measured as plaque formation on *E. coli* B, is shown by the lower curve (circles). Light absorption by free proflavine in the medium accounts for the small decline in rate at high concentrations.

The lower curve represents the rate constant for photodynamic diminution of total phage yield in the same experiment. (Although most of the subsequent work to be reported was carried out with 9-aminoacridine, the incapacity of that substance to effect photosensitization necessitates the use of a different mutagen to permit this analysis of intraphage binding.) It will be seen that the optimum fractional yield of mutants occurs at the same concentration as the maximum photosensitization of phage. The significance of the precise magnitude of this concentration for comparing the two types of experiments and *in vitro* measurements is rendered uncertain because of the unknown contributions of the environment to the binding equilibrium. That is to say, studies *in vitro* on pure DNA have shown that binding is sensitive to salt concentration, solvent, and the presence of other cationic substances, and the relative contributions of these in the comparison of binding to pure DNA with binding to the interior of a phage or to DNA in an infected *E. coli* cell is not known at present. Nevertheless, the concentration at the maximum, 2×10^{-5} M, has been shown by Peacocke and others to saturate the strong binding mode with pure DNA; this binding is now identified with the intercalated structure. Very little, if any, external binding of dye aggregates occurs unless the external concentration is much higher. Thus, the rough proportionality of the photosensitization to proflavine concentrations at low concentrations together with saturation of the sensitization at 2×10^{-5} M is compatible with the supposition that phage is sensitized by intercalated dye. There are no indications of the participation of dye aggregates, which could hypothetically either provide additional photosensitization or diminish the photosensitivity by displacing intercalated proflavine; it is conceivable that phage heads are unable to accommodate externally bound dye aggregates. For our present purpose the important feature of the mutagenicity curve is the decline in mutagenicity after the maximum has been reached. With the reasonable assumption that the binding to the DNA of whole phage is at least an approximate measure of the binding to phage DNA in the infected cell, the failure of the mutation yield to continue to increase after the intercalation limit indicates that this biological action cannot be attributed to dye that is bound in the form of externally attached aggregates. Hessler (personal communication) has found that the mutant yield given by an early exposure to proflavine followed by dilution (in order to achieve a substantial burst size) results in a concentration-dependence function that again rises to high values at about the same external concentration as shown here, but remains the same for increased concentrations, rather than declining. Thus, her data support the same conclusion—that only the intercalated acridine is effective. Unfortunately, this argument does not exclude the possibility that the binding to some other substance, for example, an enzyme, that is characterized by the same structural requirements and the same affinity as intercalation, effects the mutagenic event.

2. Cooperative Effects in Mutation

From the steepness of the increase in mutation with increasing proflavine concentration, it may be inferred that single acridine molecules do not act independently to induce a mutation. We cannot at present distinguish between various conjectures—the mutational event may require the drastic gross structural alteration that is achieved with maximal intercalation, or the local cooperative effect of two or three acridine molecules could be implicated. The decline in yield of mutants at higher concentrations could be due similarly to the displacement of intercalated acridines by externally attached stacks or additional local interactions. It is attractive to speculate that the intercalation of a small number of acridine molecules facilitates erroneous pairing between nucleotide strands that are complementary for a short region. The error could result merely from the displacement of the partners by one, two, or more steps away from the correct pairing positions, as might be probable where the sequence is repetitive. The exchange of base pairs between sequences that differ from homology by the presence of an additional pair in one could be imagined to become permissible when an intercalated acridine in the second helix brings the entire sequence into register by filling the gap. A recombinational event which included mispaired strands would generate a covalently continuous polynucleotide sequence containing an insertion or deletion of a length determined by the pairing error and at a site determined by the crossover. Since the mutations measured in these experiments must function as suppressors of mutations already carried on the chromosome of the parental phage, the specifications of the new error are more or less predetermined, and specificity with respect to the number of intercalated acridines required, depending on which particular mutant is to be suppressed, might be expected. Evidence that different mutations show a different concentration dependence in the fractional yield of revertants has been found (Morse, 1966) and will be presented elsewhere. Evidence for the participation of a recombinational event in the spontaneous, rather than acridine-induced, reversion of these mutants has been provided by Strigini (1965), who showed that the first phage carrying the new mutations is heterozygous for other nearby markers on either or both sides of the mutated site.

Acridine-induced mutations, however, occur at concentrations in which recombination between two parents in a mixed infection cannot be demonstrated. If the mechanism requires an event related to recombination, it must operate incestuously and would remain undetectable by conventional genetic experiments. Recombination between progeny and parental strands might be anticipated as a means of circumventing a local block to DNA synthesis; it has been proposed by Rupp and Howard-Flanders (1967), as the means by which sequences in which thymine dimers have been induced by ultraviolet irradiation are propagated properly.

II. Perturbation of DNA Synthesis

A. The Intermediate Forms in the Normal Replication of T4

In order to approach the direct molecular investigation of the hypotheses of this sort, we have found it desirable first to examine some of the more general properties of the system that synthesizes T4 DNA in infected *E. coli* cells and to determine the effects of the presence of aminoacridines on this system. While we have not yet reached the level of resolution that will detect that part of the synthesis of a DNA molecule that is associated with the mutational event, the results are interesting and offer some indications for the direction of further work. The kinetic studies from which some of these experiments are selected will be presented in detail elsewhere (Altman and Lerman, 1968).

1. Sedimentation Analysis

If the contents of *E. coli* cells are examined immediately after infection by T4 and later as a function of time, phage DNA can be detected in at least five different types of particles that are readily distinguished by their characteristic sedimentation coefficients; these are indicated in Table I, together with our terminology. Our observations have depended principally on the distribution of radioactively labeled material as a function of the distance of sedimentation

TABLE I

SEDIMENTATION SPECTRUM OF THYMINE LABEL
IN INTRACELLULAR T4 DNA[a]

Log $s_{20,w}$	$s_{20,w}$	Designation	Density in CsCl
1.78	60	Slow	1.7
2.52	330	Early fast	1.7
2.74	530	Late fast	1.7
2.92	830	Free phage	1.5
3.30	2000 ⎱	Bottom	1.35
3.54	3500 ⎰		

[a] Acridine absent.

after centrifugation to a certain value of force × time with density gradient stabilization of the zone, using either sucrose or heavy water.

2. The Early Slow Component

The slow component, which sediments at a rate similar to that of DNA extracted from phage by phenol, is found immediately after infection, before synthesis begins, and is composed entirely of parental DNA. Our lysis procedure does not liberate DNA from viable phage. Coincident with the beginning of synthesis, at least one-half disappears. New DNA appears in the component relatively late in development, corresponding both in time and amount produced to the new phage which also appears.

3. The Bottom Component

Concurrent with its disappearance from the slow component, parental DNA appears in the bottom component, along with the first newly synthesized DNA. As judged by its density in cesium chloride, DNA is altogether a relatively minor constituent of this material. Growth conditions or agents that tend to destabilize the bacterial cell wall diminish the amount that is found. It is relatively heterogeneous with respect to sedimentation although the maximum sedimentation coefficient is much less than that of intact cells.

4. The Fast Component

The first new DNA that is found without association with cell material, as indicated by its bouyant density, is characterized by an extremely large sedimentation coefficient and is designated here the fast component.

Under the conditions of these experiments host DNA is not labeled, it makes a negligible contribution to phage DNA; and phage DNA is synthesized continuously at constant specific radioactivity. By the use of parental phage labeled with a different isotope and the use of pulse labeling where necessary, we have been able to make reasonably reliable estimates of the kinetics of synthesis and the sequential relationships between various DNA species. The estimates of sedimentation coefficients are derived from consideration of the centrifugal field, the time of centrifugation, and the density and viscosity gradients in the medium, together with cross-checks against the sedimentation of the intact DNA extracted directly from mature phage. The kinetic data for the infection of *E. coli* strain G by an *rII* mutant is shown in Fig. 2. The time course of the distribution of both parental and newly synthesized DNA are indicated. It appears to be the product of the bottom component or perhaps a specialized sample of DNA contained in the bottom component, but kinetically it is clearly distinguishable. Its extraordinary sedimentation coefficient leads to the obvious question of its molecular weight. Extrapolation of any of the familiar molecular weight-sedimentation coefficient relations would suggest that the size of these particles is over one hundred times that of phage DNA.

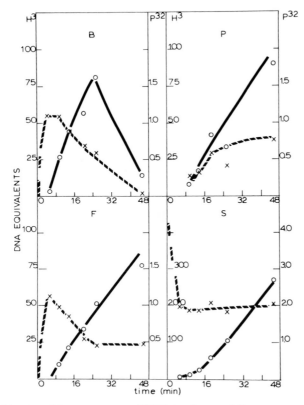

FIG. 2. The kinetics of the appearance of DNA in phage and its precursors. *E. coli* strain G were infected with an average of four ^{32}P-labeled phage (T4B carrying the proflavine-induced *rII* mutation, P3) in a medium containing salts, glucose, tryptophan, and *p*-fluoro-phenylalanine. Development was initiated at time zero by transfer to a new medium without inhibitor and containing fluorodeoxyuridine, ^{3}H-labeled thymidine, an amino acid mixture, glucose, and a lower salt concentration. At the time indicated on the abscissa the cells were lysed with lysozyme, EDTA, and sodium lauryl sulfate, and the lysates were analyzed without further treatment by zone sedimentation in 5–20% sucrose gradients containing appropriate ions and buffer concentrations. Each point represents an integrated isotope measurement from appropriate contiguous fractions in the sedimentation distribution. The letters B, P, F, and S represent the bottom, phage, fast, and slow components as defined in Table I. DNA is indicated in units equivalent to the content of one phage particle per infected cell. Newly synthesized DNA is indicated by the solid line, referring to the scale on the left, and parental DNA is indicated by the dashed line referring to the scale on the right of each graph.

Although we have no data that correspond to a conventional molecular weight estimation, it may be noted that this material is clearly detectable in our infected cells when the average amount of DNA synthesized per infected cell is not greater than the equivalent of 5–10 phage particles.

Although it might be tempting to reconcile the small amount of synthesis with the supposed high molecular weight by assuming that all of the early synthesis has occurred in only a very few cells, the development of infected cells is effectively synchronized in our experiments by the use of an amino acid analog at the time of infection, and a uniquely high rate of DNA synthesis would also have to be assumed in those cells. Other evidence for a modest value of the molecular weight of what may be the same material has been advanced by Kozinski et al. (1967). Firm attachment of both RNA and portein could conceivably modify a DNA-containing particle to give a high sedimentation coefficient together with an apparently normal bouyant density in CsCl. We have been unable to detect any phage-induced RNA in the fast component, however, using an independent tracer. Fast-sedimenting T4 DNA has also been seen by Frankel (1966a, b), but his calculation of sedimentation coefficient is somewhat different from ours. Our experiments also appear to detect an even faster component about which we know essentially nothing else. It may be identifiable with phage heads or some other partially assembled structure.

5. Phage

Since our lysis procedure does not interfere with the viability of completed phage, they are detected both by their infectivity and as a separate sedimentation peak. Some phage also appears associated with the bottom component and our kinetic data are always appropriately corrected.

6. The Late Slow Component

More rigorous kinetic analysis shows indeed that the slow component, which it will be recalled resembles phage DNA in sedimentation rate, is a by-product rather than an intermediate in phage synthesis. Although the transfer of a small part of this component back into the stream of phage synthesis is not precluded by our data, for the most part this material must be regarded as garbage. Although a constant amount of parental DNA appears to be retained in the slow component following its initial drop, other arguments indicate that the curve reflects a delayed utilization of some of the entering DNA together with a gradual accumulation of parental label in association with the accumulation representing new synthesis. The scheme of synthesis is summarized in Fig. 3.

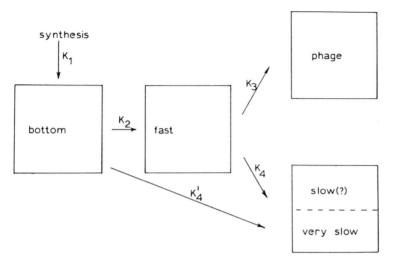

FIG. 3. A formal scheme for the sequence of intermediates in phage DNA synthesis. The components are the same as those indicated in Fig. 2 and Table I.

B. The Effects of 9-Aminoarcridine

1. Changes in Rates

In the presence of 3.5×10^{-5} M 9-aminoacridine, a concentration approaching the optimum for mutagenicity, the kinetics are drastically altered, as shown in Fig. 4. First, it can be seen that after 10 minutes the overall rate of DNA synthesis is only slightly lower than that in the absence of acridine. For a phage carrying both ac and q markers for acridine resistance, total synthesis is 0.68 as fast, and without the resistance markers it is still 0.55 as fast. Drastic changes are seen, however, in the internal kinetics of the intermediates; these can be enumerated as follows: (1) The initial rapid phase of the disappearance of parental DNA from the slow component is nearly five times as slow. (2) The amount of bottom component, as indicated by newly synthesized DNA, is diminished roughly threefold, and the amount of parental DNA it contains is similarly reduced. (3) The amount of fast component is even more drastically diminished, and its appearance is somewhat delayed. (4) The number of complete phage that appear, as judged by viable titer, is diminished about a hundredfold, so that the amount of tracer attributable to these infectious particles lies below the limit of error imposed by the spreading of neighboring peaks in the sedimentation distribution. (The amount of DNA calculated by

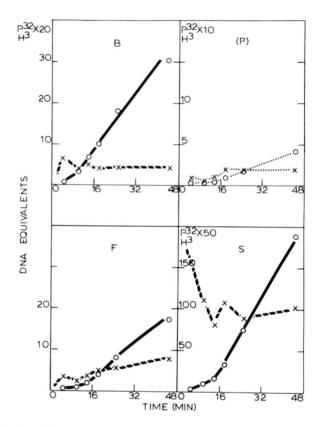

FIG. 4. Kinetics of the appearance of DNA in phage precursors in the presence of 9-amino-acridine (9AA). Except for the addition of 3.5×10^{-5} M 9AA at the time of infection, all conditions are identical to those resulting in the data of Fig. 2.

routine integration of the sedimentation patterns is indicated in the figure nevertheless and will be seen to resemble the time course of the fast component.) (5) Nearly all of the DNA that is produced can be accounted for in the slow component. Interpretation of these data in terms of the formal kinetic scheme shown in Fig. 3 is straightforward after the 10-minute point. The rates corresponding to K_1, K_2, K_3, and K_4 for phage-carrying acridine resistance markers are reduced to 0.68, 0.52, 0.16 and 1.14, respectively, of their normal value in the presence of 3×10^{-5} M 9-aminoacridine. The relative value of these changes is so much smaller than the change in the frequency of mutants (relative to the spontaneous frequency) induced by the presence of this concentration of acridine, that their relevance is not obvious.

2. The Size of Particles in the Slow Component

There is, however, a more substantial difference that is not apparent in the formal kinetics; the slow component synthesized in the presence of 9-amino-acridine sediments much more slowly than normal slow component; more extensive sedimentation studies support the inference that it is of much lower molecular weight. In addition, the sedimentation distribution changes continuously after the first appearance of new DNA in this component. Representative data are shown in Fig. 5. It can be seen that in the 10-minute lysate

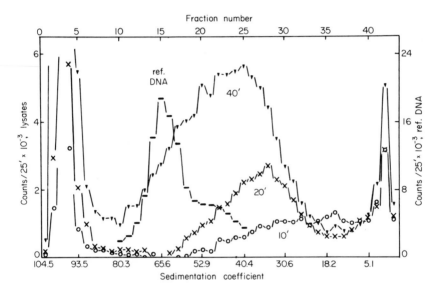

FIG. 5. The sedimentation distribution of slow component produced in the presence of 3×10^{-5} M 9AA in cells lysed 10, 20, and 40 minutes after development is initiated. Conditions were generally similar to those in the experiments of Fig. 4 except that the phage carried acridine resistance markers. The peaks at the extreme left represent a mixture of all more rapidly sedimenting components accumulated at the boundary of a very dense solution at the bottom of the centrifuge tube. Reference DNA was obtained by phenol extraction of labeled phage. Sedimentation coefficients have been calculated using a polynomial representation of the viscosity and density gradients by means of a computer program provided by Blattner (1966).

the DNA is exceedingly heterogeneous and is of very low molecular weight, corresponding to about one-tenth the size of phage DNA at the center of its distribution. The changes at later times appear to reflect the appearance of longer and longer fragments in addition to the small pieces that are produced first. Even at 40 minutes the estimated modal molecular weight is about half that of phage DNA.

3. Small Fragments and Mutation

We can speculate on the possible significance of the low molecular weight particles with respect to mutagenesis. The presence of small particles indicates the production in the infected cell of a larger number of molecular ends per unit mass DNA than are detected under normal conditions. Obviously, for each end of the helix on a small molecule there must have been either another end from which it has been broken or a new initiation point. Since we have presently no basis for identifying these—for example, they could become repaired rapidly inside the replicating system—it will be convenient merely to enumerate them by counting the small particles. If the sedimentation distribution is converted to a molecular weight distribution, the mass of DNA as indicated by the amount of tracer in each fraction can be used to calculate the number of particles, and correspondingly, the number of ends. In the normal infection we assume that both mature phage DNA and the slow component are equivalent and monodisperse and have the usual molecular weight. The comparison with acridine-induced slow component will obviously become less extreme with increasing time during infection, but it is plausible to suppose that we are interested in relatively early times on the basis of the mutation kinetics. The fractional yield of mutants among the progeny drops twofold if the proflavine is diluted at 10 minutes, although the increased burst size indicates that acridine is indeed removed. The complementary experiment in which the addition of acridine is delayed (in this case carried out with 9-aminoacridine) shows that most of the mutations are generated by acridine present before 10 minutes and 80% are generated by acridine present before 15 minutes. Accordingly, for the 10- and 20-minute samples of slow component corresponding to the central peaks in Fig. 5, we find that the number of particles is 36 and 13 times as great as the same mass of DNA would contain if divided into normal phage-size lengths. If we consider only the material that has been added between 10 and 20 minutes, we find that it consists of only 4.8 times as many particles as the same amount of phage DNA. Thus, the relative preponderance of free ends is produced quite early in the infection. (Since the low molecular weight tail of the distribution makes a major contribution to this calculation and the limit of the 10-minute sample is poorly defined, the distributions were arbitrarily terminated at the minimum before the right-hand peak, which is thought to be an artifact associated with the meniscus in collection of the sedimented material.) There are at least two obvious reasons why these values are likely to be underestimates—perhaps very seriously in error, for the numbers of ends. The smallest particles make the largest contributions, and we have not included in our calculations the material in the upper part of the gradient where the present measurements are not useful. In addition, we are estimating only ends where both strands are broken. It is conceivable that a single-strand interruption may be biologically equally significant, but measurements under conditions of strand separation, which would provide this parameter, are not yet available.

4. The Association of Recombination with the Ends of the Genome

Although any relation between ends and mutations remains purely con-jectural, there are hints that may be relevant. Mosig (1963, 1967) has shown by crosses with incomplete phage genomes that recombination is sixfold higher when a marker is within 2 % of the end of the genome as compared with the same marker in a more central position in other permuted phages. Womack (1966) draws similar conclusions on the basis of single burst analysis of multiple factor crosses (see also Doermann and Parma, 1967). Drake (1967) has pointed out that the long regions of heterozygoses observed by Strigini (1965) are more comfortably understood as terminal redundancy, rather than internal, heterozygotes. At a purely schematic level, it would seem simpler to invoke out-of-phase base pairing induced by intercalation near the end of the molecule where its termination renders unnecessary the restoration of correct phase relationships for continued pairing; Streisinger *et al.* (1966) have already presented a similar argument.

5. Comparison with the Effects of Actinomycin

These changes seen in the presence of 9-aminoacridine may be compared with the effects of actinomycin, which has been introduced into T4-infected cells by Korn (1967) by means of ethylenediaminetetraacetic acid (EDTA) treat-ment. Korn finds the accumulation of nearly all the newly synthesized DNA in a rapidly sedimenting component which is characterized by a sedimentation coefficient (when calculated by our method) closely similar to that of our early fast component. Although the acridines have been useful for the inhibition of transcription, it would appear that the principal effect here is substantially different from that of actinomycin.

III. The Relief from Acridine Inhibition

A. The Incorporation of DNA Synthesized in Acridine in New Phage

Although it is clear that the normal slow component is not a direct inter-mediate in phage synthesis, the slow component that is accumulated in the presence of acridine may serve as a direct source of phage DNA when acridine is removed. The kinetics of this process are shown in Fig. 6, where zero time

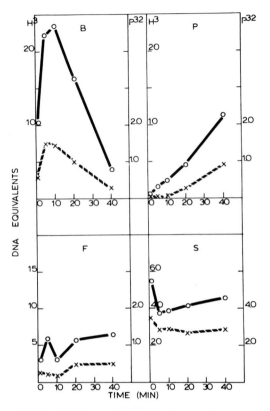

FIG. 6. Kinetics of the appearance of DNA in phage and precursors after removal of 9-aminoacridine. The experiment was started with 3×10^{-5} M 9AA, and 20 minutes later the cells were centrifuged and resuspended without 9AA. Time zero represents the beginning of incubation after drug removal.

represents the initiation of incubation in fresh medium after the infected cells have been allowed to develop for 20 minutes in 3×10^{-5} M acridine and then freed of the drug. All of the thymine label indicated in the figure was acquired during synthesis in the presence of acridine, and the new medium is devoid of any tracer. The phosphorus, as previously, indicates DNA derived from parental phage. It can be seen that there is a rapid initial drop in slow component together with a rapid initial rise in bottom component, suggesting that in the absence of acridine the fragments are taken up into the synthesizing system that is associated with the cell wall in order to be reassembled into complete molecules that can be used for phage. Within 40 minutes roughly one-half of the amount of DNA that initially was withdrawn from the slow component has appeared in new complete phage. Since there is an appreciable amount of DNA as bottom component at the end of the period in acridine, this experiment fails to demonstrate unequivocally that DNA from the slow component actually becomes part of the phage that finally appear, although it does demonstrate its entry into the synthetic apparatus. Nor does it indicate whether the labeled DNA enters as long continuous segments or in a highly dispersed state.

B. The Requirement for Further DNA Synthesis

1. Limitation of Synthesis with a Temperature-Sensitive Mutant

In order to determine how complex the process is by which the DNA synthesized in acridine is converted to phage, we have examined its dependence on DNA synthesis after acridine is removed. One experiment of this sort is shown in Fig. 7. A marker, provided by Dr. Robert Edgar, which confers temperature sensitivity on the product of gene 42, cytosine hydroxymethylase, has been introduced into our standard phage stock, retaining the other markers.

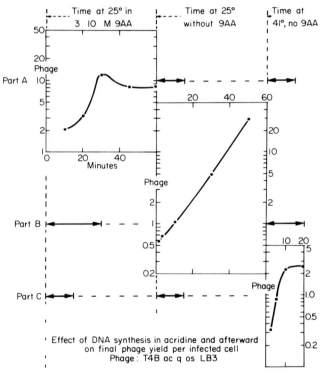

FIG. 7. The requirement for DNA synthesis after acridine removal. Cells infected with T4B P3 *ac q os* carrying a temperature sensitive mutation in gene 42 (and synchronized with fluorophenylalanine) were incubated, as indicated, at 25°C with 3×10^{-5} *M* 9AA, freed of 9AA and continued at 25°C, and then raised to 41°C before lysis and plating. The time schedule for the three steps is different in parts A, B, and C in that two steps are carried out for fixed intervals, as shown by the bold arrows, and the time for the remaining step is varied as shown by the graphs.

Because of the instability of the enzyme, hydroxymethylcytosine and, consequently, DNA, cannot be synthesized at 41°C, but there is sufficient synthesis at 25°C to permit a substantial phage burst. The infected cells are first incubated at 25°C in the presence of 3×10^{-5} M acridine, they are then washed and incubated for another time period at 25°C to allow an additional controlled amount of DNA synthesis in the presence of 9-aminoacridine, and they are then raised to 41°C to permit the completion of any maturation processes that do not depend on DNA synthesis. The experiment shown in the figure is divided into three parts; in each part the time of incubation in two of the time periods is fixed, whereas the time in the third is systematically varied in order to establish the contribution of the events occurring in that period to the final product. Part C, reading from left to right at the bottom of the figure, shows that after 30 minutes in acridine and 15 minutes without acridine at 25°C, at least an additional 10 minutes of incubation at high temperature is required for the maximum phage yield to be achieved, but no further increase in phage yield occurs with longer incubation at elevated temperature. It can be seen that ample time at 41°C has been allowed in parts A and B. Part A shows that the synthesis which occurs in the presence of acridine, if allowed to continue for at least 30 minutes, is required in order to achieve a substantial phage yield, if synthesis in the absence of acridine is severely limited. Part B, where long synthesis in acridine has been allowed, shows that virtually no phage can be produced, even allowing maturation at 41°C, unless some synthesis in the absence of acridine is permitted.

2. Limitation by Thymine Starvation

The conclusion that further DNA synthesis is necessary in order that the material made in acridine be converted to viable phage is supported also by a related experiment in which synthesis after acridine removal is restricted by limitation of the available thymidine in the presence of fluorodeoxyuridine. It is attractive to suppose that the additional DNA synthesis is necessary to establish the connections between fragments. Although connections may constitute part of the requirement, it is expected that the process is considerably more complex.

A biological effect which would appear to be associated with an assembly process is observed as an elevated recombination frequency in a two-factor cross under similar conditions (Hessler, 1966).

IV. Discussion

It would seem that any proposal for the mechanism of the generation of small molecules must take into consideration the increase in size of the fragments as time goes on. If the effect is to be attributed to a simple chemical interaction of acridine with the replicating DNA, something about the replication must differ between early and late times. If the early small fragments were to represent precocious rupture close to the starting point of replication, it is difficult to see how this process could progressively select longer fragments. In the absence of direct evidence it seems simpler to suppose that the increasing length of the fragments is due either to a diminution of the primary interaction with the acridines owing to the gradual accumulation of an interfering substance generated during the infection or that the primary interaction remains constant, continuously generating helix interruptions or their precursors, and a repair system whose rate of appearance is inhibited by the acridine, gradually comes into play. This system would presumably have to act on some disturbed structure that precedes liberation of the fragment, and not after the rupture. The existence in the infected cell of processes that might be responsible for the fragmentation even in the absence of acridine is suggested by examination of the entering DNA after infection with an amber mutant that is incapable of producing even noncovalent junctions between DNA molecules (Tomizawa et al., 1966). A broad distribution of fragments is produced, reportedly down to one-tenth the size of phage DNA. No new DNA is synthesized. One example of a model resulting in fragmentation and mutation in the presence of acridine might be imagined as follows.

Suppose that the roles of the two component strands of the parental double helix are not entirely identical, either because of dissymmetry in the replicating structure or because the enzyme site effecting 5'-addition is not identical in properties with the site providing 3'-addition. If synthesis along one template strand is delayed because of the presence of acridines, perhaps in a particularly sensitive sequence, that segment might comprise a three-strand complex in which the replicated template strand has the possibility of pairing either with its original partner or with the newly synthesized strand. If the base sequence is repetitive, the intercalated acridines would permit an out-of-phase association between the paired strands and the third strand. The branch whose replication is blocked may be nipped off or perhaps undergo an exchange of a phosphodiester connection with the termination of the newly synthesized strand (they are of the same polarity). In this way a DNA fragment is liberated, and the template helix, which now contains a replicated segment, has been reconnected; but the reconnection may have introduced an insertion or deletion.

V. Summary

Some properties of the response of T4-infected *E. coli* cells as a function of external acridine concentration suggest that mutagenesis is due to one or several intercalated acridines. The effects of 9-aminoacridine on the kinetics of

DNA synthesis in this system has been analyzed in terms of the separate inhibition of a number of sequential reactions leading to the production of mature phage. The largest effect is on the step by which a replicative intermediate form of DNA, which does not resemble normal phage DNA, is converted into mature phage. There is nevertheless a copious production of low molecular weight DNA, nearly equal to the amount of DNA normally packed into viable phage when acridines are absent; the first fragments that appear are very small, but they are produced with a constantly increasing molecular weight. A hypothesis is offered that the number of these particles reflects the number of a certain type of unusual events occuring in the presence of acridines which is part of the mutational process and may be related to recombination.

ACKNOWLEDGMENTS

Some of the experiments reported here have been carried out by Mrs. Suzanne Swanson, Mr. S. L. Allen (Lt. Col., USAF, retired), and Mrs. Ann Knapp, to whom we are deeply grateful.

REFERENCES

Altman, S., and Lerman, L. S. (1968). In preparation.
Blattner, F. (1966). Personal communication.
Doermann, A. H., and Parma, D. H. (1967). Conference on chromosome mechanics at the molecular level. Oak Ridge National Laboratory (in press).
Drake, J. (1967). Personal communication.
Frankel, F. R. (1966a). *J. Mol. Biol.* **18**, 109.
Frankel, F. R. (1966b). *J. Mol. Biol.* **18**, 127.
Hessler, A. (1966). Personal communication.
Korn, D. (1967). *J. Biol. Chem.* **242**, 160.
Kozinski, A. P. , Kozinski, B. P. and James, R. (1967) *J. Virol.* **1**, 758.
Morse, H. (1966). Ph.D. Thesis, University of Colorado School of Medicine.
Mosig. G. (1963). *Cold Spring Harbor Symp. Quant. Biol.* **28**, 35.
Mosig, G. (1967). Unpublished data.
Rupp, W. D., and Howard-Flanders, P. (1967). *Biophys. J.* **7**, 79.
Streisinger, G., Okada, Y., Emrich, J., Newton, J., Tsugita, A., Terzahgi, E., and Inouye, M. (1966). *Cold Spring Harbor Symp. Quant. Biol.* **31**, 77.
Strigini, P. (1965). *Genetics* **52**, 759.
Tomizawa, J., Anraku, N., and Iwama, Y. (1966). *J. Mol. Biol.* **21**, 247.
Womack, F. (1966). Unpublished data.
Whitfield, H. J., Martin, R. G., and Ames, B. N. (1966). *J. Mol. Biol.* **21**, 335.

Interaction of Antibiotics with Nucleic Acids*

W. KERSTEN AND H. KERSTEN

Physiologisch Chemisches Institut
Universität Münster
Münster, Germany

The complex formation of actinomycin with DNA has been extensively discussed by Reich and Goldberg (1964). Studies of the interaction of acridines with nucleic acids will be presented elsewhere in this volume. We would like to present some results of comparative studies on the complexes of DNA with two other groups of substances.† These are antibiotics referred to as chromoglycosides. The first group to be discussed are the anthracyclines (daunomycin, nogalamycin, and cinerubin) with a tetrahydrotetracenquinone ring system as chromophore. The second group includes derivatives of tetrahydroanthracenes (chromomycin, mithramycin, and olivomycin). All these antibiotics inhibit to a different extent DNA-dependent processes *in vitro* and within the cell. The question was raised whether the complexes of DNA with these two groups of antiobiotics are like the actinomycin-DNA complex or do they resemble in some respect the interaction of DNA with acridines. The physicochemical interactions of the antibiotics with DNA were studied to evaluate the similarities and differences in the molecular mechanism of action as related to the chemical structure.

I. Daunomycin, Nogalamycin, and Cinerubin

The complete structures of these substances are not yet known. The structures of the chromophore of daunomycin and the sugar daunosamin have been elucidated (Arcamone *et al.*, 1964) (Fig. 1). Nogalamycin has two sugars (Wiley, 1965) and cinerubin three sugars (Ettlinger *et al.*, 1959) attached to the chromophore.

A. Effect on Thermal Transition (T_m) of DNA

The T_m value of DNA is markedly increased upon addition of cinerubin, nogalamycin, or daunomycin (Kersten *et al.*, 1965, 1966). A shift in T_m of 10°C is achieved at a molar ratio of 1–3 to 120 for drug to DNA-P. To get a

---

*This work was supported by Deutsche Forschungsgemeinschaft.

† Abbreviations used in this chapter: DNA, deoxyribonucleic acid; DNA-P, DNA-phosphate; dAT, an alternating copolymer of deoxyadenylate and deoxythymidilate; GMP, guanosine monophosphate; dGMP, deoxy-GMP, CMP, cytidine monophosphate; dCMP, deoxy-CMP; sRNA, soluble ribonucleic acid.

comparable shift in the T_m of DNA with proflavine the concentration needed is about tenfold; for acridine orange the concentration needed is even higher. Actinomycin also shifts the T_m to higher values, whereas chromomycin and mithramycin have no influence up to a molar ratio of 25 to 120 DNA-P.

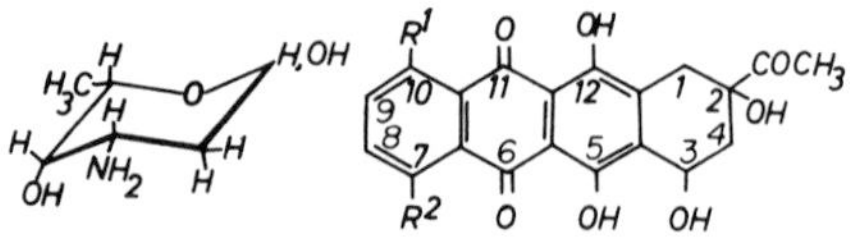

FIG. 1. Structure of daunomycin.

B. Effect on Viscosity and Sedimentation of DNA

The anthracyclines can raise the reduced specific viscosity (η/η_c) of DNA threefold. As a consequence, the sedimentation coefficient of DNA is lowered upon association with these compounds. At comparable concentrations actinomycin affects the hydrodynamic properties of DNA in the opposite way: the viscosity is lowered and the sedimentation coefficient increased. The tetrahydroanthracene glycosides have no effect on the viscosity of DNA under the same experimental conditions (Kersten *et al.*, 1966).

The effects of anthracyclines are quite analogous to those observed for the acridine dyes, indicating stiffening and/or elongation of the DNA molecule. Binding of the anthracyclines to DNA, however, is little affected by high concentrations of CsCl, whereas the acridines can be displaced by increasing CsCl concentrations. Because of the stability of the anthracycline-DNA complexes at high ionic strength one can measure the influence of the antibiotics on the buoyant density of DNA.

C. Effect on Buoyant Density of DNA

Depending on its conformation and base composition DNA exhibits a characteristic density in CsCl and $CsSO_4$ gradients. The density of DNA in CsCl becomes markedly depressed by the antibiotics in the order: nogalamycin > cinerubin > daunomycin. This effect, however, is not specific for the anthracyclines, because mithramycin, chromomycin, olivomycin, and actinomycin also decrease the buoyant density of DNA. A specific difference between both groups of antibiotics has been observed by measuring their influence on the buoyant density of dAT. Only the anthracyclines drastically depress the buoyant density of this synthetic polymer, indicating that chromomycin, mithramycin, and olivomycin need either cytosine or—like actinomycin—guanine for binding.

From the data presented it is evident that the interaction of anthracyclines with DNA in many respects resembles the interaction of acridines with DNA.

The acridines are cationic dyes which can be displaced from the DNA by high ionic salt concentrations. According to Lerman (1962) these substances intercalate between adjacent layers of nucleotide pairs. A modified model for intercalation was recently proposed by Pritchard et al. (1966). The fact that some acridines are strong mutagens can be explained by both models. The anthracyclines, however, are not mutagenic (Tabaczynski et al., 1965), and with the exception of daunomycin they do not cause chromosome breaks in human leukocytes and HeLa cells in cultures (Ostertag and Kersten, 1966). This may indicate that despite the similarities in the interaction of acridines and anthracyclines with DNA differences do exist.

II. Chromomycin and Mithramycin

The structure of chromomycinon is shown in Fig. 2. There is now agreement that in chromomycin A_3 five sugars are attached to the chromophore

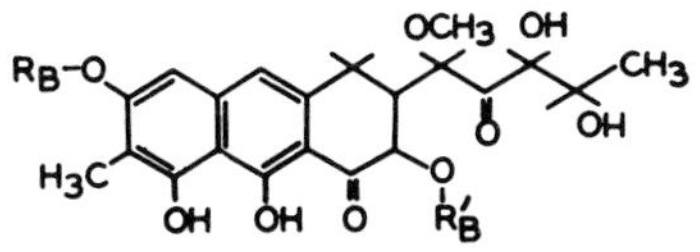

FIG. 2. Structure of chromomycinon.

(Berlin et al., 1966, Miyamoto et al., 1967). The structure of mithramycin is not yet known, but the chromophore is supposed to be identical with chromomycinon (Gause, 1966).

A. Effect of Nucleic Acids on the Spectra of the Dyes Dependent on Mg^{2+}

The interaction of chromomycin and mithramycin with nucleic acids can be shown by spectrophotometric measurements. The spectrum of chromomycin is shifted to longer wavelengths by DNA in the presence of Mg^{2+} or other divalent cations (Hartmann et al., 1964; Ward et al., 1965; Behr and Hartmann, 1965). Native DNA also changes the visible spectrum of mithramycin only in the presence of Mg^{2+} (Fig. 3).

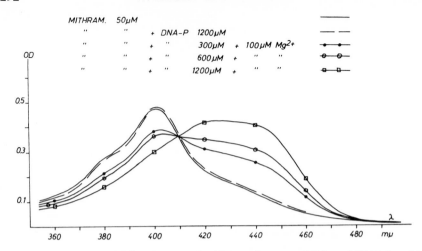

FIG. 3. Shift of the visible spectrum of mithramycin upon addition of DNA in raising concentrations in the presence of Mg^{2+}. Phosphate buffer 0.066 M, ph 7.0. The reaction mixtures were kept for 30 minutes at 37°C before measurement.

B. Effect of Nucleic Acids on the Spectra of the Antibiotics Independent of Mg^{2+}

The UV spectra of chromomycin and mithramycin are decreased by native DNA and decreased less by denatured DNA in the absence of Mg^{2+}, indicating interactions of the antibiotics with DNA in the absence of Mg^{2+} (Fig. 4). Since chromomycin and mithramycin do not interact with dAT, the

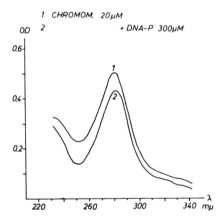

FIG. 4. Effect of DNA on the UV spectrum of chromomycin in the absence of Mg^{2+}. Experimental details given in Fig. 3.

question was raised whether these substances interact with cytosine or guanine nucleotides. The visible spectra are not altered by monoribonucleotides, ribohomopolymers, ribocopolymers, or deoxyribonucleotides (in the absence or presence of Mg^{2+}). In the UV part of the spectrum the absoprtion of the dyes is drastically reduced upon addition of dCMP and it is reduced less by dGMP (Fig. 5). Also CMP, GMP, and polynucleotides containing C slightly decrease the UV absorption of chromomycin and mithramycin. As in the case with DNA Mg^{2+} is not necessary for these interactions.

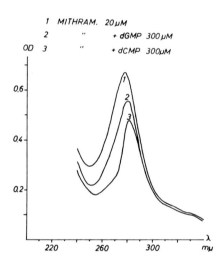

FIG. 5. Effect of dCMP and dGMP on the UV spectrum of mithramycin in the absence of Mg^{2+}. Experimental details given in Fig. 3.

C. Effect on Melting and Absorption of DNA

Mithramycin and chromomycin in low concentrations have no influence on the thermal transition of DNA in the presence or absence of Mg^{2+}. Using conditions where the molar concentration of chromomycin to DNA nucleotide is 1 : 1 and higher, Kaziro and Kamiyama (1965) have shown that the T_m value of DNA is shifted to higher temperatures. Both chromomycin and mithramycin show this effect in the absence of Mg^{2+}. No further increase in T_m of DNA is found by adding Mg^{2+} (Fig. 6).

During these experiments we noticed an influence of the antibiotics on the absorption of DNA, which is shown for chromomycin and mithramycin (Figs. 7 and 8). However, this spectral change is rather unspecific because it can also be shown with all the above-mentioned derivatives.

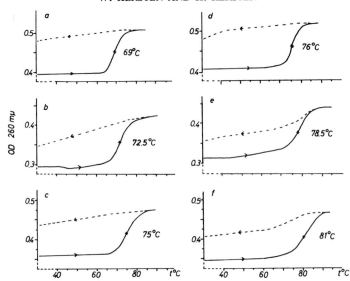

FIG. 6. Effect of mithramycin and chromomycin on thermal transition of DNA. (a) Melting of 20 μg (60 mμmoles DNA nucleotides) DNA of herring sperm/ml in 0.0016 M phosphate buffer, pH 7.0. (b) + 100 mμmoles mithramycin/ml. (c) + 100 mμmoles chromomycin/ml. (d) + 100 mμmoles Mg²⁺/ml. (e) DNA + Mg²⁺ + 100 mμmoles mithramycin/ml. (f) DNA + Mg²⁺ + 100 mμmoles chromomycin/ml.

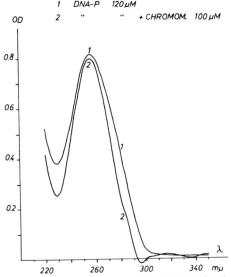

FIG. 7. Effect of chromomycin on the spectrum of DNA in the absence of Mg²⁺. Experimental details given in Fig. 3.

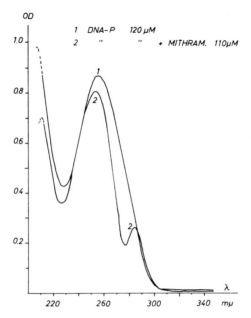

FIG. 8. Effect of mithramycin on the spectrum of DNA in the absence of Mg^{2+}. Experimental details given in Fig. 3.

D. Quantitative Measurements and Specificity

From the data presented one can conclude that chromomycin and mithramycin are strongly bound to DNA in the presence of Mg^{2+}. Weak interactions already occur without added Mg^{2+}, which is indicated by a decrease in the UV absorption of the antibiotics by DNA. We suggest that the dye molecules are oriented along the DNA at certain regions because of a weak association.

Small amounts of chromomycin (1 per 30 nucleotides) are bound to DNA in equilibrium dialysis without added Mg^{2+}. In the presence of one Mg^{2+} per dye up to one chromomycin is bound per 7.5 nucleotides. Mithramycin does not show binding to DNA in the absence of added Mg^{2+}, but in the presence of Mg^{2+} (one per dye) up to one mithramycin is bound per 10 nucleotides. The results are shown in Table I.

Although both antibiotics are very similar with respect to their interaction with DNA, certain differences may exist. During our studies on the action of both antibiotics on DNA-dependent processes *in vitro*, we have observed that mithramycin inhibits the enzymatic methylation and degradation of DNA less than chromomycin does (Table II). Also included in Table II are the effects of actinomycin and anthracyclines on DNA-dependent processes *in vitro*.

TABLE I

MOLAR BINDING RATIOS OF ANTIBIOTICS PER DNA-P IN THE ABSENCE
AND PRESENCE OF Mg^{2+} [a]

Antibiotic	nDNA		dDNA	
	$-Mg^{2+}$	$+Mg^{2+}$	$-Mg^{2+}$	$+Mg^{2+}$
Chromomycin	0.033	0.133	0.033	0.091
Mithramycin	0.0	0.1	0.0	0.066

[a] Results from equilibrium dialysis. One chamber contained 150 μM DNA-P (calf-thymus). The other chamber was filled with antibiotic solution. The concentration of the antibiotic was increased in different experiments until saturation of DNA with the antibiotic was achieved. Dialysis was performed at 4°C in phosphate buffer 0066M, pH 7.0. nDNA, native DNA; dDNA, heat-denatured DNA.

TABLE II

EFFECTS OF ANTIBIOTICS ON DNA-DEPENDENT REACTIONS *in vitro* [a]

Antibiotic	DNA→DNA [b]	DNA→RNA [c]	DNA + CH$_3$ [d]	DNase [e]	
				Exo	Endo
Actinomycin	2	0.03	1.6	5	15
Chromomycin	2	0.03	1.6	5	10
Mithramycin	7	1.0	4.4	10	15
Daunomycin	3	1.0	2.4	5	20
Nogalamycin	2.6	0.5	3.2	5	10
Cinerubin	3	1.0	0.4	5	10

[a] Concentrations ($M \times 10^{-5}$) from which 50% inhibition of DNA-dependent reactions *in vitro* are achieved.

[b] DNA-dependent sythesis of DNA, reaction mixture containing $12 \times 10^{-5} M$ DNA-P (Koschel *et al.*, 1966).

[c] DNA-dependent sythesis of RNA, reaction mixture containing $30 \times 10^{-5} M$ DNA-P (Koschel *et al.*, 1966).

[d] Methylation of DNA of *B. subtilis* with crude extracts of *E. coli*, reaction mixture containing $40 \times 10^{-5} M$ DNA-P (Kersten and Kersten, 1967).

[e] Degradation of DNA by endo- or exo-DNase, reaction mixture containing $150 \times 10^{-5} M$ DNA-P (Kersten and Kersten, 1967).

The spectral changes of the dyes in the UV region upon addition of several ribonucleotides and deoxyribonucleotides give evidence that chromomycin and mithramycin show preference for C. Both substances do not interact with dAT; they interact with dCMP and to a lesser degree with dGMP. These findings agree with the observations of Behr and Hartmann (1965) that apurinic acid binds one chromomycin per eight nucleotides, whereas with apyrimidinic acid only one chromomycin is bound per 18 nucleotides. Furthermore, Ward *et al.* (1965) showed by using dGdC as primer for RNA polymerase that the incorporation of G is inhibited far more than the incorporation of C.

III. Discussion

Specific interaction of substances with DNA very probably involves the purines or pyrimidines. Interaction with the bases can occur by various internal forces, as discussed elsewhere in this volume. In the binding of acridines to DNA interaction between the bases and the heterocyclic ring system are involved (Lerman, 1962). Besides these forces electrostatic interactions between the cationic dyes and the negatively charged phosphate groups play an important role in binding (Prichard *et al.*, 1966).

The anthracyclines contain uncharged chromophores. Probably the sugar side chain with the amino group is needed to form the complexes (Calendi *et al.*, 1965). The binding of anthracyclines with DNA also seems to involve two types of interactions. Since the anthracyclines like the acridines change the physical properties of DNA, they may also intercalate. The anthracyclines cannot be removed from DNA by CsCl in high concentrations. Under these conditions the acridines are displaced from DNA. Daunomycin is removed from DNA by Mg^{2+} (Calendi *et al.*, 1965) and nogalamycin does not interact with DNA in the presence of 1 M NaCl (Bhuyan and Smith, 1965). From these results one can argue that anthracycline interaction with DNA also comprises electrostatic forces.

The tetrahydroanthracenes apparently do not contain ionizable groups. These antibiotics form complexes with divalent cations. Whether these complexes are positively charged is subject to question. From the structure of the chromophore it is evident that the chromophore can exist in at least two mesomeric forms. One can speculate that one form preferentially interacts with DNA and is stabilized by Mg^{2+}.

From our results we are inclined to interpret the stronger interactions of mithramycin and chromomycin in the presence of Mg^{2+} as being a consequence of electrostatic interactions. Thus, the formation of complexes between the anthracene derivatives and nucleic acids may also involve two types of interactions. Since these substances do not change the physicochemical properties of DNA as the acridines do, the specific interaction with C or GC is not caused by intercalation.

298 W. KERSTEN AND H. KERSTEN

REFERENCES

Arcamone, F., Franceschi, G., Orezzi, P., Cassinelli, G., Barbieri, W., and Mandelli, R. (1964). *J. Am. Chem. Soc.* **86**, 5334.
Behr, W., and Hartmann, G. (1965). *Biochem Z.* **343**, 519.
Berlin, Y. A., Esipov, S. E., Kolosov, M. N., and Shemyakin, M. M. (1966). *Tetrahedron Letters* **15**, 1643.
Bhuyan, B. K., and Smith, C. G. (1965). *Proc. Natl. Acad. Sci. U.S.* **54**, 566.
Calendi, E., Di Marco, A., Reggiani, M., Scarpinato, B., and Valentini, L. (1965). *Biochim. Biophys. Acta* **103**, 25.
Ettlinger, L., Gäumann, E., Hütter, R., Keller-Schierlein, W., Neip, L., Prelog, V., Reusser, P., and Zähner, H. (1959). *Chem. Ber.* **92**, 1867.
Gause, G. F. (1966). *Chem. & Ind. (London)* p. 1506.
Hartmann, G., Goller, H., Koschel, K., Kersten, W., and Kersten, H. (1964). *Biochem. Z.* **341**, 126.
Kaziro, Y., and Kamiyama, M. (1965). *Biochem. Biophys. Res. Commun.* **19**, 433.
Kersten, W., and Kersten, H. (1965). *Biochem. Z.* **341**, 174.
Kersten, W., and Kersten, H. (1967). In " Wirkungsmechanismen von Fungiziden und Antibiotika ", p. 177. Akademie-Verlag, Berlin.
Kersten, W., Kersten, H., and Szybalski, W. (1966). *Biochemistry* **5**, 236.
Koschel, H., Hartmann, G., Kersten, W., and Kersten, H. (1966). *Biochem. Z.* **344**, 76.
Lerman, L. S. (1962). *Proc. Natl. Acad. Sci. U.S.* **49**, 94.
Miyamoto, M., Kawamatsu, Y., Kawashima, K., Shinohara, M., Tanaka, K., Tatsuoka, S., and Nakanishi, K. (1967). *Tetrahedron Letters* **23**, 421.
Ostertag, W., and Kersten, W. (1966). Unpublished result.
Pritchard, N. J., Blake, A., and Peacocke, A. R. (1966). *Nature* **212**, 1360.
Reich, E., and Goldberg, I. H. (1964). *Progr. Nucleic Acid Res. Molecular Biol.* **3**, 184.
Tabaczynski, M., Sheldrick, P., and Szybalski, W. (1965). *Microbial Genet. Bull.* **23**, 7.
Ward, D., Reich, E., and Goldberg, I. H. (1965). *Science* **149**, 1259.
Wiley, P. F. (1965), Personal communication.

Effect of Light on Dyes and Photodynamic Action on Biomolecules

MICHEL DELMELLE AND JULES DUCHESNE

Department of Atomic and Molecular Physics
University of Liège
Cointe-Sclessin, Belgium

I. Introduction

It is known that biomolecules, as well as living organisms, are rather insensitive to visible light. When dyes are added as photosensitizers, however, damage generally takes place. This phenomenon, which has been called "photodynamic action" was first described by Raab (13) at the beginning of the century. Since then (19), the field has grown rapidly, and its importance became evident, especially where photomutation (12, 14–16) photoinactivation (7, 21, 22) and even photocancerization (17) were concerned. Many methods have been employed to study this effect but it appears that very little use has been made of electron spin resonance (2, 18), although this seemed to be a very valuable method especially for detecting energy transfer from the dye to the biomolecule and also identifying the free radicals produced.

We chose to consider the effect of some acridines, xanthenes, and thiazines on nucleosides, DNA, and nucleoproteins. As a preliminary investigation, it was found necessary to analyze the behavior of the dyes alone when irradiated in their visible absorption bands. The second step consisted in the analysis of the energy transfer between these and the chosen biomolecules.

II. Experimental Techniques

The light source for the study of the dyes alone was a 250-watt Osram flood lamp, whereas for energy transfer, a high pressure mercury vapor lamp (Osram HBO 500) was advantageously used, because of its higher emission power in the region of the wavelengths involved. The spectrometer was a Varian type (4502-06) with a cavity V 4532 which permits irradiation during observation. The samples were introduced in quartz tubes about 4 mm in diameter and were sealed either in air or under vacuum.

III. Production of Free Radicals in Irradiated Dyes

Machmer and Duchesne (10, 11) have shown that when acridines dissolved in dimethyl sulfoxide are irradiated in the cavity, the free radicals generated are observable by electron spin resonance spectroscopy within a temperature range of $-60°$ to $-10°C$, as shown for proflavine in Fig. 1. More recently, Delmelle and Duchesne (4) used water as a solvent, with concentrations of about 10^{-2} M/liter for a series of dyes at a temperature of $-80°C$. Singlet signals were obtained, characterized by $g = 2.003$ and a width of 13 Oe, and these signals compare closely with Machmer and Duchesne's results, although they disappear at $-50°C$ instead of $-10°C$, as for dimethyl sulfoxide. Table I gives the absorption maxima in the visible region for the dyes in solution.

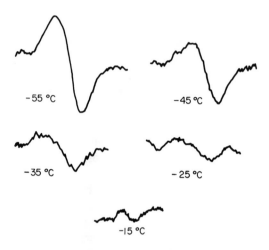

FIG. 1. ESR spectra of proflavine dissolved in dimethyl-sulfoxide and irradiated in the visible range at $-60°C$., based on temperature dependence.

TABLE I

ABSORPTION MAXIMA OF VARIOUS DYES

Substances	Wavelength of the maxima of intensity of the bands (in mμ units)	References
Proflavine	444	1
Acridine orange	492	1
9-Aminoacridine	401	1
Acriflavine	452	1
Acridine yellow	425	9
Methylene blue	662	8
Acridine red	496 and 550	This work
Pyronine	543	8

It can be seen in Fig. 2 that the intensity of the signals closely follows oxygen pressure, so that it may be concluded that the production of free radicals depends directly on this pressure and that oxyradicals are formed.

The number of free radicals are compared in Table II with the intensity of the photodynamic effects which the dyes are known to produce.

To estimate the free radical concentration, changes in the absorption coefficient in the visible region for the different dyes considered have to be taken into account. In the case of photomutation, results up to now are,

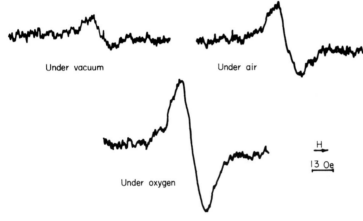

FIG. 2. ESR spectra of proflavine dissolved in water and irradiated in the visible range at −196°C. Observations at −150°C; based on dependence on atmosphere.

TABLE II[a]
PHOTODYNAMIC EFFECTS

Dyes	Number of paramagnetic centers (relative scale)	Photo-mutation of *E. coli* (12)	Photo-inactivation of *E. coli* (21)	Photo-inactivation of T5 (22)	Photo-inactivation of pneumo-coccal trans-forming DNA (7)
Methylene blue	100	+ +	+ +	+ + + +	+ + + +
Acridine orange	40	+ +	?	+ + +	+ + +
Acriflavine	30	?	?	+ +	+ +
Pyronine	10	?	+	+	?
Acridine red	—	—	?	?	+
Proflavine	60	?	?	?	?

[a] In each column, the number of + signs increases with photodynamic activity; − indicates an absence of activity, and ? an absence of comparative experimental results.

unfortunately, insufficient to make it possible to draw any definite conclusion. But, data summarized in the last three columns seem to be really gratifying, since a satisfactory parallel appears between the degree of inactivation and the relative concentration of free radicals. It will be seen that acridine red does not produce any observable radical; but on increasing the intensity of the light source by using the Hg lamp, a weak signal may be detected, in agreement with the small inactivation effect reported in the last column of Table II.

IV. Photodynamic Effects

We must now consider what occurs when biomolecules are added to the foregoing systems.

A. Effects of Light and Dyes on Deoxyribonucleosides*

Using a dye solution of 10^{-3} M/liter, a weaker concentration than that necessary to obtain a signal, the four nucleosides, deoxyadenosine, deoxycytidine, deoxyguanosine, and thymidine, were dissolved therein, at a concentration of about 10^{-2} M/liter.

For practical reasons, the systems were irradiated before placing them in the cavity of the spectrometer. The irradiations were performed in air for 2 hours at $-196°C$ and paramagnetic observations were made at $-150°C$. Under these conditions, it was gratifying to observe paramagnetic signals having shapes that were characteristic of each nucleoside. This is illustrated in Fig. 3 for the systems light-proflavine-nucleosides. Of these, thymidine is immediately recognized because of its well-known multiplet of eight lines. In the three remaining cases, the widths of the central peaks, which amount, respectively, to 14, 20, and 16 Oe, the distribution of the satellite lines, and the "g" values of 2.004 permit identification. As for the dyes alone, a decreasing air pressure, as shown in Fig. 4, produces a weakening of the paramagnetic spectra. Figure 5 represents the evolution of the line intensities in terms of temperature for the system proflavine-deoxyadenosine.

B. Effects of Light and Dyes on DNA and Nucleoproteins*

Acriflavine and proflavine were the only dyes used. Temperatures were the same as for nucleosides, whereas the dye and the calf thymus DNA, highly polymerized and undenaturated, were at the respective concentrations of

---

* Cf. Delmelle and Duchesne (5).

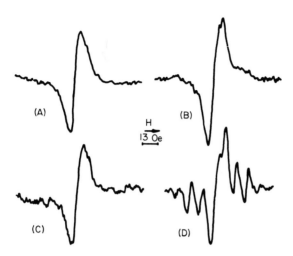

FIG. 3. ESR spectra of proflavine-deoxyribonucleosides in aqueous solution irradiated in the visible range at −196°C. Observations at −150°C. (A) Proflavine-deoxyadenosine, (B) proflavine-deoxycytidine, (C) proflavine-deoxyguanosine, and (D) proflavine-thymidine.

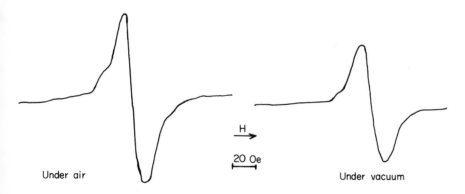

FIG. 4. ESR spectra of proflavine-deoxycytidine in aqueous solution irradiated in the visible range at −196°C. Observations at −150°C; based on dependence on atmosphere.

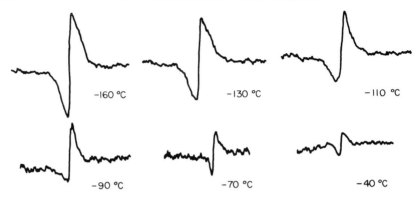

FIG. 5. ESR spectra of proflavine-deoxyadenosine in aqueous solution irradiated in the visible range at $-196°C$; based on temperature dependence.

3×10^{-4} M/liter and 0.1 %. Figure 6 shows the paramagnetic spectrum of the irradiated system proflavine-DNA and this is characterized, apart from some still undetermined satellite lines, by a central peak about 20 Oe wide and a "g" value of 2.003. This compares favorably with the signal of DNA when irradiated directly with X-rays at low temperature. The results for acriflavine were quite similar. As illustrated in Fig. 7, the line intensity weakens with temperature and disappears at $-60°C$.

Further studies have been performed on nucleohistone of calf thymus origin (concentration 0.18 %, corresponding to a DNA concentration of 0.08 %). The spectrum obtained is shown in Fig. 8, along with the spectra of DNA, histone, and nucleohistone (20), as obtained from samples directly irradiated by X-rays, for comparison. It is immediately seen that the central peaks of all the spectra have almost identical widths, and each, except histone, has a superimposed narrow singlet. It may therefore be strongly suggested that for nucleoproteins the energy transferred is localized in the DNA moiety.

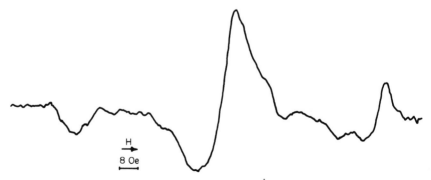

FIG. 6. ESR spectrum of proflavine-DNA in aqueous solution irradiated in the visible range at $-196°C$. Observations at $-150°C$.

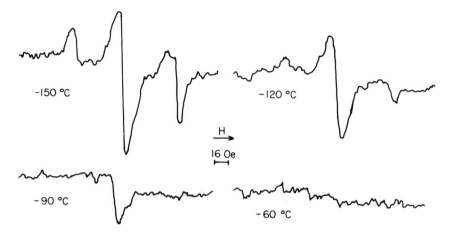

FIG. 7. ESR spectra of proflavine-DNA in aqueous solution irradiated in the visible range at $-196°C$; based on temperature dependence.

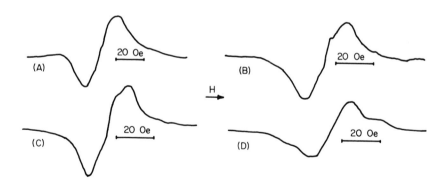

FIG. 8. (A) ESR spectrum of proflavine-nucleoprotein in aqueous solution irradiated in the visible range at $-196°C$. Observations at $-150°C$. (B) ESR spectrum of nucleohistone irradiated with X-rays at $-150°C$. (C) ESR spectrum of DNA irradiated with X-rays at $-150°C$. (D) ESR spectrum of histone irradiated with X-rays at $-150°C$.

C. Protection effect by Cystamine*

It seemed of interest to take advantage of the simplicity of the method to extend the study to another interrelated question—protection phenomena at the molecular level. With this in view, we adopted the system proflavine-DNA to which the well-known protector (3), cystamine, was added at a concentration of 1 gm/liter. These three components in aqueous solution were irradiated for $3\frac{1}{2}$ hours under the same conditions as before.

As hoped, we observed the spectrum of cystamine rather than that of DNA, as shown in Fig. 9. The identification was made on the basis of the spectrum obtained after having irradiated an aqueous solution of cystamine in the ultraviolet region. On decreasing the protector concentration to $2 \times 10^{-2}\%$,

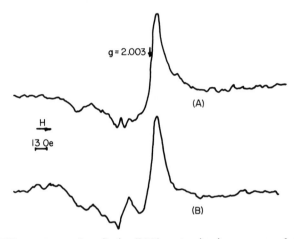

FIG. 9. (A) ESR spectrum of proflavine-DNA-cystamine in aqueous solution irradiated in the visible range at $-196°C$. Observations at $-150°C$. (B) ESR spectrum of cystamine irradiated in the ultraviolet range at $-196°C$. Observations at $-150°C$.

which means 20% with respect to DNA, the signal obtained does not seem to arise only from cystamine and appears to be superimposed by a DNA component. This would indicate the point from which the protector does not act with full efficiency.

V. General Conclusions

It cannot now be denied that electron spin resonance is a very valuable tool by which to approach the fundamental problem of photodynamic action, including those areas in biophysics. Indeed, if spectral, chromato-graphic, and ultracentrifugation methods have shown themselves to be very

---

* Cf. Delmelle and Duchesne (6).

useful by giving results which drew attention to the marked alterations produced by light and dyes either on DNA, proteins, and enzymes or in the area of genetics, it is now quite clear that the fact that free radicals may be identified in such systems constitutes a significant advance. It should be pointed out that the parallelism found by us between the free radicals and the intensity of photoinactivation shows that these radicals play an important role in such phenomena. This opinion is strengthened by the fact that irradiated dyes and irradiated dyes-DNA must be kept in air in order that free radicals may be observed, so that the second mechanism is derived directly from the first. Should this be so, then it would be expected that photomutation itself would at least partially be governed by the number of free radicals induced in biological systems, and this is now being investigated in our laboratory.

It is now possible to ascertain that where nucleoproteins are concerned, free radicals localize themselves in the DNA moiety and not in the protein constituent. It was also very gratifying to find that the method was most efficient in the analysis of the protection effects. In relation to this, it is to be noted that the system consisting only of proflavine and cystamine does not give rise to any energy transfer. The fact that the free radicals are concentrated in cystamine when DNA is added evidently supposes a specific interaction between the latter and the protector.

REFERENCES

1. Albert, A. (1966). "The Acridines," Arnold, London.
2. Azizova, O. A., Kayushin, L. P., and Pulatova, M. K. (1966). *Biofizika* **11**, 970.
3. Bacq, Z. (1965). "Chemical Protection Against Ionizing Radiation," Thomas, Springfield, Illinois.
4. Delmelle, M., Depireux, J., and Duchesne, J. (1966). *Compt. Rend.* **D263**, 1625.
5. Delmelle, M., and Duchesne, J. (1967). *Compt. Rend.* **D264**, 138.
6. Delmelle, M., and Duchesne, J. (1967). *Compt. Rend.* **D264**, 1651.
7. Fujita, H., Moriguchi, E., Yamagami, H., Suzuki, K., and Wada, A. (1963). *Ann. Rept. Natl. Inst. Radiation Sci. Japan*, p. 23.
8. "International Critical Tables of Numerical Data, Physics, Chemistry and Technology" (E. W. Washburn, C. West, and N. Dorsey, eds.) (1930). Vol. 7. McGraw-Hill, New York.
9. Löber, G. (1965). *Photochem. Photobiol.* **4**, 607.
10. Machmer, P., and Duchesne, J. (1966). *Compt. Rend.* **D262**, 307.
11. Machmer, P., and Duchesne, J. (1966). *Compt. Rend.* **D262**, 705.
12. Nakai, S., and Saeki, T. (1964). *Genet. Res.* **5**, 158.
13. Raab, O. (1900). *Z. Biol.* **39**, 524.
14. Ritchie, D. A. (1964). *Genet. Res.* **5**, 168.
15. Ritchie, D. A. (1965). *Genet. Res.* **6**, 474.
16. Ritchie, D. A. (1965). *Biochem. Biophys. Res. Commun.* **6**, 720.
17. Santamaria, L., Giordano, G. G., Alfisi, M., and Cascione, F. (1966). *Nature* **210**, 824.
18. Santamaria, L. (1962). *Bull. Soc. Chim. Belges* **71**, 889.

19. Spikes, J. D., and Ghiron, C. A. (1964). *In* "Physical Processes in Radiation Biology" (L. G. Augenstein, R. Mason, and B. Rosenberg, eds.), p. 309. Academic Press, New York.
20. Van de Vorst, A. (1967). Personal communication.
21. Wacker, A., Dellweg, H., Träger, L., Kornhauser, A., Lodemann, E., Türck, G., Selzer, R., Chandra, P., and Ishimoto, M. (1964). *Photochem. Photobiol.* **3**, 369.
22. Yamamoto, N. (1958). *J. Bacteriol.* **75**, 443.

Charge-Transfer Complexes in Biological Oxidations

G. CILENTO AND K. ZINNER

Department of Chemistry, Faculdade de Filosofia
Ciências e Letras, Universidade de São Paulo
São Paulo, Brazil

I. Introduction

Mulliken (1952) predicted that charge-transfer forces would open new ways for understanding intermolecular interactions in biological systems. This belief has received considerable support from Szent-Györgyi (1960) and is now widespread.

We shall analyze here the effects resulting from charge-transfer association and their possible occurrence in connection with the electron-transport chains. In this connection the results of the quantum mechanical calculations by Pullman and Pullman (1963) of the uppermost filled and lowest vacant orbitals in biologically important compounds are of fundamental importance. As a matter of fact, one can now even predict whether two biochemical species or moieties are prone to form a charge-transfer complex.

Charge-transfer forces may—as do other types of forces—lead or contribute to the association of two or more entities. In the same species a preferred conformation may result, which, in turn, may confer biological activity. Likewise, an allosteric effector may add to the critical center by way of charge-transfer forces.

Similar to other types of association, charge-transfer complexing may influence equilibrium and reactivity. For instance, a difference in complexing ability of the ionized and neutral forms of a molecule with a certain species will result in a change in the pK value (Cilento and Berenholc, 1963). Kinetically, if a reactant is a better charge-transfer complex former (either electron donor or acceptor) when in the transition state than in the ground state, then a nonreactive species (electron acceptor or donor) may accelerate the reaction. Alternately, if the ground state is a better complex former, the reaction will be inhibited (Colter *et al.*, 1964). Often, but not always, the result of this enthalpy effect is to slow down the reaction.

The rate can also be modified by entropy effects, inasmuch as charge-transfer association between reactants or between a reactant and a foreign species may properly approximate the reacting groups.

Other effects are also feasible. Consider a molecule which can act as both electron acceptor and donor. This duality of character might conceivably lead to some interesting effects. Thus, extensive stacking of the molecules might occur under proper conditions.

In electron-transfer complexes in which the ionization potential of the donor is low enough an electron may be completely transferred to the acceptor; hence, in a biological aggregate the transferred electron and/or the hole may become completely delocalized and independent functions may be performed by them (Kearns and Calvin, 1961).

Clearly, listing these possible effects of charge-transfer complexing does not mean that they are completely independent of each other.

II. Charge-Transfer Association

A. Internal Interaction

The aromatic enzyme side chains may act as π donors; some other side chains, such as the lysyl residues, as n donors. The only protein side chain which might be able to act as an acceptor is the protonated histidine residue (Shinitzky et al., 1966). For enzymes involved in biological oxidations this internal donor-acceptor interaction has never been reported.

Internal charge-transfer interaction is conceivable for the oxidized form of the pyridine and flavine coenzymes, inasmuch as their functional moiety is a good acceptor and the purine ring is a donor. Although it is likely that charge-transfer forces do not contribute to the folding of flavine adenine dinucleotide (FAD) (G. Weber, 1965; Tsibris et al., 1965; Strittmatter, 1966; Wilson, 1966), it may well be that they operate in the case of nicotinamide adenine dinucleotide (NAD). For, not only a new absorption is found in a system of pyridinium salts and adenosine or adenosine diphosphate (ADP) (Cilento and Schreier, 1964; Anderson and Reynolds, 1966), but even more important, splitting of the 3-isoadenosine analog of NAD produces a hypochromic effect in the 300–330 mμ region (Leonard and Laursen, 1965).

B. Enzyme-Coenzyme Interaction

Theoretically, interaction can take place between a donor side chain of the enzyme and the oxidized coenzyme or between the reduced coenzymes and a protonated imidazole group of the enzyme.

The donor side chains of greatest interest are the indolyl and, next, the tyrosyl groups. Indole seems to be even more efficient than expected probably because of more localized interaction (Green and Malrieu, 1965). The ionized tyrosyl group may be a better complex former than the nondissociated group if hydrogen bonding is not involved in the association. The high pK value, however, makes it unlikely that the ionized form can participate in a charge-transfer complex.

Charge-transfer interaction between an enzyme and an oxidized pyridine nucleotide is most likely to occur in the case of the 3-phosphoglyceraldehyde dehydrogenase. Accordingly, the addition of NAD to this enzyme produces a new absoprtion band which, as pointed out by Kosower (1956), is reminiscent of charge-transfer interaction. A similar band could be reproduced in a system of NAD or pyridinium salts and indole derivatives (Cilento and Giusti, 1959; Alivisatos *et al.*, 1960, 1961; Cilento and Tedeschi, 1961; Alivisatos, 1961), or even more efficiently if the pyridinium ring and the indole nucleus are incorporated in the same molecule (Shifrin, 1964a,b).

To the best of our knowledge, there is no report of charge-transfer association between a flavine and the apoenzyme in flavoenzymes, as demonstrated by the appearance of a charge-transfer band. A possible case is provided by the " old yellow enzyme," which when treated with ammonium sulfate at high pH values becomes green and a band reminiscent of charge-transfer interaction appears in the spectrum (Rutter and Rolander, 1957).

Nevertheless, association of the flavine with a tyrosyl group of the apoenzyme at least partially by charge-transfer forces has been considered likely (Harbury and Foley, 1958; Fleischman and Tollin, 1965a) on the grounds that charge-transfer interaction occurs between flavines and phenols. Moreover, the formation of hydrogen bond may not be essential (Tsibris *et al.*, 1966). Iodination of the phenolic group, however, while enhancing association with flavines (Cilento and Berenholc, 1965), may interfere, in the case of an apoenzyme, with flavine binding (Nygaard and Theorell, 1955; Strittmatter, 1961). Therefore, either charge-transfer forces between a tyrosyl group of the apoenzyme and the flavine do not contribute to the association or the bulky iodine group sterically hampers binding.

Strong interaction, probably of the charge-transfer type, occurs *in vitro* between indole or derivatives and flavines (Isenberg and Szent-Györgyi, 1958; Wilson, 1966). It seems possible that under suitable conditions a complete electron transfer takes place (Isenberg, 1964).

A naturally occurring charge-transfer interaction between an enzyme and a quinone does not seem to have been described. Using chloranil as a model for ubiquinone, Birks and Slifkin (1963) observed n,π charge-transfer interaction with the amino group of amino acids and proteins. With the aromatic amino acids, π,π complexation is expected to occur. Charge-transfer association between methylindoles and quinones has been reported by Foster and Hanson (1964).

One may presume that a tyrosyl group of an enzyme might activate the reduction of a quinone. This stemms from the observation that under special conditions, phenols by associating with quinones may catalyze their reduction (Braud *et al.*, 1954). In several experiments performed in our laboratories, however, employing a variety of solvents, quinones and phenols gave always negative results. Nevertheless, since a quinonelike structure is part of the isoalloxazine ring, it is interesting to note that a tyrosyl group may interact with FAD in microsomal NADH-cytochrome b_5 reductase (Strittmatter, 1965).

Because porphyrins in aqueous solutions form molecular complexes with a variety of molecules, it has been suggested by Mauzerall (1965) that such complexes occur in heme proteins as a part of the phosphorylating mechanism. Charge-transfer forces are believed to contribute to the association with *sym*-tri-nitrobenzene (Gouterman and Stevenson, 1962); if so, the observation of no new bands could be due either to the low porphyrin concentration or to its occurrence in the near infra-red region.

C. Coenzyme-Coenzyme Interaction

The reduced form of a coenzyme of an electron-transport chain may form a charge-transfer complex with the oxidized form of any of the coenzymes of the chain, including its own oxidized form.

The pyridinium cation and 1,4-dihydronicotinamides, used as models for the natural coenzymes, do form charge-transfer complexes (Cilento and Schreier, 1964; Ludowieg and Levy, 1964). The association between the natural coenzymes appears to be very weak *in vitro* presumably because of internal charge-transfer association in NAD (Cilento and Schreier, 1964).

Isenberg and Szent-Györgyi (1959) believe that charge-transfer association occurs between the reduced pyridine nucleotides and flavine mononucleotide (FMN), but a characteristic absorption band was not observed. The formation of a NADH-flavine complex has also been postulated in the mechanism of oxidative phosphorylation (Kosower, 1962; Grabe, 1964). Yet, according to Radda and Calvin (1964) no complexation occurs; this, however, might be due to the low concentrations employed.

On the other hand, the charge-transfer association between reduced flavines and NAD or models has been reported by Massey and Palmer (1962) and Sakurai and Hosoya (1966).

The possibility of charge-transfer complexation between the pyridine nucleotides and flavine coenzymes may prove to be especially significant in the mechanism of the NADH-cytochrome b_5 reductase, thoroughly investigated by Strittmatter (1965). It may well be that hydrogen transfer occurs reversibly between the NADH·FAD and NAD·FADH$_2$ charge-transfer complexes specifically bounded to the apoenzyme.

It must, however, be stressed that a more complete description of these intermediate complexes is warranted (Strittmatter, 1965). This is also true for the red intermediates in flavoprotein oxidoreductions (A. Pullman, 1964; Hemmerich *et al.*, 1965). In the presence of arsenite, however, NAD and

NADP do form a charge transfer complex with the fully reduced flavoenzymes, lipoyl dehydrogenase and glutathione reductase, respectively, with an absorption maximum at 720 mμ (Massey and Williams, 1965). Yet these complexes are probably of no biological significance.

Interaction between oxidized and reduced flavines is so favorable that the charge-transfer complex absorbs in the infrared; from this complex, formation of semiquinone also occurs (Gibson et al., 1962), but the primary step is one of hydrogen transfer (Swinehart, 1966; Fox and Tollin, 1966a).

It is interesting that flavine semiquinone can complex with NAD (Massey and Williams, 1965), phenols (Fleischman and Tollin, 1965b), and the oxidized and reduced flavines (Ehrenberg et al., 1964; Gibson et al., 1962). Of special significance may be the stabilization of the flavine free radical by purines (Slifkin, 1965; Fox and Tollin, 1966a,b).

Let us now consider the interaction between 1,4-dihydropyridines and quinones. The reaction may proceed by hydrogen transfer (Wallenfels and Gellrich, 1959), but electron transfer also seems to occur as shown by appearance of the semiquinone absorption band (Cilento and Zinner, 1966). Whether the expected charge-transfer complexation precedes these transfers or whether it is only abortive is an open question. It is, however, interesting that the p-semiquinone can be readily oxidized by oxygen associated to an o-diphenol as reported in Section II, E.

D. Substrate — Coenzyme Interaction

According to Massey et al. (1965) in the action of D-amino acid oxidase, the intermediates are in the sequence: enzyme-substrate complex, a complex of flavine and amino acid radicals, and a charge-transfer complexes of the reduced flavine and the imino acid.

E. Oxygen Complexes

Oxygen is an electron acceptor and forms charge transfer complexes with a variety of donors (Tsubomura and Mulliken, 1960) including amino acids (Slifkin, 1962). From work carried out in our laboratories it appears that in some of these complexes, oxygen becomes a much better oxidant. It may be added that very recently Carlsson and Robb (1966) have found it necessary to postulate an initial oxygen-hydrocarbon charge-transfer complex as precursor to the thermal initiation step of autoxidation of indene and tetralin.

To date we have discovered three classes of donors acting as catalysts in autoxidation reactions: (1) the monoanion of catechol and catecholamines (Cilento and Zinner, 1966, 1967a), (2) the monoprotonated form of p-phenylenediamines (Cilento and Zinner, 1967b), and (3) the iodide ion (Cilento and Zinner, 1967c).

The catalytic effect of catechol and catecholamines has been observed in the autoxidation of *p*-hydroquinone, *p*-phenylenediamines, and catechol itself, usually by both spectrophotometric and manometric techniques. In the case of catechol the catalytic effect was inferred from the second-order dependence of the oxygen uptake upon the substrate when the latter is present in low concentrations (Joslyn and Branch, 1935).

A representative example of *o*-diphenol catalysis is illustrated in Fig. 1.

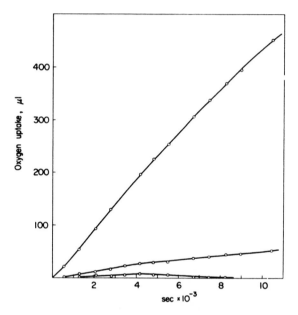

FIG. 1. Upper line: the catalytic effect of 20 m*M* norepinephrine on the oxygen consumption by 11 m*M* *p*-hydroquinone; solvent, 0.1 *M* phosphate buffer, pH 7.50-methanol (1 : 1, v/v). Middle line: *p*-hydroquinone alone; lower line: norepinephrine alone.

The *o*-diphenol effect is detected in very low concentrations and does not need the cooperation of metal ions. Catechol also seems able to catalyze the autoxidation of *p*-semiquinones, as dihydropyridines are oxidized in the presence of substoichiometric amounts of quinones (Fig. 2), even vitamin K_3.

NADH   Quinone H_2O + catechol

NADH-*e*   *p*-Semiquinone $O_2 \cdot$ catechol

Oxygen activation by monoprotonated *p*-phenylenediamines was inferred from the maximum observed in the pH dependence of the rate of autoxidation of the base. The monoprotonated *N*,*N*,*N'*,*N'*-tetramethyl-*p*-phenylenediamine appears to be active in concentrations as low as 10^{-5} *M*. (Fig. 3).

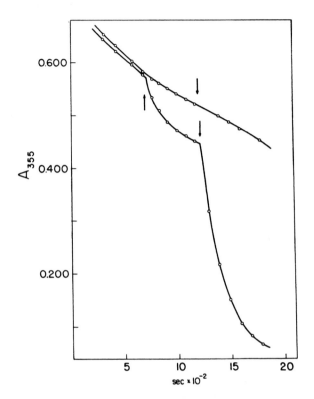

FIG. 2. The catalytic effect of 2×10^{-2} M catechol upon the dehydrogenation of 9.0×10^{-5} M 1-benzyl-1,4-dihydronicotinamide in the presence of 3.0×10^{-5} M chloranil. Initially, the cells were carefully evacuated at 0°C. Upper curve: no catechol; the upward arrow indicates catechol addition; the downward arrow indicates when air was admitted. Solvent; 0.02 M phosphate buffer, pH 6.9-methanol (1 : 1, v/v); optical path, 1 cm.

Iodide catalysis has up to now only be observed in the autoxidation of *p*-phenylenediamines and is quite specific (Fig. 4). Kinetic studies indicate that the effect is modest because it is exerted on a step which is only partially rate-determining.

F. Other Complexes

Polycyclic hydrocarbons, including carcinogenic ones, form electron-transfer complexes with the pyridinium ring (Cilento and Sanioto, 1965) and quinones (Cilento and Sanioto, 1963). No correlation is apparent between carcinogenic activity and complexing.

Not only tyrosine, but also its halogenated derivatives, form charge-transfer

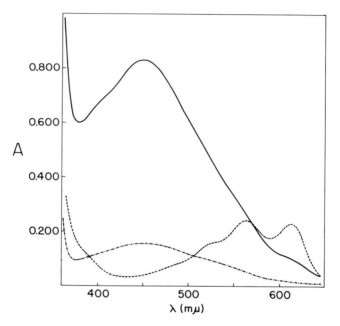

FIG. 3. The catalytic effect of the monoprotonated N,N,N',N'-tetramethyl-p-phenylenediamine ((TMPDH$^+$) upon the autoxidation of p-phenylenediamine (PPDA). Absorption spectra: Dashed-dotted line: 4.28×10^{-2} M PPDA, pH 8.49, after 3410 seconds; dashed line: 5.44×10^{-3} M TMPD, pH 9.13, after 3040 seconds; solid line: 4.28×10^{-2} M PPDA $+ 5.44 \times 10^{-3}$ M TMPD, pH 9.07, after 2700 seconds. Solvent, methanol-water (3 : 1, v/v); optical path, 1 cm. The differences in pH between the mixture and controls are responsible for only an insignificant part of the effect.

complexes with the pyridinium ring, flavines, and quinones (Cilento and Berenholc, 1965). The charge-transfer band is observed with the phenolate forms; in the latter the donor ability is enhanced. No doubt, complexing should also occur with the thyroid hormones.

2,4-Dinitrophenol, as expected, complexes with tryptophan (Fujimori, 1960).

G. Complexes of Electronically Excited Species

The possibility that electronically excited states are functional in biological systems has recently been analyzed (Cilento, 1965). Further important additional evidence has been provided by Stauff (1964), who observed chemiluminescence in breathing mitochondria and electron-transfer particles, especially after addition of acridine orange. It is, therefore, justified to consider here also charge-transfer complexes of excited molecules.

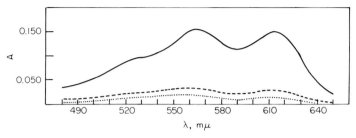

FIG. 4. The effect of 0.076 M NaI (upper curve) and 0.105 M NaBr (middle curve) upon the autoxidation spectrum of 5.1×10^{-4} M N,N,N',N'-tetramethyl-p-phenylenediamine (TMPD). Lower curve; 5.1×10^{-4} M TMPD. Solvent, methanol; spectra were taken after 5400 seconds; optical path, 1 cm.

Flavines are likely to appear in excited states (McGlynn *et al.*, 1964; Cilento, 1965) and to mediate oxidations (Steele, 1963).* According to Radda and Calvin (1963) triplet FMN can form charge-transfer complexes with donors. FMN in the triplet state is a much better acceptor than in its ground state. As the result of complex formation, the FMN triplet (Shiga and Piette, 1964) and very probably that of riboflavine (Tomita, 1967) may be quenched by tryptophan.

Excited indole forms charge-transfer complexes with polar solvents, the "exciplex," closely resembling the excimers between two aromatic molecules (Walker *et al.*, 1966). The indole exciplex is the immediate donor of electrons to the solvent.

III. Discussion

Caution is required in asserting that a charge-transfer association may be biologically significant, not only because the occurrence of a charge-transfer complex between biochemicals *in vitro* is by no means proof of its functional occurrence *in vivo*, but also because in several asserted instances the characteristic charge-transfer band has not been observed. Yet, the potential importance of the available data in the context of biological oxidations is beyond any question.

Interaction within the oxidized pyridine and flavine nucleotides or between the functional moiety of these nucleotides and enzyme side chains will certainly influence the redox properties of the coenzymes and, consequently, the electron flux. It may be added that the folded conformation of FAD will also allow stabilization of the radical form of the functional moiety by complexation with the purine ring (Fox and Tollin, 1966a).

---

* This statement has been misunderstood and interpreted (Fox and Tollin, 1966a) in the sense that dark reactions of flavines may proceed through a triplet transition state.

For two reasons the enzyme 3-phosphoglyceraldehyde dehydrogenase deserves special consideration. First, we have seen that addition of NAD results in charge-transfer complexing between the pyridinium ring and an enzyme side chain, probably an indole nucleus; here the coenzyme presumably acts as an allosteric effector (Kirschner *et al.*, 1966). Second, this enzyme is the only dehydrogenase which maintains NAD in the folded conformation (Velick, 1961). This arrangement could favor the oxidation of the reduced coenzyme, because the purine moiety may stabilize the incipient pyridinium ring in the transition state (Cilento and Sanioto, 1965).

Actually, this type of catalysis may prove to be of more general significance in biological oxidations and will positively operate in the direction of the coenzyme oxidation, provided a donor group is properly available to stabilize the incipient form in the transition state. A possible example occurs in the NADH-cytochrome b_5 reductase system. If an ionized mercapto group is very close to the pyridine coenzyme (Strittmatter, 1965), it is conceivable that this group will favor the oxidation of NADH.

Coenzyme-coenzyme charge-transfer interaction is of multiple potential interest.

First, it could influence reactivity. For instance, in the system of the NADH-cytochrome b_5 reductase the easy hydrogen transfer between the NADH·FAD and NAD-FADH$_2$ complexes may be the result of a favorable entropy of activation, inasmuch as the reacting groups might be properly oriented.

Second, but still in the realm of reactivity, an association such as that between NADH and the isoalloxazine ring might favor ATP generation (Kosower, 1962; Grabe, 1964).

Third, coenzyme-coenzyme charge-transfer association may, in special cases, give information concerning the stereochemistry of the system (Cilento and Schreier, 1964). Consider the NAD-NADH exchange catalyzed by the symmetrical transhydrogenase (San Pietro *et al.*, 1955); consider also the postulates of Kosower (1956) and Burton and Kaplan (1963) that a charge-transfer complex may be formed between coenzyme and substrate in the enzyme. Then, if NAD and NADH are in the unfolded conformation, complexing will occur between the pyridine moieties; these—being presumably parallel to each other—must have the carboxamide groups pointing in opposite directions in order to keep the same stereospecificity.

Likewise, in the case of the energy-linked transhydrogenases in which a hydrogen is transferred from the *A* locus of NADH to the *B* locus of NADP (Lee *et al.*, 1965; Roberton and Griffiths, 1965), the pyridine moieties must have the carboxamide groups pointing in the same direction if a complex is formed.

Perhaps it is in connection with oxygen itself that charge-transfer complexation may be of the utmost importance. The resultant catalytic effect in autoxidation reactions might be connected with the unusual property that, in oxygen, entrance of one electron will favor acceptance of a second electron (George, 1964). Therefore, oxygen may be a better electron acceptor in the transition state of autoxidation reactions than in the ground state; hence, suitable donors may stabilize the transition state more than the ground state and, thus, reduce the activation energy.

Several important inferences can be made.

Catalysis by catechol and catecholamines suggests that one of the primary functions of the very important, naturally occurring *o*-diphenols is to activate oxygen and, thus, to promote electron transfer.

The *o*-diphenol catalysis makes feasible other sequences of biological electron transport; thus, in the presence of quinones, the aerobic oxidation of 1,4-dihydropyridines may proceed in one-electron steps. The enzyme polyphenol oxidase—which can form both the *o*-diphenol and the quinone— might play a role in electron transport. Also, the enzymatic formation of naphthosemiquinone radicals (M. M. Weber *et al.*, 1965) acquires further significance, as these radicals may shuttle electrons between the reduced pyridine coenzymes and *o*-diphenol-activated oxygen.

This catalysis by *o*-diphenols readily explains the Greenstein-Riley phenomenon (Riley, 1958) and related effects, at the root of which was the observation that abnormal oxygen consumption occurs in a system of *p*-phenylenediamine as substrate and melanoma extracts. This increased oxidation is the result of oxygen activation by DOPA-like compounds present in great excess in the tumor. For this reason *p*-phenylenediamine is less toxic to melanoma-bearing mices than to their nontumor controls.

The observed catalysis of autoxidations by monoprotonated *p*-phenylenediamines raises the possibility that in studies of biological electron transport employing *p*-phenylenediamines as artificial mediators between the substrate and the respiration-phosphorylation chain, part of the substrate oxidation bypasses the chain. This may be a contributing factor to the lowering of the P : O ratio observed as a result of high N,N,N',N'-tetramethyl-*p*-phenylenediamine concentration (Howland, 1963; Lee *et al.*, 1964; Packer and Mustafa, 1966).

Because a *p*-phenylenediamine type of structure is part of the tetrahydropteridine nucleus, it is conceivable that the latter, being indeed an excellent donor, could participate in oxygen activation. The initial rate of oxygen uptake in a system of a tetrahydropteridine and NADPH or cysteine (Mager and Berends, 1965) is higher than expected on the basis of the cyclic oxidation-reduction role of the tetrahydropteridine; hence it may well be that a fraction of NADPH and cysteine is directly oxidized by the tetrahydropteridine-oxygen complex.

A further important inference can be made from the iodide catalysis. Since iodide is a major metabolite from L-thyroxine (Nunez and Jacquemin, 1964; Reinwein and Rall, 1966) and iodotyrosines (Choufoer and Querido, 1964), a link might exist between deiodination and biological oxidations.

In conclusion, catalysis of autoxidation reactions by charge-transfer complexing with oxygen is likely to be both important and common. It may well be that it operates in those oxidizing enzyme systems in which no activator is required, the substrate itself fulfilling the role of activator as suggested by Hayaishi (1964). Other facts, hitherto unexplained, may conceivably find an explanation in this context, e.g., the cytochrome c catalyzed oxidation of epinephrine (Carlisle, 1949):

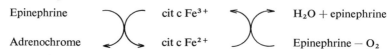

Finally, a few words are in order concerning the effect of uncouplers and inhibitors of the chain. Whether they act as electron donors or acceptors it is conceivable that in some cases their effect is primarily dependent on charge-transfer complexing, e.g., with a coenzyme (Yagi *et al.*, 1960; Liao and Williams-Ashman, 1962; Devartanian and Veeger, 1964) or an allosteric site of an enzyme.

IV. Concluding Remarks

At present, evidence of charge-transfer complexing in biological oxidation exists, but apart from a few cases, it is more inferential than factual. Yet the hypothesis is so promising and conditions in mitochondria conceivably so favorable, that it will be, indeed, surprising if additional and more conclusive evidence is not forthcoming.

REFERENCES

Alivisatos, S. G. A. (1961). *Biochem. Biophys. Res. Commun.* **4**, 292.
Alivisatos, S. G. A., Mourkides, G. A., and Jibril, A. (1960). *Nature* **186**, 718.
Alivisatos, S. G. A., Ungar, F., Jibril, A., and Mourkides, G. A. (1961). *Biochim. Biophys. Acta* **51**, 361.
Anderson, B. M., and Reynolds, M. L. (1966). *Arch. Biochem. Biophys.* **114**, 299.
Birks, J. B., and Slifkin, M. A. (1963). *Nature* **197**, 42.
Braude, E. A., Jackman, L. M., and Linstead, R. P. (1954). *J. Chem. Soc.* p. 3548.
Burton, R. M., and Kaplan, N. O. (1963). *Arch. Biochem. Biophys.* **101**, 139.
Carlisle, E. (1949). Thesis, University of Wisconsin. We thank Prof. H. Lardy for this information.
Carlsson, D. J., and Robb, J. C. (1966). *Trans. Faraday Soc.* **62**, 3403.
Choufoer, J. C., and Querido, A. (1964). *Exposes Ann. Biochim. Med.* **25**, 213.
Cilento, G. (1965). *Photochem. Photobiol.* **4**, 1243.

Cilento, G., and Berenholc, M. (1963). *J. Phys. Chem.* **67**, 1159.
Cilento, G., and Berenholc, M. (1965). *Biochim. Biophys. Acta* **94**, 271.
Cilento, G., and Giusti, P. (1959). *J. Am. Chem. Soc.* **81**, 3801.
Cilento, G., and Sanioto, D. L. (1963). *Ber. Busenges. Physik. Chem.* **67**, 426.
Cilento, G., and Sanioto, D. L. (1965). *Arch. Biochem. Biophys.* **110**, 133.
Cilento, G., and Schreier, S. (1964). *Arch. Biochem. Biophys.* **107**, 102.
Cilento, G., and Tedeschi, P. (1961). *J. Biol. Chem.* **236**, 907.
Cilento, G., and Zinner, K. (1966). *Biochim. Biophys. Acta* **120**, 84.
Cilento, G., and Zinner, K. (1967a). *Biochim. Biophys. Acta* **143**, 88.
Cilento, G., and Zinner, K. (1967b). *Biochim. Biophys. Acta* **143**, 93.
Cilento, G., and Zinner, K. (1967c). *Arch. Biochem. Biophys.* **120**, 244.
Colter, A. K., Wang, S. S., Megerle, G. H., and Ossip, P. S. (1964). *J. Am. Chem. Soc.* **86**, 3106.
Devartanian, D. V., and Veeger, C. (1964). *Biochim. Biophys. Acta* **92**, 233.
Ehrenberg, A., Eriksson, G., and Hemmerich, P. (1964). *Intern. Symp. Oxidases and Related Oxidation-Reduction Systems, Amherst, Mass., 1964*, p. 179.
Fleischman, D. E., and Tollin, G. (1965a). *Biochim. Biophys. Acta* **94**, 248.
Fleischman, D. E., and Tollin, G. (1965b). *Proc. Natl. Acad. Sci. U.S.* **53**, 237.
Foster, R., and Hanson, P. (1964). *Trans. Faraday Soc.* **60**, 2189.
Fox, J. L., and Tollin, G. (1966a). *Biochemistry* **5**, 3865.
Fox, J. L., and Tollin, G. (1966b). *Biochemistry* **5**, 3873.
Fujimori, E. (1960). *Biochim. Biophys. Acta* **40**, 251.
George, P. (1964). *Intern. Symp. Oxidases and Related Oxidation-Reduction Systems, Amherst, Mass., 1964*, p. 3.
Gibson, Q. H., Massey, V., and Atlerton, N. M. (1962). *Biochem. J.* **85**, 369.
Gouterman, M., and Stevenson, P. E. (1962). *J. Chem. Phys.* **37**, 2267.
Grabe, B. (1964). *J. Theoret. Biol.* **7**, 112.
Green, J. P., and Malrieu, J. P. (1965). *Proc. Natl. Acad. Sci. U.S.* **54**, 659.
Harbury, H. A., and Foley, K. A. (1958). *Proc. Natl. Acad. Sci. U.S.* **44**, 662.
Hayaishi, O. (1964). *Proc. 6th Intern. Congr. Biochem., New York, 1964*, I.U.B. Vol. 33, p. 31.
Hemmerich, P., Veeger, C., and Wood, H. C. S. (1965). *Angew. Chem. Intern. Ed. Engl.* **4**, 671.
Howland, J. L. (1963). *Biochim. Biophys. Acta* **77**, 419.
Isenberg, I. (1964). *Physiol. Rev.* **44**, 487.
Isenberg, I., and Szent-Györgyi, A. (1958). *Proc. Natl. Acad. Sci. U.S.* **44**, 857.
Isenberg, I., and Szent-Györgyi, A. (1959). *Proc. Natl. Acad. Sci. U.S.* **45**, 1229.
Joslyn, M. A., and Branch, G. E. K. (1935). *J. Am. Chem. Soc.* **57**, 1779.
Kearns, D. R., and Calvin, M. (1961(. *J. Am. Chem. Soc.* **83**, 2110.
Kirschner, K., Eigen, M., Bittman, R., and Voigt, B. (1966). *Proc. Natl. Acad. Sci. U.S.* **56**, 1661.
Kosower, E. M. (1956). *J. Am. Chem. Soc.* **78**, 3497.
Kosower, E. M. (1962). "Molecular Biochemistry," p. 20. McGraw-Hill, New York.
Lee, C., Nordenbrand, K., and Ernster, L. (1964). *Intern. Symp. Oxidases and Related Oxidation-Reduction Systems, Amherst, Mass., 1964*, p. 960.
Lee, C., Simard-Duquesne, N., Ernster, L., and Hoberman, H. D. (1965). *Biochim. Biophys. Acta* **105**, 397.
Leonard, N. J., and Laursen, R. A. (1965). *Biochemistry* **4**, 365.
Liao, S., and Williams-Ashman, H. G. (1962). *Federation Proc.* **21**, No. 2, 51 (abstr.).
Ludowieg, J., and Levy, A. (1964). *Biochemistry* **3**, 373.

McGlynn, S. P., Smith, F. J., and Cilento, G. (1964). *Photochem. Photobiol.* **3**, 269.
Mager, H. I. X., and Berends, W. (1965). *Rec. Trav. Chim.* **84**, 1329.
Massey, V., and Palmer, G. (1962). *J. Biol. Chem.* **237**, 2347.
Massey, V., and Williams, C. H., Jr. (1965). *J. Biol. Chem.* **240**, 4470.
Massey, V., Palmer, G., Williams, C. H., Jr., Swoboda, B. E. P., and Sands, R. H. (1965). *In* "Symposium on Flavins and Flavoprotiens." Elsevier, Amsterdam (quoted from Strittmatter, 1966).
Mauzerall, D. (1965). *Biochemistry* **4**, 1801.
Mulliken, R. S. (1952). *J. Am. Chem. Soc.* **74**, 811.
Nunez, J., and Jacquemin, C. (1964). *Exposes Ann. Biochim. Med.* **25**, 75.
Nygaard, A. P., and Theorell, H. (1955). *Acta Chem. Scand.* **9**, 1587.
Packer, L., and Mustafa, M. G. (1966). *Biochim. Biophys. Acta* **113**, 1.
Pullman, A. (1964). *J. Chim. Phys.*, p. 1666.
Pullman, B., and Pullman, A. (1963). "Quantum Biochemistry," Wiley (Interscience), New York.
Radda, G. K., and Calvin, M. (1963). *Nature* **200**, 464.
Radda, G. K., and Calvin, M. (1964). *Biochemistry* **3**, 384.
Reinwein, D., and Rall, J. E. (1966). *J. Biol. Chem.* **241**, 1636.
Riley, V. (1958). *Proc. Soc. Exptl. Biol. Med.* **98**, 57.
Robertson, A. M., and Griffiths, D. E. (1965). *Biochem. J.* **94**, 30P.
Rutter, W. J., and Rolander, B. (1957). *Acta Chem. Scand.* **11**, 1663.
Sakurai, T., and Hosoya, H. (1966). *Biochim. Biophys. Acta* **112**, 459.
San Pietro, A., Kaplan, N. O., and Colowick, S. P. (1955). *J. Biol. Chem.* **212**, 941.
Shifrin, S. (1964a). *Biochim. Biophys. Acta* **81**, 205.
Shifrin, S. (1964b). *Biochemistry* **3**, 829.
Shiga, T., and Piette, L. H. (1964). *Photochem. Photobiol.* **3**, 213.
Shinitzky, M., Katchalski, E., Grisaro, V., and Sharon, N. (1966). *Arch. Biochem. Biophys.* **116**, 332.
Slifkin, M. A. (1962). *Nature* **193**, 464.
Slifkin, M. A. (1965). *Biochim. Biophys. Acta* **103**, 365.
Stauff, J. (1964). *Ber. Bunsenges. Physik. Chem.* **68**, 773.
Steele, R. H. (1963). *Biochemistry* **2**, 527.
Strittmatter, P. (1961). *J. Biol. Chem.* **236**, 2329.
Strittmatter, P. (1965). *J. Biol. Chem.* **240**, 4481.
Strittmatter, P. (1966). *Ann. Rev. Biochem.* **35**, 125.
Swinehart, J. H. (1966). *J. Am. Chem. Soc.* **88**, 1056.
Szent-Györgyi, A. (1960). "Introduction to a Submolecular Biology." Academic Press, New York.
Tomita, G. (1967). *Experientia* **23**, 25.
Tsibris, J. C. M., McCormick, D. B., and Wright, L. D. (1965). *Biochemistry* **4**, 504.
Tsibris, J. C. M., McCormick, D. B., and Wright, L. D. (1966). *J. Biol. Chem.* **241**, 1138.
Tsubomura, H., and Mulliken, R. S. (1960). *J. Am. Chem. Soc.* **82**, 5966.
Velick, S. F. (1961). *In* "Light and Life" (W. D. McElroy and B. Glass, eds.), p. 108. Johns Hopkins Press, Baltimore, Maryland.
Walker, M. S., Bednar, T. W., and Lumry, R. (1966). *J. Chem. Phys.* **45**, 3455.
Wallenfels, K., and Gellrich, M. (1959). *Ann. Chem.* **621**, 149.
Weber, G. (1965). *In* "Symposium on Flavins and Flavoproteins." Elsevier, Amsterdam. (quoted from Strittmatter, 1966).
Weber, M. M., Hollocher, T. C., and Rosso, G. (1965). *J. Biol. Chem.* **240**, 1776.
Wilson, J. E. (1966). *Biochemistry* **5**, 1351.
Yagi, K., Ozawa, T., and Nagatsu, T. (1960). *Biochim. Biophys. Acta* **43**, 310.

Charge-Transfer Complexes in Enzyme-Coenzyme Models

SIDNEY SHIFRIN

National Cancer Institute
National Institutes of Health
Bethesda, Maryland

The possible importance of charge-transfer complexes in biological systems was first suggested by Mulliken in 1952 and more recently by Szent-Györgyi (1960). Within recent years there has been an increase in the study of electron donor acceptor properties of molecules which are of biological interest.

Theoretical calculations of purines, pyrimidines, and amino acids (Pullman and Pullman, 1958) had predicted that indole should be the most effective electron donor of the amino acids. The prediction was supported by several experimental studies (Isenberg and Szent-Györgi, 1958; Fujimora, 1959; Harbury *et al.*, 1959; Cilento and Tedeschi, 1961). One of the many electron acceptors which has been examined in a study of charge-transfer complexes is the coenzyme nicotinamide adenine dinucleotide (NAD^+) or its model compound 1-alkyl-3-carbamoyl-pyridinium halide. For example, indole was found to form an intensely yellow complex when mixed with concentrated solutions of NAD^+ or with 1-alkylpyridinium salts (Cilento and Giusti, 1959; Alivisatos *et al.*, 1961). The electron-acceptor properties of pyridinium salts were examined extensively by Kosower (1960) using the iodide ion or other inorganic ions as the electron donor; however, the electron-donor properties of amino acid side chains have not been examined with the coenzyme model as the acceptor.

Since the term "charge-transfer complex" has been employed to account for otherwise inexplicable results in many biochemical systems, an experimental examination of the donor properties of a number of amino acid side chains should provide a more scientific foundation as a basis on which to implicate charge-transfer interactions.

One of the difficulties which has prevented an examination of the donor properties of amino acids is the proximity of the charge-transfer transition to the absorption maximum of either the donor or the acceptor. Since high concentrations of material are required, the weak transition arising from charge-transfer complex formation is overshadowed by the strong absorption contributed by either or both components. One of the best electron acceptors which was made available in recent years is tetracyanoethylene (TCNE) and it was employed by Foster and Hanson (1965) in an examination of the donor properties of indole. Although a long wavelength band which was assigned to an intermolecular charge-transfer transition was initially formed, this transient intermediate gradually disappeared and 3-tricyanovinyl indole was isolated from the mixture.

In order to examine the absorption spectra of charge-transfer complexes very close to the locally excited transitions, the donor and acceptor moieties were incorporated into the same molecule and were insulated from one

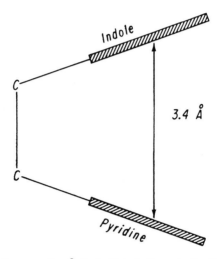

FIG. 1. Schematic diagram of α, β-disubstituted ethane in the "eclipsed" configuration. The distance between the geometric centers of the two aromatic structures was determined from molecular models.

another by two methylene groups. Thus, passage of an electron from the donor to the acceptor molecule must take place through space while propagation through the saturated hydrocarbon groups is minimized. The model compound which was used to study the interaction between indole and 3-carbamoylpyridinium chloride is shown by formula I. The two groups on

(I)

the disubstituted ethylene may assume many orientations, one of which could be the "eclipsed" conformation such that the two rings form a sandwichlike structure in which the geometric centers are separated by 3.4 Å (Fig. 1). This particular orientation does not appear to be essential for charge transfer to take place since W. N. White (1959) demonstrated that formation of an intra-molecular charge-transfer complex between the *p*-aminophenyl system as the donor and the *p*-nitrophenyl system as the acceptor was independent of the geometrical arrangement of the two rings.

I. Indole-Pyridinium Ion Interaction*

The crystalline compound and methanolic solutions of indolylethylnico-tinamide are characterized by their intensely yellow color. The absorption spectrum of a methanolic solution of this compound from 240 to 500 mμ is shown in Fig. 2. The individual contribution of the absorbancies of indole

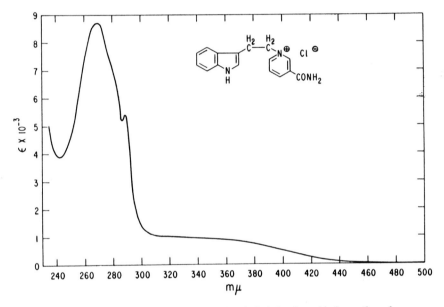

FIG. 2. Absorption spectrum of indolylethylnicotinamide in methanol.

and the pyridinium ion were subtracted from the spectrum of the complex by difference spectroscopy to give the spectrum shown in Fig. 3.

The most prominent feature of the difference spectrum is the long, diffuse absorption band which extends from 300 to 450 mμ with an apparent maxi-mum in the region between 310 and 330 mμ and a maximum extinction coefficieint of 1000. The absorption properties of this compound obey Beer's law over a concentration range from 0.10 to 10^{-7} M. Thus, there does not appear to be any significant contribution from an *inter*molecular charge-

---

* Cf. Shifrin (1964a).

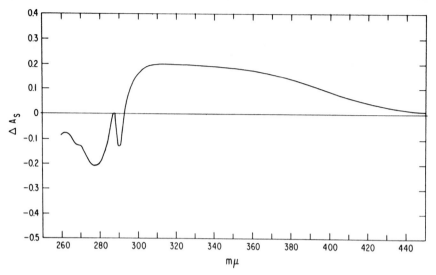

Fig. 3. Difference spectrum obtained by subtracting the individual contributions of the indole and 1-alklynicotinamide chromophores from the spectrum of indolylethylnicotinamide. The concentration of the model compound as well as that of tryptamine hydrochloride and 1-ethyl-3-carbamoylpyridinium perchlorate was $2 \times 10^{-4} M$.

transfer complex and the new band found in indolylethylnicotinamide has been assigned to an *intra*molecular charge-transfer transition.

Since fluorescence is a sensitive detector of changes in the molecular environment of the chromophore, the emission spectrum of the synthetic complex was compared with tryptamine which has its fluorescence maximum at 340 mμ (A. White, 1959). Excitation of tryptamine at 280 mμ resulted in the expected behavior, while emission from indole of the intramolecular complex was completely quenched. Equimolar mixtures of tryptamine hydrochloride and N-methylnicotinamide perchlorate showed the same fluorescence intensity at 340 mμ as given by tryptamine alone. In an effort to enhance any fluorescence which may have been emitted but was of such low intensity as to be undetectable, the complex was prepared in glycerol solution. Neither the increased viscosity nor lowering the temperature of methanolic solutions of the complex has any effect on increasing the fluorescence of the indole moiety. The absence of any detectable fluorescence may simply result from reabsorption of the 340 mμ emission by the charge-transfer band which overlaps the fluorescence band. Nevertheless, Orgel (1954) has reported that the order of quenching efficiency of electron acceptors closely parallels their effectiveness as charge-transfer acceptors.

The absorption and fluorescence properties of indolylethylpyridinium chloride were compared with the behavior of NAD$^+$ in the presence of glyceraldehyde-3-phosphate dehydrogenase. Racker and Krimsky (1952) noted that the broad absorption band of the enzyme-coenzyme complex had an apparent absorption maximum at 360 mμ with an extinction coefficient of 1000. Removal of the coenzyme resulted in the disappearance of the visible absorption band.

In addition to the similarities in the absorption properties of the model system and glyceraldehyde-3-phosphate dehydrogenase : NAD$^+$, the fluorescence behavior of both systems are analogous. Thus, Velick (1958) determined the dissociation constant of the enzyme-coenzyme complex by following quenching of apoenzyme fluorescence after addition of NAD$^+$. The results of absorption and emission behavior are consistent with the view that a tryptophan residue is located in the vicinity of the nicotinamide moiety of NAD$^+$ shown schematically in Fig. 4.

Sigman and Blout (1967) have recently reported that alkylation of chymotrypsin with α-bromo-4-nitroacetophenone results in the appearance of an intense absorption band (λ_{max} 350 mμ, ε 7.55 $\times$ 10^3 M^{-1} cm^{-1}) which is not

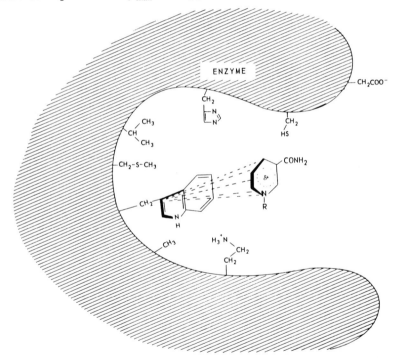

FIG. 4. Schematic diagram of a hypothetical situation in which the nicotinamide moiety of NAD$^+$ is located in the vicinity of the indole of a tryptophan residue such that electron donation from the C-3 of indole to the pyridinium ring may produce a charge-transfer transition.

shown by either the enzyme or the alkylating agent alone. On the basis of studies carried out with indole, cresol, and toluene, which are representative of the side chains of tryptophan, tyrosine, and phenylalanine, respectively, with the alkylating agent, it was concluded that the donor in the eznyme charge-transfer complex was the tryptophan residue. The characteristic tryptophan fluorescence of chymotrypsin was also 60% quenched in the alkylated enzyme.

Szent-Györgyi et al. (1961) had suggested that electron transfer from the indole moiety is localized on one of the atoms (particularly the C-3) and does not come from the overall π-system. More recently, Green and Malrieu (1965) have carried out quantum chemical studies on indoles and indole derivatives with results which are in agreement with the earlier suggestions. Isolation of 3-tricyanovinylindole from the reaction between indole and TCNE (Foster and Hanson, 1965) supports the conclusion that the C-3 position of indole has the greatest electron density and is localized site from which electrons are donated. The electron-acceptor site in the pyridinium ion may also be localized. Theoretical calculations (B. Pullman and Pullman, 1959) indicate that the most positive carbon of the 3-carbamoylpyridinium ring is C-4.

Some modifications were made in the indole moiety in an effort to examine the dependence of the charge-transfer band: (1) on the length of the hydrocarbon chain which separates the donor and acceptor groups and (2) on the position of the indole nucleus at which the alkyl side chain is substituted.

Three methylene groups were placed between indole and the pyridinium ion instead of the usual ethylene linkage which is used throughout this investigation. The additional methylene group had no effect whatever on the apparent absorption maximum of the charge-transfer transition or on its intensity. An examination of molecular models of this compound shows that the two rings are not held at restricted distances from one another, but may actually be closer to one another than in the lower homolog. Further studies are underway in the synthesis of compounds in which the connecting hydrocarbon bridge is lengthened.

The indole nucleus was substituted on the nitrogen instead of in the C-3 position as in tryptophan. The change in position of substitution did not seem to have any effect on the apparent absorption maximum of the charge-transfer band, but the intensity of the transition was 50% of the value found with 3-substituted indole.

II. Indole: 1,4-Dihydronicotinamide Interaction

Chemical reduction of the pyridinium moiety with sodium hydrosulfite results in the addition of a hydride at C-4 accompanied by removal of the positive charge of the electron-acceptor moiety. The absorption spectrum of

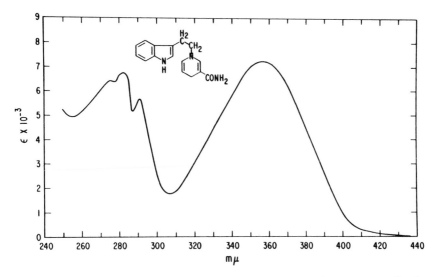

FIG. 5. Absorption spectrum of a methanolic solution of indolylethyl-1, 4-dihydronicotinamide as well as an equimolar mixture of tryptamine and 1-methyl-1 4-dihydronicotinamide.

indolylethyl-1,4-dihydronicotinamide in methanol (Fig. 5) is completely coincident with the combined spectra of equimolar solutions of tryptamine and 1,4-dihydronicotinamide, i.e., there are no detectable charge-transfer transitions or spectral shifts.

The sensitivity of fluorescence, however, to minor alterations in the environment of the emitting species would be expected to show evidence of interaction. Indole and dihydronicotinamide are both fluorescent. The former has its emission maximum at 340 mμ and the reduced nicotanimide fluoresces maximally at 450–465 mμ. When a methanolic solution of indolylethyldihydronicotinamide was excited at 280 mμ where more than 90 % of the light is absorbed by the indole moiety, the only detectable fluorescence was emitted by the dihydronicotinamide ring (λ_{fl} 460 mμ). Transfer of electronic excitation energy ("sensitized fluorescence") is most probable when the emission band of the donor (indole) overlaps the absorption band of the acceptor (dihydronicotinamide). The solid curve in Fig. 6 represents the excitation spectrum of indolylethyldihydronicotinamide which demonstrates that all of the energy absorbed by indole is transferred to the fluorescent dihydronicotinamide. The excitation spectrum of 1-benzyl-1,4-dihydronicotinamide is given by the dashed curve in Fig. 6. Transfer of excitation energy from protein tryptophan residues to bound NADH has been reported for a large number of pyridine coenzyme-linked dehydrogenases (Shrifin and Kaplan, 1960).

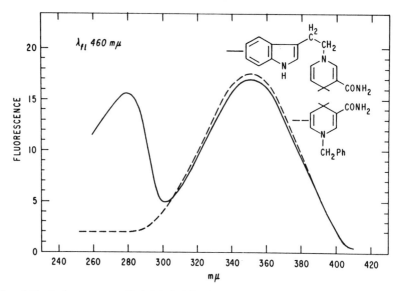

FIG. 6. Excitation spectra of indolylethyldihydronicotinamide (solid curve) and 1-benzyl-1, 4-dihydronicotinamide (dashed curve) in methanol keeping the emission wavelength at the maximum of dihydronicotinamide fluorescence. The absorbancy of both solutions was adjusted to 0.100 at 356 mμ for the study. Fluorescence intensity is in arbitrary units and the spectra were corrected for variations in the intensity of the exciting source.

III. Amino Acid-Pyridinium Interactions*

Examination of indole-pyridinium ion interactions was facilitated by the long wavelength band of the charge-transfer complex which was far removed from the locally excited transitions. As the electron availability of the amino acid side chain decreases, however, the position of the charge-transfer transition comes closer to the locally excited chromophores and the assignment of the apparent maximum to the new band is less accurate. We may tentatively assume that amino acid side chains, however, form charge-transfer complexes with the pyridinium ion and see if the properties of the complexes follow the behavior expected of other donor : acceptor pairs.

The apparent absorption maxima obtained from difference spectra of β-substituted ethyl-3-carbamoylpyridinium chlorides for the side chains of methionine, tyrosine, histidine, and phenylalanine are summarized in the first column of Table I. The apparent absorption maxima are listed in order

---

* Cf. Shifrin (1964b).

TABLE I

Absorption and Emission Properties of β-Substituted
1-Ethyl-3-Carbamoylpyridinium Chlorides

R Side chains	Difference spectrum		Cyanide adduct		Dithionite reduction		
	λmax (mμ)	(ϵ)	λmax (mμ)	(ϵ)	λmax (mμ)	(ϵ)	Emission intensity at 450 mμ
indolyl	325	(1000)	340	(7000)	356	(7000)	1.00
CH_3S-	300	(900)	337	(6100)	355	(7000)	0.95
$HO-C_6H_4-$	296.5	(1000)	341	(6800)	356.5	(7000)	1.10
imidazolyl	294.5	(900)	337.5	(7900)	355	–	1.00
phenyl	282.5	(1000)	338	(7000)	355	(7100)	1.12
$(CH_3)_3\overset{+}{N}-$	–	–	330	(6000)	340	–	0.02

of decreasing wavelength which should coincide with decreasing electron availability of the donors. Immediately below indole is the thiomethyl group of methionine in which the nonbonding lone pair n electrons are shown to be more efficient electron donors than the π electrons of aromatic phenol. Although the source of electrons in indole complexes was found to be localized at C-3, the n-donor action of the sulfur atom is even more localized. The donor properties of sulfur have been studied by several authors using I_2 as the acceptor molecule (Tsubomura and Lang, 1961; Niedzielski et al., 1964; Wayland and Drago, 1964).

There have been several unsuccessful attempts in this laboratory to synthesize a model compound containing a free sulfhydryl group so that its non-covalent interaction with the pyridinium ring could be examined. The initial studies on the broad absorption band in gylceraldehyde-3-phosphatede-hydrogenase : NAD$^+$ complexes (Racker and Krimsky, 1952) suggested that the protein sulfhydryl group was involved in production of the new band, since the latter was lost in the presence of p-mercuribenzoate, a reagent which reacts with the sulfhydryl group. It might appear reasonable from the present studies of n donors that the ionized mercaptide group of the protein may serve as an excellent donor with the nicotinamide side chain of NAD$^+$. The assignment of the indole side chain of tryptophan, however, in the role of electron donor in the enzyme-coenzyme complex is consistent with both absorption and emission behavior.

Interaction of phenol with the pyridinium ion when examined in methanol or aqueous solution at pH 7 results in the appearance of a new band with an apparent maximum at 296 mμ. Ionization of the phenolic hydroxyl group results in an increase in the electron-donor properties of the aromatic system and a long wavelength shift of the intramolecular charge-transfer transition. Indeed, the difference spectrum of p-hydroxyphenylethylnicotinamide in 0.1 N NaOH shows an apparent maximum at 320 mμ with considerably higher absorbancy between 330 and 460 mμ than that found in the un-ionized sample (Fig. 7).

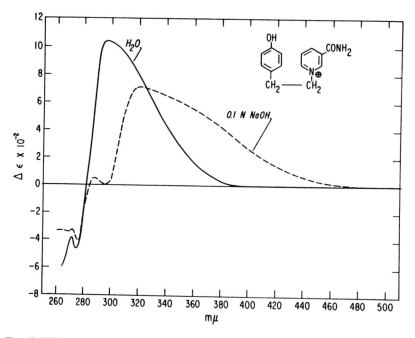

FIG. 7. Difference absorption spectrum of p-hydroxyphenylethylnicotinamide in water at pH 7 (solid curve) and in 0.1 N NaOH (dashed curve).

The imidazole side chain of histidine contains two nitrogen atoms with a lone pair of electrons which function as the n donor to the pyridinium ion giving rise to a broad transition which extends from 280 to 360 mμ. Protonation of the imidazole nitrogens with 0.1 N HCl abolished the charge-transfer transition. This is further support for the assignment of an n donor to the imidazole group.

The weakest electron donor among the π or n donors was the benzene side chain of phenylalanine. Here the electrons would be the least localized and the apparent maximum of the new transition at 282 mμ is very close to the locally excited bands.

The extinction coefficient of all of the new bands was consistently around 1000 which suggests that the intensity of the charge-transfer transition is associated with a feature common to all the synthetic compounds. The most apparent factors are (1) the presence of the 3-carbamoylpyridinium ion and (2) the orientation of the donor and acceptor molecules. McConnell et $al.$ (1953) had suggested that the intensity of the charge-transfer transition is related to the acceptor moiety of the complex, whereas Orgel and Mulliken (1957) have indicated that the apparent extinction coefficient is a sensitive indicator of the orientation of the two groups.

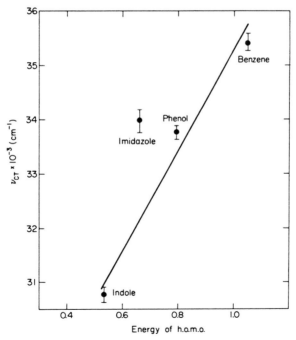

FIG. 8. Relationship between the frequency of the apparent absorption maximum of the intramolecular charge-transfer transition and calculated energies of the highest occupied molecular orbitals (h.o.m.o.) for several amino acid side chains.

Since the electron-acceptor moiety and the ethylene bridge are constant features of the series, the wavelength of the new transition is dependent upon the ionization potential of the electron donor. Although the ionization potentials of indole or imidazole are not available from the literature, related properties have been calculated by B. Pullman and Pullman (1958) for the energy of the highest occupied molecular orbital (h.o.m.o.). These calculated values are plotted against the frequency of the charge-transfer transition (Fig. 8). The linear relationship is relatively good except for imidazole in which case we have been informed that changes are being made in the quantum chemical calculations (B. Pullman, 1964).

IV. Amino Acid Side Chains—1,4-Dihydronicotinamide Interactions

Chemical reduction of the nicotinamide moiety with sodium dithionite gave the corresponding 1,4-dihydronicotinamide derivative. The maxima of the longest wavelength absorption band together with their corresponding extinction coefficients are summarized above in Table I. With the exception of the trimethylammonium derivative which is a model for the lysine side chain, the spectral properties of the reduced models are identical with the combined spectra of the amino acid side chains and 1-ethyl-1,4-dihydronicotinamide, i.e., there is no indication from the absorption spectra that the two moieties interact to produce a new transition nor are there any detectable shifts in the absorption maxima.

The observation that imidazole interferes with electron transport in mitochondrial oxidative phosphorylation was interpreted in terms of the ability of the imidazole ring to form a charge-transfer complex with the dihydronicotinamide moiety of NADH (Estabrook et al., 1963). The results obtained from model compounds in these studies does not support such an hypothesis. There is a possibility that the dihydronicotinamide moiety of NADH may serve as an electron donor with the positively charged imidazolium ion as the electron acceptor; however, creation of an imidazoliumion by the addition of acid would destroy the dihydronicotinamide in the model compounds described in these studies. Recent reports have shown that the imidazolium ion may, indeed, function as an electron acceptor in the presence of indole (Shinitzky et al., 1966).

The high sensitivity of fluorometric methods indicates that the energy used to excite the phenol moiety of p-hydroxyphenylethyl-1,4-dihydronicotinamide electronically is transferred quantitatively to the fluorescent dihydronicotinamide group. The absorbancy of all the solutions was adjusted to 0.100 at the longest wavelength band and the emission intensities and maxima were compared with a solution of 1-methyl-1,4-dihydronicotinamide. The values shown in the last column of Table I indicate that the fluorescent behavior of the di-hydronicotinamide moiety is unaffected by all of the amino acid side chains except the trimethylammonium ion which is the model for the charged ε-amino group of lysine.

The absorption maximum of the cyanide adduct and the dithionite reduction product of trimethylaminoethylnicotinamide is shifted by about 15 mμ to the blue compared with maxima of other model compounds. In addition to the hypsochromic shift in the absorption maxima, the fluorescence intensity of the dihydronicotinamide moiety is quenched by the nearby positively charged group.

Binding of NADH to horse liver alcohol dehydrogenase was accompanied by a shift in the absorption maximum of the reduced coenzyme from 340 to 325 mμ (Theorell and Bonnichsen, 1951). There has been a considerable amount of speculation in an attempt to predict the environment of the bound coenzyme which could account for this very large spectral shift. From theoretical considerations and by analogy with published spectra Kosower and Remy (1959) calculated that a positively charged group placed 3.1 Å from the dihydronicotinamide ring nitrogen could account for the increased energy required to raise the reduced coenzyme to an excited singlet state. The increased energy is reflected by the hypsochromic shift in enzyme-co-enzyme systems and in the model compound.

Boyer and Theorell (1955) reported that the binding of NADH to liver alcohol dehydrogenase resulted in a marked enhancement in the intensity of NADH fluorescence. Similar behavior has now been reported for a large number of NAD$^+$-linked dehydrogenases (Shifrin and Kaplan, 1960). In the present study, however, dihydronicotinamide fluorescence is markedly quenched by the neighboring positively charged nitrogen in contrast to the behavior of the enzyme-coenzyme complex. Nevertheless, a more thorough understanding of the fluorescence behavior of proteins may shed light on the emission properties of NADH-enzyme complexes.

V. Substituted Benzene-Pyridinium Interactions*

Although the new absorption bands observed in spectra of amino acid side chain-pyridinium models were tentatively assigned to intramolecular charge-transfer transitions, the relationship between the frequency of the new transition and the ionization potentials of the electron donors would be most conveniently studied using substituted benzene derivatives whose ionization potentials are accurately known from photoionization data (Watanabe, 1959).

The following groups were employed as electron donors in the β-substituted

---

* Cf. Shifrin (1965).

ethyl-3-carbamoyl pyridinium salts: *p*-aminophenyl, *p*-methoxyphenyl, *p*-hydroxyphenyl, *p*-tolyl, *p*-chlorophenyl, and unsubstituted phenyl. Some representative difference spectra for the *p*-amino-, *p*-hydroxy-, and *p*-methyl-substituted derivatives are shown in Fig. 9. The half-band width of the *p*-hydroxyphenyl derivative is representative of the majority of the new transitions ($v_{1/2} = 4.4$–4.8×10^3 cm^{-1}). The *p*-tolyl derivative is the narrowest band in this series ($v_{1/2} = 3.1 \times 10^3$ cm^{-1}) and the *p*-aminophenyl group

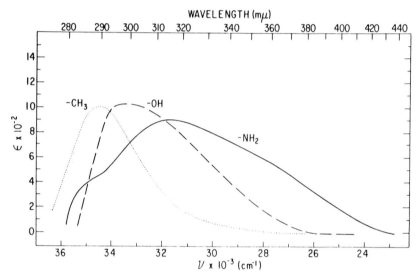

FIG. 9. Difference spectra for *p*-tolylethylnicotinamide (dotted curve, —CH$_3$), *p*-hydroxyphenylethylnicotinamide (dashed curve, —OH), and *p*-aminophenylethylnicotinamide (solid curve, —NH$_2$).

($v_{1/2} = 7.7 \times 10^3$ cm^{-1}) gives rise to a band whose half-band width is nearly double that of the others. The new transition formed between aniline and nicotinamide can be clearly seen to be composed of at least two separable bands. One of the transitions may result from the electron-donor properties of the amino group alone (*n* donor) and the other from the π system of the aromatic ring. Since the nonbonding electrons on the aniline nitrogen are conjugated with the aromatic ring, the results of the present study are inconclusive as to the source of the donor electrons. Although Mukherjee and Chandra (1964) discuss the difficulty in classifying aniline as an *n* donor or a π donor with chloranil, they eventually conclude that the electrons come from the nitrogen lone pair. While a similar conclusion was made by Tsubomura (1960) in a study of molecular complexes between *N,N*-dialkylaniline and iodine, the results of a more recent study of aniline : chloranil interaction (Carper *et al.*, 1965) are inconclusive as to the source of electrons. Studies of aniline or alkylaniline with the excellent electron-acceptor properties of tetracyanoethylene are complicated by the irreversible formation of tri-cyanovinly derivatives of the aniline (Rappoport and Horowitz, 1964; Isaacs, 1966).

The spectral properties of the substituted benzene derivatives are sum-marized in Table II. The extinction coefficients of the new transition are

TABLE I

SPECTRAL PROPERTIES OF THE INTRAMOLECULAR CHARGE-TRANSFER
TRANSITION OF N-(β-p-X-PHENYLETHYL)-3-CARBAMOYLPYRIDINIUM HALIDES

X	$\nu_{ct}(cm^{-1})$	ε_m	$\Delta\nu_{1/2}(cm^{-1})$	f
—NH$_2$	31750 ± 150	910	7750	0.030
—OH	33600 ± 150	1040	4850	0.022
—OCH$_3$	33850 ± 150	1160	4700	0.023
—CH$_3$	33500 ± 50	1010	3100	0.013
—Cl	35000 ± 80	950	4870	0.020
—H	35400 ± 70	1080	4450	0.026

relatively constant ($\varepsilon \approx 1000$) as had been observed with the amino acid side chains as electron donors.

In addition to the vast literature which demonstrates the linear relationship between kinetic parameters and Hammett's sigma functions for substituted benzene derivatives, there have been some applications to regularities in electronic absorption spectra (Gerson and Heilbronner, 1959; Kosower et al., 1962). The correlation between the frequency of the intramolecular charge-transfer transition and the Hammett sigma functions for the p-substituted benzene derivatives employed as electron donors is shown in Fig. 10. The p-chlorophenyl derivative was found to be a better electron donor than unsubstituted benzene in the present studies, although the Hammett functions would have predicted the opposite.

Of greater significance to a study of charge-transfer transitions is the relationship between the frequency of the new transition and the ionization potential of the electron donor. Within the rather narrow range of ionization potential values of the benzene derivatives employed in this study the relation-ship should be linear. The frequency of the new band is plotted against the ionization potentials of the electron donors in Fig. 11. The linear relationship is represented by the expression shown in Fig. 11.

In summary, it has been demonstrated that charge-transfer transitions in biological systems where NAD$^+$ may be serving as the electron acceptor should behave in the same predictable manner as other charge-transfer complexes.

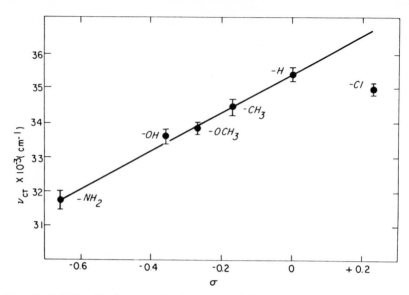

FIG. 10. Relationship between the frequency of the apparent absorption maxima of
p-phenyl derivatives of the model compounds and the Hammet sigma functions (σ) for
the corresponding benzene derivatives.

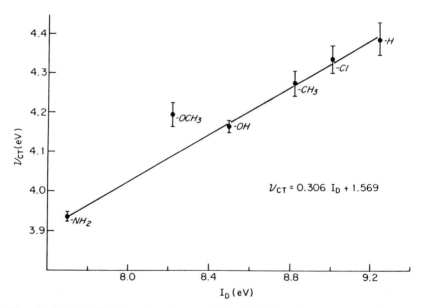

FIG. 11. Relationship between the frequency of the apparent absorption maxima of
p-phenyl derivatives of the β-substituted ethylpyridinium salts and the ionization potential
of the substituted benzene derivatives.

VI. Purine-Nicotinamide Interactions

The linear relationship between the frequency of the intramolecular charge-transfer transition and the ionization potential of the electron donor in β-substituted ethyl-3-carbamoylpyridinium halides should permit an experimental evaluation of the ionization potentials of some purines and pyrimidines. Quantum chemical calculations of the purines and pyrimidines have been reported (Veillard and Pullman, 1963; A. Pullman and Rossi, 1964), although there are difficulties encountered with the lone pairs of electrons on the nitrogen and oxygen atoms.

The synthetic methods of preparing the desired purine and pyrimidine derivatives of N-ethylnicotinamide are considerably more involved than that which was required for the amino acid side chains. Adenylethylnicotinamide (II) was synthesized although in very poor yield.

(II)

The absorption spectrum of (II) was compared with the spectral properties of NAD^+ in water and methanolic potassium cyanide. The spectral curves for the two compounds are almost identical. Adenylethylnicotinamide should serve as an excellent model for examining interactions in the coenzyme. Cilento and Schreier (1964) have demonstrated that NAD^+ exists in a folded conformation which allows a charge-transfer band to be formed as a result of the electron-donor properties of the adenine. In the present study an equimolar mixture of 9-(2'-chloroethyl) adenine and N-ethyl-3-carbamoylpyridinium chloride was compared with the spectrum of (II). There were no new transitions present in (II) that were far removed from the locally excited bands. The half-band width of (II) is considerably greater than that given by the equimolar mixture which suggests that if there is a charge-transfer transition, it lies very close to the main bands of the individual chromophores. A more accurate assignment of the position of the new band awaits the synthesis of larger quantities of (II) which will also permit more extensive purification.

On the basis of quantum chemical calculations of the energy of the highest occupied molecular orbital of purines, pyrimidines, and amino acids B. Pullman and Pullman (1958) would have predicted that adenine would be a better electron donor than indole, although this does not appear to be supported in the present studies. Further studies are in progress however, in an examination of the donor efficiency of purines and pyrimidines.

REFERENCES

Alivisatos, S. G. A., Unger, R., Jibril, A., and Mourkides, G. A. (1961). *Biochim. Biophys. Acta* **51**, 361.

Boyer, P. D., and Theorell, H. (1955), *Acta Chem. Scand.* **10**, 447.

Carper, W. R., Hedges, R. M., and Simpson, H. N. (1965), *J. Phys. Chem.* **69**, 1707.

Cilento, G., and Giusti, P. (1959). *J. Am. Chem. Soc.* **81**, 3801.

Cilento, G., and Schreier, S. (1964). *Arch. Biochem. Biophys.* **107**, 102.

Cilento, G., and Tedeschi, P. (1961). *J. Biol. Chem.* **236**, 907.

Estabrook, R. W., Gonze, J., and Nissley, S. P. (1963). *Federation Proc.* **22**, 1071.

Foster, R., and Hanson, P. (1965). *Tetrahedron* **21**, 255.

Fujimora, E. (1959). *Proc. Natl. Acad. Sci. U.S.* **45**, 133.

Gerson, F., and Heilbronner, E. (1959). *Helv. Chim. Acta* **42**, 1877.

Green, J. P., and Malrieu, J. P. (1965). *Proc. Natl. Acad. Sci. U.S.* **54**, 659.

Harbury, H. A., Lanoue, K. F., Loach, P. A., and Amick, R. M. (1959). *Proc. Natl. Acad. Sci. U.S.* **45**, 1708.

Isaacs, N. S. (1966). *J. Chem. Soc. B, Phys. Org.* p. 1053.

Isenberg, I., and Szent-Györgyi, A. (1958). *Proc. Natl. Acad. Sci. U.S.* **44**, 857.

Kosower, E. M. (1960). *In* "The Enzymes" (P. D. Boyer, H. Lardy, and K. Myrbäck, eds.), Vol. 3, p. 171. Academic Press, New York.

Kosower, E. M., and Remy, D. C. (1959). *Tetrahedron* **5**, 281.

Kosower, E. M., Hofmann, D., and Wallenfels, K. (1962). *J. Am. Chem. Soc.* **84**, 2755.

McConnell, H., Ham, J. S., and Platt, J. R. (1953), *J. Chem. Phys.* **21**, 66.

Mukherjee, D. C., and Chandra, A. K. (1964), *J. Phys. Chem.* **68**, 477.

Mulliken, R. S. (1952), *J. Am. Chem. Soc.* **74**, 811.

Niedzielski, R. J., Drago, R. S., and Middaugh, R. L. (1964). *J. Am. Chem. Soc.* **86**, 1694.

Orgel, L. E. (1954). *Quart. Rev. (London)* **8**, 422.

Orgel, L. E., and Mulliken, R. S. (1957). *J. Am. Chem. Soc.* **79**, 4839.

Pullman, A., and Rossi, M. (1964). *Biochim. Biophys. Acta* **88**, 211.

Pullman, B. (1964). Personal communication.

Pullman, B., and Pullman, A. (1958). *Proc. Natl. Acad. Sci. U.S.* **44**, 1197.

Pullman, B., and Pullman, A. (1959). *Proc. Natl. Acad. Sci. U.S.* **45**, 136.

Racker, E., and Krimsky, I. (1952). *J. Biol. Chem.* **198**, 731.

Rappoport, Z., and Horowitz, A. (1964). *J. Chem. Soc.* p. 1348.

Shifrin, S. (1964a). *Biochim. Biophys. Acta* **81**, 205.

Shifrin, S. (1964b). *Biochemistry* **3**, 829.

Shifrin, S. (1965). *Biochim. Biophys. Acta* **96**, 173.

Shifrin, S., and Kaplan, N. O. (1960). *Advan. Enzymol.* **22**, 337.

Shinitzky, M., Katchalski, E., Grisaro, V., and Sharon, N. (1966). *Arch. Biochem. Biophys.* **116**, 332.

Sigman, D. S., and Blout, E. R. (1967). *J. Am. Chem. Soc.* **89**, 1747.

Szent-Györgyi, A. (1960). "Introduction to a Submolecular Biology," Academic Press, New York.

Szent-Györgyi, A., Isenberg, I., and McLaughlin, J. (1961). *Proc. Natl. Acad. Sci. U.S.* **47**, 1089.

Theorell, H., and Bonnichsen, R. K. (1951). *Acta Chem. Scand.* **5**, 127.

Tsubomura, H. (1960). *J. Am. Chem. Soc.* **82**, 40.

Tsubomura, H., and Lang, R. P. (1961). *J. Am. Chem. Soc.* **83**, 2085.

Veillard, A., and Pullman, B. (1963). *J. Theoret. Chem.* **4**, 37.

Velick, S. F. (1958). *J. Biol. Chem.* **233**, 1455.

Watanabe, K. (1959). "Final Report on Ionization Potentials of Molecules from Photo-ionization Data."

Wayland, B. D., and Drago, R. S. (1964). *J. Am. Chem. Soc.* **86**, 5240.

Weber, G. (1960). *Biochem. J.* **75**, 335.

White, A. (1959). *Biochem. J.* **71**, 217.

White, W. N. (1959). *J. Am. Chem. Soc.* **81**, 2912.

Charge-Transfer Interactions in Certain Physiological Processes

M. A. SLIFKIN AND J. G. HEATHCOTE

Department of Pure and Applied Physics
and Department of Chemistry
The University
Salford, England

I. Introduction

Charge-transfer forces are now accepted as being of interest in biological systems; one aspect which has rarely been studied is the effect of pH on such systems. To the authors' knowledge the only systems studied at different pH values have been amino acids, proteins and chloranil (Birks and Slifkin, 1963), and proline and chloranil (Slifkin and Heathcote, 1967). It would appear important to study pH effects on biological charge-transfer complexes if any knowledge of their physiological importance is to be gained. In this chapter results are presented on two types of systems of complexes—those involving cyanocobalamin (vitamin B_{12}) and hydroxocobalamin (vitamin B_{12b}) and those involving hematoporphyrin.

II. Vitamin B_{12}

A. Introduction

A brief account of the isolation and properties of the various cobalamins is given in the monograph by Smith (1960). Our interest in vitamin B_{12} stems from its chemotherapeutic properties in the treatment of pernicious (megaloblastic) anemia. A good deal of controversy exists concerning the mode of absorption of the vitamin in the intestine. The early theory (due to Castle, 1953) was that the vitamin combined in the stomach with a high molecular weight protein which facilitated its absorption. Apart from the fact that the vitamin does not occur naturally in a form free to combine with protein, Heathcote and Mooney (1958) have pointed out that proteolytic breakdown (not synthesis) is essential before the vitamin can be released for absorption. These authors have suggested that low molecular weight peptides (Milhaud, 1961) and even amino acids (Heathcote and Mooney, 1962; Mooney and Heathcote, 1963, 1965) may facilitate the absorption of the vitamin.

Chemical studies on hydroxocobalamin (vitamin B_{12b}) give some evidence of association with other molecules. Adler *et al.* (1966) have shown that some thiols interact with hydroxocobalamin to form 1 : 1 complexes, the interaction occurring with the sulfhydryl group. Randall and Alberty (1966) have studied the kinetics of thiocyanate ion binding to vitamin B_{12b} as a function of pH. Forward and backward rate constants as well as equilibrium coefficients have been given.

A purely theoretical study of cyanocobalamin has been made by Veillard and Pullman (1965) and these authors have concluded that it is a good electron acceptor.

B. Study of the Interactions of Hydroxocobalamin (Vitamin B_{12b})

The addition of amino acids and amines to hydroxocobalamin (vitamin B_{12b}) in water or 50% ethanol, buffered or unbuffered, causes the following spectral changes in each case. The main absorption bands of vitamin B_{12b} at 28.2 and 18.8 kK shift toward 27.4 and 18 kK, the amount of shift in wavelength toward the red being proportional to the concentration of the amino acid. Isosbestic points are observed at 18.4, 25.6, 28, and 30.8 kK. Figure 1 illustrates these changes. If the difference spectra are observed, then the same effects are shown as positive peaks at 27.4 and 18 kK and negative peaks at

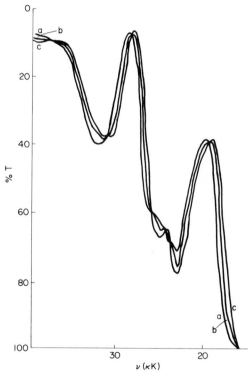

Fig. 1. Absorption spectra of glycine in vitamin B_{12b}; pH 7 buffer. (a) Vitamin B_{12b} alone; (b) plus 6.8×10^{-2} M glycine; and (c) plus 1.27×10^{-1} M glycine.

28.2 and 18.8 kK as in Fig. 2. The same isosbestic points are, of course, seen where the spectra cross the 100% transmission line. On heating solutions of amino acids and vitamin B_{12b} both the negative peaks and the positive peaks decrease, i.e., move toward the 100% transmission line (Fig. 3). These effects are, in general, fully reversible on cooling. The spectra of several amines and amino acids have been studied as a function of concentration and temperature. Application of the Benesi-Hildebrand equation (1949) to the data for glycine and vitamin B_{12b} in phosphate buffer at pH 7 yields equilibrium constants (K_c) of 76.8 liters/mole at 25°C, 72.4 liters/mole at 35°C, and 65.5 liters/mole at 45°C. This gives an enthalpy of dissociation, ΔH, of -1.52 kcal/mole (Briegleb, 1961).

The aliphatic amines give spectra very similar to those of the amino acids except for some increased absorption in the region above 29 kK as in Fig. 4. Equilibrium constants for the triethylamine-vitamin B_{12b} interaction at pH 7 are 141 liters/mole at 25°C, 100 liters/mole at 35°C, and 62.4 liters/mole at 45°C, yielding an enthalpy of dissociation of -7.64 kcal/mole. Cysteine-HCl, an amino acid containing a sulfhydryl group, initially reacts with vitamin B_{12b} to give a purple solution with positive peaks at 26.5 and 17.5 kK and negative peaks at 28.6 and 18.8 kK. With time, irreversible changes occur and the solutions go to a pale peach color and spectral detail is lost in the ultraviolet and visible regions. A white precipitate comes down after 12 hours. Details of the aliphatic amines and amino acids which have been studied are given in Table I.

TABLE I

INTERACTIONS OF VITAMINS B_{12} AND B_{12b}

(a) Materials interacting with vitamin B_{12b}	
glycine ($\Delta H = -1.52$ kcal/mole)	bovine serum albumen
triethylamine ($\Delta H = -7.64$ kcal/mole)	caffeine
8-azaadenine ($\Delta H = -26.8$ kcal/mole)	6-methylaminopurine
tryptophan	xanthine
valine	purine
serine	8-azaguanine
lysine	hypoxanthine
glutamic acid-HCl	methylamine
aspartic acid	diethylamine
arginine-HCl	cysteine-HCl (sulfyhydryl bonded)
p-aminobenzoic acid	alloxan
m-aminobenzoic acid	
(b) Materials interacting with vitamin B_{12}	
alloxan	chloranil
glutaric acid	2,5-dichlorobenzoquinone
iodine	glutamic acid-HCl

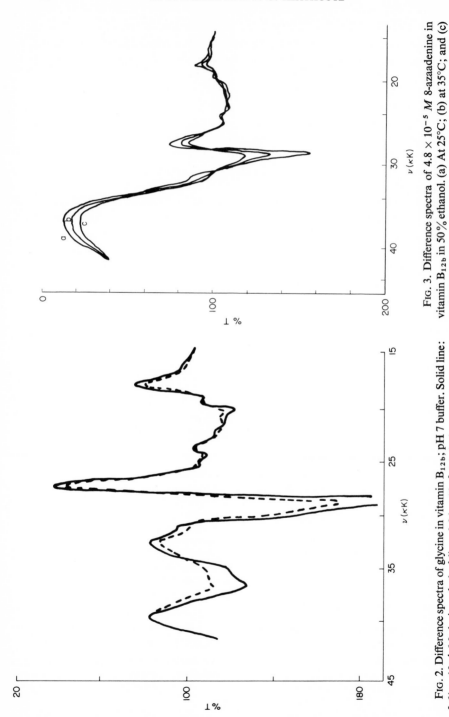

FIG. 3. Difference spectra of $4.8 \times 10^{-5}\ M$ 8-azaadenine in vitamin B_{12b} in 50% ethanol. (a) At 25°C; (b) at 35°C; and (c) at 45°C.

FIG. 2. Difference spectra of glycine in vitamin B_{12b}; pH 7 buffer. Solid line: $3.61 \times 10^{-1}\ M$ glycine; dashed line: $3.34 \times 10^{-2}\ M$ glycine.

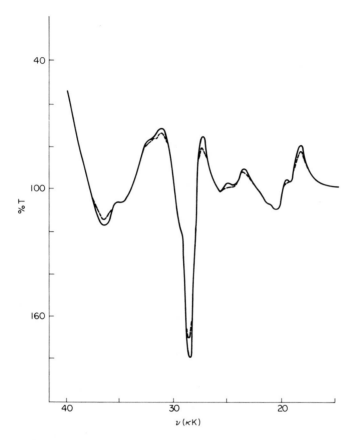

FIG. 4. Difference spectra of 10^{-1} M triethylamine in vitamin B_{12b} in pH 7 buffer. Solid line: at 25°C; dashed line: at 35°C.

The interaction of nucleic acid bases with vitamin B_{12b} has also been studied. Similar results to those described for the amino acids are obtained. One noteworthy point is the particularly strong interaction of 8 azaadenine. The equilibrium constants in 50% ethanol are 214 liters/mole at 25°C, 51 liters/mole at 35°C, and 24.6 liters/mole at 40°C, yielding an enthalpy of dissociation of −26.8 kcal/mole. Details of the nucleic acid bases which have been studied are also given in Table I.

A compound which gave interesting results was alloxan (a purine). The spectra of fresh alloxan-vitamin B_{12b} mixtures resembled those of glycine, but changes occurred in them with time, as with cysteine, which resulted in a reduction in color to that of pale peach and the loss of all spectral detail in the ultraviolet and visible regions.

The effect of pH on the mixtures was studied. First, the spectrum of vitamin B_{12b} in water is independent of pH over the range 2 to 12 provided that the measurements are taken fairly rapidly. The difference spectra were studied as a function of pH by adding concentrated NaOH or HCl to change the pH. The only change which was observed for both the amino acids and the nucleic acid

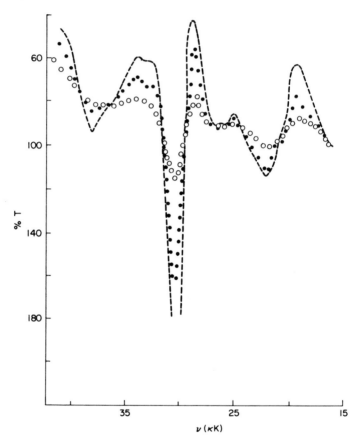

FIG. 5. Difference spectra of 2.26×10^{-2} M glutamic acid-HCl in vitamin B_{12b} at (a) pH 5, circled line; (b) pH 8, dotted line and (c) pH 12, dashed line.

bases listed in Table I was a very slight intensification of the spectra in alkaline solutions. No changes in the shapes of the spectral diagrams were observed. An exception was glutamic acid-HCl which showed a fourfold intensification in spectrum in going from pH 6 to 8 but without any change in the shape of the spectrum. This is illustrated in Fig. 5. An attempt was made to study the rates of these various interactions with B_{12b} but the spectra changed too quickly, equilibrium being reached in a few minutes.

C. Study of the Interactions of Cyanocobalamin (Vitamin B₁₂)

The addition of amino acids or nucleic acid bases to vitamin B_{12} in water or 50% ethanol, buffered or unbuffered, causes no change at all in the spectrum whether observed directly or as a difference spectrum. The addition of glutaric acid or alloxan causes the spectrum to move to a higher frequency. This shows up in the difference spectrum (Fig. 6) as two negative peaks at 27.4 and 18 kK

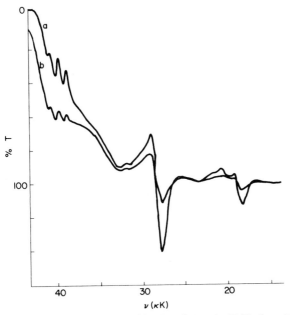

Fig. 6. Difference spectra of glutaric acid in vitamin B_{12} in 50% ethanol. (a) 2.61×10^{-1} M and (b) 8.7×10^{-2} M glutaric acid.

and positive peaks at 28.2 and 18.8 kK, which is the inverted mirror image of the amino acid-vitamin B_{12b} difference spectrum. At room temperature the optical density of the spectrum as a function of concentration of added alloxan or glutaric acid is linear, but heating the solutions causes partially irreversible increases in the difference spectrum of each complex. Therefore, equilibrium constants have only been derived for freshly made solutions at 25°C. They are 3.16 liters/mole for glutaric acid and 219 liters/mole for alloxan. The effect of pH on the glutaric acid-vitamin B_{12} difference spectrum was examined rapidly after adding very small amounts of concentrated HCl or NaOH to the aqueous solution. As the pH increases from 2 to 12, the intensity of the difference spectrum decreases by about a third but with no overall change in shape (Fig. 7). The spectrum of vitamin B_{12} in water shows no change with pH over

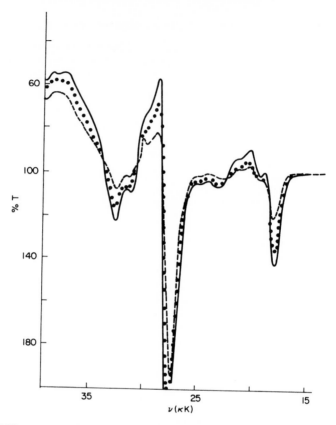

FIG. 7. Difference spectra of 2.65×10^{-2} M glutaric acid in vitamin B_{12} in water at (a) pH 5, solid line; (b) pH 8, dotted line; and (c) pH 12, dashed line.

this region. The addition of the electron acceptors, iodine, chloranil, and 2,5-dichlorobenzoquinone, to B_{12} in water causes positive peaks to appear at 28.6 kK and negative peaks at 18 and 27.4 kK. Owing to the marked coloration of iodine solutions as well as to their instability (Slifkin, 1965a) and also to the very low solubility of the other acceptors in water, however, no quantitative studies were carried out.

D. Discussion of the Observed Interactions of Vitamins B_{12} and B_{12b}

The spectra of mixtures of vitamin B_{12b} and amino acids or nucleic acid bases are consistent with the formation of 1 : 1 molecular complexes. The isosbestic points, the temperature reversibility, and the agreement with the Benesi-Hildebrand equation are all indicative of such formation. Furthermore, the amino acids and nucleic acid bases are known electron donors

(Birks and Slifkin, 1963; Slifkin, 1965b). The nature of the binding would appear to be that between an electron donor and an electron acceptor. This type of complexing is probably the same as that described by Mulliken (1950) as a charge-transfer complex. One criterion suggested by Mulliken is that a new absorption band corresponding to the transfer of an electron from the donor to the acceptor should be seen. Such a band has not been observed here or, indeed, in other similar complexes (Birks and Slifkin, 1963; Slifkin, 1965b,c), but there are various explanations for this. Firstly, Mulliken derived his theory for very weak complexes in which the amount of charge transfer in the ground state is negligible. In stronger complexes, still bound together by similar forces, the ground state can be quite ionic in character so that absorption of light might not result in any strong further donation of charge to the acceptor. Indeed, some complexes have been shown to be almost completely ionic in the ground state so that absorption by the complex results in excitation to the bound negative ion of the acceptor (Slifkin, 1964). Again, an earlier investigation of some biological complexes showed that the charge-transfer band was observed in organic solvents but not in aqueous solvents (Birks and Slifkin, 1963). Finally, the charge-transfer band might occur, but it might be lost under other absorption bands or appear in regions not accessible to the instrument used. In the case of the aliphatic amino acids and amines the donated electron must be one of the lone-pair electrons located on the nitrogen atom in the amino group. The interaction of cysteine-HCl with vitamin B_{12b} gives a different spectrum from the other amino acids and has been shown by Randall and Alberty (1966) to be due to sulfhydryl bonding. The white precipitation which forms later is probably due to the aerial oxidation of uncomplexed cysteine to cystine which is very insoluble. The spectrum of the complexed vitamin B_{12b} is identical with that of vitamin B_{12}, cyanocobalamin. This can be interpreted on the electronic level quite simply. The addition of an electron to cobalamin$^+$ either by the chemical bonding of an electron donor like CN^- or by the formation of a weak complex with an electron donor gives rise to the cobalamin entity with its own distinctive spectrum. Conversely, the removal of an electron from cobalamin either by the physical removal of the electron-donating CN^- or by the formation of a complex with an electron acceptor causes the spectrum to revert to that of cobalamin$^+$.

The spectra arising from the electron acceptors, chloranil, dichlorobenzoquinone, and iodine, are thus easily explained. However, the interactions of alloxan and glutaric acid are not as clear. The interactions are partially, if not completely, irreversible. Acids should be able to split off cyanogen quite easily in aqueous solution and this explains the behavior of glutaric acid and alloxan with vitamin B_{12}. The decreased reaction of glutaric acid at alkaline pH values is probably due to the neutralizing action of the alkali.

The behavior of alloxan is remarkable. With vitamin B_{12b} it behaves in a similar manner to the electron donors and with vitamin B_{12} it behaves in a similar manner to electron acceptors and acids. Alloxan contains both electron-abundant nitrogen atoms and electron-deficient carbonyl groups and seems to be one of the few known molecules which can act both as a donor and an acceptor. The nature of the interaction between alloxan and vitamins B_{12} or B_{12b} is unclear as in both cases irreversible changes occur, but initially both reactions appear to involve electron-transfer processes. Finally, it is inferred from the experiments with amino acids that peptides are also capable of forming complexes with vitamin B_{12b} but not with vitamin B_{12}.

III. Hematoporphyrin

A. Introduction

Hematoporphyrin is another molecule of physiological interest because it is a precursor in the synthesis of hemoglobin. It has certain structural affinities to cobalamin and it is interesting to contrast its electron-donor and electron-acceptor properties with those of vitamins B_{12} and B_{12b}.

B. Study of the Interactions of Hematoporphyrin

The addition of amino acids or aliphatic amines to hematoporphyrin in water or weakly buffered solutions results in quite distinctive difference spectra with well-marked isosbestic points. Heating the solutions causes reversible decreases in the difference spectra. From a study of absorbance vs. concentration of amine for a fixed concentration of hematoporphyrin, values of apparent equilibrium constants and enthalpies of dissociation can be obtained. They are listed in Table II. Figure 8 shows a typical difference spectrum. Similarly, the addition of one of the charge acceptors, tetracyanoethylene (TCNE), glutaric acid, or glutamic acid-HCl, to hematoporphyrin gives rise to a distinct difference spectrum together with well-marked isosbestic points. Again equilibrium constants have been derived as well as an enthalpy of dissociation (Table II). A typical spectrum is shown in Fig. 9.

On measuring the pH of these solutions, however, it was found that in the case of the donors, the solutions had become more alkaline and, in that of the acceptors, more acidic. Furthermore, identical results can be obtained simply by adding HCl or NaOH. The variation in absorbance of a $1 \times 10^{-4} M$ solution of hematoporphyrin at different pH values is purely a function of pH and is independent of the additive producing this change. The absorbance of hematoporphyrin as a function of pH is shown in Fig. 10 and it is seen to vary

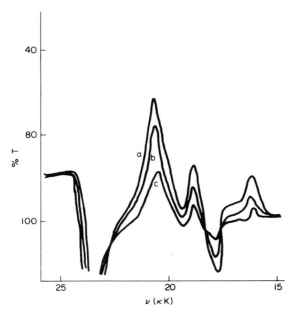

FIG. 8. Difference spectra of proline in 7.45×10^{-5} M hematoporphyrin. (a) 3.45×10^{-1} M, (b) 8.69×10^{-2} M, and (c) 1.739×10^{-2} M proline.

TABLE II

APPARENT COMPLEXES WITH HEMATOPORPHYRIN IN UNBUFFERED SOLUTION

Compounds	Equilibrium constants (liters per mole)	Temperature (°C)	Frequency of determination (kK)	Enthalpy of dissociation (kcal/mole)
Triethylamine	7250	23	20.4	—
Diethylamine	4420	23	20.4	—
	2500	38	20.4	−16.0
	685	50	20.4	—
	226	62	20.4	—
Proline	10.2	23	20.4	—
Glycine	110	23	20.4	—
Valine	36.6	23	20.4	—
TCNE	126	23	18.3	—
	114	38	—	−1.7
	103	50	—	—
	96.5	62	—	—
	86	74	—	—
Glutaric acid	6.86	23	18.3	—
Glutamic acid-HCl	206	23	18.3	—

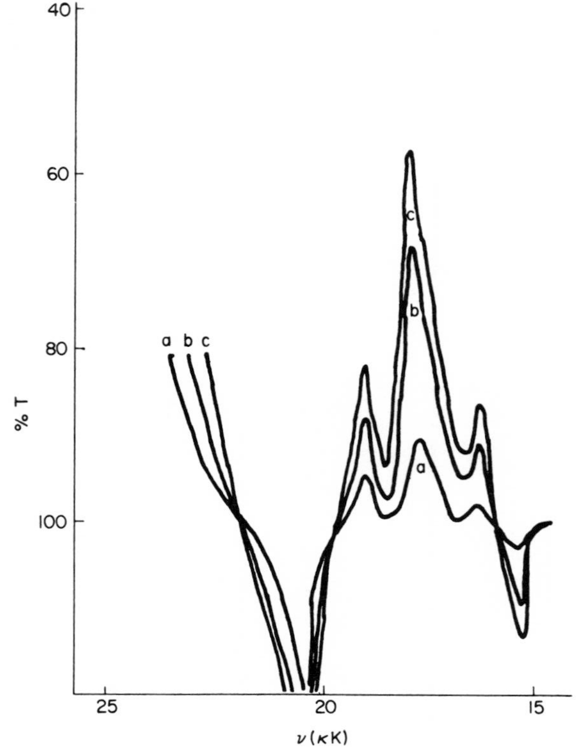

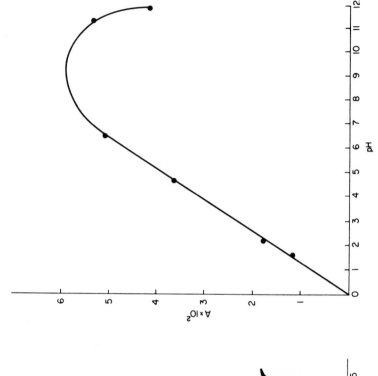

FIG. 9. Difference spectra of glutaric acid in 7.45×10^{-5} M hematoporphyrin in 50% ethanol. (a) 1.51×10^{-2} M, (b) 1.23×10^{-1} M, and (c) 2.45×10^{-1} M glutaric acid.

FIG. 10. Relation between absorbance at 20.4 kK and pH (for 10^{-5} M hematoporphyrin in 50% ethanol).

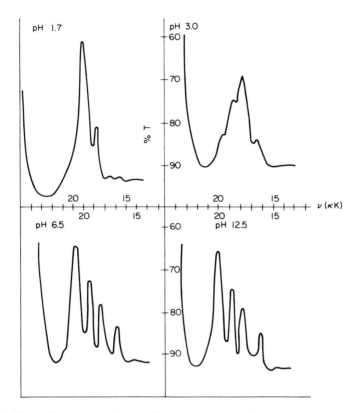

FIG. 11. Absorption spectra of ca. 10^{-5} M hematoporphyrin at various values of pH.

linearly over the range pH 2 to 7, but it is almost independent at pH 9. Figure 11 shows the spectrum of hematoporphyrin at different pH values.

A study has been made of the interaction between hematoporphyrin in phosphate buffer (pH 9) and various electron donors. The difference spectrum shown in Fig. 12 is quite different from those seen in unbuffered solutions. Well-marked isosbestic points occur together with temperature reversibility.

Values of equilibrium constants and enthalpies of dissociation have been derived and are given in Table III. The interaction of hematoporphyrin with nondissociative electron acceptors, e.g., trinitrobenzene and chloranil, in unbuffered solutions gives rise to difference spectra which are quite different from those of TCNE and are illustrated in Fig. 13. These are temperature reversible; they obey the Benesi-Hildebrand equation and contain well-defined isosbestic points.

Equilibrium constants and some enthalpies of dissociation are given in Table III.

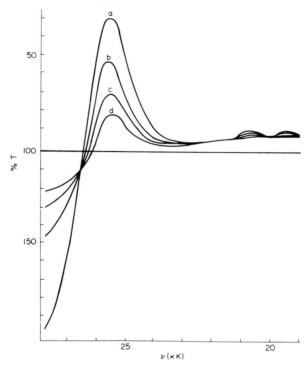

FIG. 12. Difference spectra of tryptophan in 2.09×10^{-5} M hematoporphyrin in pH 9 buffer. (a) 6.57×10^{-2} M, (b) 3.29×10^{-2} M, (c) 1.97×10^{-2} M, and (d) 1.31×10^{-2} M tryptophan.

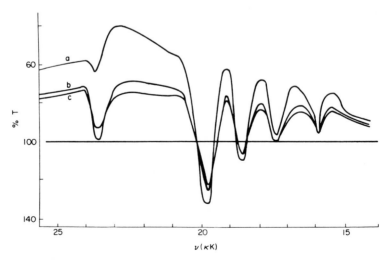

FIG. 13. Difference spectra of trinitrobenzene in 8.08×10^{-5} M hematoporphyrin in 20% ethanol/80% toluene. (a) 4.14×10^{-2} M, (b) 1.24×10^{-2} M, and (c) 0.83×10^{-2} M trinitrobenzene.

TABLE III

INTERACTIONS OF HEMATOPORPHYRIN

Compound	Solvent system	ΔH(kcal/mole)
Amino acids	pH 9 buffer	
Tryptophan		−5.2
6-Aminocaproic acid		−5.5
Arginine		−5.0
Histidine		Reacts but too insoluble to measure
Phenylalanine		Very weak interaction
Other materials	pH 9 buffer	
Caffeine		−10.8
6-Methylaminopurine		− 3.5
Pyridine		Very weak interaction
Cytosine		—
Diethylamine		—
Methyl cyanide		—
dimethyl sulfoxide		—
sym-Trinitrobenzene	20% ethanol/80% toluene	−5.2
sym-Trinitrobenzene	50% ethanol	−1.7
Chloranil		[a]

Materials showing no interaction in pH 9 *buffer*

Glycine	Lysine (hydrochloride)	Indole
Alanine	Asparagine	Tyrosine
Sodium glutamate	Ammonia	Proline
Alloxan	Serine	Ornithine (hydrobromide)
Glutaric acid	Cysteine	Allantoin
Cysteine (hydrochloride)	Citrulline	

[a] K_c at 23°C = 588 liters/mole.

C. Discussion

The spectrum obtained with donors in unbuffered solutions is caused by an increase in pH, resulting in deprotonation of the porphyrin since all the donors are basic. Hematoporphyrin in alkaline solution contains four absorption bands in the visible region, but in acidic solution it contains only two as shown in Fig. 11. At the pH at which the measurements were taken, pH 4 to 6, hematoporphyrin is undergoing change from one form to another. What is being measured is the acidity of the donors and the alkalinity of the dissociable acceptors. This can, however, be related to the donor abilities as shown by Ilmet and Krasij (1966), who noted that plots of dissociation constants of azanaphthalenes in water against the equilibrium constants of the complexes formed with iodine in nonpolar solvents were linear. As can be seen from Table II, the apparent equilibrium constant is, in reality, a measure of the alkalinity; hence, dissociation constants of the donors correlate quite well with their donor properties. The aliphatic amines would be expected to be much stronger electron donors than the amino acids.

The interaction in pH 9 buffer solution ensured that the formation of complexes of hematoporphyrin was due to electron donor-acceptor forces. Particularly noteworthy are the enthalpies of dissociation for the amino acids which are all very similar. This is corroborative evidence for the bonding being due to the lone-pair electron on the nitrogen of the amino group. The fact that tryptophan has a ΔH value very similar to those of the other amino acids and that indole does not appear to react with hematoporphyrin shows that tryptophan is behaving as an n-electron donor and not as a π donor from its indole ring. Those amino acids which do not appear to react with hematoporphyrin at pH 9 have their carboxyl groups adjacent to the amino groups and perhaps some form of steric hindrance occurs. The interactions with the electron acceptors, chloranil and trinitrobenzene (even in unbuffered solutions), are due to charge-transfer interaction. The difference spectra are very similar to those obtained by Gouterman and Stevenson (1962) for complexes of etioporphyrin and coproporphyrin with trinitrobenzene, as are the enthalpies of dissociation and K_c values. The interaction of TCNE in unbuffered solution, on the other hand, is identical with that of HCl and is probably due to the dissociation of TCNE to HCN giving rising to hydrogen ions. The alloxan does not interact with hematoporphyrin. This can be contrasted with its behavior with vitamins B_{12} and B_{12b} where it acts as an acceptor with the former, but as a donor with the latter. Alloxan contains both electron-positive and electron-negative moieties so that this result is not too surprising. Pullman and Pullman (1963) have shown on theoretical grounds that alloxan should tend to be a good acceptor. The presence of the electron-abundant nitrogen atoms, however, should enable it to be an electron donor in appropriate circumstances such as the present.

IV. General Conclusions

The formation of complexes between vitamins B_{12b} and various biological electron donors and their lack of dependence on pH have been shown. The formation of complexes of hematoporphyrin with some biological donors and organic acceptors has also been demonstrated. The behavior of hematoporphyrin complexes is very different from those of vitamin B_{12} complexes. Only in buffered solutions do electron donor-acceptor complexes appear to be formed with dissociable donors or acceptors. In unbuffered solutions the effect of adding the dissociable donor or acceptor is simply to change the concentration of hydrogen ion and hence the equilibrium between the oxidized and reduced hematoporphyrin since no evidence of complexing is found. The behavior of complexes of vitamin B_{12b} and hematoporphyrin should be compared to that of the complexes of amino acids and proteins with chloranil (Birks and Slifkin, 1963). Here, the effect of complexing is to shift the main absorption band of chloranil from 34 kK to the region of 30 to 27 kK depending on the pH. In the case of chloranil, however, complexes are still formed in unbuffered solutions because the chloranil peak is not shifted by acid or alkali alone.

It has thus been shown that pH can be very important in biological complexes and that any physiological discussion of complexes must take this into account. It is obvious that the aliphatic amines and amino acids can only be n-electron donors. Many discussions of biological charge transfer seems to disregard n donation. It is evident that n donation can play a very important role in biological charge-transfer processes and should not be disregarded. Furthermore, any such discussion must take into account the effect of physiological pH.

ACKNOWLEDGMENTS

The spectrophotometer used in this study was the gift of the Medical Research Council. We would like to acknowledge the technical help of Mr. G. J. Hill and Mr. P. Rothwell.

REFERENCES

Adler, N., Medwick, T., and Posnanski, T. J. (1966). *J. Am. Chem. Soc.* **88**, 5018.
Benesi, H. A., and Hildebrand, J. H. (1949). *J. Am. Chem. Soc.* **71**, 2703.
Birks, J. B., and Slifkin, M. A. (1963). *Nature* **197**, 42.
Briegleb, H. (1961). "Elektronen-Donator-Accepter-Komplexen." Springer, Berlin.
Castle, W. B. (1953), *New Engl. J. Med.* **249**, 603.
Gouterman, M., and Stevenson, P. E. (1962), *J. Chem. Phys.* **37**, 2266.
Heathcote, J. G., and Mooney, F. S. (1958). *Lancet* **I**, 982.
Heathcote, J. G., and Mooney, F. S. (1962). *Nature* **193**, 380.
Ilmet, I., and Krasij, M. (1966). *J. Phys. Chem.* **70**, 3755.
Milhaud, G. (1961). *Nature* **189**, 33.
Mooney, F. S., and Heathcote, J. G. (1963). *Nature* **199**, 289.
Mooney, F. S., and Heathcote, J. G. (1965). *Nature* **205**, 393.
Mulliken, R. S. (1950). *J. Am. Chem. Soc.* **72**, 600.
Pullman, B., and Pullman, A. (1963). "Quantum Biochemistry." Wiley (Interscience), New York.
Randall, W. C., and Alberty, R. A. (1966). *Biochemistry* **5**, 3189.
Slifkin, M. A. (1964). *Spectrochim. Acta* **20**, 1543.
Slifkin, M. A. (1965a). *Spectrochim. Acta* **21**, 1391.
Slifkin, M. A. (1965b). *Biochim. Biophys. Acta* **103**, 365.
Slifkin, M. A. (1965c). *Biochim. Biophys. Acta* **109**, 617.
Slifkin, M. A., and Heathcote, J. G. (1967). *Spectrochim. Acta* **23A**, 2893.
Smith, E. L. (1960). "Vitamin B_{12}." Methuen, London.
Veillard, A., and Pullman, B. (1965). *J. Theoret. Biol.* **8**, 307.

Complexes between Indole and Imidazole Derivatives of the Charge-Transfer Type

MEIR SHINITZKY AND EPHRAIM KATCHALSKI

Department of Biophysics
The Weizmann Institute of Science
Rehovoth, Israel

Indole as well as tryptophan derivatives have been shown to act as good electron donors (Pullman and Pullman, 1963), capable to form charge-transfer complexes with characteristic acceptors, such as tetracyanoethylene, chloranil, and trinitrobenzene (Szent-Györgyi *et al.*, 1961). Of biological interest is the charge-transfer complex formed between indole and 1-alkyl-3-carboxamide pyridinium, the latter acting as electron acceptor (Cilento and Giusti, 1959; Shifrin, 1964). It is this type of complex which seems to determine, to a large extent, the forces binding NAD^+ with the protein moiety of the corresponding dehydrogenases (Kosower, 1956).* The imidazole ring when protonated closely resembles the pyridinium ring, since both contain a system of six conjugated π electrons that is positively charged. One should thus expect that the imidazolium ion will have electron-acceptor properties similar to those of the pyridinium ion.

In this chapter we describe *inter-* as well as *intra*molecular complexes of the charge-transfer type between indole and imidazolium derivatives. Intermolecular complexes were shown to form in aqueous solutions containing both types of electron donors and acceptors. Intramolecular complexes were found to exist in some peptides and proteins containing tryptophan and histidine residues.

I. Intermolecular Indole-Imidazolium Complexes in Aqueous Solution

The formation of complexes between indole and imidazolium derivatives in aqueous solution was demonstrated by two thermodynamic and two optical methods (Shinitzky *et al.*, 1966). With the aid of the thermodynamic methods it could be demonstrated that the solubility in water of indole and 3-methylindole is markedly increased in the presence of imidazolium salts. The optical methods allowed the detection of marked alterations in the fluorescence and absorbance of indole derivatives in the presence of imidazolium salts.

---

* NAD, nicotinamide adenine dinucleotide.

A. The Thermodynamic Methods

The solubility of indole in water is highly temperature dependent. When a hot aqueous solution of indole (0.04 to 0.07 M) is cooled gradually a temperature is reached at which the solution suddenly turns into a milky suspension because of the low melting point of indole (m.p. 51°C). This temperature can be denoted as the saturation temperature. Preliminary experiments have shown that the saturation temperature of indole in water decreases in the presence of imidazolium salts, whereas in the presence of inorganic salts the reverse effect is usually observed. Assuming that the observed increase in solubility is due to the formation of a highly water-soluble indole-imidazolium complex, the derivation of the association constant K ($K = $ [Ind-ImH$^+$]/ [Ind][ImH$^+$]) at different temperatures from solubility data was attempted. Figure 1 shows two typical saturation temperature curves of indole in water, the first in the presence of imidazolium perchlorate (curve B), the second in the presence of sodium perchlorate at the same concentration (curve A). The horizontal lines joining the two curves give the molar concentration of the complex formed at the corresponding temperature, whereas the concentration of the free indole is determined by curve A. Curves A and B thus allow the evaluation of the equilibrium constant K for any given temperature in the

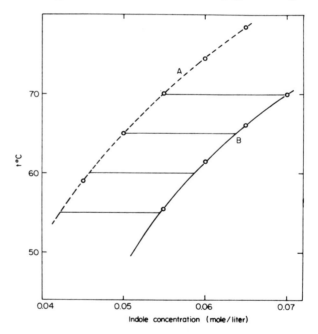

FIG. 1. Saturation temperature curves for indole in water. (A) In the presence of 0.2 M NaClO$_4$; (B) in the presence of 0.2 M imidazolium perchlorate.

range of 55°–80°C. The enthalpy of complex formation, ΔH, and the value of K at 22°C were obtained by plotting $\ln K$ vs. $1/T$ according to the well-known van't-Hoff equation. The recorded values of K and ΔH for the indole-imidazolium system are given in Table I.

TABLE I

ASSOCIATION CONSTANT K AND THE ENTHALPY OF FORMATION OF SOME INDOLE-IMIDAZOLIUM COMPLEXES AS OBTAINED BY SATURATION TEMPERATURE AND PHASE DISTRIBUTION METHODS[a]

Mixture	K (22°C) (liter mole^{-1})		ΔH (kcal mole^{-1})	
	Saturation temperature method	Distribution method	Saturation temperature method	Distribution method
In-Im·HClO$_4$	2.2 ± 0.2	1.6 ± 0.1	-1.8 ± 0.4	-3.2 ± 0.2
In-His·HClO$_4$	2.0 ± 0.2	1.5 ± 0.1	-2.0 ± 0.4	-3.2 ± 0.2
3-Methyl In-Im·HClO$_4$	—	1.8 ± 0.1	—	-3.2 ± 0.2
3-Methyl In-His·HClO$_4$	—	1.4 ± 0.1	—	-3.4 ± 0.2

[a] In, Im, and His stand for indole, imidazole and histidine respectively.

An independent determination of the complex association constant, K, for indole and imidazolium perchlorate was obtained by measuring the distribution of indole (at 0° to 40°C) between methylcyclohexane and aqueous solutions 0.1 or 0.2 M in NaClO$_4$ or imidazolium perchlorate. The aqueous phase containing the imidazolium salt was found to contain a considerably higher concentration of indole than the aqueous phase containing NaClO$_4$. The association constant, K, at 22°C, given in Table I was calculated from the experimental data assuming absence of imidazolium or indole-imidazolium complex in the equilibrated organic phase. The enthalpy of association, ΔH, was obtained from a linear plot of $\ln K$ vs. $1/T$. Analogous experiments were carried out with indole and histidinium perchlorate using methylcyclohexane as the organic phase or with 3-methylindole and imidazolium perchlorate and histidinium perchlorate using cyclohexane as the organic phase. The results obtained for these systems are also included in Table I.

B. Indirect Spectrophotometric Estimation of the Binding of Indole with Imidazolium Derivatives

All the indole-imidazolium mixtures tested showed no new absorption bands in the visible or the near UV, when compared with the absorption spectra of the constituents. This is not surprising since it has been shown that the characteristic absorption band of the charge-transfer complex formed between indole and alkylnicotinamide ion has a maximum around 320 mμ (Shifrin, 1964), and imidazolium is a considerably poorer acceptor than charged nicotinamide (see below). Because of the above findings we were unable to detect the presence of indole-imidazolium complexes by the usual spectroscopic techniques (Briegleb, 1961), and an indirect spectrophotometric method was employed.

The indirect method developed is based on the observation that the addition of imidazolium to a solution containing a charge-transfer complex of indole and methylnicotinamide ion leads to a redistribution of the indole between both acceptors, and since it is only the indole-nicotinamide complex which has a characteristic absorption band in the region of 330–450 mμ, one can evaluate the concentration of the indole-imidazolium complex in the system from the decrease in the optical density (O.D.) at wavelengths longer than 330 mμ. The O.D. of the characteristic band at 330–450 mμ appearing in aqueous mixtures of indole and 1-methylnicotinamide ion (Fig. 2), which might be attributed to an *inter*molecular charge-transfer complex, resembling the corresponding *intra*molecular complex investigated by Shifrin (1964), did not obey the Beer-Lambert law. It was thus possible to derive the association constant for the charge-transfer complex, $K_1 = 3.0$ liters mole^{-1}, from the change in absorption intensity, as measured at 330, 340, and 350 mμ, with the concentration of the components. An association constant $K_1 = 4.2$ liters

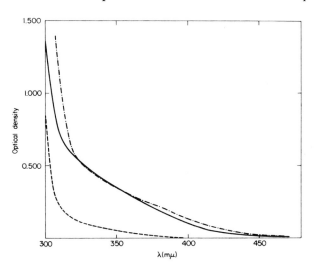

FIG. 2. Absorption spectra of aqueous solutions containing indole and pyridinium derivatives, each at a concentration of 10^{-2} M. Dashed line: Indole + 1-methy-pyridinium chloride; solid line: indole + 1-methyl-3-carboxamide pyridinium chloride; dashed, dotted line: N-acetyl-L-tryptophan amide + 1-methyl-3-carboxamide pyridinium chloride.

mole^{-1} was obtained in aqueous solution for the system N-acetyl-L-trypto-phan amide/1-methylnicotinamide. Addition of different amounts of imidazo-lium salts, up to a final concentration of 0.8 M, to an aqueous solution 0.1 M in 1-methylnicotinamide and 0.001 M in indole or N-acetyl-L-tryptophan amide caused a simultaneous decrease, up to 50%, in the O.D. of the charge-transfer band at 330–450 mμ. The association constant K_2 for the charge-transfer complex between indole and imidazolium may be derived with the aid of Eq. 1, which holds for a system containing a large excess of both acceptors. In this equation (O.D.)$_0$ and (O.D.)

$$\frac{(O.D.)_0}{(O.D.)} = 1 + \frac{K_2[ImH^+]}{1 + K_1[N^+]} \qquad (1)$$

are the otpical densities of the mixture in the absence and the presence of the imidazolium salt, respectively, $[ImH^+]$ is the concentration of the imidazolium salt, $[N^+]$ is the concentration of the nicotinamide salt, and K_1 is the association constant for the indole-nicotinamide complex determined as above. (O.D.)$_0$ was measured as a rule in the presence of a concentration of NaCl equivalent to the imidazolium chloride used in the final experiment. Table II gives the

TABLE II

ASSOCIATION CONSTANTS K OF INDOLE AND N-ACETYLTRYPTOPHAN AMIDE WITH SOME IMIDAZOLIUM DERIVATIVES[a]

Donor	Acceptor	K (liter mole^{-1})
Indole	Imidazole·HCl	0.5 ± 0.1
Indole	Histamine·2HCl	0.6 ± 0.1
Indole	L-Histidine·HCl	0.6 ± 0.1
N-Acetyl-L-tryptophan amide	Imidazole·HCl	0.7 ± 0.1
N-Acetyl-L-typtophan amide	Histamine·2HCl	0.9 ± 0.1
N-Acetyl-L-tryptophan amide	L-Histidine·HCl	0.8 ± 0.1

[a] As derived from competition experiments with 1-methyl-3-carboxamide pyridinium chloride.

association constants obtained by using the above procedure for some representative indole-imidazolium complexes.

C. Imidazolium and Pyridinium Salts as Quenchers of the Fluorescence of Indole

Imidazole hydrochloride as well as some of its derivative (see Table III), in a concentration range of 10^{-3} to 10^{-2} M, was found to quench the emission at 345 mμ of indole, at a concentration of 10^{-6} M, when excited at 285 mμ.

A 10^{-6} M aqueous solution of indole was quenched even to a greater extent by pyridine hydrochloride and some of its derivatives (Table III), in the concentration range of 10^{-3} to 10^{-2} M.

TABLE III

APPARENT STERN-VOLMER CONSTANT K_{SV} FOR QUENCHING OF THE
FLUORESCENCE OF INDOLE BY IMIDAZOLIUM AND PYRIDINIUM DERIVATIVES[a]

Quencher	K_{SV} (liter mole^{-1})
Imidazole·HCl	13
Histamine·2HCl	17
α, N-Acetyl-L-histidine amide·HCl	28
Pyridine·HCl	75
1-Methylpyridinium chloride	90
1-Methyl-3-carboxamide pyridinium chloride	142

[a] In aqueous solution.

The variation of the intensity of the fluorescence of indole as a function of the concentration of the ionized quencher (see Table III) could be described by the Stern-Volmer equation

$$F_0/F = 1 + K_{SV}[Q] \qquad (2)$$

where F_0 and F are the fluorescence intensities in the absence and in the presence of the quencher, respectively, $[Q]$ is the concentration of quencher, and K_{SV} is the apparent Stern-Volmer constant. The values of K_{SV} obtained for several imidazolium and pyridinium chlorides are given in Table III.

In a previous article (Shinitzky *et al.*, 1966) it has been shown that the Stern-Volmer constant, K_{SV}, at relatively low concentrations of quencher is given by

$$K_{SV} = K + k\tau \qquad (3)$$

where K is the association constant of the indole-imidazolium (or pyridinium) complex τ is the lifetime of the excited indole, and k is the collision rate constant determining the rate of quenching of excited indole molecules by quencher molecules. In all of the cases dealt with in Table III, K_{SV} is considerably greater than K (the values of K are recorded in Tables I and II). The quenching of indole by the imidazolium and pyridinium derivatives seems, therefore, to be caused mainly by quencher molecules colliding with excited indole molecules. The rate of collision is diffusion controlled. The efficiency of each collision, however, is determined by a set of characteristic parameters determining the nature of the intermediate complex. In this connection it is pertinent to note that the K_{SV} values given in Table III increase with the electron affinity of the quencher. It may thus be tentatively concluded that the excited indole-quencher intermediate is a complex of the charge-transfer type.

In a highly viscous solvent such as glycerol $K \gg k\tau$ and the Stern-Volmer constant, K_{SV}, equals the association constant K. Some of the values of K obtained by the fluorescence quenching technique for several indole-imidazo-lium complexes in 95% glycerol, are given in Table IV. The enthalpy of complex formation, ΔH, was calculated graphically from the variation of K with temperature in the range of 0–30°C.

Finally, it should be noted that Eq. (2) can be used to describe the variation with pH of the fluorescence, F, of a solution containing indole and imidazole (see Fig. 3). For this purpose, however, one has to substitute [Q] by the

TABLE IV

ASSOCIATION CONSTANT K AND THE ENTHALPY OF FORMATION OF SOME
INDOLE-IMIDAZOLIUM COMPLEXES IN 95% GLYCEROL[a]

Mixture	K (liter mole^{-1})	ΔH (kcal mole^{-1})
In-Im · HClO$_4$	2.4 ± 0.2	−3.0 ± 0.4
In-His · HClO$_4$	2.3 ± 0.2	−3.0 ± 0.4
3-Methyl In-Im · HClO$_4$	2.4 ± 0.2	−3.1 ± 0.4
3-Methyl In-His · HClO$_4$	2.2 ± 0.2	−3.1 ± 0.4

[a] In, Im, and His stands for indole, imidazole, and histidine, respectively.

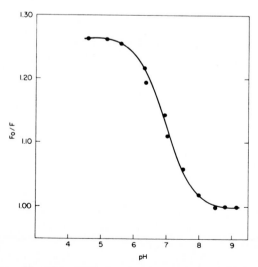

FIG. 3. The change with pH in F_0/F of an aqueous solution 0.02 M in imidazole and 10^{-5} M in indole. F_0 and F are the fluorescence intensities in the absence and presence of imidazole, respectively.

concentration of the protonated imidazole at any of the pH values studied, given by the expression $K_i[\text{Im}]_0[\text{H}^+]/(1 + K_i[\text{H}^+])$, in which K_i is a characteristic constant defining the basicity of imidazole, and $[\text{Im}]_0$ is the total concentration of the imidazole. The extent of quenching of the fluorescence of indole by imidazole as a function of pH is given in Fig. 3. The data presented show that maximum quenching occurs at pH values below pH 5.5 at which the imidazole ring is completely protonated. At pH values above 8.5 at which imidazole appears as a free base it does not affect the fluorescent of indole.

II. Intramolecular Indole-Imidazolium Complexes in Model Compounds

To investigate the possible formation of intramolecular indole-imidazolium complexes the following linear and cyclic histidine- and tryptophan-containing peptides were synthesized, and some of their optical properties investigated: cyclo(L-His-L-Try),   cyclo(L-His-D-Try), α,N-acetyl-L-His-L-Try-OMe, α,N-acetyl-L-His-L-Try, and (L-His-L-Try)$_3$ (Shinitzky et al., 1967).

In all the above model compounds it was found that the fluorescence intensity of the tryptophan residue decreases markedly on protonation of the adjacent histidyl residue. A typical fluorometric titration curve of cyclo-(L-His-L-Try) is given in Fig. 4. For comparison the figure also contains a

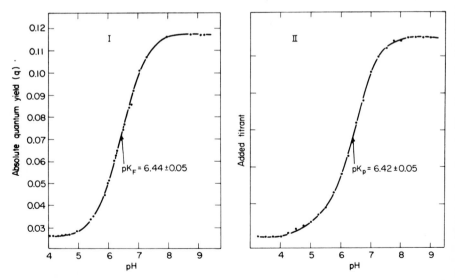

FIG. 4. The fluorometric titration curve (I) and the potentiometric titration curve (II) of cyclo(L-His-L-Try) in aqueous solution.

normal potentiometric titration curve of the histidyltryptophan diketopiper-azine. It should be noted that both titration curves practically overlap and yield the same apparent pK values for the histidine side chain at the inflection point.

The fluorometric titration curves of the other model compounds synthesized closely resembled in shape the one recorded for cyclo(L-His-L-Try). The maximal and minimal quantum yields, q_{max} and q_{min}, respectively, differed, however, from compound to compound (see Table V). Close agreement was

TABLE V

CHARACTERISTIC DATA ON THE FLUOROMETRIC TITRATION IN AQUEOUS SOLUTION OF SOME HISTIDINE- AND TRYPTOPHAN-CONTAINING COMPOUNDS[a]

Compound	q_{min}	q_{max}	$\dfrac{\Delta q}{q_{max}}$	pK_F	pK_p
Cyclo(L-His · L-Try)	0.026	0.118	0.78	6.44 ± 0.05	6.42 ± 0.05
Cyclo(L-His · D-Try)	0.042	0.090	0.53	6.20 ± 0.1	6.2 ± 0.1
α,N-Ac-L-His · L-TryOMe	0.033	0.057	0.42	6.58 ± 0.05	6.5 ± 0.1
α,N-Ac-L-His · L-Try	0.036	0.075	0.52	6.80 ± 0.05	6.8 ± 0.1
(L-His · L-Try)$_3$	0.024	0.083	0.71	6.75 ± 0.05	6.8 ± 0.1
N,Ac-L-Try	0.286	0.286	~ 0	—	—
N,Ac-L-TryNH$_2$	0.216	0.216	~ 0	—	—
N,Ac-L-TryOMe	0.094	0.094	~ 0	—	—

[a] The quantum yields at the lower and upper plateaus of the fluorometric titration curve (at pH 4.4–4.6 and 9.0–9.2, respectively) are denoted as q_{min} and q_{max}, respectively. pK_F and pK_p define the basicity of the imidazole side chains as determined fluorometrically and potentiometrically. $\Delta q = q_{max} - q_{min}$.

found between the pH values corresponding to the midpoint of the fluoro-metric and the potentiometric titration curves. It is thus possible to determine the apparent imidazolium dissociation constants of the model compounds synthesized fluorometrically (pK_F values in Table V) or potentiometrically (pK_p values in Table V). The fluorometric titration could obviously be carried out at high dilution ($\sim 10^{-6}$ M) and were practically independent of the nature and concentration of buffer.

In an attempt to explain the fluorometric as well as the potentiometric titration of the model compounds discussed, in which an intramolecular complex can be formed between the indole and the imidazole moieties of the tryptophan and histidine residues, respectively, we adopted the following scheme to describe the equilibria occurring in the system:

$$A + H^+ \rightleftharpoons AH^+ \rightleftharpoons B^+ \qquad (4)$$

A denotes un-ionized molecules, AH^+ denotes protonated molecules in which the histidinium residues have not reacted with adjacent tryptophan residues, and B^+ denotes protonated molecules in which the histidinium residues have formed a complex of the type discussed with the adjacent tryptophan residues. Both protonated forms, AH^+ and B^+, are in equilibrium and thus fulfill Eq. (5)

$$\frac{[B^+]}{[AH^+]} = \beta \tag{5}$$

where β is a constant. The relation between the concentration of uncharged molecules and charged ones, is given by

$$\frac{[AH^+] + [B^+]}{[A][H^+]} = K_a = K_i(1 + \beta) \tag{6}$$

where K_i is a constant defining the intrinsic basicity of the imidazole moeity ($K_i = [AH^+]/[A][H^+]$).

Equation (6) shows that the potentiometric titration of the model compounds in which a complex of the type B^+ can be formed is determined by the constant K_a, which is $(1 + \beta)$ times greater than the intrinsic imidazolium constant K_i. In the following it will be shown that the constant K_a also determines the course of the fluorometric titration of the compounds under consideration.

The assumptions specified below, which might be justified by the findings described, enabled the calculation of the intensity of fluorescence emitted by a system containing A, AH^+ and B^+. (a) The three forms, A, AH^+, and B^+, possess the same absorbancies at the excitation wavelengths. (b) The fluorescence intensity of A is given by $F_A = pq_A[A]$, where q_A is the corresponding quantum yield, and p is a proportionality factor. (c) B^+ is void of fluorescence. (d) Excited AH^+ molecules can lose their excitation energy either by a fluorescence decay mechanism analogous to that of A determined by the radiation rate constant k_1, or as a result of intramolecular collisions between the excited tryptophan residues and the adjacent histidynium side chains, the rate of which is determined by the first-order rate constant k_2. AH^+ will thus emit light with a quantum yield q_{AH^+} given by $q_{AH^+} = k_1 q_A/(k_1 + k_2)$. In accord with these assumptions one may deduce for the fluorescence intensity, F, of the above system the expression:

$$F = \frac{p[A]_0}{1 + K_a[H^+]} \left[q_A + (q_{AH^+})\left(\frac{K_a[H^+]}{1 + \beta}\right) \right] \tag{7}$$

where $[A]_0$ stands for the total concentration of the histidine- and tryptophan-containing peptide.

Equation(7) shows that the intensity of fluorescence reaches a maximum value of

$$F_{max} = pq_A[A]_0 \tag{8}$$

when $K_a[H^+] \ll 1$. When $K_a[H^+] \gg 1$ the fluorescence intensity reaches a minimum value given by

$$F_{min} = (pq_{AH^+})[A]_0/(1 + \beta) \tag{9}$$

Substracting Eq. (9) from Eq. (7) one obtains

$$F - F_{min} = \frac{p[A]_0}{1 + K_a[H^+]} \left(q_A - \frac{q_{AH^+}}{1 + \beta} \right) \tag{10}$$

Since $q_A - q_{AH^+}/(1 + \beta)$ is a constant, Eq. (10) describes a function closely resembling the corresponding potentiometric titration of the system.

When $[H^+] = 1/K_a$, $F - F_{min} = (F_{max} - F_{min})/2$. The midpoint of the fluorometric titration curve thus gives the apparent acidic dissociation constant of the histidinium residue in the model compound under investigation.

The absorption spectra of cyclo(L-His-L-Try) at pH 4.0 and 9.0 are given in Fig. 5. The data presented show that the protonation of the imidazole ring leads to hypochromicity of about 6% in the region of 253–300 mμ. In the

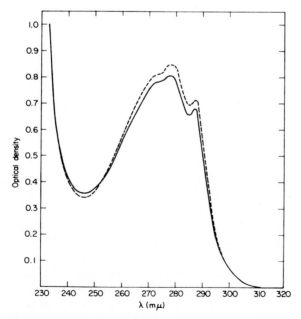

FIG. 5. Absorption spectra of aqueous solutions of cyclo(L-His-L-Try) at pH 4.0 (solid line) and pH 9.0 (dashed line).

range 240–253 mμ, however, a hyperchromic effect appears. The latter is most likely due to a characteristic charge-transfer band of the indole-imidazolium complex which overcomes the general decrease of indole absorbancy in the complex. Hypochromic and hyperchromic effects similar to the ones detected in cyclo(L-His-L-Try) were found also in the other model compounds on the ionization of the histidine residue.

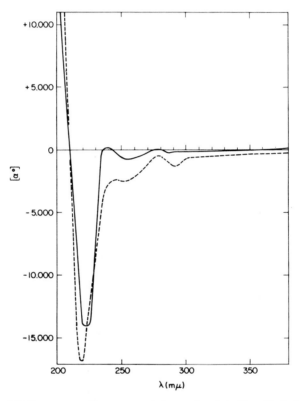

FIG. 6. The O.R.D. spectra of aqueous solutions of cyclo(L-his-L-Try) at pH 4.0 (solid line) and pH 9.0 (dashed line).

Figure 6 gives the optical rotatory dispersion of cyclo(L-His-L-Try) at pH 4.0 and 9.0. Protonation of the imidazole ring diminishes the magnitude of the characteristic Cotton effect of tryptophan around $\lambda_0 = 280$ mμ and leads to a marked decrease in the magnitude of the specific rotation around 245 mμ. These changes might be correlated with the hypo- and hyperchromic effects observed in the same region (see Fig. 5). Changes similar to those recorded for cyclo(L-His-L-Try) were observed when the histidine residues of the other model compounds were protonated.

III. Fluorometric Detection of Histidine-Tryptophan Complexes in Proteins

The detection of intramolecular complexes in the model compounds containing tryptophan and histidine induced the search for similar complexes in proteins (Shinitzky and Goldman, 1967). A survey of the change in fluorescence intensity at 330–350 mμ, in the pH range 5 to 8.5, of a considerable number of proteins, when excited at 285–290 mμ, revealed in some of the cases a behaviour resembling that of the model peptides (see Table VI). A striking example suggesting the existence of a well-defined complex between tryptophan and histidine residues was found in the case of papain (see Fig. 7). A papain molecule (molecular weight 22,000) contains five tryptophans and two histidines; it is thus remarkable that the fluorescence of papain decreases

TABLE VI

FLUOROMETRIC TITRATION DATA OF PROTEINS AND NATURAL PEPTIDES[a]

Peptide or protein	q_{min}	q_{max}	$\Delta q/q_{max}$	pK_F	Residues per molecule	
					His	Try
α-MSH	0.130	0.154	0.15	6.4 ± 0.05	1	1
ACTH	0.131	0.147	0.11	6.3 ± 0.05	1	1
Glucagon	0.145	0.145	~0	—	1	1
Papain	0.080	0.121	0.34	6.6 ± 0.05	2	5
Activated papain[b]	0.062	0.145	0.57	7.22 ± 0.05	2	5
Chymopapain	0.056	0.121	0.54	6.2 ± 0.1	—	—
Activated chymopapain[c]	0.056	0.121	0.54	6.2 ± 0.1	—	—
Ficin	0.051	0.080	0.36	7.95 ± 0.05	—	—
Bromelain	0.024	0.038	0.37	6.2 ± 0.1	2	8
Pepsinogen	0.222	0.222	~0	—	3	6
Pepsin	0.218	0.194	<0	—	1	6
Chymotrypsinogen	0.095	0.111	0.14	6.1 ± 0.1	2	8
Chymotrypsin	0.118	0.140	0.16	5.8 ± 0.1	2	8
Carboxypeptidase	0.145	0.161	0.10	6.0 ± 0.1	8	6
Hen egg-white lysozyme	0.073	0.085	0.14	6.25 ± 0.05	1	6
Goose egg-white lysozyme	0.248	0.248	~0	—	3–4	2–3

[a] In 0.1 M NaCl at 23°C. The notations are those of Table V. The pK_F is taken as the pH at which $q = q_{min} + \Delta q/2$.

[b] In the presence of 0.005 M cysteine and 0.002 M ethylenediaminetetraacetic acid (EDTA).

[c] In the presence of 0.01 M cysteine.

markedly as the pH is decreased from pH 9 to 5. Moreover, a marked alteration occurred in the fluorometric titration curve on activation of the enzyme with cysteine. In view of the considerations forwarded above, one may assume that the imidazole which forms a complex with tryptophan in the nonactive enzyme has a pK of 6.60, whereas in the activated enzyme a pK of 7.22 was recorded. This difference in the fluorometric titration might be explained by a conformational change which occurred as a result of enzyme activation which had altered the pK of the bound imidazole as well as the

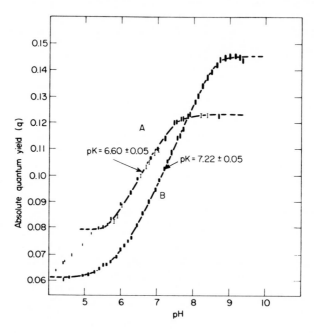

FIG. 7. Fluorometric titration curves of papain (A) and activated papain (B) in aqueous solutions containing 0.1 M NaCl.

microenvironment to which the complex is exposed. The shift in the imidazole pK suggests a more stable complex in the activated enzyme than in the nonactivated one. Marked changes in fluorescence with pH were observed also with chymopapain, ficin, and bromelain (see Table VI).

The difference in behavior of various proteins is illustrated by the finding that glucagon, pepsinogen, pepsin, and goose egg-white lysozyme show, in contradistinction to papain, no change in their fluorescence intensity in the pH range 5 to 8.5. No interaction seems, therefore, to prevail in these proteins between their tryptophan and histidine residues.

IV. Discussion

The existence of *intra-* as well as *inter*molecular complexes between indole and imidazolium derivatives has been established above by thermodynamic and spectroscopic techniques. No final proof as to the nature of the complex has been obtained so far. The finding described strongly suggest, however, that it is a complex of the charge-transfer type resembling that between indole and pyridinium (Shifrin, 1964).

Indole and trytophan derivatives are known electron donors (Pullman and Pullman, 1963). Preliminary spectroscopic experiments in which the charge-transfer band formed on complexing indole or various tryptophan derivatives with classic acceptors such as trinitrobenzene or chloranil revealed ionization potentials of 8.05 ± 0.05 and 7.90 ± 0.05 eV for indole and tryptophan derivatives, respectively (Shinitzky, 1967). These values indicate that indole derivatives are fairly good electron donors and, as such, may play the role of the donor in the suitable charge-transfer complexes.

The structure of the imidazolium ring resembles in many features the pyridinium ring; one would, therefore, expect a priori that imidazolium derivatives should act as electron acceptors as do pyridinium derivatives (Kosower, 1960). Molecular orbital calculations, using the Hückel approximation, resulted in values of -0.75 and -0.36 for the energy coefficients of the lowest empty molecular orbitals of imidazolium and 1-alkyl-3-carboxamide pyridinium, respectively (Pullman, 1967). These show that both compounds may act as electron acceptors, although the nicotinamide ion is a considerably stronger electron acceptor than the imidazolium ion. Supporting evidence for the ability of imidazolium to act as an electron acceptor comes from the absorption spectra of the iodides of several imidazolium derivatives (Shinitzky *et al.*, 1966). All the iodides studied when dissolved in organic solvents, such as ethanol and butanol-cyclohexane, did not obey the Beer-Lambert law in the range of 250–300 mμ, probably as a result of the formation of ion pairs of the charge-transfer type in which the imidazolium derivatives act as acceptors.

No prediction can be made at this stage as to the role of the complexes of the charge-transfer type between tryptophan and histidine residues in determining the structure and function of proteins. It should be noted, however, that the binding forces between the indole and imidazole moieties are weak and strongly dependent on pH and orientation. Amino acid residues other that those of tryptophan and histidine could act, in principle, in proteins as electron donors and electron acceptors. Tyrosine, phenylalanine, un-ionized histidine, methionine and cystine might be expected to act as donors, whereas arginine and protonated terminal α-amino groups could act as acceptors. Bonds between the latter potential donors and acceptors should be weaker than those between tryptophan and histidine; they might nevertheless markedly affect the structure of proteins by a cooperative effect. An imidazole-imidazolium complex has been recently detected in ribonuclease (Crestfield *et al.*, 1963) and chymotrypsin (Bender and Kézdy, 1964).

REFERENCES

Bender, M. L. and Kézdy, F. J. (1964). *J. Am. Chem. Soc.* **86**, 3704.
Briegleb, G. (1961). " Electronen-Donator-Acceptor-Komplexe," Springer, Berlin.
Cilento, G., and Giusti, P. (1959), *J. Am. Chem. Soc.* **81**, 3801.
Crestfield, A. M., Stein, W. H., and Moore, S. (1963). *J. Biol. Chem.* **238**, 2421.
Kosower, E. M. (1956). *J. Am. Chem. Soc.* **78**, 3497.
Kosower, E. M. (1960), *In* " The Enzymes" (P. D. Boyer, H. Lardy, and K. Myrbäck, eds.), Vol. 3, p. 171. Academic Press, New York.
Pullman, B. (1967). Private communication.
Pullman, B., and Pullman, A. (1963). "Quantum Biochemistry." Wiley (Interscience), New York.
Shifrin, S. (1964). *Biochim. Biophys. Acta* **81**, 205.
Shinitzky, M. (1967). Unpublished results.
Shinitzky, M., Katchalski, E., Grisaro, V., and Sharon, N. (1966). *Arch. Biochem. Biophys.* **116**, 332.
Shinitzky, M., Fridkin, M., and Katchalski, E. (1967). To be submitted for publication.
Shinitzky, M., and Goldman, R., (1967). *European J. Biochem.* In press.
Szent-Györgyi, A., Isenberg, I., and McLaughlin, J. (1961). *Proc. Natl. Acad. Sci. U.S.* **47**, 1089.

Nature of the Intramolecular Complex of Flavine Adenine Dinucleotide

DONALD B. McCORMICK

Section of Biochemistry and Molecular Biology
and Graduate School of Nutrition
Cornell University
Ithaca, New York

I. Introduction

A. Historical Considerations

Numerous studies indicate a close association of the riboflavine and adenine ring systems within the coenzyme, flavine adenine dinucleotide (FAD), a planar representation of which is given in Fig. 1. Early investigations on the light absorption of flavines (Warburg and Christian, 1938a,b) showed that the molar extinction coefficient of FAD at 260 mμ is lower than the sum of molar extinction coefficients of riboflavine and adenosine at this wavelength. This was confirmed (Whitby, 1953) by demonstrating that after hydrolysis of FAD, the absorption rises to a level approximating the sum of the individual spectra of flavine mononucleotide (FMN) and 5'-adenylic acid (AMP). Interaction of flavines with purines was shown to lead to 1 : 1 complexes (Harbury and Foley, 1958). Spectral changes suggest that such complexes are formed both with oxidized and reduced flavine, including FAD (Wilson, 1966). Recent studies with polarized light have revealed greater anomalies in the optical rotatory dispersion of FAD than in its component parts (Gascoigne and Radda, 1965; Simpson and Vallee, 1966) and a marked effect of solvent on such spectra (Listowsky *et al.*, 1966). Investigations on the fluorescence emission of flavines revealed that purines quench the fluorescence (Weil-Malherbe, 1946; Burton, 1951), which supports the view that the internal quenching that occurs within FAD reflects an interaction between flavine and adenine moieties (Bessey *et al.*, 1949; Weber, 1950). The equilibrium between fluorescent and nonfluorescent forms of FAD was shown to be influenced by solvent, salt, and hydrogen ion concentrations (Walaas and Walaas, 1956; Cerletti and Siliprandi, 1958). It has been suggested that water plays an obligatory role in the dark complex (Weber, 1966). Additional evidence for the internal complex of FAD was supplied by observations on its greater light stability relative to riboflavine or FMN, wherein photolysis of the ribityl chain occurs at a more rapid rate (Bessey *et al.*, 1949). Also, the internal quenching in FAD makes it less suitable as a transmitter of light energy, as substantiated by its decreased photodynamic action (Frisell *et al.*, 1959).

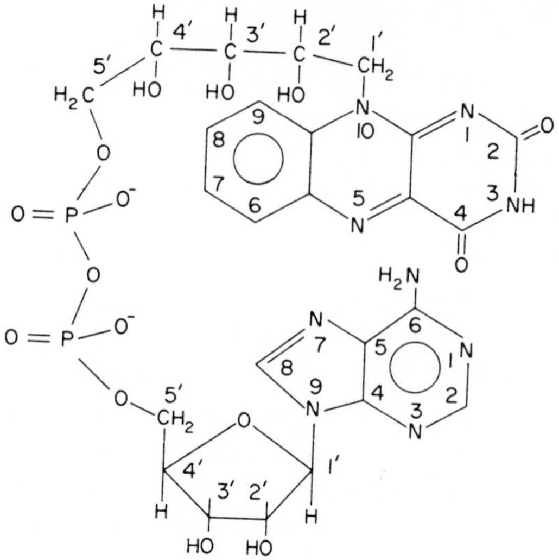

FIG. 1. Planar representation of FAD with positions numbered.

B. Current Aims

Certainly, considerable evidence for flavine-purine complexes had accumulated, and some progress toward understanding the characteristics of these complexes had resulted. Much remained to be delineated concerning their nature, however, including that of the biologically important FAD. A more detailed understanding of the role and localization of interactive forces was especially needed. The limits for length and extent of hydroxylation of the side chain of flavine and also the contribution from the side group of the adenylate moiety required assessment. The particular function of the 6-amino group of adenine within the FAD complex was unclear, although its intimate participation is reflected by the pH profile for fluorescence (Weber, 1950) and by molar rotation at 262 mμ (Simpson and Vallee, 1966). Possible interplay of this purine group with the 2- and 4-keto functions in the isoalloxazine part of the flavine had not been thoroughly studied. For these reasons, we have attempted a systematic examination, largely using the sensitive technique of measuring fluorescence quenching, of both inter- and intramolecular complexes of flavines with purines. The effects of alterations in both side groupings and ring systems have been investigated. Results obtained (Tsibris *et al.*, 1965; Chassy and McCormick, 1965; Roth and McCormick, 1967), although not yet complete, seem to aid in further elucidation of the nature of FAD.

II. Experimental Findings

A. Theoretical Relationships

The reversible interaction of flavine, F, with quencher, Q, to form a $1:1$ molecular complex, FQ, is related to the equilibrium constant for association K by

$$K = [FQ]/([F][Q]) \tag{1}$$

When the complex is nonfluorescent, the fluorescence of flavine in the presence of quencher, I, must reflect only the free flavine, i.e.,

$$[F] = 1 \tag{2}$$

The flavine involved in the complex is then reflected by the difference between the fluorescence of total flavine, I_0, and that which is free as

$$[FQ] = I_0 - I \tag{3}$$

Since the concentration of flavine is usually quite small when compared to the concentration of quencher, the latter is considered to be that added, and substitution of the fluorescence intensities into the equation for the association constant gives

$$K = (I_0 - I)/(I[Q]) = [(I_0/I) - 1]/[Q] \tag{4}$$

The last expression can be rearranged to a linear form

$$I_0/I = K[Q] + 1 \tag{5}$$

Therefore, a plot of the ratio of fluorescence intensities against the quencher concentration will produce a straight line, the slope of which is the association constant.

B. Materials and Methods

Flavines were synthesized by procedures already outlined in the literature. The monophosphate ester of each was prepared by phosphorylation of the terminal hydroxymethyl group with dichlorophosphoric acid (Flexser and Farkas, 1952). FAD analogs were made by reaction of pyridinium salts of flavine phosphates with the appropriate 4-morpholine N,N'-dicyclohexyl-carboxamidinium nucleoside-5'-phosphoromorpholidates (Moffatt and Khorana, 1958). Purine analogs were generally the best commercial preparations.

Measurements of fluorescence were made with an Aminco-Bowman spectrophotofluorometer which was equipped with a xenon lamp, a IP21 photomultiplier tube, and slit arrangement 2 or 3. Activation wavelength was set at the optimum for each flavine at 445 to 470 mμ and fluorescence emission read at an optimum of 520 to 535 mμ. The cell compartment was specially thermostated by fitted brass blocks which were bored to allow passage of water at a desired temperature maintained by a Haake Model F constant temperature circulator.

III. Alterations in the Flavine Moiety

A. Side Group at Position 10

As shown by the data in Fig. 2, the fluorescence of tetraacetylriboflavine is less effectively quenched by comparable concentrations of adenosine than is the fluorescence of riboflavine. Thus, the complexing affinity of riboflavine with adenosine is decreased by acetylation of the ribityl chain at N-10 of the flavine. The tetraacetyl derivative was also observed to have a more intense fluorescence than riboflavine in dimethylformamide. Acetyl groups may hinder formation of the intermolecular complex in part by a steric effect, but they also prevent hydroxyl groups in the chain from an intramolecular interaction with

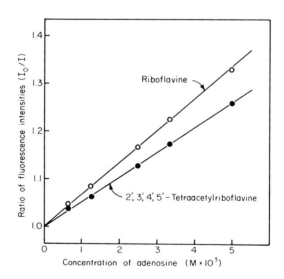

FIG. 2. Change in the ratio of fluorescence intensities of riboflavine and its tetraacetyl derivative as influenced by varying concentrations of adenosine. Flavines were 6.25×10^{-6} M in 0.01 M sodium phosphate buffer, pH 7, at 30°C.

the isoalloxazine system, which normally occurs in riboflavine. The self-quenching due to secondary hydroxyls in glycityl chains is lost in the more fluorescent analogs with ω-hydroxyalkyl chains.

The influence of changing pH on the intensities of fluorescence of ribityl-replaced analogs of FAD is illustrated in Fig. 3. As with flavines and flavine phosphates, the isoalloxazine portions of these FAD analogs exist mainly as nonfluorescent species in the low pH range where shortening of the mean life

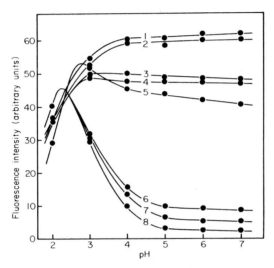

FIG. 3. Effect of pH change on the intensities of fluorescence of ribityl-replaced FAD analogs. Flavines were 10^{-6} M in 0.2 M buffers of glycine-HCl, pH 2–3; sodium acetate, pH 4–5; sodium phosphate, pH 6–7. Curves correspond to: D-allo-FAD (1), DL-glycero-FAD (2), 3′-hydroxypropyl-FAD (3), D-erythro-FAD (4), 2′-deoxyribo-FAD (5), 6′-hydroxyhexyl-FAD (6), 5′-hydroxypentyl-FAD (7), and 4′-hydroxybutyl-FAD (8).

of the excited molecules occurs (Weber, 1950). Similar to FAD, optima for these analogs are evident from pH 2.5 to 3, and quenching caused by intra-molecular complexing occurs in the pH region above the optima. The ability of 4′-hydroxybutyl, 5′-hydroxypentyl, and 6′-hydroxyhexyl analogs of FAD to form intramolecular complexes indicates that secondary hydroxyl groups in the ribityl portion of FAD play no obligatory role, and some variation in the length of this connecting portion is allowed. The absence of significant internal quenching of fluorescence with 3′-hydroxypropyl-FAD and DL-glycero-FAD must be due to insufficient length of the three-carbon chains to allow adenine and isoalloxazine systems to interact. Proof that no restriction other than a steric one prevents intramolecular complexing in these cases is proferred by the observation that fluorescence quenching of their FMN analogs occurs by intermolecular complexing with AMP.

B. Isoalloxazine Ring System

Listed in Table I are the association constants for complexes of 2-substituted riboflavines with adenosine. The values indicate some decrease in complexing from that of the β-hydroxyethylamino analog, which is approximately the same as for riboflavine, through methylmercapto- and morpholinoriboflavine, to the weakly basic anilinoriboflavine. These findings would seem to rule out the possibility that the 2-keto group of the flavine simply is reacting with the 6-amino group of adenosine through hydrogen bonding. In fact, the 2-substituted flavines complex better with dimethylamino- than methylamino- than aminopurine ribosides.

TABLE I

ASSOCIATION CONSTANTS OF COMPLEXES OF 2-SUBSTITUTED
RIBOFLAVINES WITH ADENOSINE[a]

2-Substituted riboflavine	$K_{assoc.}$
β-Hydroxyethylamino	79
Methylmercapto	61
Morpholino	45
Anilino	27

[a] Determined by fluorescence quenching of 8×10^{-6} M 2-β-hydroxyethylamino- or 2-methylmercaptoriboflavine and 4×10^{-6} M 2-morpholino- or 4×10^{-5} M 2-anilinoriboflavine with varying concentrations of adenosine.

Change in the ratio of fluorescence intensities of 2-substituted analogs of riboflavine as influenced by varying pH in the presence of constant adenosine is shown in Fig. 4. The complexing of these analogs by adenosine, as with complexing of riboflavine by this base, exhibits grossly similar ratios of fluorescence intensities throughout the pH range, is maximal above pH 5, and decrease rapidly at lower pH values where adenosine becomes protonated ($pK_a = 3.5$).

Relative fluorescence and pH profiles of riboflavine and its 2-substituted analogs are shown in Fig. 5. The effect of the 2-substituents in internally decreasing the fluorescence of the flavine is seen in the following order: β-hydroxyethylamino < methylmercapto < morpholino < anilino. When intermolecular interaction of riboflavine with the corresponding amines was examined at pH 7, aniline was found to be the only effective quencher. Aniline exists predominantly as the free base at this pH, whereas morpholine and β-hydroxyethylamine are largely protonated and hence unable to complex. The 2-substituted riboflavines as well as riboflavine exhibit decreased fluorescence at high acidity where the mean life of the excited state species may

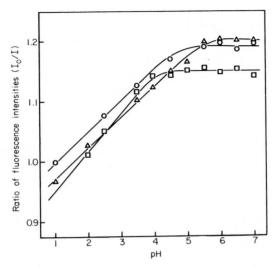

FIG. 4. Change in the ratio of fluorescence intensities of 2-substituted analogs of ribo-flavine as influenced by varying pH in the presence of constant adenosine. Adenosine was 2.77×10^{-3} M in 0.1 M buffers of KCl-HCl, pH 1–2; glycine-HCl, pH 2.5–3.5; sodium acetate, pH 4–5; sodium phosphate, pH 5.5–7. Flavines were 8×10^{-6} M methylmercapto (○), β-hydroxyethylamino (△), and morpholino (□).

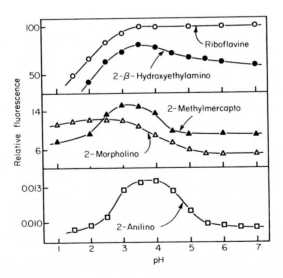

FIG. 5. Relative fluorescence and pH profiles of riboflavine and its 2-substituted analogs. Buffers were as given under Fig. 4.

be shortened. In addition, at pH values below 1, protonation of the isoalloxa-zine portion becomes significant (Michaelis *et al.*, 1936) and may contribute to the population of nonfluorescent flavine molecules. The 2-substituted analogs, however, but not riboflavine, also show decreased fluorescence at pH values above an optimum characteristic for each such flavine in the moderately acid range. The behavior of these analogs thereby indicates the internal

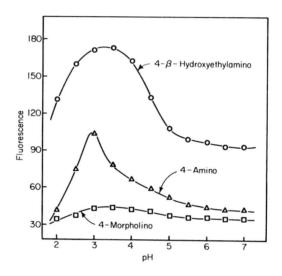

FIG. 6. Fluorescence and pH profiles of 4-substituted analogs of riboflavine. Buffers were as given under Fig. 4.

quenching of fluorescence that results from removal of a proton from the 2-substituent at a pH value above an optimally fluorescent species which is protonated on the 2-substituent but not in the isoalloxazine portion.

Data on the fluorescence of 4-substituted flavines at different pH values are presented in Fig. 6. Again, unlike riboflavine and FMN, but like FAD, optima for fluorescence in the weakly acid range are seen with the absolute fluorescence of β-hydroxyethylamino > amino > morpholino. Apparently, flavines with amine functions in either positions 2 or 4 may have similarities in the ionic species which account for their fluorescence optima and which may afford models for FAD, wherein the amine portion is covalently bonded at a greater distance from the isoalloxazine moiety, but can be brought into close contact in the complex.

The influence of changing pH on the intensities of fluorescence of 3-methyl-ribo- and 7,8-dichlororibo-FAD is illustrated in Fig. 7. The 3-methyl analog has a pH optimum and general profile for fluorescence quite similar to FAD.

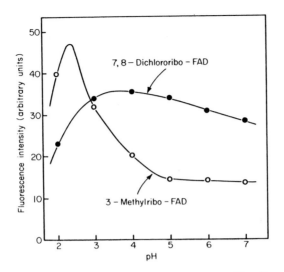

FIG. 7. Effect of pH change on the intensities of fluorescence of isoalloxazine-substituted FAD analogs. Flavines were 10^{-6} M in buffers as given under Fig. 3.

Hence, methylation at the 3-imino position in the isoalloxazine has no significant effect upon its ability to form an intramolecular complex with the adenosine portion. Formation of intermolecular complexes with 3-methylflavines had already been demonstrated and cited as evidence against hydrogen-donor action from position 3 (Harbury and Foley, 1958; Harbury *et al.*, 1959). The meaning which can be attributed to the curve found with the 7,8-dichloro analog is not clear, but may reflect some charge perturbation in formation of an intramolecular complex. The fluorescence of 7,8-dichloro-FMN is, however, quenched by high concentrations of adenosine.

IV. Alterations in the Adenosine Moiety

A. Side Group at Position 9

As shown by the different slopes to the lines in Fig. 8, the nature of the side groupings on N-9 of the adenine ring considerably influences the ability of the purine to form an intermolecular complex with riboflavine. The same was found with tetraacetylriboflavine. Flavine complex formation follows the order: adenine > adenosine > AMP. Both increased size and polarity of the side groups, especially the anionic phosphate ester group of AMP, are detrimental to complexing. Only very slight change is caused by absence of the 2′-hydroxy function, as the behavior of adenosine vs. deoxyadenosine and AMP vs. deoxy-AMP are similar.

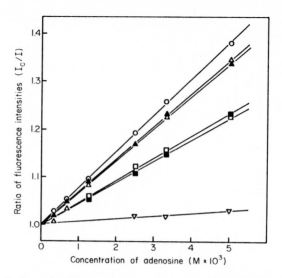

FIG. 8. Change in the ratio of fluorescence intensity of riboflavine as influenced by varying concentrations of adenine and its 9-substituted derivatives. Riboflavine was 6.25×10^{-6} M in 0.01 M sodium phosphate buffer, pH 7, with adenine (○), adenosine (△), deoxyadenosine (▲), AMP (□), and deoxy-AMP (■), and in dimethylformamide with adenosine (▽), all at 30°C.

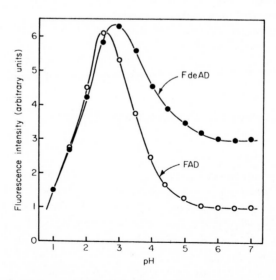

FIG. 9. Effect of pH change on the intensities of fluorescence of FAD and its deoxyadenosyl analog. Flavines were 1.44×10^{-6} M at 30°C in buffers as given under Fig. 3.

The influence of changing pH on the intensities of fluorescence of FAD and its deoxyadenosyl analog is illustrated in Fig. 9. The internal complexing of ribo and deoxyribo compounds are similar as expected. This, too, was noted with other analogs which varied only in the ribosyl portion.

B. Purine Ring System

Listed in Table II are the association constants for complexes of riboflavine with purines and pyrimidines. In general, purines are better complexers than

TABLE II

ASSOCIATION CONSTANTS OF COMPLEXES OF RIBOFLAVINE
WITH PURINES AND PYRIMIDINES[a]

Compound	$K_{assoc.}$
Purines	
2-Amino-6-methylaminopurine	172
Guanine	141
Xanthine	130
2-Amino-6-methylpurine	88
8-Azaadenine	87
8-Azaxanthine	79
Adenine	76
2,6-Diaminopurine	65
2-Aminopurine mononitrate	64
Hypoxanthine	39
6-Methylpurine	5
Purine	2
Pyrimidines	
2,4,6-Triaminopyrimidine	56
2,5-Diaminopyrimidine	48
2-Aminopyrimidine	27
Thymine	24
4-Amino-5-aminomethyl-2-methylpyrimidine	20
Cytosine	18
Uracil	5
6-Azathymine	5
6-Azauracil	1

[a] Determined by fluorescence quenching of 6.25×10^{-6} M riboflavine with approximately 5×10^{-3} M analog in 0.01 M sodium phosphate buffer, pH 7, at 30°C.

pyrimidines. The gain in complex stability through addition of the purine N-7-C-8-N-9 system to the pyrimidine structure is often quite large, e.g., xanthine > uracil. Further gain in stability results from substitution of a methyl group into a pyrimidine—thymine complexes more avidly with riboflavine than does uracil—or into the pyrimidinoid portion of a purine—6-methylpurine complexes more strongly than purine. Some of these effects may be explained by an increase in hydrophobicity of the purine with the resultant tendency to associate with flavine in an aqueous medium, but additional causes are in operation with certain of the substituted purines and pyrimidines. For examples, keto and particularly amino functions at positions 2 and 6 seem to increase the tendency to complex with flavine, whereas a 6-aza function detracts. These effects may reflect electronic interactions which mainly alter the character of the purine.

The association constants of complexes of riboflavine with a rather extensive group of 6-substituted purines and purine ribosides are listed in Table III. The more basic diethylamino and dimethylamino substituents enhance complex stability more than their less basic monoalkylamino analogs, which, in

TABLE III

ASSOCIATION CONSTANTS OF COMPLEXES OF RIBOFLAVINE WITH
6-SUBSTITUTED PURINES AND PURINE RIBOSIDES[a]

Compound	$K_{assoc.}$
Purines	
Seleno	92
β-Hydroxyethylamino	85
Amino	82
Bis(β-hydroxyethyl)amino	66
Hydroxylamino	44
Hydroxy	37
Methoxy	31
Purine ribosides	
Diethylamino	346
Dimethylamino	153
Ethylamino	101
Methylamino	96
n-Propylamino	95
Methylmercapto	82
Isopropylamino	80
Di-n-propylamino	79
Amino	77
Iodo	57
Mercapto	49
Hydroxy	48
Bromo	37
Chloro	3

[a] Determined by fluorescence quenching of 4×10^{-6} M riboflavine with various concentrations of compounds in 0.01 M sodium phosphate buffer, pH 7.

turn, are stronger complexers than adenosine. These effects from *N*-alkylation also rule out hydrogen bonding directly from the 6-amino group, since replacement of the hydrogens from the amino nitrogen abolishes this group as a potential hydrogen donor. Moreover, bulkiness of the di-*n*-propylamino substituent decreases complex stability to below that found with the *n*-propylamino analog. The bis(β-hydroxyethyl)amino substituent, comparable in size

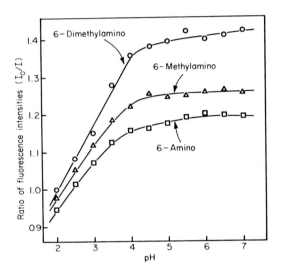

Fig. 10. Change in the ratio of fluorescence intensity of riboflavine as influenced by pH and the nature of the 6-substituents of purine ribosides. Riboflavine was 4×10^{-6} M with 2.77×10^{-3} M riboside in buffers as given under Fig. 4.

to the di-*n*-propylamino group, is also a less effective complexer than the β-hydroxyethylamino analog. A similar though less marked optimum can be seen among the monoalkylamino derivatives where a progressive decrease in complex stability occurs with sizes of the *N*-alkyl group increasing past ethyl. Within the halogenated series, the decrease in electronegativity of the halo group follows an increase in the ability of the purine to complex—iodo > bromo > chloro.

Change in the ratio of fluorescence intensity of riboflavine as influenced by pH and the nature of the 6-substituents of purine ribosides is shown in Fig. 10. Maximal complexing of riboflavine occurs toward neutral pH, and the order for efficiency of complexing with the 6-substituted purine ribosides is again seen to be dimethylamino > methylamino > amino. Decrease in the ratio of fluorescence intensities is seen toward acid values, where the quenching propensities of the purine ribosides is curtailed upon protonation.

The influence of changing pH on the intensities of fluorescence of adenosyl-replaced analogs of FAD is illustrated in Fig. 11. The intramolecular complexing, which is observed best above pH 4, is considerably weaker with the analogs shown than with FAD. Some internal complex, however, seems to exist with the hypoxanthine-containing analog, FHD; also the cytosine-containing analog, FCD, is more quenched than the thymine-containing FTD.

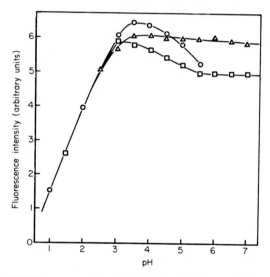

FIG. 11. Effect of pH change on the intensities of fluorescence of adenosyl-replaced FAD analogs. Flavines were 1.44×10^{-6} M at 30°C in buffers as given under Fig. 3. Symbols used for curves: FCD (O), FTD (△), and FHD (□).

The last behaves nearly like free riboflavine or FMN. FA(1-N-oxide)D was found to be poorly complexed intramolecularly. Thus, there again appears to be at least rough agreement between findings for inter- and intramolecular complexes of flavines.

V. Shape of Flavine Complexes

The shape of any flavine complex can be ascertained with certainty only by X-ray analysis. Unfortunately, the crystalline materials so used may not represent the appearance of a complex in an aqueous environment. Some progress, however, is to be expected from such studies and may bear at least indirectly upon the elucidation of reasonable three-dimensional structures for such a complex as FAD. Recent X-ray analysis of 1,3,10-trimethylisoalloxazinium iodide (Kierkegaard, 1967) has established a relationship of flavine to the iodide anion such as is shown by the partial structure illustrated in the top of

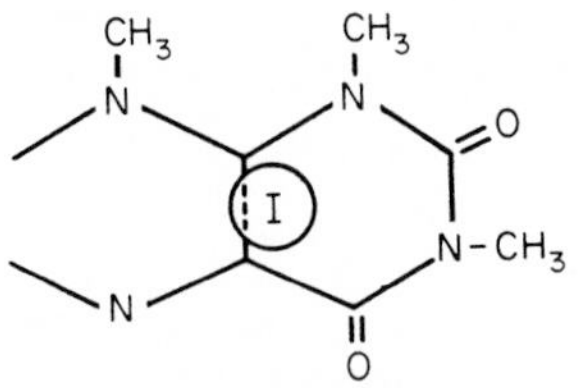

1,3,10 - Trimethylisoalloxazinium iodide

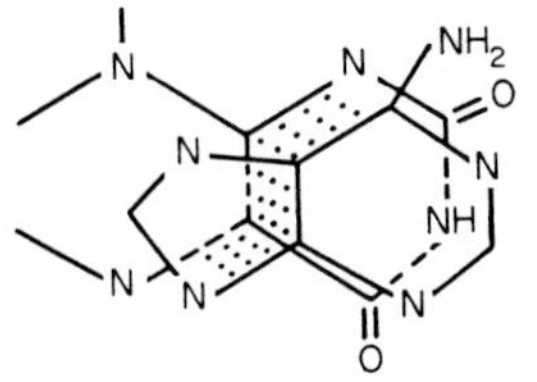

Flavine - adenine

FIG. 12. Structures of complexes considered probable for flavinium iodide (top) and possible for flavine adenine (bottom).

Fig. 12. Although the nature of the flavine-purine complex is quite different in many respects, it is at least possible that the molecular overlay in flavine-adenine or FAD may be as conceived in the bottom of this figure. As with charge donation from iodide into the acceptor region of the flavine, predicted earlier on the basis of calculations of the net charge distribution (Tsibris *et al.*, 1965), both flavine and adenine may be mutually polarized by a weak inter-action between the overlapping molecules. This does *not* mean that the flavine adenine type of complex is best characterized by charge transfer. As we stated earlier (Tsibris *et al.*, 1965), "The complexes which result show only small changes in the flavin absorption spectrum and therefore are probably not of the classic charge-transfer type wherein electronic perturbation between complexing halves is sufficient to exhibit an additional absorption band absent in either separate partner." Certainly, contribution from even partial charge transfer to the binding energy of such complexes is negligible. The acid com-plexes of flavine with phenol have been shown to have a charge-transfer character, but this does not contribute to the strength of intermolecular binding (Fleischman and Tollin, 1965). Those factors which affect the fluores-cence of many flavine complexes and contribute to their stability may be hydrophobic bonding forces and/or London dispersion forces such as may predominate in nucleotide stacking (DeVoe and Tinoco, 1962; Hanlon, 1966). Hydrophobic regions between the aromatic rings would be expected to facilitate the operation of dispersion forces. The role of water may also be considered essential in the mutual orientation of partners in the flavine com-plex (Weber, 1966) of greatest biological significance, FAD. Further study must be made to delineate and quantitate these factors.

ACKNOWLEDGMENTS

Many of the foregoing results were obtained by Drs. J. C. M. Tsibris and B. M. Chassy and Mr. J. A. Roth to whom thanks are due.

These investigations were supported by PHS Research Grant AM-04585 from the National Institutes of Health. Presentation of the work was facilitated by the tenure of a Guggenheim Memorial Foundation Fellowship.

REFERENCES

Bessey, O. A., Lowry, O. H., and Love, R. H. (1949). *J. Biol. Chem.* **180**, 755.

Burton, K. (1951). *Biochem. J.* **48**, 458.

Cerletti, P., and Siliprandi, N. (1958). *Arch. Biochem. Biophys.* **76**, 214.

Chassy, B. M., and McCormick, D. B. (1965). *Biochemistry* **4**, 2612.

DeVoe, H., and Tinoco, I., Jr. (1962). *J. Mol. Biol.* **4**, 500.

Fleischman, D. E., and Tollin, G. (1965). *Proc. Natl. Acad. Sci. U.S.* **53**, 38.

Flexser, L. A., and Farkas, W. G. (1952). U.S. Patent 2,610,179; *Chem. Abstr.* **47**, 8781g (1952).

Frisell, W. R., Chung, C. W., and Mackenzie, C. G. (1959). *J. Biol. Chem.* **234**, 1297.

Gascoigne, I. M., and Radda, G. K. (1965). *Chem. Commun.* No. 21, p. 534.

Hanlon, S. (1966). *Biochem. Biophys. Res. Commun.* **23**, 861.

Harbury, H. A., and Foley, K. A. (1958). *Proc. Natl. Acad. Sci. U.S.* **44**, 662.

Harbury, H. A., LaNoue, K. F., Loach, P. A., and Amick, R. M. (1959). *Proc. Natl. Acad. Sci. U.S.* **45**, 1708.

Kierkegaard, P. (1967). Private communication.

Listowsky, I., England, S., Betheil, J. J., and Seifter, S. (1966). *Biochemistry* **5**, 2548.

Michaelis, L., Schubert, M. P., and Smythe, C. V. (1936). *J. Biol. Chem.* **116**, 587.

Moffatt, J. G., and Khorana, J. (1958). *J. Am. Chem. Soc.* **80**, 3756.

Roth, J. A., and McCormick, D. B. (1967). *Photochem. Photobiol.* **6**, 657.

Simpson, R. T., and Vallee, B. L. (1966). *Biochem. Biophys. Res. Commun.* **22**, 712.

Tsibris, J. C. M., McCormick, D. B., and Wright, L. D. (1965). *Biochemistry* **4**, 504.

Walaas, E., and Walaas, O. (1956). *Acta Chem. Scand.* **10**, 122.

Warburg, O., and Christian, W. (1938a). *Biochem. Z.* **296**, 294.

Warburg, O., and Christian, W. (1938b). *Biochem. Z.* **298**, 150.

Weber, G. (1950). *Biochem. J.* **47**, 114.

Weber, G. (1966). *In* "Flavins and Flavoproteins" (E. C. Slater, ed.), p. 15. Elsevier, Amsterdam.

Weil-Malherbe, H. (1946). *Biochem. J.* **40**, 363.

Whitby, L. G. (1953). *Biochem. J.* **54**, 437.

Wilson, J. E. (1966). *Biochemistry* **5**, 1351.

Molecular Complexes of Flavines*

GORDON TOLLIN

Department of Chemistry
University of Arizona
Tucson, Arizona

I. Introduction

Flavines are known to be quite avid molecular complex formers.† Among the compounds which will interact in this manner with flavine in its oxidized state are the following: phenols (Fleischman and Tollin, 1965a, b; Harbury and Foley, 1958; Yagi and Matsuoka, 1956; Yagi et al., 1959a), indoles (Harbury and Foley, 1958; Isenberg and Szent-Györgyi, 1958; Pereira and Tollin, 1967; Wilson, 1966), purines (Burton, 1951; Tsibris et al., 1965; Weber, 1950; Wilson, 1966; Wright and McCormick, 1964), quinine (Burton, 1951), chlortetracycline (Yagi et al., 1956), chlorpromazine (Yagi et al., 1959b), and benzoic acid (Harbury and Foley, 1958). All of these would be expected to function as electron donors in donor-acceptor (charge-transfer) type complexes. It is interesting that oxidized flavine can also form complexes with electron acceptors, e.g., tetracyanoethylene and dichlorodicyano-*p*-benzoquinone (Matsunaga, 1967). In the fully reduced state, flavine is capable of forming molecular complexes with, for example, NAD^+ (Sakurai and Hosoya, 1966), oxidized flavine (Beinert, 1956; Gibson et al., 1962), and caffeine (Harbury et al., 1959). Finally, flavine at the semiquinone level of reduction will complex with fully reduced flavine (Kuhn and Ströbele, 1937), NADH (Mahler and Brand, 1961), HI (Fleischman and Tollin, 1965c), and phenols (Fleischman and Tollin, 1965c). Thus, complexing ability would appear to be a characteristic property of flavine in all of its redox forms.

The following discussion will concern itself primarily with certain of the properties of flavine molecular complexes as well as with the modification of flavine properties caused by complex formation.

---

* The work described in this paper was supported in part by the U.S. Atomic Energy Commission [Contract No. AT(11-1)908] and the Air Force Cambridge Research Laboratory [Contract No. AF19(628)4376].

† Abbreviations used in this chapter: F, flavine; FMN, flavine mononuculeotide; FAD, flavine adenine dinucleotide; NAD^+, nicotinamide adenine dinucleotide; NADH, NAD^+ reduced form.

II. pH Dependence of Complex Formation

Flavine can participate in the following protonic equilibria in its various redox states:

$$FH_2^+ \; \underset{\longleftarrow}{\overset{pK_a \sim 0}{\longrightarrow}} \; FH \; \underset{\longleftarrow}{\overset{pK_a \sim 10}{\longrightarrow}} \; F^- \qquad \text{(oxidized form)}$$

$$FH_3^+\cdot \; \underset{\longleftarrow}{\overset{pK_a \sim 1\text{--}3}{\longrightarrow}} \; FH_2\cdot \; \underset{\longleftarrow}{\overset{pK_a \sim 6.5}{\longrightarrow}} \; FH^-\cdot \qquad \text{(semiquinone form)}$$

$$FH_4^+ \; \underset{\longleftarrow}{\overset{pK_a < 0}{\longrightarrow}} \; FH_3 \; \underset{\longleftarrow}{\overset{pK_a \sim 6.2}{\longrightarrow}} \; FH_2^- \qquad \text{(fully reduced form)}$$

Thus far, complex formation has been observed with FH, FH_2^+, $FH_2^+\cdot$, and FH_3.

In many cases, one can isolate these complexes as nicely crystalline, highly colored solids. In Fig. 1, some examples of these are shown. Neutral complexes are invariably orange to orange-red, whereas the acid complexes range from green to deep red or black.

III. Complex Stoichiometry

For the most part, 1 : 1 stoichiometries have been observed with flavine complexes. Table I presents some representative results. It should be noted that in some solid state complexes flavine seems able to associate with two species simultaneously.

TABLE I

STOICHIOMETRIES OF FLAVINE MOLECULAR COMPLEXES

Components	Mole ratios	State	Reference
$FMNH_2$–N–alkyl pyridinium salts	1 : 1	Solution	Sakurai and Hosoya (1966)
$FH_3^+\cdot$–I^-–2,7–naphthalenediol	1 : 1 : 1	Crystal	Fleischman and Tollin (1965a)
$FH_3^+\cdot$–I^-	1 : 2	Crystal	Fleischman and Tollin (1965a)
FH_2^+–phenols	1 : 1	Solution	Fleischman and Tollin (1965b)
FH_2^+–Cl^-–phenols	1 : 1 : 1	Crystal	Ray et al. (1965)
FH–naphthalenediols	1 : 1	Solution	Fleischman and Tollin (1965c)
FH–naphthalenediols	1 : 2	Crystal	Fleischman and Tollin (1965c)
FH_2^+–indoles	1 : 1	Solution	Pereira and Tollin (1967)
FH_2^+–Cl^-–indoles	1 : 1 : 1	Crystal	Pereira and Tollin (1967)
FH–indoles	1 : 1	Crystal	Pereira and Tollin (1967)

IV. Spectral Properties of Complexes

Complexes of FH in solution generally show a broadening and flattening of the long wavelength absorption band of the flavine and a more or less pronounced tailing of this band toward the red. Some typical examples are

FIG. 1. Crystals of flavine molecular complexes. (a) Lumiflavine-carbazole; neutral. (b) Lumiflavine-tryptophan; neutral. (c) Riboflavine-2,3-naphthalenediol; neutral. (d) 9-Methyl isoalloxazine-1,4-naphthalenediol; acid.

shown in Fig. 2. In some cases, spectra of crystals of these materials show evidence of new absorption bands (Fig. 3). There is some correlation between the band position and the donor ability of the phenol (i.e., the better the donor, the further to the red is the absorption) (Fleischman and Tollin, 1965b), and thus it is possible that these represent charge-transfer transitions.

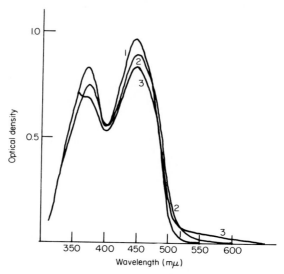

FIG. 2. Optical absorption spectra of neutral flavine complexes in aqueous solution. (1) FMN; (2) FMN-4-methyl catechol; and (3) FMN-1,4-naphthalenediol.

Solution spectra of complexes of FH_2^+ with most monobenzenoid phenols also show a broadening and tailing of the long wavelength flavine absorption (Fig. 4). With naphthalenediols and trimethylhydroquinone, however, new bands are seen to the red of the flavine absorption (Fig. 5). Again, the position of these bands roughly correlates with the predicted donor ability of the phenol, and thus they are probably charge transfer in origin. Similar new absorptions, strongly red-shifted as compared to solution, are found in the spectra of crystalline FH_2^+-phenol complexes (Fig. 6).

Absorption spectra of complexes of FH_2^+ with indoles in solution do not show any new bands, but merely show the broadening and tailing effects. The crystal spectra, on the other hand, have additional absorption bands to the red of the flavine absorption. Since these are less clearly defined than in the case of the phenol complexes, it is difficult to know whether to assign these to charge-transfer transitions.

Spectra of flavine semiquinone complexes have been obtained with crystalline systems consisting of $FH_3^+\cdot$, HI, and phenols. These show a broadening and slight shifting of the semiquinone absorption; no indications of charge-transfer transitions are found.

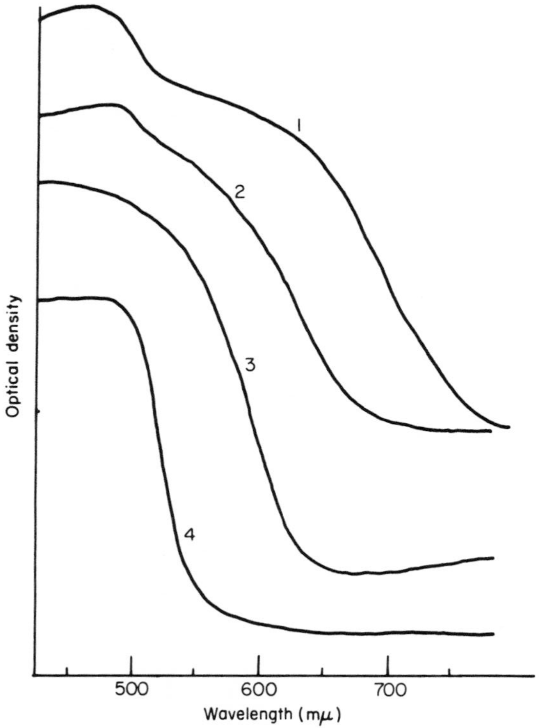

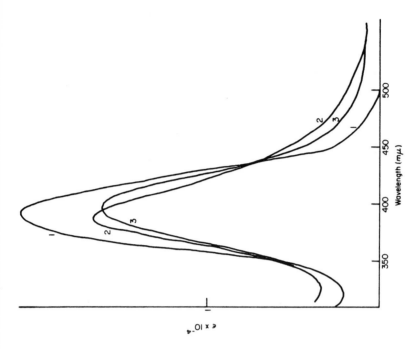

FIG. 4. Optical absorption spectra of flavine complexes in 6 N HCl. (1) Riboflavine; (2) riboflavine-pyrogallol, and (3) riboflavine-4-methyl catechol.

FIG. 3. Optical absorption spectra of suspensions of microcrystals of neutral flavine complexes. (1) Lumiflavine-1,4-naphthalenediol; (2) lumiflavine-1,7-naphthalenediol; (3) lumiflavine-2,3-naphthalenediol; and (4) lumiflavine. Baselines have been shifted for clarity.

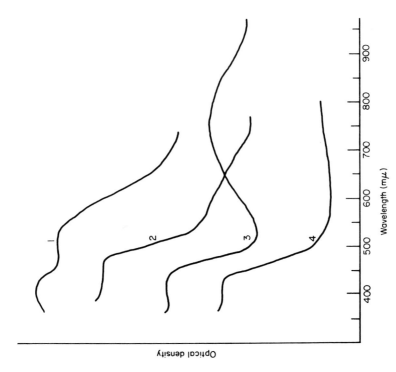

Fig. 6. Optical absorption spectra of suspensions of microcrystals of acid flavine complexes. (1) 9-Methyl isoalloxazine-4-chlorocatechol; (2) 9-methyl isoalloxazine-1,2-naphthalenediol; and (4) 9-methyl isoalloxazine hydrochloride.

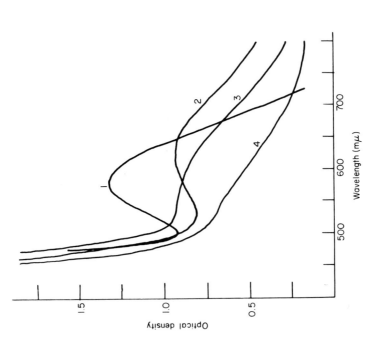

Fig. 5. Optical absorption spectra of flavine complexes in 6 N HCl. (1) 9-Methyl isoalloxazine-1,5-naphthalenediol; (2) 9-methyl isoalloxazine-1,4-naphthalenediol; (3) 9-methyl iso-alloxazine-1,2-naphthalenediol; and (4) 9-methyl isoalloxazine-trimethyl hydroquinone.

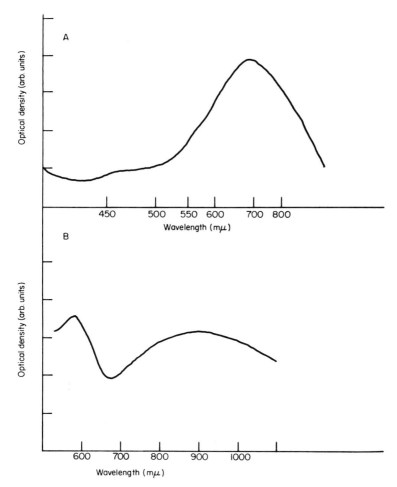

Fig. 7. Optical spectra of fully reduced flavine complexes in aqueous solution. (A) $FMNH_2$-N-methyl-4-carbomethoxy pyridinium chloride. After Sakurai and Hosoya (1966). (B) $FMNH_2$–FMN. After Gibson et al. (1962).

Two fairly well-characterized examples of FH_3 complexes exist: FH—FH_3 (Gibson et al., 1962) and N-alkylpyridinium ion—FH_3 (Sakurai and Hosoya, 1966). In both cases, new broad long wavelength absorption bands are present in solution spectra which are probably charge transfer in nature (Fig. 7). In the latter system, the position of the new band was shown to be a linear function of the energy of the lowest molecular orbital of the pyridinium salt.

V. Thermodynamics of Complex Formation

In Tables II and III are given stability constants for a variety of flavine complexes. A number of generalizations emerge from these data.

TABLE II

STABILITY CONSTANTS OF SOME FLAVINE MOLECULAR COMPLEXES IN AQUEOUS SOLUTION

Components	Temperature (°C)	$K(M^{-1})$	References
Riboflavine-adenosine	17	120	Weber (1950)
Riboflavine-caffeine	17	91	Weber (1950)
FAD-caffeine	17	83	Burton (1951)
Riboflavine-chlorpromazine	—	1000	Yagi et al. (1959a)
FMN-tryptophan	—	59	Isenberg and Szent-Györgyi (1958)
FMN-tryptophan	25	92	Wilson (1966)
FMN-catechol	25	10.4	Fleischman and Tollin (1965b)
FMN-2,3-naphthalenediol	25	242	Fleischman and Tollin (1965b)
FMN (protonated)-catechol	25	0.68	Fleischman and Tollin (1965b)
FMN (protonated)-2,3-naphthalenediol	25	68	Fleischman and Tollin (1965b)
Riboflavine (protonated)-hydroquinone	25	2.9	Fleischman and Tollin (1965b)
Riboflavine (protonated)-catechol	25	3.9	Fleischman and Tollin (1965b)
Riboflavine (protonated)-1,3,5-trihydroxybenzene	25	7.4	Fleischman and Tollin (1965b)
Riboflavine (protonated)-3,5-dihydroxytoluene	25	9.2	Fleischman and Tollin (1965b)
Riboflavine (protonated)-trimethylhydroquinone	25	10	Fleischman and Tollin (1965b)
Riboflavine (protonated)-1,4-naphthalenediol	25	55	Fleischman and Tollin (1965b)
Riboflavine (protonated)-1,7-naphthalenediol	25	98	Fleischman and Tollin (1965b)
Riboflavine (protonated)-2,7-naphthalenediol	25	102	Fleischman and Tollin (1965b)
Riboflavine (protonated)-1,5-naphthalenediol	25	111	Fleischman and Tollin (1965b)
Riboflavine (protonated)-2,3-naphthalenediol	25	162	Fleischman and Tollin (1965b)

(1) Stability constants range between 1–1000 M^{-1}.

(2) The strongest complexes are formed with polycyclic compounds. This is probably an indication that the polarizability and/or the "electron richness" of the donor are important in stabilizing the complex.

(3) Among the naphthalenediol complexes of FH_2^+, no correlation between complex stability and donor ability exists. Further, although FH_2^+ is a better electron acceptor than FH, as evidenced by the fact that charge-transfer bands can be observed with the former and not with the latter, the FH complexes are appreciably more stable than the FH_2^+ complexes. This indicates little, if any, charge-transfer stabilization of the ground states of these complexes.

(4) Among the phenol complexes of FH_2^+, the addition of methyl groups to the phenolic component seems to be particularly effective in increasing stability (e.g., 1,3,5-trihydroxybenzene vs. 3,5-dihydroxytoluene). This could be interpreted in terms of a role of hydrophobic forces in complex stabilization, particularly in view of statement (3) above.

(5) Indole complexes of FH_2^+ are generally more stable than are the phenol complexes. As we have noted above, however, many of the phenol complexes have easily observable charge-transfer absorptions, whereas the indole complexes do not. Also, with the indole complexes, as with the phenol complexes, no correlation exists between donor ability and complex stability. These facts provide further evidence for the lack of charge-transfer stabilization of ground states.

(6) With the FH-naphthalenediol complexes in aqueous ethanol, the order of decreasing stabilities is approximately the same as the order of decreasing ionization potentials of the phenols (Fleischman and Tollin, 1965b) (an exception to this is the 1,7-naphthalenediol complex, which is less stable than would be predicted on this basis). This suggests that, in this case, charge-transfer forces may be involved in ground-state stabilization. It is interesting to note here that for the FH complexes, the indole systems are *less* stable than the phenol systems, in contrast to statement (5) above.

(7) Addition of ethanol to the solvent invariably destabilizes the complex. Further, *no* complex formation can be detected in pure ethanol. Wilson (1966) has also observed this with FH-indole complexes and Weber (1966) has found, using pyridine-water solvents, that the presence of water is necessary in order for complex formation to occur. These facts further suggest the participation of solvent structure in complex stabilization. Additional support for this, at least in the case of the FH_2^+-indole complexes, can be obtained from a consideration of the thermodynamic parameters for complex formation (Table IV). Thus, when there is no charged group present in the donor molecule, we see that the entropy change is uniformly positive. Further, one would expect that the presence of ionized groups in the complex would

TABLE III

STABILITY CONSTANTS OF SOME FLAVINE MOLECULAR COMPLEXES IN AQUEOUS ETHANOL[a]

Components	Solvent[b]	$K(M^{-1})$	Components	Solvent[b]	$K(M^{-1})$
Riboflavine-2-methylindole	(60% Abs. ethanol–40% 12 N HCL)	603	FMN-3-methylindole	(30% Abs. ethanol–70% 0.1 M phosphate buffer, pH 6.8)	48.8
Lumiflavine-1,2-dimethylindole	(60% Abs. ethanol–40% 12 N HCL)	284	FMN-1,2-dimethylindole	(30% Abs. ethanol–70% 0.1 M phosphate buffer, pH 6.8)	45.2
Riboflavine-1,2-dimethylindole	(60% Abs. ethanol–40% 12 N HCL)	268	Riboflavine-1,5-naphthalenediol	(60% Abs. ethanol–40% 12 N HCL)	22.0
Riboflavine-3-methylindole	(5% Abs. ethanol–95% 12 N HCL)	150	Riboflavine-2-3,naphthalenediol	(60% Abs. ethanol–40% 12 N HCL)	14.9
Lumiflavine-3-methylindole	(60% Abs. ethanol–40% 12 N HCL)	67.1	Riboflavine-2,7-naphthalenediol	(60% Abs. ethanol–40% 12 N HCL)	6.1
FMN-3-methylindole	(60% Abs. ethanol–40% 12 N HCL)	60–70[c]	FMN-1,4-naphthalenediol	(30% Abs. ethanol–70% 0.1 M phosphate buffer, pH 6.8)(30% Abs. ethanol–	230

Complex	Solvent system[b]	Value
FMN-tryptophan	(60% Abs. ethanol–40% 12 N HCL)	30–40[c]
Riboflavine-3-methylindole	(60% Abs. ethanol–40% 12 N HCL)	35.5
Lumiflavine-tryptophan	(60% Abs. ethanol–40% 12 N HCL)	28.2
Riboflavine-tryptophan	(60% Abs. ethanol–40% 12 N HCL)(30% Abs. ethanol–	15.3
FMN-tryptophan	70% 0.1 M phosphate buffer, pH 6.8 (30% Abs. ethanol–	98.4
FMN-2-methylindole	70% 0.1 M phosphate buffer, pH 6.8	89.6
FMN-1,5-naphthalenediol	70% 0.1 M phosphate buffer, pH 6.8 (30% Abs. ethanol–	191
FMN-2,3-naphthalenediol	70% 0.1 M phosphate buffer, pH 6.8 (30% Abs. ethanol–	176
FMN-2,7-naphthalenediol	70% 0.1 M phosphate buffer, pH 6.8 (30% Abs. ethanol–	124
FMN-1,7-naphthalenediol	70% 0.1 M phosphate buffer, pH 6.8 (30% Abs. ethanol–	73
FMN-resorcinol	70% 0.1 M phosphate buffer, pH 6.8	33.7

[a] Pereira and Tollin (1967).
[b] Mixture prepared by volume.
[c] Some uncertainty indicated due to the fact that FMN was not completely soluble in the solvent system used.

TABLE IV

THERMODYNAMIC PARAMETERS FOR FORMATION OF FLAVINE-INDOLE
COMPLEXES IN ACIDIC ETHANOL[a]

Complex	K (M^{-1})	ΔF^0_{300} (kcal/mole)	ΔH^0_{300} (kcal/mole)	ΔS^0_{300} (cal/mole-deg)
Riboflavine-2-methylindole	603	−3.82	−1.02	+9.3
Lumiflavine-1,2-dimethylindole	284	−3.37	−1.26	+7.0
Lumiflavine-3-methylindole	67.1	−2.51	−1.99	+1.7
Riboflavine-3-methylindole	35.5	−2.13	−1.92	+0.7
Lumiflavine-tryptophan	28.2	−1.99	−2.60	−2.0
Riboflavine-tryptophan	15.3	−1.63	−2.22	−2.0

[a] Solvent in each case was a mixture (by volume) of 40% 12N HCL–60% absolute ethanol [from Pereira and Tollin (1967)].

tend to reduce stability and possibly also to change the nature of the stabilizing forces. The results with tryptophan as a donor support this. With the FMN-indole (Wilson, 1966) and the flavine-tyrosine (Cilento and Berenholc, 1965) complexes, the entropy changes are invariably negative suggesting that hydrophobic interactions are relatively unimportant here. In both these systems, however, charged groups are present in either the donor or the acceptor.

VI. Solid-State Properties of Complexes

We will be concerned here particularly with electrical and magnetic properties. These are of interest with respect to models for charge and energy storage and transport systems in biology (Szent-Györgyi, 1960). For a detailed description of experimental techniques and results see Ray et al. (1965).

At room temperature, specific resistivities of solid $FH_2{}^+$-phenol complexes in the dark are in the range 10^6–10^{10} ohm cm. These are comparable to values found for many other organic molecular complexes and should be compared with a resistivity for pure FH of greater than 5×10^{12} ohm cm. Current may be carried by both positive and negative species. Illumination of the crystals with visible light produces a reversible increase in conductivity with unimolecular rate constants for rise and decay (in the case of the riboflavine-hydroquinone complex) of the order of 0.1 $\sec^{-1}$ (these vary with the conditions of measurement). Although these constants are virtually temperature-independent, the dark conductivity and the steady-state photoconductivity have activation energies of approximately 1 eV.

The kinetic picture which emerges from these studies is as follows. Illumination leads to electron transfer from donor (phenol) to acceptor (flavine).

A steady state with respect to charge carriers in traps and in the conduction band is reached quickly compared to the rate of carrier migration to the electrodes. Thus, the steady-state photocurrent (and the dark current as well) is determined by the depth of the trapping centers. This accounts for the 1eV activation energy. The rate of rise and decay, however, is determined by carrier migration.

Crystals of the complexes are paramagnetic. This paramagnetism seems to be associated with imperfection sites in the crystal at which complete electron transfer from donor to acceptor occurs, leading to deeply trapped, unpaired electrons (these do not contribute to the conductivity). These can be generated during the crystallization process or, reversibly, by removing solvent from fully formed crystals. The maximum unpaired spin concentration which can be produced corresponds to about 1 mole % of the complex.

VII. Modification of Flavine Properties by Complex Formation

A. Quenching of Excited States

Weber (1950) has observed that phenols and purines quench riboflavine fluorescence by virtue of ground-state complex formation. He has also attributed the much lower fluorescence yield of FAD relative to riboflavine to an intramolecular complex between the isoalloxazine and adenine rings.

Using EPR techniques, Shiga and Piette (1965) have shown that tryptophan diminishes the photoinduced triplet-state yield for FMN in rigid solutions at 77°K, presumably via ground-state complex formation.

Radda and Calvin (1963; Radda, 1966) have shown that a wide variety of compounds (including phenols, indoles, and purines) inhibit the photobleaching of flavines in neutral solution at room temperature. This has been attributed to complex formation between the flavine triplet state and the quencher, although the evidence for this is indirect. Similarly, Pereira and Tollin (1967) have demonstrated that phenols and indoles quench flavine free radical formation by light in acid solution (Fig. 8). The triplet state is probably involved here also.

B. Chemical Reactivity

Very little information is available on the modification of the dark chemistry of flavins by complex formation. Table V summarizes much of the present knowledge in this area. The intramolecular complexing in FAD seems to decrease slightly the overall reactivity, to stabilize the flavine radical, and to destabilize the F—FH$_2$ complex. The latter probably occurs because of

TABLE V

EFFECT OF COMPLEXATION ON DARK CHEMISTRY OF FLAVINE[a]

Reaction system	Overall reaction rate	Rate of flavine radical formation	Rate of F–FH$_2$ complex formation
1. FMN + NADH	50	1.0	2.4
FAD + NADH	32	2.2	0.53
2. FMN + NADH	38	3.4	2.5
+2,3-Naphthalenediol	20	2.0	0.85
+Tryptophan	33	2.4	1.95

[a]Numbers represent stopped-flow measurements of reaction rates in anaerobic systems. Limits of error are ±2 for overall reaction rates, ±0.12 for radical formation rate, and ±0.06 for complex formation rate. Data given in arbitrary units. From Fox and Tollin (1966).

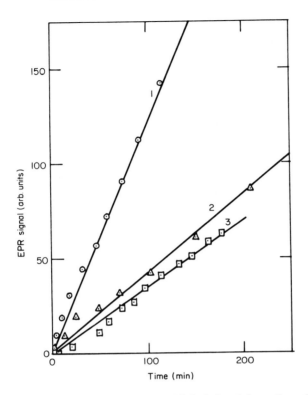

FIG. 8. Effect of complexing agents on rate of light-induced formation of flavine semi-quinone in acid solution. (1) Riboflavine; (2) riboflavine plus 2,3-naphthalenediol; and (3) riboflavine plus tryptophan.

steric effects. Phenols and indoles, on the other hand, decrease all measured reaction rates to approximately the same extent, presumably by virtue of an inhibitory effect upon the initial, rate-limiting, hydrogen-transfer reaction.

VIII. Biochemical Implications

There is evidence that, at least in some flavoenzymes, tyrosine residues are involved in flavine binding (Strittmatter, 1961; Yagi and Ozawa, 1960). This could involve complex formation of the type we have been discussing here. In addition, although no evidence is yet available, tryptophan residues could participate as well. Interactions such as these could be partly responsible for the low fluorescence yield of most FMN enzymes (Tsibris *et al.*, 1966), as well as for the modification of flavine chemistry which occurs upon binding to an apoenzyme, e.g., stabilization of flavine semiquinone, change in redox potential, etc.

A number of flavoenzymes have been shown to develop broad, long wave-length, absorption bands upon addition of various aromatic compounds (including, in some cases, competitive inhibitors) that have been attributed to charge-transfer complex formation (Massey and Ganther, 1965; Massey and Palmer, 1962; Veeger et al., 1966). Thus, it is possible that flavine-substrate interactions may involve complexing as a component. The ability of flavine to complex simultaneously with two donors (Table I) may be of significance here, e.g., with a group in the protein and with the substrate. Also, this latter property could conceivably result in an influence of flavine upon the tertiary structure of flavoproteins, e.g., by complexation with two amino acid residue side chains in different regions of the main polypeptide chain.

Several highly physiologically active indole derivatives are known (serotonin, indole acetic acid). One should keep in mind the possibility that complex formation with flavoenzymes may be involved in the biological action of these materials.

REFERENCES

Beinert, H. (1956). *J. Am. Chem. Soc.* **78**, 5323.
Burton, K. (1951). *Biochem. J.* **48**, 458.
Cilento, G., and Berenholc, M. (1965). *Biochim. Biophys. Acta* **94**, 271.
Fleischman, D., and Tollin, G. (1965a). *Biochim. Biophys. Acta* **94**, 248.
Fleischman, D., and Tollin, G. (1965b). *Proc. Natl. Acad. Sci. U.S.* **53**, 38.
Fleischman, D. E., and Tollin, G. (1965c). *Proc. Natl. Acad. Sci. U.S.* **53**, 237.
Fox, J. L., and Tollin, G. (1966). *Biochemistry* **5**, 3865 and 3873.
Gibson, Q. H., Massey, V., and Atherton, N. M. (1962). *Biochem. J.* **85**, 369.
Harbury, H. A., and Foley, K. A. (1958). *Proc. Natl. Acad. Sci. U.S.* **44**, 662.
Harbury, H. A., LaNoue, K. F., Loach, P. A., and Amick, R. M. (1959). *Proc. Natl. Acad. Sci. U.S.* **45**, 1708.
Isenberg, I., and Szent-Györgyi, A. (1958). *Proc. Natl. Acad. Sci. U.S.* **44**, 857; **45**, 1229 (1959).
Kuhn, R., and Ströbele, R. (1937). *Chem. Ber.* **70**, 753.
Mahler, H. R., and Brand, L. (1961). *In* "Free Radicals in Biological Systems" (M. S. Blois *et al.*, eds.), p. 163. Academic Press, New York.
Massey, V., and Ganther, H. (1965). *Biochemistry* **4**, 1161.
Massey, V., and Palmer, G. (1962). *J. Biol. Chem.* **237**, 2347.
Matsunaga, Y. (1967). Personal communication.
Pereira, J. F., and Tollin, G. (1967). *Biochim. Biophys. Acta* **143**, 79.
Radda, G. K. (1966). *Biochim. Biophys. Acta* **112**, 448.
Radda, G. K., and Calvin, M. (1963). *Nature* **200**, 464.
Ray, A., Guzzo, A. V., and Tollin, G. (1965). *Biochim. Biophys. Acta* **94**, 258.
Sakurai, T., and Hosoya, H. (1966). *Biochim. Biophys. Acta* **112**, 459.
Shiga, T., and Piette, L. H. (1965). *Photochem. Photobiol.* **4**, 769.
Strittmatter, P. (1961). *J. Biol. Chem.* **236**, 2329.
Szent-Györgyi, A. (1960). "Introduction to a Submolecular Biology." Academic Press, New York.

Tsibris, J. C. M., McCormick, D. B., and Wright, L. D. (1965). *Biochemistry* **4**, 504.

Tsibris, J. C. M., McCormick, D. B., and Wright, L. D. (1966). *J. Biol. Chem.* **241**, 1138.

Veeger, C., DerVartanian, D. V., Kalse, J. F., DeKok, A., and Koster, J. F. (1966). *In* "Flavins and Flavoproteins" (E. C. Slater, ed.), p. 242. Elsevier, Amsterdam.

Weber, G. (1950). *Biochem. J.* **54**, 114.

Weber, G. (1966). *In* "Flavins and Flavoproteins" (E. C. Slater, ed.), p. 15. Elsevier, Amsterdam.

Wilson, J. E. (1966). *Biochemistry* **5**, 1351.

Wright, L. D., and McCormick, D. B. (1964). *Experientia* **20**, 501.

Yagi, K., and Matsuoka, Y. (1956). *Biochem. Z.* **328**, 138.

Yagi, K., and Ozawa, T. (1960). *Biochim. Biophys. Acta* **42**, 381.

Yagi, K., Okuda, J., Ozawa, T., and Okada, K. (1956). *Science* **124**, 273.

Yagi, K., Ozawa, T., and Okada, K. (1959a). *Biochim. Biophys. Acta* **35**, 102.

Yagi, K., Ozawa, T., and Nagatsu, T. (1959b). *Nature* **184**, 982.

Les Forces de Van der Waals-London dans les Complexes Dits de Transfert de Charge

MARIE-JOSÉ MANTIONE

Service de Biochimie Théorique
Institut de Biologie Physico- chimique
Paris, France

I. Introduction

Certaines confusions semblent exister dans le domaine des complexes dits de transfert de charge, en particulier au sujet de la nature des forces d'interaction entre les deux constituants.

En effet de nombreux auteurs attribuent a priori aux éventuels transferts d'électrons entre les deux constituants la totalité ou la quasi totalité de la stabilité de ces complexes. Cette conception est également répandue en biochimie, où un rôle prépondérant est fréquemment donné, dans des associations inter- ou intramoléculaires, à des « forces de transfert de charge » et cela même lorsqu'un tel transfert est fort hypothétique.

Or dans les problèmes impliquant des transferts de charge, deux aspects sont à distinguer. L'apparition d'une bande caractéristique dans le spectre ultraviolet de deux composés associés est interprétée comme un transfert de charge (Mulliken, 1952), c'est-à-dire une excitation électronique de l'un des partenaires du complexe vers l'autre. Un tel transfert correspond au passage aux *états excités* du complexe. Cela est à distinguer des aspects concernant le complexe dans son *état fondamental*. En particulier, la stabilité d'un complexe est relative à son état fondamental, et il n'y a aucune raison d'attribuer cette stabilité à l'existence de transferts de charge entre les deux constituants.

Bien plus, la théorie même des interactions moléculaires, si elle est correctement faite, c'est-à-dire sous une forme générale, permet de prévoir qu'aucune distinction de nature fondamentale n'existe entre les complexes avec ou sans transfert de charge, en ce qui concerne leur énergie d'interaction (donc leur stabilité).

II. Expression de l'Énergie d'Interaction Intermoléculaire

Considérons en effet le système constitué par deux molécules A et B susceptibles d'interagir. Dans le cadre d'un traitement par perturbations, le

* Ce travail a bénéficié de la subvention 67-00-532 de la Délégation Générale à la Recherche Scientifique et Technique (Comité de Biologie Moléculaire).

potentiel perturbateur U représente précisément cette interaction. L'énergie de stabilisation (ou d'interaction) du complexe formé entre A et B peut être obtenue par un développement au second ordre.

Avec les notations de Dirac cette énergie peut s'écrire sous la forme :

$$E = \langle \Psi_0{}^A \Psi_0{}^B | U | \Psi_0{}^A \Psi_0{}^B \rangle + \sum_{I,J} \left[\frac{|\langle \Psi_0{}^A \Psi_0{}^B | U | \Psi_I{}^A \Psi_0{}^B \rangle|^2}{E_0{}^A - E_I{}^A} \right.$$

$$\left. + \frac{|\langle \Psi_0{}^A \Psi_0{}^B | U | \Psi_0{}^A \Psi_J{}^B \rangle|^2}{E_0{}^B - E_J{}^B} + \frac{|\langle \Psi_0{}^A \Psi_0{}^B | U | \Psi_I{}^A \Psi_J{}^B \rangle|^2}{E_0{}^A + E_0{}^B - (E_I{}^A + E_J{}^B)} \right]$$

Dans cette expression, $\Psi_0{}^A$ et $\Psi_0{}^B$ représentent respectivement les états fondamentaux de A et de B, d'énergie $E_0{}^A$ et $E_0{}^B$. Le terme du premier ordre correspond donc à l'interaction électrostatique entre les répartitions de charges de A et B dans leurs états fondamentaux.

Dans les termes au second ordre, $\Psi_I{}^A$ et $\Psi_J{}^B$ représentent les états excités de A et B d'énergie $E_I{}^A$ et $E_J{}^B$. Il est important de souligner que ces termes incluent, en toute rigueur, l'ensemble complet des états excités, et en particulier les excitations « lointaines » de A et B. Or, la possibilité d'excitations lointaines de la molécule A par exemple, signifie que la densité électronique peut devenir non-négligeable dans une région relativement éloignée du squelette de la molécule A. Si une autre molécule B est précisément dans le voisinage de A, une certaine densité électronique due à la molécule A pourra être suffisamment importante pour interagir avec B. Tout se passe alors comme si la molécule B avait capté une fraction de charge provenant de la molécule A, c'est-à-dire comme si un transfert de charge partiel s'effectuait entre A et B.

Ainsi l'expression théorique de l'énergie d'interaction à l'état fondamental, d'un complexe intermoléculaire, inclut sous sa forme la plus générale, des contributions de type transfert de charge (de poids variable selon les cas) et rien ne permet de distinguer formellement ces contributions parmi l'ensemble des interactions.

On peut signaler d'ailleurs que les interactions entre deux molécules voisines, désignées sous le nom d'interactions de Van der Waals-London, devraient correspondre, en toute rigueur, à l'ensemble complet de ces contributions, y compris celles du type transfert de charge.

Mais cet aspect théorique, rigoureux de l'expression des interactions moléculaires, est un peu différent de l'aspect pratique sous lequel on traite actuellement ce problème. Il est impossible, en effet, de considérer réellement l'ensemble complet des états excités possibles. On se limite pratiquement à un choix de configurations appropriées au cas étudié. Cela revient à dire qu'on utilise une « *base incomplète.* » Dans ces conditions, mais dans ces conditions seulement, on fait apparaître dans les termes du second ordre un certain nombre de contributions distinctes, auxquelles on peut attribuer des significa-tions physiques particulières.*

C'est ainsi que, pour un complexe au sens le plus général, c'est-à-dire où l'on veut tenir compte des contributions dues aux transferts de charge entre les partenaires, on peut distinguer, pratiquement, parmi les termes du second ordre :

—des termes de polarisation électrique de chacune des molécules par l'autre, du type :

$$\sum_{i,j*} \frac{|\langle \Psi_0{}^A \Psi_0{}^B | U | \Psi_{i \to j*}^A \Psi_0{}^B \rangle|^2}{E_0{}^A - E_{i \to j*}^A}$$

(Il existe évidemment le terme symétrique.) L'état excité $\Psi_{i \to j*}^A$ d'énergie $E_{i \to j*}^A$, correspond à l'excitation électronique de l'orbitale occupée i de A vers l'orbitale vide j de la *même* molécule.

—un terme de dispersion :

$$\sum_{i,j*,i',j'*} \frac{|\langle \Psi_0{}^A \Psi_0{}^B | U | \Psi_{i \to j*}^A \Psi_{i' \to j'*}^B \rangle|^2}{E_0{}^A + E_0{}^B - (E_{i \to j*}^A + E_{i' \to j'*}^B)}$$

où interviennent simultanément les états excités $\Psi_{i \to j*}^A$ et $\Psi_{i' \to j'*}^B$ de chacune des deux molécules.

—un terme supplémentaire, devant représenter plus spécifiquement les interactions entre les formes ioniques, c'est-à-dire les contributions propres aux transferts de charge entre les partenaires. (Ce terme correspond, approxi-mativement, aux «excitations lointaines» qui ne figurent pas dans les deux précédentes sommations, pour lesquelles la base d'états excités est *incomplète*.) Ce terme est de la forme

$$\sum_{i,j'*} \frac{|\langle \Psi_0{}^A \Psi_0{}^B | U | \Psi_i^{A+} \Psi_{j'*}^{B-} \rangle|^2}{E_0{}^A + E_0{}^B - (E_i^{A+} + E_{j'*}^{B-})}$$

et il correspond, par exemple, aux transferts électroniques des orbitales occu-pées i de A vers les orbitales vides $j'*$ de B.

Comme on ne tient pas compte de cette dernière contribution, dans les évaluations actuelles des énergies intermoléculaires, on appelle communément *interactions de Van der Waals-London* l'ensemble des autres contributions,

---

* En fait, dans notre actuel procédé de calcul des interactions, nous introduisons les polarisabilités de liaison expérimentales. On pourrait donc considérer que nous tenons compte de l'ensemble des états excités possibles (y compris les «excitations lointaines»), c'est-à-dire que la base utilisée forme une base complète. Mais le fait que chaque liaison soit simplement assimilée à un dipôle électrique entraîne que les excitations lointaines ne sont pratiquement pas représentées.

c'est-à-dire les interactions électrostatique, de polarisation et de dispersion. Mais en réalité il serait plus correct de les appeler *interactions de Van der Waals-London approchées*, puisqu'elles n'incluent pas tous les types d'interactions possibles.

Ainsi, on voit combien il serait artificiel d'établir une distinction fondamentale entre la nature des forces d'interaction dans les complexes avec ou sans transfert de charge.

Ce premier point établi, on peut, en outre, prévoir que, même dans des complexes avec transfert de charge, la contribution apportée par les formes ioniques (c'est-à-dire le terme spécifique du transfert de charge) à l'énergie du complexe dans son état fondamental, doit avoir, dans la généralité des cas, un « poids » relativement beaucoup plus faible que les autres termes. Cela peut se montrer en développant les expressions contenues dans les numérateurs des différents termes et en comparant les ordres de grandeur des types d'intégrales auxquelles on aboutit.

Toutefois, dans certains cas particuliers de transfert de charge, la situation est certainement différente. Il s'agit en premier lieu des cas où il y a formation de deux ions véritables A^+ et B^- (Anex et Hill, 1966); en solution ces ions sont d'ailleurs vraisemblablement solvatés. Dans d'autres cas, on est sans doute en présence d'associations à caractère fortement ionique, pour lesquelles on peut avoir un moment dipolaire très important, et dans lesquelles les partenaires sont à des distances très courtes, nettement inférieures à la somme des rayons de Van der Waals (c'est le cas, par exemple, des complexes formés entre certains composés aminés et des halogènes) (Malrieu, 1967).

Pour ces cas particuliers, il est probable soit que les contributions des formes ioniques aient un poids important, soit même qu'on doive renoncer à traiter les interactions dans ces complexes par une méthode de perturbation.

Pour tous les autres cas de complexes avec transfert de charge, il est prévisible que le terme spécifique du transfert de charge apporte à l'énergie totale, donc à la stabilité expérimentale du complexe, une contribution beaucoup plus faible que l'ensemble des autres termes (interactions électrostatique, de polarisation et de dispersion appelées communément interactions de Van der Waals-London).

Ce point mérite d'être souligné car il semble que beaucoup d'auteurs aient négligé ou sous-estimé le rôle que devaient jouer les interactions de Van der Waals-London dans les complexes avec transfert de charge.

Peut-être cela est-il dû au fait qu'on cherche généralement à relier la stabilité expérimentale des complexes avec les grandeurs relatives à la bande de transfert de charge (énergie d'excitation hv ou longueur d'onde λ). Cela revient au point de vue théorique à relier l'énergie de stabilisation au terme spécifique des transferts de charge; en effet, les quantités $E_0^A + E_0^B - (E_i^{A^+} + E_{j'*}^{B^-})$ qui figurent au dénominateur de ce terme peuvent être ramenées, en première approximation, à l' *énergie d'excitation hv* de la bande de transfert de charge. Dans beaucoup de cas, ces relations apparaissent comme très peu satisfaisantes. Néanmoins, il peut arriver pour des complexes formés entre un accepteur constant et des composés d'une même famille, que la stabilité varie approximativement en raison inverse de l'énergie hv de la bande caractéristique. Cela peut provenir, en effet, d'une variation parallèle du terme d'interaction spécifique du transfert de charge et des autres termes du second ordre (polarisation et dispersion). L'énergie d'excitation hv suit en général, dans une famille de complexes, les variations du potentiel d'ionisation du donneur. Dans certains groupes de composés apparentés, ces potentiels d'ionisation peuvent refléter, dans une certaine mesure les énergies de transition. Or les termes d'interaction de polarisation et de dispersion dépendant fortement des énergies de transition qui figurent à leur dénominateur. Il peut en résulter, si toutefois les numérateurs de tous ces termes ne varient pas trop, que le terme spécifique du transfert de charge varie dans le même sens que les autres termes au second ordre. Cela se traduira expérimentalement par l'existence d'un certain parallélisme entre la stabilité du complexe et l'inverse de l'énergie d'excitation hv.

Mais, même dans le cas où une telle relation est obtenue, il serait injustifié de conclure sans autre vérification à la prépondérance du terme de transfert de charge par rapport à l'ensemble des autres interactions. D'autre part, il n'y a aucune raison de croire à la généralité de semblables relations.

Cette dernière conception a été récemment exposée par Dewar et Thompson (1966) qui ont également pressenti le rôle important que devaient jouer les forces de Van der Waals-London dans les complexes avec transfert de charge. Ces auteurs ont étudié une série de complexes particulièrement caractéristiques formés entre le tetracyanoéthylène (TCNE) et une famille d'hydrocarbures aromatiques. Ils ont montré qu'ils n'obtenaient pas de relation de linéarité entre la longueur d'onde de la bande et la stabilité expérimentale de ces complexes (ou plus précisément leur énergie libre $\Delta F = -RT \log K$, où K est la constante d'association). Dewar et Thompson en concluent que le rôle des termes propres au transfert de charge n'est sans doute pas prépondérant.

Il restait à vérifier que les termes correspondant aux interactions de Van der Waals-London approchées, pouvaient rendre compte de la stabilité expérimentale de tels complexes avec transfert de charge.

III. Calcul des Interactions de Van der Waals-London

Nous disposons actuellement pour le calcul de ces interactions de méthodes de calcul, progressivement améliorées au cours de ces dernières années, en particulier dans notre laboratoire (Pullman, Claverie et Caillet, 1966;

Claverie et Rein, 1967; Rein, Claverie et Pollak, 1967; Claverie, 1967, ce volume).

Le calcul des énergies de Van der Waals-London comprend, sous sa forme actuelle, quatre contributions: l'interaction électrostatique E_{el} entre les charges nettes atomiques, l'énergie de polarisation électrique E_{pol}, l'énergie de dispersion E_{disp} et un terme de répulsion E_{rep}.

Les trois premières contributions sont calculées dans l'approximation dite « monopôles-polarisabilités de liaisons. »

Quant au terme de répulsion, son origine n'apparaît pas explicitement dans les expressions des termes d'interactions telles qu'elles sont écrites précédemment. En effet dans ces termes nous avons considéré les simples produits de fonctions d'onde. Or comme on travaille dans une base incomplète, il est nécessaire d'antisymétriser ces produits, tout au moins lorsque la distance entre les molécules devient très courte. Il apparaît alors des termes d'échange et de recouvrement (correspondant au premier ordre), donnant une répulsion entre les atomes. Actuellement ces termes de répulsion sont évalués d'après un ajustement de la formule semi-empirique de Kitaygorodski.

A. Complexes Formés entre le TCNE et une Famille d'Hydrocarbures Aromatiques

Nous avons commencé l'étude des interactions de Van der Waals-London par la famille de complexes observée précisément par Dewar et Thompson, c'est-à-dire formés entre le TCNE et un ensemble d'hydrocarbures aromatiques. Parmi ces derniers nous avons dû exclure, toutefois, ceux qui ne sont pas plans.

Pour chaque couple de molécules la configuration la plus favorable, correspondant au maximum d'interactions, est déterminée par un ensemble de rotations, et éventuellement translations, des deux molécules, l'une par rapport à l'autre, dans deux plans parallèles.

Pour ces configurations les plus favorables, les variations de l'énergie totale d'interaction dans cette série de complexes, en fonction de la distance entre les plans des deux molécules, sont représentées sur la Fig. 1. L'allure très satisfaisante des courbes obtenues justifie l'introduction des termes de répulsion. Le minimum d'énergie obtenu correspond à une distance d'équilibre très voisine de 3.4 Å.

Le Tableau I rassemble, pour les configurations les plus favorables et à la distance d'équilibre, les valeurs des quatre contributions calculées E_{el}, E_{pol}, E_{disp} et E_{rep}, de leur somme E_{tot} et de la valeur « corrigée » E_{corr}. Il s'agit là d'une correction faite sur la somme des deux contributions E_{pol} et E_{disp}, qui font intervenir les polarisabilités des liaisons des molécules. En effet, le procédé de calcul actuel ne tient pas compte du fait que, dans les cycles

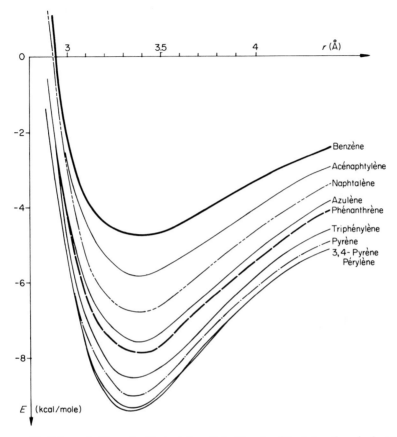

FIG. 1. Variations de l'énergie d'interaction E de Van der Waals-London, incluant les répulsions, en fonction de la distance intermoléculaire, pour les complexes formés entre le TCNE et une série d'hydrocarbures aromatiques.

aromatiques, les polarisabilités (ou plus précisément la contribution π de ces polarisabilités) varient en raison inverse des énergies d'excitation, c'est-à-dire dans ce cas, à peu près parallèlement aux potentiels d'ionisation. Ainsi les polarisabilités de liaisons des cycles aromatiques doivent être corrigées, relativement à celles du benzène, par un facteur multiplicatif égal au rapport des potentiels d'ionisation $I_{(\text{benzène})}/I$. C'est pourquoi nous avons multiplié les sommes $E_{\text{disp}} + E_{\text{pol}}$ par ce facteur.

Les valeurs des énergies d'interaction calculées sont comparées, sur le Tableau I, aux énergies libres expérimentales $\Delta F = -RT \log K$. En réalité, la grandeur thermodynamique correspondant aux énergies calculées est l'enthalpie ΔH, et non l'énergie libre. Comme on a la relation $\Delta F = \Delta H - T \Delta S$, la comparaison des variations de l'énergie libre avec celles de l'énergie

TABLEAU I

INTERACTIONS CALCULÉES DE VAN DER WAALS-LONDON ET ÉNERGIES
LIBRES EXPÉRIMENTALES DANS LES COMPLEXES FORMÉS ENTRE UNE SÉRIE
D'HYDROCARBURES AROMATIQUES ET LE TCNE

Complexe formé avec le TCNE		Energies d'interaction (kcal/mole)						ΔF (kcal/mole)
		E_{el}	E_{pol}	E_{disp}	E_{rep}	E_{tot}	E_{corr}	
Benzène		-2.47	-1.35	-3.20	2.23	-4.80	-4.80	+0.76 ± 0.12
Naphtalène		-2.94	-1.76	-4.03	2.70	-6.03	-6.76	-0.02 ± 0.06
Phénanthrène		-2.81	-2.18	-4.76	2.98	-6.75	-7.85	-0.46 ± 0.05
Pyrène		-3.09	-2.45	-5.32	3.62	-7.25	-8.94	-0.61 ± 0.12
Acénaphtylène		-2.96	-2.07	-4.33	2.96	-6.40	-7.63	-0.62 ± 0.04
Triphénylène		-2.37	-2.80	-5.84	3.55	-7.46	-8.57	-0.75 ± 0.07
Pérylène		-2.28	-2.71	-5.17	3.49	-6.67	-9.21	-0.83 ± 0.05
3,4-Benzopyrène		-2.90	-3.03	-6.49	5.28	-7.14	-9.37	-0.97 ± 0.07
Azulène		-3.16	-1.80	-3.77	2.90	-5.83	-7.13	-1.59 ± 0.03

calculée, dans une série de complexes, n'a de validité que si les variations
d'entropie ΔS ne sont pas trop irrégulières.

On constate l'existence d'un très net parallélisme entre les énergies calculées
et les énergies libres expérimentales, tout au moins pour la série des sept
hydrocarbures alternants. La Fig. 2 illustre la relation linéaire obtenue entre
ces deux quantités.

Ainsi les variations de l'énergie de Van der Waals-London dans cette
série de complexes rendent très bien compte des variations de leur stabilité
expérimentale, alors qu'aucune relation n'existait entre cette dernière grandeur
et la longueur d'onde de la bande de transfert de charge. Le rôle joué par les
forces de Van der Waals-London apparaît donc tout à fait prépondérant par
rapport à la contribution de type transfert de charge.

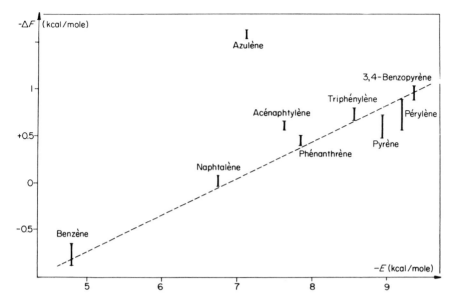

Fig. 2. Relation entre l'énergie libre expérimentale ΔF et l'énergie d'interaction calculée E pour les complexes formés entre le TCNE et une série d'hydrocarbures aromatiques.

Quant aux deux hydrocarbures non-alternants: acénaphtylène et azulène, les interactions calculées sont visiblement sous-estimées. Cela provient sans doute de plusieurs facteurs d'origine différente. Les uns sont liés au calcul lui-même. En effet la répartition des charges sur ces deux corps est fort mal connue, aucune méthode ne donnant un diagramme de charges capable de reproduire correctement le moment dipolaire expérimental. D'autre part, les transitions électroniques π–π^* de ces deux corps, et principalement de l'azulène, sont beaucoup plus basses que celles d'un hydrocarbure alternant qui aurait le même potentiel d'ionisation. Il en résulte que, dans notre procédé actuel de calcul, où interviennent non pas l'ensemble des énergies de transition comme il serait correct de le faire, mais les potentiels d'ionisation, les énergies calculées de polarisation et de dispersion sont considérablement sous-estimées, par rapport à un hydrocarbure alternant. Enfin, d'autres facteurs peuvent être dûs à l'effet du solvant, qui peut présenter une discontinuité pour ces deux corps polaires.

B. Complexes Formés entre le TCNE et une Famille de Composés à Hétérocycles Pentagonaux

Nous avons également calculé les interactions de Van der Waals-London dans une autre série de complexes avec transfert de charge, formés entre le TCNE et une famille d'hétérocycles pentagonaux: furanne, thiophène, pyrrole, et leurs homologues benzèniques. Cooper et coll. (1966) ont en effet mesuré les constantes d'association K de ces complexes.

Nous avons utilisé pour ces composés hétérocycliques les répartitions des charges σ et π obtenues récemment dans notre laboratoire, qui correspondent à des moments dipolaires en bon accord avec les valeurs expérimentales (Berthod et Pullman, 1965, 1966).

L'étude de la variation de l'énergie totale d'interaction (y compris le terme de répulsion), en fonction de la distance entre les plans des molécules indique que la distance d'équilibre est, ici également, voisine de 3.4 Å.

A cette distance et pour les configurations les plus favorables, les valeurs des quatre contributions calculées, de leur somme, et de la valeur «corrigée,» de façon semblable au cas des hydrocarbures, sont rassemblées dans le Tableau II.

TABLEAU II

INTERACTIONS CALCULÉES DE VAN DER WAALS-LONDON ET ÉNERGIES LIBRES EXPÉRIMENTALES DANS LES COMPLEXES FORMÉS ENTRE UNE SÉRIE DE COMPOSÉS À HÉTÉROCYCLE PENTAGONAL ET LE TCNE

Complexe formé avec le TCNE	Energies d'interaction (kcal/mole)						ΔF (kcal/mole)
	$E_{él}$	E_{pol}	E_{disp}	E_{rep}	E_{tot}	E_{corr}	
Furanne	−2.23	−1.29	−2.45	1.73	−4.24	−4.48	0.73±0.02
Thiophène	−2.10	−1.91	−3.53	2.24	−5.30	−5.33	0.43±0.02
Dibenzofuranne	−1.29	−2.00	−4.13	2.64	−4.78	−5.32	−0.03±0.02
Benzofuranne	−2.13	−1.76	−3.64	2.48	−5.05	−5.46	−0.04±0.02
Benzothiophène	−1.77	−2.22	−4.47	3.04	−5.42	−6.02	−0.04±0.02
Dibenzothiophène	−0.82	−2.61	−5.02	3.21	−5.24	−6.27	−0.17±0.02
Pyrrole	−3.20	−1.71	−2.80	1.87	−5.84	−6.38	−
Indole	−2.65	−2.05	−3.82	2.65	−5.87	−6.70	−0.92±0.03
Carbazole	−2.59	−2.14	−4.33	3.08	−5.98	−6.98	−0.96±0.02

On constate l'existence d'un parallélisme entre l'énergie d'interaction calculée et l'énergie libre expérimentale, parallélisme moins bon toutefois que dans le cas des hydrocarbures alternants. Sur la Fig. 3, qui représente les variations de ces deux grandeurs, on constate en effet une certaine dispersion des points de part et d'autre de la droite.

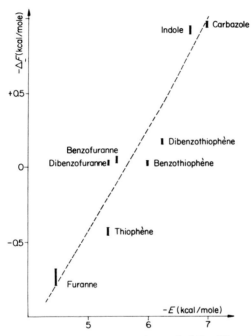

Fig. 3. Relation entre l'énergie libre expérimentale ΔF et l'énergie d'interaction calculée E pour les complexes formés entre le TCNE et une série de composés à hétérocycle pentagonal.

Il est possible que cette dispersion provienne en partie du calcul lui-même, qui comporte évidemment une marge d'incertitude un peu plus grande que dans le cas d'hydrocarbures. Mais on peut penser que, pour cet ensemble de corps de taille et de polarité différentes, les variations de l'entropie, sont beaucoup moins régulières que dans le cas des hydrocarbures aromatiques alternants: alors, les variations de l'enthalpie peuvent ne pas suivre exactement celles de l'énergie libre. Comme les énergies d'interaction calculées correspondent à l'enthalpie, il serait ici particulièrement important de connaître les valeurs expérimentales de cette dernière quantité.

Or nous disposons des valeurs de l'enthalpie pour trois complexes de la série considérée: Cooper et coll. ont pu déterminer ces valeurs pour le furanne et le thiophène, et une valeur pour le complexe indole-TCNE était connue antérieurement (Foster et Hanson, 1965). Si nous comparons ces valeurs de l'enthalpie avec celles des énergies calculées, les points obtenus (Fig. 4) sont bien situés sur une droite, passant par l'origine, et on remarque que la dispersion qui existait sur la Fig. 3 disparaît complètement. Ainsi, bien qu'il soit difficile de conclure avec certitude puisque nous n'avons que trois points, il apparaît cependant probable que la dispersion constatée sur la Fig. 3 provienne éssentiellement de la variation du terme d'entropie, et qu'une bien meilleure relation serait obtenue entre l'enthalpie et les interactions calculées de Van der Waals-London.

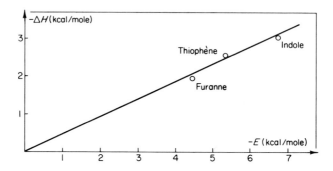

FIG. 4. Relation entre l'enthalpie expérimentale ΔH et l'énergie d'interaction calculée E pour les complexes formés entre le TCNE et les trois composés: furanne, thiophène et indole.

D'autre part, nous voyons sur la Fig. 4 que les valeurs numériques des énergies calculées sont d'un ordre de grandeur voisin du double des valeurs expérimentales. Cela n'a rien de surprenant: les énergies calculées correspondent en effet aux interactions dans le *vide* (on ne tient aucunement compte du solvant).

De ces résultats on peut dégager une remarque d'intérêt biochimique. En effet, parmi les trois familles de composés hétérocycliques étudiés, c'est-à-dire dans lesquelles l'hétéroatome est respectivement l'oxygène, le soufre et l'azote, ce sont les composés azotés qui donnent les interactions les plus fortes. (La constante K du complexe pyrrole-TCNE n'a pas pu être mesurée, mais nous avons calculé son énergie d'interaction et la valeur obtenue se place entre le dibenzothiophène et l'indole.) La forte tendance à s'associer que possèdent donc les composés azotés par rapport à leurs analogues oxygénés ou soufrés n'est probablement pas sans relation avec le rôle privilégié qu'ils jouent en biologie: l'indole en particulier intervient dans de nombreuses associations biochimiques.

C. Associations Observées entre la Flavine et une Série de Purines

Enfin nous avons étudié un troisième groupe de complexes, de nature biochimique et dans lesquels il n'est pas certain qu'il se produise réellement un transfert de charge. Il s'agit des associations formées entre la riboflavine et les bases puriques, associations dont Tsibris et coll. (1965) ont récemment pu mesurer les constantes d'association.

L'existence d'un complexe interne dans le flavine-adénine-dinucléotide (FAD), entre l'adénine et la 6,7-diméthylisoalloxazine est connue depuis plusieurs années (Weber, 1950); cette association provoque en effet une extinction de la fluorescence de la flavine. Plus récemment furent étudiés les analogues de ce dinucléotide dans lesquels l'adénine était remplacée par des composés tels que les purines, les pyrimidines ou les dérivés de l'indole (Tsibris et coll., 1965; Slifkin, 1965; Wilson, 1966). L'existence d'un transfert de charge véritable dans ces complexes internes est fort discutée. En fait il n'apparaît jamais de bande nouvelle caractéristique dans le spectre des constituants. Cependant l'allure du spectre est légèrement modifiée (surtout pour les composés indoliques).

Nous avons calculé les interactions de Van der Waals-London entre la 6,7-diméthylisoalloxazine et un certain nombre de purines. Mais le calcul de ces interactions comporte ici une assez grande part d'incertitude. On ne dispose d'aucun moyen susceptible de tester la validité des répartitions de charge pour ces composés, puisqu'on ne connait pas les moments dipolaires expérimentaux. D'autre part, la taille et la complexité de ces molécules rendent fort difficile la détermination du minimum d'énergie, c'est-à-dire de la configuration la plus favorable (le nombre de rotations et translations nécessaires devient très grand).

Par ailleurs, il est assez hasardeux de comparer les énergies d'interaction calculées avec les stabilités expérimentales de ces dinucléotides, car la présence de la chaîne sucre-phosphate qui unit la flavine et la base purique doit certainement modifier les interactions entre les deux constituants, et peut-être favoriser certains types de configurations.

Le Tableau III indique les valeurs obtenues des énergies d'interaction calculées, et celles des énergies libres expérimentales obtenues par Tsibris et coll. (1965).

On constate que, grosso modo, les interactions obtenues sont d'autant plus fortes que le dinucléotide est plus stable, mais la corrélation n'est pas parfaite, comme on pouvait le craindre.

Tsibris et ses collaborateurs ont proposé, pour le FAD, une configuration représentée sur la Fig. 5a analogue à celle que Karreman (1962) avait suggérée pour le complexe flavine-indole. Le choix de cette configuration est basé sur la complémentarité de charges π entre les atomes des régions N_1–N_{10} de la

TABLEAU III

INTERACTIONS CALCULÉES DE VAN DER WAALS-LONDON ET ÉNERGIES LIBRES
EXPÉRIMENTALES DANS LES ASSOCIATIONS INTRAMOLÉCULAIRES ENTRE LA
RIBOFLAVINE ET LES PURINES

Complexe formé avec la 6,7-diméthylisoalloxazine	Energies d'interaction (kcal/mole)					Energies libres ΔF (kcal/mole)
	E_{el}	E_{pol}	E_{disp}	E_{rep}	E_{tot}	
Purine	−3.44	−0.86	−5.85	3.11	−7.04	1.02
6-Méthylpurine	−5.09	−1.01	−6.00	3.61	−8.49	0.42
Hypoxanthine	−5.46	−1.14	−5.41	2.69	−9.32	−0.89
2-Aminopurine	−4.58	−0.99	−5.42	2.52	−8.46	−1.12
2,6-Diaminopurine	−4.63	−1.17	−5.91	2.72	−9.00	−1.13
Adénine	−4.77	−1.05	−5.53	2.52	−8.83	−1.22
Xanthine	−6.27	−1.47	−4.78	2.20	−10.31	−1.58
Guanine	−7.13	−1.43	−5.35	2.66	−11.25	−1.63

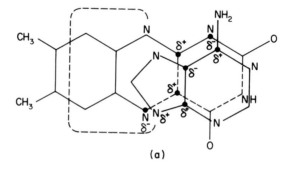

(a)

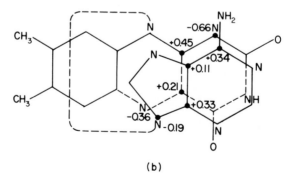

(b)

FIG. 5. (a) Configuration du FAD proposée par Tsibris *et al.* (1965). (b) Valeurs des charges nettes totales $\sigma + \pi$.

flavine et C_6–N_9 de l'adénine. Mais on peut remarquer que, dans l'interaction électrostatique entre deux molécules il est plus correct de considérer l'ensemble des charges nettes σ et π, et on voit sur la Fig. 5b qu'alors la complémentarité de charges disparaît; et surtout on ne doit pas considérer une région isolée des molécules, mais l'*ensemble* des atomes: en réalité, avec les répartitions de charges utilisées, cette configuration correspond à une énergie électrostatique répulsive (les moments dipolaires calculés sont en effet presque parallèles et de même sens) et, par suite, à une énergie totale d'interaction assez faible.

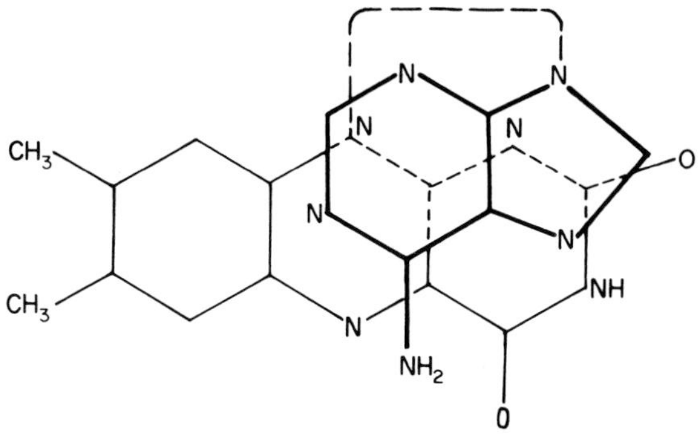

FIG. 6. Configuration correspondant à un maximum d'interactions calculées de Van der Waals-London entre la 6,7-diméthylisoalloxazine et l'adénine.

Le type de configuration pour lequel nous obtenons un maximum d'interactions entre la 6,7-diméthylisoalloxazine et l'adénine est indiqué sur la Fig. 6. Il est bien évident, en raison des diverses causes d'incertitude précédemment citées, qu'on ne peut pas être assuré que cette configuration correspond bien à la réalité.

IV. Conclusion

Nous espérons avoir mis en évidence, sur ces différents exemples, la contribution apportée à l'énergie de stabilisation des complexes avec transfert de charge, par les termes d'interaction électrostatique, de polarisation et de dispersion—contribution certainement *prépondérante* par rapport au terme spécifique du transfert de charge.

Cette étude a pu ainsi contribuer à élucider partiellement le problème de la nature des forces qui assurent la stabilité des complexes avec transfert de charge, tout au moins pour la catégorie de ces complexes où il est justifié d'étudier leurs interactions par un traitement de perturbation.

REMERCIEMENTS

L'auteur remercie le Professeur Pullman qui lui a suggéré l'idée de ce travail et MM. Pierre Claverie et Jean-Paul Malrieu pour toutes les remarques et discussions dont ils l'ont fait bénéficier.

BIBLIOGRAPHIE

Anex, B. G., et Hill, E. B. (1966). *J. Am. Chem. Soc.* **88**, 3648.
Berthod, H., et Pullman, A. (1965). *J. Chim. Phys.* p. 942.
Berthod, H., et Pullman, A. (1966). *Compt. Rend.* **262**, 76.
Claverie, P., et Rein, R. (1967). *Intern. J. Quantum Chem.* sous presse,
Cooper, A. R., Crowne, C. W. P., et Farrell, P. G. (1966). *Trans. Faraday Soc.* **62**, 18.
Dewar, M. J. S., et Thompson, C. C. (1966). *Tetrahedron Suppl.* **7**, 37.
Foster, R., et Hanson, P. (1965). *Tetrahedron* **21**, 255.
Karreman, G. (1962). *Ann. N.Y. Acad. Sci.* **96**, 1029.
Malrieu, J. P. (1967). Deuxième sujet de thèse, Paris.
Mulliken, R. S. (1952). *J. Am. Chem. Soc.* **84**, 811.
Pullman, B., Claverie, P., et Caillet, J. (1966). *Proc. Natl. Acad. Sci. U.S.A.* **55**, 905.
Rein, R., Claverie, P., et Pollak, M. (1967). Article soumis à *Intern. J. Quantum Chem.*
Slifkin, M. A. (1965). *Biochim. Biophys. Acta* **103**, 365.
Tsibris, J. C. M., McCormick, D. B., et Wright, L. D. (1965). *Biochemistry* **4**, 504.
Weber, G. (1950). *Biochem. J.* **47**, 114.
Wilson, J. E. (1966). *Biochemistry* **5**, 1351.

Solvent Effects on Molecular Associations

OKTAY SINANOĞLU*

*Sterling Chemistry Laboratory and
Department of Molecular Biophysics
Yale University
New Haven, Connecticut*

I. Introduction

Molecular associations of the type

$$A + B \rightleftarrows (AB) \tag{1}$$

or isomerizations or conformation changes

$$A \rightleftarrows C \tag{2}$$

are affected by solvents. The association may be chemical with (AB) a product as in the dimerization of cyclopentadiene (1),

$$\tag{3}$$

or physical with (AB) a "complex" held together by van der Waals, electrostatic, charge transfer, and "solvophobic" type forces. Equation (2) may also be a chemical rearrangement or a conformation change which may involve an intramolecular association, a "stacking," or a "folding up." Presumably TpT [thymidylyl-(3′ → 5′)-thymidine] exhibits such an equilibrium between an open, "linear," and a folded form in solution.

We shall be concerned with solvent effects† on equilibrium constants, K, but the same considerations apply to the rates of associations where reactions are not diffusion controlled and where we can invoke an activated complex and treat it as in quasi-equilibrium with the reactants, e.g.,

$$A + B \underset{}{\overset{K^{\ddagger}}{\rightleftarrows}} (AB)^{\ddagger} \overset{k}{\rightarrow} D, \qquad k \propto K^{\ddagger} \tag{4}$$

$$A \underset{K^{\ddagger}}{\rightleftarrows} (C) \overset{}{\underset{k}{\rightarrow}} C$$

Equations (1) and (2) may refer to biopolymers as well, in which case water is an especially favorable solvent mainly due to a difference in degree rather

---

* Alfred P. Sloan Fellow.

† The solvent is presumed not to enter the reaction, nor to catalyze it; it should act merely as a physical environment.

than kind when compared to other solvents (2). The DNA double helix is stable in water, but not in other solvents.* Protein side groups tend to associate in an aqueous medium. An analysis of solvent effects is also revealing with regard to what is involved in such "hydrophobic" phenomena.

Finally, a study of solvent effects combined with experiment would yield information on the intrinsic properties of the reaction itself such as the forces between A and B alone in (AB) with no solvent around.

II. Theory

A. Definition of Solvent Effect

Consider reaction Eqs. (1) and (2) for molecules first, rather than polymers. For Eq. (1) in solution we have

$$A + B \rightleftarrows (AB); \qquad \Delta F = 0 \tag{5}$$
$$x_A \quad x_B \quad x_{AB}$$

The x_i is the mole fraction of species $i \equiv A$, B, or (AB) in solution. Each species is assumed to be in the Henry's Law region so that the standard states are $x_i{}^0 \equiv 1$, a hypothetical state of a molecule i in solution. Then

$$F_i = F_i{}^0 + RT \ln x_i \tag{6}$$

and since $\Delta F = 0$, the standard free energy change $\Delta F^0 = F^0_{(AB)} - F_A{}^0 - F_B{}^0$ equals

$$\Delta F^0 = -RT \ln K_{\text{soln.}} \tag{7}$$

$$K_{\text{soln.}} = \frac{x_{(AB)}}{x_A x_B} \tag{8}$$

The standard free energy change with these standard states is also called the "unitary free energy change" (3).

$$\Delta F^0 \equiv \Delta F^0_{\text{(unitary)}} \tag{9}$$

In the gas phase (even where the reaction would be hypothetical),

$$A + B \rightleftarrows (AB) \qquad \text{(standard states: } p_i{}^0 \equiv 1 \text{ atm ideal gas)} \tag{10}$$
$$p_A \quad p_B \quad p_{(AB)}$$

$$F_{i(\text{gas})} = F^0_{i(\text{gas})} + RT \ln p_i$$

$$\Delta F^0_{\text{gas}} = -RT \ln K_{\text{gas}}; \qquad K_{\text{gas}} = \frac{p_{(AB)}}{p_A p_B}$$

---

* The reader is referred to Sinanoğlu and Abdulnur (2) for a detailed study of the role of water in the DNA double-helix case with a semiquantitative theory.

Subtracting Eq. (10) from Eq. (5), we have the definition of solvent effect:

$$\Delta F^0_{\text{solv. eff.}} \equiv \Delta F^0_{\text{soln.}} - \Delta F^0_{\text{gas}} = RT \ln \frac{k^H_{(AB)}}{k_A^H k_B^H} \tag{11}$$

where k_i^H are Henry's Law constants, e.g.,

$$p_A = k_A^H x_A \tag{12}$$

The right-hand side of Eq. (11) is a constant independent of concentrations, dependent only on T. This is true in the Henry's Law region and with smaller molecules. With polymers, the concentration effects are not given simply by Eq. (6) even in dilute solution (2, 4).

B. Single Solute in Solution

Consider the equilibrium

$$A \text{ (soln.)} \rightleftarrows A \text{ (gas)}; \qquad k_A^H = \frac{p_A}{x_A} \tag{13}$$

$$\overset{x_A}{} \qquad \overset{p_A}{}$$

$$\Delta F_A^0 = F_A^0(\text{gas}; p_A^0 \equiv 1) - F_A^0(\text{soln.}; x_A^0 \equiv 1) = -RT \ln k_A^H$$

in the region where Eqs. (5)–(11) apply. The reverse process of taking a solute molecule A from the gas at $p_A^0 \equiv 1$ atm to a fixed state in solution ($x_A^0 \equiv 1$; "unitary" state) may be imagined as taking place in two hypothetical steps. This is a mathematical separation. In the first step a hole is prepared in the liquid. In step 2, A is placed in the hole. The molecule A interacts with its new environment. The ΔF_A^0 in Eq. (13) is thus given by

$$\Delta F_A^0 \approx -F^A_{\text{int.}} - F_c^A - RT \ln \frac{kT/p_0}{v_1} \tag{14}$$

The F_c^A is the free energy required to make the cavity with the appropriate walls in the liquid; $F^A_{\text{int.}}$ is the free energy of interaction of A with its environment in step 2. The last term is an entropy effect, in a rough approximation taking care of the "free volume" of A in liquid l whose average molecular volume is v_1 (5). The p_0 is 1 atm. From Eqs. (14) and (13), also

$$k_A^H \approx \frac{kT}{v_1} \exp\left(F_c^A + F^A_{\text{int.}}/RT\right) \tag{15}$$

Equation (14) may be separated into its enthalpy and entropy parts:

$$\Delta H_A^0 \approx -e^A_{\text{int.}} - e_c^A + RT(1 - T\mathscr{A}_1) \tag{16}$$

$$\Delta S_A^0 \approx -s^A_{\text{int.}} - s_c^A + R \ln \frac{kT}{v_1} + R(1 - T\mathscr{A}_1) \tag{17}$$

where

$$\mathscr{A}_1 \equiv \left(\frac{1}{v_1} \frac{\partial v_1}{\partial T} \right)_p$$

is the coefficient of thermal expansion of solvent. In Eq. (16), $e_{int.}^A$ and e_c^A are energies of interaction and cavity formation. Here a term of the order of $p^0 v_1$ has been omitted since it is usually negligible.

1. The Cavity Term

The energy required to make a cavity that will accommodate an A molecule comes from the surface energy of such a cavity. It takes work to separate the solvent molecules from one another. The surface that is created is of microscopic dimensions* and it is highly curved and not flat. The energy part of the macroscopic, ordinary surface tension is given by

$$\gamma_1 \left[1 - \frac{\partial \ln \gamma_1}{\partial \ln T} \right] \tag{18}$$

Based on thermodynamic formulas (7) the curvature correction at molecular dimensions does not appear appreciable (8). These formulas, however, most likely underestimate the effect. More quantitatively the microscopic cavity properties may be extracted (9) for use in solvent effect theory from the thermodynamic properties of pure liquids and dilute solutions. We have

$$e_c^A = \kappa_1^e (\varphi_{1A}^{-1/3}) \times 4.836 v_A^{2/3} \gamma_1 \left[1 - \frac{\partial \ln \gamma_1}{\partial \ln T} - \frac{2}{3} \mathscr{A}_A T \right] \tag{19}$$

v_A is the average molecular volume of pure A in liquid form (at the same reduced temperature as the solvent) and $\mathscr{A}_A$ its liquid coefficient of thermal expansion. If A does not exist as a liquid, v_A is estimated from density of similar model compounds or from molecular models. γ_1 is the pure solvent ordinary surface tension; $\kappa_1^e (\varphi_{1A}^{-1/3})$ is a constant dependent on the volume fraction $\varphi_{1A} = v_1 / v_A$. For large A in a solvent of much smaller molecules, κ_1^e should approach 1.0. It is about 0.6 for pure nonpolar liquids and somewhat above 1.0 for water and other polar liquids (9) (i.e., for single molecule of liquid in a cavity of its own kind).

---

* For theoretical discussions of surface tension down to microscopic dimensions we shall refer the reader to the writings of Buff (6) and other papers.

For F_c^A itself, we have

$$F_c^A = \kappa_1^e(\varphi_{1A}^{-1/3}) \times 4.836 v_A^{2/3} \gamma_1 (1 - W_{1A}) \tag{20a}$$

$$W_{1A} \equiv (1 - \eta_{1A})\left(\frac{\partial \ln \gamma_1}{\partial \ln T} + \frac{2}{3}\mathscr{A}_A T\right) \tag{20b}$$

where

$$\eta_{1A} = \frac{\kappa_1^s(\varphi_{1A}^{-1/3})}{\kappa_1^e(\varphi_{1A}^{-1/3})} \tag{20c}$$

The κ_1^s is a constant similar to that in Eq. (19), but occurring in S_c^A instead. Thus it measures the deviation of the cavity entropy from the macroscopic surface entropy. Especially for polar liquids κ_1^s, in general, will differ from κ_1^e, but if $\kappa_1^s \approx \kappa_1^e$, then $\eta_{1A} \approx 1$ and W_{1A} vanishes.

2. The Interaction Term

Consider now the interaction of a solute molecule like benzene or the side group of a biopolymer such as a purine base of a single polynucleotide strand interacting with the solvent around it. The interaction will contain a van der Waals (vdw) part which includes all the nonbonded, nonelectrostatic attractions *and* repulsions and also an "electrostatic" part by which we mean solute-solvent interactions involving electrostatic charge distributions including various multipole-multipole and multipole-induced multipole forces. Thus*

$$F_{\text{int.}}^A = F_{\text{vdw}}^A + F_{\text{e.s.}}^A \tag{21}$$

The F_{vdw}^A results from the interaction of A in its cavity with an average distribution of all the solvent molecules. It may be obtained by integrating an effective intermolecular (solute-solvent molecule) potential function $V_{1A}^{\text{eff}}(R)$ over such a distribution. We shall not go into the details of this procedure, but simply indicate the nature of $V_{1A}^{\text{eff}}(R)$ and the form of the result.

$V_{1A}^{\text{eff}}(R)$ is an effective pair potential since the dispersion forces between two molecules are affected by as much as 10–40% when the two molecules are surrounded by a liquid environment (10). A detailed theory of such effective potentials dependent on the relative sizes of solute-solvent molecules, density of the medium, etc., has recently been presented with a number of examples such as the CCl_4—CCl_4 potential in liquid benzene (11). For quasi-spherical

---

* There is also a smaller term due to induced moments in the solute for nonpolar solutes in polar liquids.

molecules or side groups, such as C_6H_6, CCl_4, $—C_2H_5$, purines, pyrimidines, etc., the nonpolar parts of the potential, $V_{1A}^{eff}(R)$, are

$$V_{1A}^{eff}(R) \approx V_{A1}^K(R) \times B_{A11} \qquad (22)$$

with $V_{A1}^K(R)$ a gas-phase two-molecule semiempirical Kihara potential, and B_{A11} the correction factor due to the liquid environment (11). The detailed form of Eq. (22) is shown in the Appendix. Integration over a first discrete solvation layer, followed by a solvent continuum around A, yields

$$F_{vdw}^A \approx -f(\varphi_{1A}, l_1)\Delta_{1,A}D_1 D_A \bar{B}_{1A} \qquad (23)$$

where

$$\Delta_{1,A} = \mu \frac{I_1 I_A}{I_1 + I_A}; \qquad 0 < \mu < 2, \quad \mu \approx 1.4 \qquad (23a)$$

I_1, I_A are the ionization potentials of solvent molecule and solute molecule.

Since ionization potentials of molecules of interest here range between 9–12 eV,

$$\Delta_{1,A} \approx I_1 \mp 20\%$$

D_1 and D_A are the Clausius-Mosotti functions for pure solvent and pure liquid A, respectively:

$$D_i \equiv \frac{n_i^2 - 1}{n_i^2 + 2} \qquad (23b)$$

where n_i is the refractive index of pure liquid i.

$$\bar{B}_{1A} \approx 1 - \tfrac{1}{2} \times \tfrac{3}{4} \times D_1 \times \bar{L}_{1A} \qquad (23c)$$

with $\bar{L}_{1A} \approx 2$ (cf. Appendix).

Finally, the $f(\varphi_{1A}, l_1)$ is a dimensionless function of the relative solute-solvent sizes ($\varphi_{1A} \equiv v_1/v_A$) and of l_1, the solvent molecular "core size." The latter may be estimated from macroscopic solvent properties such as the "acentric factor" (12) as discussed in the Appendix.

D_1, D_A vary little for most solvent and solute systems of interest here. Thus, the remarkable thing about Eq. (23) is that aside from the geometric factor f, it is roughly constant for a variety of solutes and solvents. The factor f will be about the same for solutes and solvents of comparable molecular sizes. In general, its dependence on v_1/v_A and l_1 may have to be taken into account however. This is a point that requires further investigation.

The "electrostatic" part of $F_{int.}^A$, i.e., $F_{e.s.}^A$ in Eq. (21) depends, of course, more directly on the nature of the solute. A given solute such as a guanine base has a charge distribution (giving rise to sizable dipole and higher multipoles) which interacts with the solvent dielectric and through the reaction fields it induces. A quantitative calculation of $F_{\text{e.s.}}^{A}$ would be very difficult. The dipole terms alone would not suffice for the sizable solutes one has, and the macroscopic static solvent dielectric constant ε_s is not entirely appropriate. Near the solute, the solvent dielectric exhibits saturation effects with a lower effective dielectric constant (in polar solvents like water)—an effect difficult to estimate. One may nevertheless see the general behavior $F_{\text{e.s.}}^{A}$ with respect to changes in solute and solvent properties, for example, from the Onsager reaction field (13). According to this

$$F_{\text{e.s.}}^{A} \approx -\frac{1}{2}\left(\frac{\mu_A^2}{R_A^3}\right)\mathscr{D}_s' \times \frac{1}{1 - \mathscr{D}_s'\left(\dfrac{\alpha_A}{R_A^3}\right)} + \text{higher multipoles} \qquad (24)$$

where $\mathscr{D}_s'$ is a function of the static solvent dielectric constant ε_s,

$$\mathscr{D}_s' \equiv \frac{2(\varepsilon_s - 1)}{(2\varepsilon_s + 1)} \qquad (24a)$$

μ_A and α_A are the dipole moment and average polarizability of solute A, R_A the average solute radius (from pure liquid A). The analogous higher solute multipole terms which should follow the first term in Eq. (24) are not expected to be small. One reasonable way to bring them in may be by applying Eq. (24) to regional dipoles in A.

$F_{\text{e.s.}}^{A}$ has roughly the form*

$$F_{\text{e.s.}}^{A} \approx -\mathscr{M}(\mu_A, Q_A, \ldots; v_A)\mathscr{F}(\varepsilon_s) \qquad (25)$$

$$\approx -\frac{1}{2}\left(\frac{\mu_A^2}{v_A}\right)\mathscr{F}'(\varepsilon_s) + \cdots$$

where $\mathscr{M}$ is some function of solute charge distribution and volume, and $\mathscr{F}$ a function of solvent ε_s. The solvent dependence appears mainly through $\mathscr{D}_s'(\varepsilon_s)$ [Eq. (24a)] which varies little from solvent to solvent. For example, from benzene to water covering an ε_s range of 2.3 to 77, $\mathscr{D}_s'$ varies only between 0.5 and 1.0. The saturation effects would diminish this variation even more.

Thus, in comparing the solvation interactions of DNA bases with solvents, not only F_{vdw}^{A}, but also $F_{\text{e.s.}}^{A}$ were found quite insensitive to solvent (2, 8). On the other hand, Eqs. (24) and (25) show a crucial dependence on solute average molecular volume v_A as well as multipole moments, μ_A, Q_A, etc.

---

* The factor $1/[1 - \mathscr{D}_s' \times (\alpha_A/R_A^3)]$ does not vary that much from solvent to solvent.

Now putting together Eqs. (13), (14), (20a), (21), (23), and (25), we have the form of ΔF_A^0 for the "unitary" free energy of taking an A from solution out into the gas:

$$\Delta F_A^0 \approx a + b\frac{\mu_A^2}{v_A} - cv_A^{2/3}\gamma_1 - RT \ln \frac{kT}{v_1} + \cdots \tag{26}$$

The "constants" a, b, and c, in principle, depend on both solute and solvent, but as above considerations show, for a wide variety of polar and nonpolar common solvent and solute systems they will be very roughly "constants." Then ΔF_A^0 exhibits a linear dependence on γ_1, the macroscopic solvent surface tension, a slight dependence on v_1 which should often be negligible, and a more complicated dependence on solute molecular volume v_A. Note that for large and/or nonpolar solutes the surface term $v_A^{2/3}\gamma_1$ dominates—hence the "solvophobic effect" ("hydrophobic" in the most drastic case with the highest γ_1, i.e., water) on bulky groups. For smaller polar solutes the $1/v_A$ term gains importance (similarly with small ions an analogous "solvation energy" term dominates). There should be a certain "critical" solute size for each solvent and fixed μ_A at which the two effects are balanced. For example, for nonpolar solutes, where the second term in Eq. (26) is zero, one would have

$$\Delta F_A^0 = 0 \qquad \text{when } v_A^{\text{crit.}} \approx \left(\frac{a - RT \ln \dfrac{kT}{v_1}}{c\gamma_1}\right)^{3/2} \tag{27}$$

Thus, smaller solutes ($v_A < v_A^{\text{crit.}}$) will be soluble, larger ones quite insoluble from the gas.

3. The "Reduction" Terms

An equation like Eq. (26) applies to A, B, and also the (AB) "complex" taken as a unit. The $\Delta F_{(AB)}^0$, however, has one more term in it. The A-B interaction itself is reduced from its value in the gas phase owing to the direct effect of the medium on (A-B) van der Waals, electrostatic, and charge transfer forces.

Thus, for (A-B) in Eq. (10) and

$$\text{(A-B)(soln.)} \underset{x_{AB}}{\overset{}{\rightleftarrows}} \text{(A-B)(gas)}; \qquad \underset{p_{AB}}{\Delta F_{(A-B)}^0} \tag{13a}$$

we have

$$\Delta F_{(A-B)}^0 \approx -F_{\text{int.}}^{(A-B)} - F_c^{(A-B)} - RT \ln \frac{kT}{v_1} - F_{\text{red.}}^{(A-B)} \tag{14a}$$

where the reduction term has

$$F_{\text{red.}} \approx F_{\text{red.}}(\text{vdw}) + F_{\text{red.}}(\text{ind.}) + F_{\text{red.}}(\text{e.s.}). \tag{28}$$

If $E_{(A-B)}$ is the direct interaction energy of A and B in the "complex" or product, in the gas phase including chemical, H-bonding, van der Waals, electrostatic (e.s.), and charge transfer (c.t.) forces,

$$E_{(A-B)}(gas) = E_{(A-B)}(chem.) + E_{(A-B)}(vdw) + E_{(A-B)}(e.s.) + \cdots$$

and

$$E_{(A-B)}(soln.) = E_{(A-B)}(chem.) + E_{(A-B)}(vdw)(1 - K_{AB,1})$$
$$+ E_{(A-B)}(ind.) \times \frac{1}{(n_1{}^4)} + E_{(A-B)}(e.s.) \times \frac{1}{\varepsilon_1{}^E} + \cdots$$

then

$$F_{red.} \approx -E_{(A-B)}(vdw)K_{AB,1} - E_{(A-B)}(ind.)\left(1 - \frac{1}{n_1{}^4}\right)$$
$$- E_{(A-B)}(e.s., c.t.)\left(1 - \frac{1}{\varepsilon_1{}^E}\right) + \cdots \quad (29)$$

The multipole-induced multipole forces between A and B are reduced by $1/(\varepsilon_1{}^\infty)^2$ (solvent high-frequency dielectric constant), i.e., $1/n_1{}^4$. The A-B van der Waals potential is reduced by $K_{AB,1}$ as in Eq. (23c) [See Appendix and Sinanoğlu (11)]:

$$K_{AB,1} \approx \tfrac{3}{4}D_1\bar{L}_{AB} \approx \tfrac{3}{2}D_1 = \frac{3}{2}\left(\frac{n_1{}^2 - 1}{n_1{}^2 + 2}\right) \quad (30)$$

Again a rough idea, especially for comparing the effect of different solvents, is obtained on $F_{red}(e.s., c.t.)$ using the Kirkwood-Westheimer (14) (K.W.) formulas, e.g., for A-B dipole-dipole interaction. Clearly the e.s. A-B interactions cannot be obtained by the dipole approximation and the monopole or other methods are more appropriate. Nevertheless, the effective microscopic dielectric constant $\varepsilon_1{}^E$ of K.W. indicates, e.g., the reductions of e.s. base-base interactions of DNA in different solvents (2) to be relatively small and quite constant over a range of solvents such as water, glycol, formamide, ethanol, etc.

C. The Overall Solvent Effect Equations

The solvent effects on molecular associations are now given by Eqs. (11), (14), (20), (21), (23), (25), and (28). Combining these we have

$$\Delta F^0_{solv.\,eff.} = \Delta F_{int.} + \Delta F_c + \Delta F_{red.} + \Delta\left(RT \ln \frac{kT}{v_1}\right) + \Delta F_{mix.} \quad (31)$$

The Δ on the right-hand side refers to [(A-B)-A-B] for associations [Eq. (1)]; for example,

$$\Delta F_{int.} \equiv F^{(A-B)}_{int.} - F^A_{int.} - F^B_{int.} \quad (32)$$

or to a conformation change [Eq. (2)] when Δ means [(C)-A], e.g.,

$$\Delta F_{\text{int.}} = F_{\text{int.}}^{(C)} - F_{\text{int.}}^{A}. \tag{33}$$

For Eq. (1), $\Delta(RT \ln kT/v_1)$ gives $- RT \ln kT/v_1$, but for Eq. (2) it is zero. We have also added a free energy of mixing, correction term, $\Delta F_{\text{mix.}}$. This occurs with large molecules, with the association of biopolymers (e.g., double-helix formation) or biopolymer side chains (e.g., in protein folding). In these cases, the Henry's law dependence of free energy and entropy on mole fraction x_i [Eq. (6)] is no longer valid. More complicated entropies of mixing [volume fractions, polymer statistics; see, e.g., Hill (4)] must be used. In the association of two single strands into a polynucleotide double helix, for example, an approximate formula was given for $\Delta F_{\text{mix.}}$ in Sinanoğlu and Abdulnur (2) which, though not negligible, did not vary much from solvent to solvent. Also with side-chain associations $\Delta(RT \ln kT/v_1)$ term is zero.

Similarly, from Eqs. (16) and (17), the enthalpy and entropy parts of $\Delta F_{\text{solv. eff.}}^0$ are

$$\Delta H_{\text{solv. eff.}}^0 \approx \Delta e_{\text{int.}} + \Delta e_c + \Delta[RT(1 - T\mathscr{A}_1)] + \Delta e_{\text{red.}} \tag{34}$$

$$\Delta S_{\text{solv. eff.}}^0 \approx \Delta S_{\text{int.}} + \Delta S_c + \Delta\left[R \ln \frac{kT}{v_1} + R(1 - T\mathscr{A}_1)\right] + \Delta S_{\text{mix.}} \tag{35}$$

Equations (21), (23), and (25) have been assumed to be split into their enthalpy and entropy parts; the entropy part of $\Delta F_{\text{red.}}$ [Eq. (28)] has been assumed negligible. One may note again from Eqs. (5)–(11) that, e.g., for an association equilibrium, $\Delta H_{(\text{exp})}^0$ would be

$$\Delta H_{(\text{exp})}^0 = \Delta H_{\text{gas}}^0 + \Delta H_{\text{solv. eff.}}^0 \tag{36}$$

where ΔH_{gas}^0 is the enthalpy of formation of the A-B complex in the dilute gas phase. Where it is not possible to study a reaction in the gas phase, one may be able to obtain the hypothetical ΔH_{gas}^0 from an evaluation of $\Delta H_{(\text{exp})}^0$ in a number of solvents.

III. The Solvophobic Force

As in Eq. (26), the total (measured) standard free-energy change of an association $A + B \rightleftarrows (AB)$, ΔF^0 in a given solvent, has very roughly the form:

$$\Delta F^0 = - RT \ln K = \Delta F_{\text{gas}}^0 + a' - b'\Delta\left(\frac{\mu_i^2}{v_i}\right) + c'\Delta(v_i^{2/3})\gamma_1$$

$$- RT \ln \frac{kT}{v_1} + \Delta F_{\text{red.}}(n_1, \varepsilon_s) \tag{37}$$

$$\Delta F^0 \approx a'' - b'\Delta\left(\frac{\mu_i^2}{v_i}\right) + c'\Delta(v_i^{2/3})\gamma_1 \tag{38}$$

In Eq. (38), we have lumped a number of the terms into an a'' for the case where these change little from solvent to solvent for a given reaction.

According to these forms, a given association reaction acquires an additional driving force due to solvent, given by the last term of Eq. (38). $\Delta(v_i^{2/3})$ is proportional to the *surface area change* in the reaction, which for most associations is negative. This effect may be referred to as "*solvophobic force.*" It is greatest in water, water having the highest surface tension of any common, inert solvent. The solvophobic driving force is competed with by "solvation energy" [second and part of first terms in Eq. (38)], especially if the reactants are polar. Only the dipole effect has been written, but with a given charge distribution, the electrostatic term will still involve some type of inverse volume dependence.

For a given association, if one plots $\Delta F^0(\exp) = -RT \ln K$ or, in the case of reaction rates of the type in Eq. (4), $\ln k$ versus $\Delta F^0_{\text{solv. eff.}}$ calculated at least semiquantitatively from Eq. (31) and previous equations, for a number of pure solvents, one should get a straight line. In some cases where a'' and b' are quite insensitive to solvent, a plot of $\ln K$ or $\ln k$ vs. just the surface tension γ_1 should yield a rough straight line as well.

One may also keep the solvent fixed, but look at associations of related but larger and larger "reactants" and examine the *inverse effective volume* $\Delta(\mu_i^2/v_A)$ and *effective surface area* $c'\Delta(v_i^{2/3})$ change of reaction, and the dependence of Eq. (38) on these.

Similar correlations apply to ΔH^0 and ΔS^0 separately; e.g.,

$$\Delta H^0 \approx \alpha - \beta \Delta \left(\frac{\mu_i^2}{v_i} \right) + \Delta(v_i^{2/3})\gamma_1 \tag{39}$$

Above considerations should apply to pure solvents. In general with mixed solvents, the problem is complicated owing to local, inhomogeneous concentration effects, although in some cases a satisfactory correlation with the macroscopic γ of the mixed solvent at different compositions may be found.

The particularly strong tendency of water to associate nonpolar groups has been called "hydrophobic bonding" and attributed mainly to entropy decrease, "structuring" in water around such groups (3, 15, 16). This effect no doubt contributes, though it may or may not be the dominant one. The last term in Eq. (39), the cavity surface energy change alone, already makes water much more "solvophobic" compared to other solvents (2, 8). Water has a great affinity for itself.

Equation (39) indicates that solvophobic forces [and "hydrophobic" taken in the sense of Eq. (39)] will act on polar groups as well as on nonpolar ones. However, the second (e.s.) term in Eq. (39) diminishes the association tendency for the same size groups if the groups happen to be polar (compare phenyl vs. pyrimidine groups).

Solvent structuring effect in solvents like water, formamide, etc., is also, in principle, contained in the more detailed form of the solvent effect equations, namely, in the entropy part $F_c{}^i$ in Eq. (20). For a solute A in solvent I, for instance, this part (cf. Eqs. (20) and (19)) is

$$S_c^A = -\kappa_1{}^s(\varphi_{1A}^{-1/3}) \times 4.836v_A^{2/3}\gamma_1\left(\frac{\partial \ln \gamma_1}{\partial T} + \frac{2}{3}\mathscr{A}_1\right) \qquad (40)$$

The $\kappa_1{}^s(\varphi_{1A}^{-1/3})$, a function of the relative solute-solvent molecule sizes, corrects the macroscopic surface entropy to the microscopic one in the "prepared cavity" [in step 1, above Eq. (14)] appropriate to solute A. For nonpolar solvents, $\kappa_1{}^s$ and $\kappa_1{}^e$ are close together (9), but in polar solvents $\kappa_1{}^s$ should differ and reflect the "structuring" effect. In the semiquantitative version of the solvent effect equations [Eqs. (38) and (39) above], some average κ's are implied. In rough estimates made previously, they have been taken as $\kappa \approx 1$ (2). The greater "solvophobic" power of water compared to other solvents based on Eq. (39), i.e., cavity surface enthalpy effect, would be enhanced even further by a lowered cavity surface entropy. Estimates of these entropies were obtained for small solutes like Ar, Kr, CH_4, N_2, and C_2H_6 in water (2). For larger solutes* $(\varphi_{1A}^{-1/2} \gg 1)$ these entropies should become more positive gradually.

IV. Discussion of Some Applications

A. p-Benzoquinone–Hydroquinone

Cassidy and Moser (17) tested the above considerations on the p-benzoquinone–hydroquinone (quinhydrone) association. They measured the equilibrium constant in different solvents at the same temperature and pressure. The results showed no consistent relationship between the degree of association in solution and the dielectric constant, dipole moment, Y values (18), or Z values (19) [see Wiberg (1)] of the solvents. A quite linear relationship, on the other hand, is obtained if the experimental ΔF^0's are plotted against calculated $\Delta F_{\text{solv. eff.}}^0$ for the pure solvents.† In this particular case, as the calculations show (9), the $\Delta F_{\text{e.s.}}$, electrostatic terms [as obtained from Eq. (24); the second term in Eq. (38)] contribute negligibly, so that there is a quite satisfactory correlation even with just γ_1 alone (17).

A plot of calculated $\Delta F_{\text{solv. eff.}}^0$ values versus $\Delta F_{\text{exp.}}^0$ would be expected to correlate only for pure solvents. With mixed solvents local concentrations on

---

* For theoretical estimates around protein side chains see Némethy and Scheraga (16).

† Discussion of these calculations which compare the magnitudes of $\Delta F_{\text{solv. eff.}}^0$ for different solvents will be found in Halıcıoğlu and Sinanoğlu (9).

the cavity walls will, in general, differ not only from the bulk composition, but also from the macroscopic surface concentration as well. The ΔF^0(exp.) vs. solvent composition, compared with surface tension of the mixed solvent vs. composition, for systems like aqueous tetrahydrofuran (17) is indicative of this effect. Among the mixed solvents studied, glycol and water have the surface tensions and dielectric constants closest to each other, with glycol behaving as the least surface-active substance. Thus, as a rather special case, $\Delta F^0_{solv.\ eff.}$ and γ_1 obtained from bulk aqueous glycol correlate with $\Delta F^0_{exp.}$ on a line with other pure solvents (9). But such a simple behavior is not expected for any mixed solvent in general.

B. Thymine Photodimerization—TpT

Semiquantitative calculations with more approximate forms of Eqs. (20a) and (23), etc., were also carried out on thymidylyl-(3′ → 5′)-thymidine (TpT) (2). An equilibrium between open ("linear") and folded ("stacked") conformations is assumed.*

$$
\begin{matrix} T \\ \diagdown \\ p{-}T \ \ (l.) \end{matrix} \ \rightleftarrows \ \begin{matrix} T \\ \diagup \\ p \ \ (s.) \\ \diagdown \\ T \end{matrix} \tag{41}
$$

The calculated $\Delta F^0_{solv.\ eff.}$, albeit with the less detailed, less quantitative form of theory (2, 8), places the pure solvents on a *solvophobic force sequence* in decreasing order with respect to tendency to cause "stacking" of thymines: water > (glycerol, formamide) > glycol > (n-butanol, methanol, n-propanol, ethanol) > tert-butanol. The solvents which are shown in parentheses are too close together to be ordered.

If it is now assumed that the covalent thymine photodimers which are formed under the action of ultraviolet light (20) will form when a TpT is in a properly "stacked" form, the extent of such photodimerization in different solvents, under otherwise equal conditions, should reflect the above sequence.† Wacker and Lodemann (21) carried out such experiments and observed a satisfactory correlation with the predicted $\Delta F^0_{solv.\ eff.}$ sequence.

The semiquantitative calculations of $\Delta F^0_{solv.\ eff.}$ for comparing different solvents involve experimental solute and solvent properties, such as ε_s, γ_1, α, μ, etc. (2). The only things that can be adjusted are the solute cavity sizes

---

* The sugar groups, of course, also participate in the folding process as a space-filling model will show.

† We thank Drs. P. O. Ts'o and S. Y. Wang for pointing out that this assumes that the solvents mainly affect the degree of "stacking" and do not affect the intrinsic UV absorption as much.

or geometry. The sizes are usually obtained from corresponding pure substance densities or from those of model compounds. The sizes and geometry being the main things that can be varied, it is interesting to examine how the solvent sequence is affected by this variation. In reference (2), such a study, in which the effective solute radii in $\Delta F_{int.}$ and ΔF_c were varied separately, was carried out and the sequence was not affected.* More recently (9), further calculations have been made with a different geometric model which contains only one variable size for both the surface and volume forces. Each wing of the TpT is assumed to be sitting in a cylindrical cavity with a fixed height-to-base-diameter ratio estimated from a molecular model. In the " stacked " form the two cylinders sit on top of each other forming a single cylinder with the same base, but having twice the height. The sugars may be considered partly in these cylinders as well. Where the $F_{int.}$ terms require spherical cavity radii, these are obtained as the radii of equivalent spheres whose volumes equal those of the corresponding cylinders. Thus, the cylinder base radius, R, is the only variable. The $\Delta F^0_{solv.\ eff.}$ as a function of this R again preserves the same solvophobic sequence over a wide range.

C. Solvent Denaturation of DNA

Semiquantitative calculations have also been carried out (2, 8) for the association of bases on single strands into a double helix. DNA double helix is stable in water at 25°C, pH 7, and 0.1 ionic strength, but it is denatured upon addition of a wide variety of solvents like glycol, ethanol, formamide, etc. $\Delta F^0_{solv.\ eff.}$ calculations in a number of *pure solvents*, though very crude on an absolute basis for a given solvent, indicated the relative magnitudes of the different $\Delta F^0_{solv.\ eff.}$ components. The special role played by water in aiding the helix stability appeared due in large part to its high surface tension and to the large reduction in the cavity surface area energies around bases in going from single strands to helix. The same pure solvent " solvophobic " sequence is obtained for A–T as well as G–C pairs, even though G and C have considerably larger dipole moments than A and T. Among other effects, the solvent dependence of the phosphate oxygen e.s. repulsions has recently been studied in more detail by Abdulnur (22), who found that they did not affect the above qualitative conclusions.

For the reasons described in previous sections, it is difficult to predict the solvophobic sequence of mixed solvents like the ones used in solvent denaturation experiments (23–25). In such experiments solvents are ordered with

---

* We thank Professor B. Pullman for calling to this writer's attention the fact that the surface area change used in reference (2) appeared too large. Mr. Abdulnur confirms that two sides of the folded TpT were not taken into account. This makes the absolute magnitudes of $\Delta F^0_{solv.\ eff.}$ larger, but the relative solvent sequence is unaffected.

respect to their aqueous concentrations which will cause denaturation. This order need not bear any resemblance, of course, to the pure solvent solvophobic sequence. Abdulnur (22) has also explored the mixed solvents with the simple theory using macroscopic, bulk surface tensions and other properties of the mixtures. As expected, on account of local concentration effects, correlation with experimental denaturing concentrations is obtained only for similar classes of solvents (like a number of alcohols) or for dimethyl formamide, formamide, etc., mixed with water.

D. Actinomycin

Recently, Crothers, Müller, and their co-workers (26, 27) have studied the dimerization of actinomycin in aqueous solutions and the monomer in a number of pure solvents. They used equilibrium centrifugation and optical rotatory dispersion (ORD). For dimerization they find (26) a $\Delta H^0 = -15$ kcal/mole and $\Delta S^0 = -38$ e.u. and argue that these values seem consistent with the above picture of solvophobic interactions. Also consistent with these hypotheses, they find they can correlate changes in the ORD spectrum of monomer in the pure solvents methanol, dioxane, glycol, formamide, glycerol, and water with the surface tensions of these solvents attributing the ORD changes to monomeric conformation changes.

V. Conclusion

The association of molecular groups to form "complexes," dimers, more compact polymeric conformations, double helices, ..., is greatly affected by solvents. Water is especially strong in providing an additional driving force to aid such associations above and beyond any intrinsic intermolecular forces that would exist between such groups even in the no-solvent case. A semi-quantitative theory was presented previously (2, 8) that examined the solvent contributions to association free energy and concluded on the importance of cavity surface area creation energies. For specific types of associations, the solvents are ordered into decreasing "solvophobic force" sequences. In the present treatment, a more quantitative theory has been developed which may aid in the extraction of such properties of an association as molecular exposed surface area changes and intrinsic ΔH^0 of associations. The treatment indicates mainly two usually opposing solvent effects which may be classified as *solvation* or *inverse volume forces* and *surface forces*. "Solvophobic forces" with "hydrophobic" as an especially strong special case arise from the surface forces and act not only on nonpolar, hydrocarbonlike groups, but also on polar groups (like purine, pyrimidine bases). It is just that if the groups are polar, they may act in a way to counteract some of the surface effect. For small groups solvation or *inverse volume forces* may dominate, but with larger ones *surface forces* take over.

Experiments determining the ΔF^0 and ΔH^0 of suitable associations in a number of pure solvents would test the theory more quantitatively. The experimental ΔF^0's and ΔH^0's should yield direct correlations with $\Delta F^0_{solv.\ eff.}$ and $\Delta H^0_{solv.\ eff.}$ calculated as above using macroscopic or semiempirical properties of liquids and simple solutions. In some cases, where the "solvation forces" do not change much from solvent to solvent, ΔF^0 and ΔH^0 are expected to correlate with surface forces alone, hence with solvent surface tensions.

The more quantitative theory outlined in this chapter has been applied to a number of association equilibria and rates (9) beyond the qualitative or semi-quantitative applications discussed above.

VI. Appendix. Nonpolar Part of the van der Waals Potential between Two Molecular Groups Immersed in a Liquid

If $v_{AB}(R)$ is a semiempirical potential energy function giving the nonpolar parts of the interactions between two molecules like CCl_4, benzene, or similar molecular groups, in the absence of liquid, the same potential in liquid l environment becomes:

$$v_{AB}^{eff}\left(R;\rho_1,\frac{R_A}{R_1},\frac{R_B}{R_1}\right) = v_{AB}(R) \times B_{AB1}\left(R,\rho_1,\frac{R_A}{R_1},\frac{R_B}{R_1}\right) \qquad (A1)$$

(vdw)   (vdw)

The $v_{AB}^{eff}(R;\rho_1, R_A/R_1, R_B/R_1)$ is now an *effective intermolecular potential* between molecules A and B, distance R apart, and depends also on the density ρ_1 of the surrounding liquid as well as on the sizes of A and B (radii R_A, R_B) relative to the average solvent molecule size (average solvent radius R_1).

$v_{AB}(R)$ includes both dispersion-type attractions and nonbonded repulsions; (vdw) it depends not only on polarizabilities, but on molecular sizes as well. Note that this being a semiempirical function of the kind say that yields good agreement with the thermodynamic properties of a gas such as CCl_4, C_6H_6, C_2H_6, etc., it is independent of the usual (dipole, no overlap) type approximations that go into the quantum mechanical calculation of London dispersion forces.

A useful semiempirical potential for use between molecular groups about the size of CCl_4, C_6H_6, purines, pyrimidines, etc., is summarized below along with ways of estimating its parameters. The potential is only for the nonpolar parts of the interaction; with polar molecules electrostatic charge distribution forces must be added.

For the no-solvent A-B interaction we have an approximate Kihara potential (28):

$$v_{AB}(R) \approx v_{AB}^K(R) \approx \mathscr{C}_{AB}\left[\frac{\sigma_{AB}^6}{(R - l_{AB})^{12}} - \frac{1}{(R - l_{AB})^6}\right] \quad (A2)$$

(vdw)

In the liquid this is multiplied by the factor (11),

$$B_{AB1} \approx 1 - \frac{1}{m} \cdot \frac{\Delta' D_1' L_{AB}}{1 - \sigma_{AB}/(R - l_{AB})^6} \approx 1 - \frac{1}{m} \cdot \frac{3}{2} \cdot D_1' \quad (A3)$$

$$\Delta' \approx \frac{3}{4}; \quad D_1' \approx \frac{D_1}{1 + D_1}; \quad D_1 = \frac{n_1^2 - 1}{n_1^2 + 2}$$

n_1 equals the refractive index of solvent; m equals 3 for $A = B = 1$, pure liquid or solvent-solvent interaction; 2 for $A \neq B = 1$, solute-solvent molecule interaction; or 1 for $A = B \neq 1$ or $A \neq B \neq 1$, solute-solute interaction in solvent.

$$L_{AB} \equiv L_{AB}\left(R, \frac{R_A}{R_1}, \frac{R_B}{R_1}\right)$$

equals a dimensionless function of R and of relative solute-solvent sizes. A detailed and solvation structure-dependent form of L_{AB} is derived and given in Sinanoğlu (11). L_{AB} varies between 0 and 2, usually

$$L_{AB} \approx 2$$

$$\frac{R_A}{R_1} = \left(\frac{\rho_1}{\rho_A}\right)_r^{1/3}; \quad \frac{R_B}{R_1} = \left(\frac{\rho_1}{\rho_B}\right)^{1/3}$$

ρ_A (or ρ_B) is the number densities of a pure liquid made of A (or B) at the same reduced temperature T/T_c as the pure solvent.

In Eq. (A2):

$$\mathscr{C}_{AB} = \frac{3}{2}\alpha_A\alpha_B\frac{\delta_A\delta_B}{\delta_A + \delta_B}; \quad \alpha_i \equiv \text{polarizability of } i$$

$$\delta_i = \mu I_i; \quad 1 < \mu < 2; \quad I_i = \text{ionization potential of } i$$

$$\mathscr{C}, \sigma, l = \text{gas phase quasi-spherical Kihara parameters}$$

$$\sigma_{AB} \approx (\sigma_A + \sigma_B)/2; \quad l_{AB} \approx (l_A + l_B)/2$$

These "core" size parameters may be evaluated from the semiempirical relations

$$l_A = \frac{1}{\beta} \cdot \left(\frac{3}{4\pi\rho_A}\right)^{1/3} \cdot \left(\frac{1}{3.24 + 7\omega_A}\right); \quad \beta \approx 1.15 \quad (A4)$$

and

$$\sigma_A = \frac{1}{2^{1/6} \cdot \beta} \cdot \left(\frac{2.24 + 7\omega_A}{3.24 + 7\omega_A}\right) \cdot \left(\frac{3}{4\pi\rho_A}\right)^{1/3} \qquad \text{(nonpolar only)} \qquad \text{(A5)}$$

In deriving these, we have used the results of Sinanoğlu (11) and Danon and Pitzer (29). For nonpolar A, ω_A is the "acentric factor" (12) of A or a model compound in pure liquid form:

$$\omega_A \equiv -\log P_r - 1$$

where P_r is the reduced vapor pressure P/P_c of pure liquid A at a reduced temperature of $T/T_c = T_r = 0.7$.

For a polar molecule, ω_A is not a measure of a van der Waals "core size" l_A alone; polar effects are mixed in it. So for such substances one may use the ω_A of an analog, nonpolar substance (e.g., benzene for pyrimidine).

For the interaction parameters of a number of liquids and details of the effective potential the reader is referred to Sinanoğlu (11).

ACKNOWLEDGMENTS

It is a pleasure to thank Mr. Timur Halıcıoğlu for many helpful discussions during the course of this work. It has also been very enjoyable and beneficial to discuss the various aspects of the above topic with Professors H. Cassidy, D. Crothers, M. Fixman, L. Onsager, B. Pullman, F. M. Richards, J. Sturtevant, and K. Wiberg, and with Mr. C. Eckhardt. This work was supported by a grant from the National Science Foundation.

REFERENCES

1. For solvent effects on this and other reactions, see for example, Wiberg, K.B. (1964). "Physical Organic Chemistry." Wiley, New York.
2. Sinanoğlu, O., and Abdulnur, S. (1965). *Federation Proc.* **24**, No. 2, Part III, S-12.
3. Kauzmann, W. (1959). *Advan. Protein Chem.* **14**, 34.
4. See, for example, Hill, T. L. (1960). "Statistical Thermodynamics." Addison-Wesley, Reading, Massachusetts.
5. Fowler, R., and Guggenheim, E. A. (1956). "Statistical Thermodynamics." Cambridge Univ. Press, London and New York.
6. Buff, F. P. (1955). *J. Chem. Phys.* **23**, 419.
7. Wakeshima, H. (1961). *J. Phys. Soc. Japan* **16**, 6.
8. Sinanoğlu, O. and Abdulnur, S. (1964). *Photochem. Photobiol.* **3**, 333.
9. Halıcıoğlu, T. and Sinanoğlu, O. (to be published).
10. Kestner, N. R. and Sinanoğlu, O. (1963). *J. Chem. Phys.* **38**, 1730.
11. Sinanoğlu, O. (1967). *In* "Intermolecular Forces" (J. O. Hirschfelder, ed.). Wiley (Interscience), New York, Vol. 12, pp. 283ff.
12. Pitzer, K. S. (1955). *J. Am. Chem. Soc.* **77**, 3427; and other papers. See Lewis, G. N., Randall, M., Pitzer, K. S., and Brewer, L. (1961). "Thermodynamics," 2nd ed., pp. 606–629. McGraw-Hill, New York.
13. Onsager, L. (1936). *J. Am. Chem. Soc.* **58**, 1486.

14. Kirkwood, J. G. and Westheimer, F. H. (1938). *J. Chem. Phys.* **6**, 506.
15. Frank, H. S., and Evans, M. W. (1945). *J. Chem. Phys.* **13**, 507.
16. Némethy G., and Scheraga, H. A. (1962). *J. Phys. Chem.* **66**, 1773.
17. Moser, R. E., and Cassidy, H. G. (1965). *J. Am. Chem. Soc.* **87**, 3463.
18. Grünwald, E., and Winstein, S. (1948). *J. Am. Chem. Soc.* **70**, 846.
19. Kosower, E. M. (1958). *J. Am. Chem. Soc.* **80**, 3253.
20. The following are some of the papers that deal with different aspects of thymine photodimers: Beukers, R., Ijlstra, J., and Berends, W. (1960). *Rec. Trav. Chim.* **79**, 101; Wacker, A., Dellweg, H., and Jacherts, D. (1962). *J. Mol. Biol.* **4**, 410; Wang, S. Y. (1963). *Nature* **200**, 879; Mantione M., and Pullman, B. (1964). *Biochim. Biophys. Acta* **91**, 387.
21. Wacker, A., and Lodemann, E. (1965). *Angew. Chem. Intern. Ed. Engl.* **4**, 150; (1965). *Angew. Chem.* **77**, 133.
22. Abdulnur, S. (1966). Ph.D. Thesis, Chem. Dept., Yale University.
23. Herskovits, T. T. (1962). *Arch. Biochem. Biophys.* **97**, 474.
24. Levine, L., Gordon, J. A., and Jencks, W. P. (1963). *Biochemistry* **2**, 168.
25. Ts'o, P. O., Helmkamp, G. K., and Sander, C. (1962). *Biochim. Biophys. Acta* **55**, 584.
26. Crothers, D. M., Sabol, S. L., Ratner, D. I., and Müller, W. "Studies Concerning the Behavior of Actinomycin in Solution" Preprint (to be published).
27. Müller, W., and Emme, I. (1965). *Z. Naturforsch.* **20b**, 835.
28. Kihara, T. (1963). *Advan. Chem. Phys.* **5**, 147.
29. Danon, F. and Pitzer, K. S. (1962). *J. Chem. Phys.* **36**, 425.

Solvent Polarity and Molecular Associations

P. DOUZOU

Service de Biospectroscopie
Institut de Biologie Physico- chimique
Paris, France

I. Introduction

Aromatic molecules can give dimer and polymer associations without any covalent bond. Such associations with ordered structures occur in water and are dependent on both concentration and temperature. These parameters act upon the stability of molecular associations which results from van der Waals-London and dispersion forces according to molecular dipole moments (dipole-dipole, dipole–induced-dipole interactions, etc.) and could involve a charge-transfer contribution according to the electron-donor or electron-acceptor character of the molecules (1).

Some parameters other than concentration and temperature influence the associations and, in fact, their stability. These include ionic strength, pH, and alcohol, and their main effect (decrease of polymerization) would be to lower the dielectric constant of medium (as the temperature does).

In order to verify eventually this hypothesis and to try to find accurate conditions for which molecular associations are not too "flickering," we experimented with hydroalcoholic mixtures, the polarity P (and then the dielectric constant ε) of which widely varies according to $\varepsilon = f(1/T)$ in a wide range of temperatures without freezing the samples.

Under these conditions, the dielectric constant and viscosity increase progressively in proportion as Brownian motion decreases. These effects should be able to increase the stability of molecular associations in a still fluid medium.

Moreover, we can determine such reversible processes without changing the global chemical composition of the medium. We change only the reciprocal geometric distribution of water and alcohol molecules around the solutes in order to study the microscopic phase transformations on solute associations.

Among aromatic molecules, many artificial dyes have a tendency toward self-association and this phenomenon is characterized by "magnificent" spectral effects: appearance of new and specific absorption and fluorescence bands, self-quenching, formation of gels, insoluble polymers, etc. (2).

We experimented with such a dye and established that in hydroalcoholic solutions, temperature effects were equivalent to concentration effects in aqueous solution at room temperature.

Then, we studied in the same way the behavior of nucleic components and other aromatics of biological interest.

In most of the cases, we found spectral proofs of associations in proportion to changes in temperature.

This chapter deals principally with experiments on "couples" of electron-donor and electron-acceptor molecules that are biologically active, the interaction of which could involve charge-transfer contribution and related physicochemical behavior. The electron-acceptor molecule is flavine mononucleotide (FMN) and the donor ones include various indole derivatives (methoxyindole, tryptophan, adenine, etc.) and p-phenylene diamine, a strong and "synthetic" donor. Moreover, we studied coenzymatic dinucleotides, flavine adenine dinucleotide (FAD) and nicotinamide adenine dinucleotide (NAD), in the presence of the preceding electron donors.

II. Experiments and Results

In the equation $\varepsilon = f(1/T)$ are present the electronic-atomic polarization term P_0 and the molecular polarization (orientation) term $4\pi N\mu^2/9k$ (N, Avogadro number; μ, dipole moment of solvent molecules; k, Boltzmann's constant). Using different (binary and ternary) hydroalcoholic mixtures, we can experiment in a wide range of temperatures (e.g., $50°C \rightleftharpoons 140°C$) and dielectric constants ($20 \rightleftharpoons 120$) without changing the global chemical composition of the medium or freezing it (3).

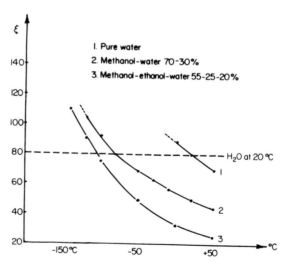

FIG. 1. Dielectric constant vs. temperature for water, methanol-water, and methanol-ethanol-water.

In Fig. 1 three curves for dielectric constant vs. temperature are shown for water, a binary, and a ternary hydroalcoholic solvent system. Of course, we can use various other mixtures and then get new scales for the dielectric constant as a function of temperature.

As the temperature drops, the dielectric constant as well as the viscosity rises; the Brownian motion also decreases. These three effects should be conjugated and we will have to evaluate their respective contribution. But the most important part of our study was to try to obtain molecular association and dispersion by using our reversible temperature process. We started these experiments with artificial dyes in order to determine the best experimental conditions and to verify that temperature effects can spontaneously build or destroy noncovalent polymers.

The principal results were published recently. They concern absorption and fluorescence recording spectra made in cryostatic cells and a special cryometric device that are described elsewhere (4).

A. Spectroscopic Behavior of Polymerizable Dyes

It is well known that the absorption spectrum of many dyes undergoes a continuous change as their concentration increases in aqueous solutions and that such a change is connected with new and specific fluorescence processes by an unmistakable parallelism (5, 6). In dilute solutions (10^{-5}–10^{-4} M) these dyes are responsible for "molecular" (M) absorption bands. As their concentration is increased, some new bands appear which are called "dimer" (D) and "polymer" (P) bands. At the same time, one can record a self-quenching of fluorescence (for many D associations) or the appearance of a new fluorescence (for many P associations) (7, 8).

At concentrations higher than 10^{-3}–10^{-2} M, "gel" formation and even crystallization of polymers occur.

Using the following dye (due to Dr Bourdon, from Kodak-Pathé-France) one can record, in dilute aqueous solutions (10^{-5} M), an absorption spectrum consisting of a band culminating at 510 mμ.

Increasing the concentration, we can verify that the dye does not follow Beer's law and that a new band appears and culminates at 590 mμ. A sharp fluorescence band is associated "in resonance" to this absorption.

The first absorption band can be considered a molecular (M) band and the second one a polymer (P) band.

Using hydroalcoholic solvents in various proportions it is possible to observe that alcohol inhibits the polymerization. The dye concentration has to be increased in order to get the polymerization.

In the mixture methanol-water (in proportion 1 : 1) we get the so-called P band (at 590 mμ) and the corresponding fluorescence only when we drop the temperature from $-30°$ to $-60°$C according to the solute concentration. So, these experiments demonstrate that $1/T$ (and, thus, polarity and viscosity) can have the same associative effect on hydroalcoholic solutions that an increase of dye concentration has on aqueous solutions. This polymerization effect under the influence of $1/T$ is quite thermoreversible with a beautiful "hysteresis" effect recorded both by absorption and fluorescence measurements (Fig. 2).

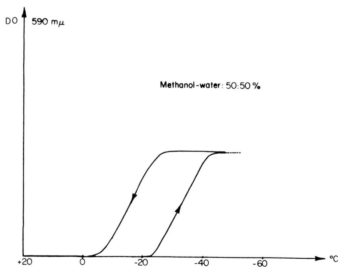

FIG. 2. Effect of temperature on the P band at 590 mμ in a methanol-water system.

Using hydroalcoholic solvents with an increasing percentage of alcohol in binary and ternary mixtures (where alcohol represents more than 50%) we can observe at room temperature and as a function of temperature a new spectroscopic behavior of the dye which raises the question of a specific solvent effect upon the association type. This effect is reported and discussed elsewhere (4).

B. Spectroscopic Behavior of Electron Acceptor and Donor Solutes

1. Hydroalcoholic Solutions of Flavine Mononucleotide (FMN)

In the ternary solvent (methanol-ethanol-water, 55 : 25 : 20) and at various concentrations (between 10^{-5} and 10^{-4} M), FMN undergoes a continuous and reversible spectroscopic evolution in the following range of temperatures: $20°C \rightleftharpoons -140°C$.

From the absorption and fluorescence spectra reported in Fig. 3a and b we can observe that the normal effects of temperature on optical density and fluorescence intensity are associated with phenomena like the blue shift recorded between $-50°$ and $-80°C$ influenced by the FMN concentration; we can also observe a hysteresis effect recorded by fluorescence for the reverse (thermal) reaction.

Such phenomena can be attributed to associative processes between FMN molecules as a function of temperature, solvent composition, and solute concentration.

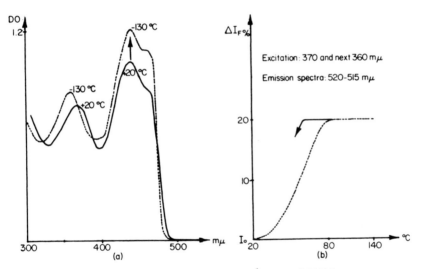

FIG. 3. Absorption and fluorescence spectra of FMN.

2. Hydroalcoholic Solutions of FMN and Indole Derivatives

a. Experimental Results. The indole derivatives are electron donors and FMN is an electron acceptor.

In order to obtain association and, if possible, charge-transfer processes between them, we studied concentrations 10^{-2}–10^{-3} M indole derivatives and 10^{-4}–10^{-5} M FMN, according to their k values.

Using these concentrations and the same ternary solvent we recorded absorption and fluorescence phenomena reported in Fig. 4 for the mixture tryptophan-FMN. From Fig. 4, it appears that after a hyperchromic effect (from 20° to −60° or −80°C according to solute concentration) there is a hypochromic effect reaching its maximum near −130° to −140°C. A blue shift occurs at −60° or −80°C (370 → 360 mμ). Below these temperatures, a small absorption appears between 490 and 550 mμ with an isosbestic point at 480 mμ.

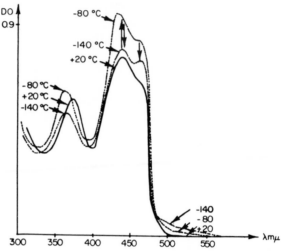

FIG. 4. Effect of temperature on spectra of tryptophan-FMN mixture.

We observed under similar conditions the evolution of fluorescence which consists initially in an enhancement between 20° and −80°C (15–20%) and, next, a quenching (40–50%) between −80° and −120°C. So there is a parallelism between the reversible evolution of absorption and fluorescence. By increasing the temperature we can observe (by recording of fluorescence) a hysteresis effect of the reverse reaction.

Several other indole derivatives gave the same qualitative results and in Fig. 5, absorption spectra for 5-methoxyindole (10^{-3} M) and FMN (10^{-5} M) are reported.

  b. Comments. We have spectroscopic evidence of molecular associations under the combined effects of low temperature, viscosity, and the dielectric constant of the solvent medium. These parameters shift the following equilibrium:

$$n\,D, A \underset{T,\,\eta,\,\epsilon}{\overset{1/T,\,\eta,\,\epsilon}{\rightleftharpoons}} (D \cdots A)n \quad (n = 2, 3, \ldots)$$

where D is the indole derivative and A represents FMN.

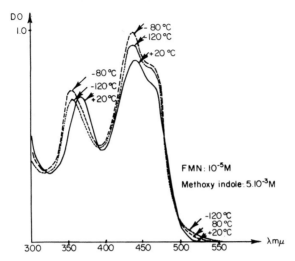

FIG. 5. Absorption spectra of FMN and methoxyindole.

This equilibrium is also concentration dependent. The temperature effect $(1/T)$ is influenced by absolute concentration in D and A and also by the solvent composition.

The hysteresis effect easily observed in fluorescence spectra when warming the solutions is a supplementary proof of the molecular association and is due to the overall stability of ordered solute molecules.

At present it is impossible to determine even approximately the energy represented in such associations. The main source of the stability should reside in the nonspecific interaction between stacked molecules, due to van der Waals-London (dipole-dipole, dipole-induced dipole) and dispersion forces (9, 10). Moreover, according to the electron-donor and electron-acceptor character of the indole derivatives (D) and FMN (A), charge-transfer forces could be also operative between these superposed molecules (9).

We know that such a type of interaction is to be expected for parallel-oriented aromatic molecules at close distance (3–4 Å) and that its quantum explanation involves a wave function Ψ which describes the complex between D and A in the following terms:

$$\Psi = a\Psi(D\ldots A) + b\Psi(D^+-A^-) \qquad \text{where } a > b$$

$\Psi(D\ldots A)$ denotes the so-called "no band" wave function involving a binding by intermolecular forces like the van der Waals-London, dispersion, and hydrogen-bonding forces, whereas $\Psi(D^+-A^-)$ denotes the so-called "dative bond" wave function resulting from a charge-transfer process (10–13).

The electronic transition from the ground to the first excited state of such a complex involving a charge-transfer contribution can be associated with the appearance of a new (additional) absorption band located generally toward longer wavelengths. The corresponding wave function of the molecular (dimer, polymer) complex for the excited state is written as:

$$\Psi_{S^*} = b^*\Psi(D \ldots A) - a^*\Psi(D^+ - A^-) \qquad \text{where } a^* > b^*$$

and the charge-transfer contribution should be thus increased. From Figs. 4a and 5, we can observe the appearance of a new absorption, quite small (its extinction coefficient is at maximum 20–30), which could represent a charge-transfer contribution involving in most cases less than 1 % of the whole complexes. This effect is quite different from previous observations (in frozen aqueous solutions, for instance) where the charge-transfer process can be considered a reversible and one-step oxidoreduction (14–16). In the present case, it was impossible to obtain an enhancement of the additional absorption by optical excitation. We simply observed destroying of FMN without any transient photoprocesses.

3. Hydroalcoholic Solutions of FMN and *p*-Phenylene Diamine as "Donor"

a. Experiments. In hydroalcoholic solutions of FMN and *p*-phenylene diamine studied as a function of $1/T$, a new absorption band appears culminating at 600 mμ and reaching its optimum at $-140°$C. This process is thermoreversible (Fig. 6).

But we can obtain the same absorption (in lower proportions) at room temperature. Such an absorption is dependent on temperature, oxygen, and time.

Using optical excitation at low temperature (for instance, at 120°C when the visible absorption band reaches its optimum intensity) we get a pronounced enhancement of the band (Fig. 7).

Such a photoprocess is only partially thermoreversible. When returning to room temperature, we get a continuous visible absorption (a "mahogany" coloration), probably due to photo end products.

Then, we also studied associations of these derivatives and coenzymatic dinucleotides (FAD, NAD) under the same conditions and observed thermoreversible photoprocesses at low temperature in still fluid solutions.

As an example, we indicate the spectroscopic evolution of the mixture NAD-serotonine (Fig. 8).

b. Comments. From these experiments, we can write a new equilibrium equation:

$$n\,D,A \underset{1/T}{\overset{}{\rightleftharpoons}} (D \cdots A)_n \underset{kT}{\overset{h\nu}{\rightleftharpoons}} (D \cdots A)_n^* \longrightarrow (D \cdots A, D^+ \cdots A^-) \downarrow O_2$$

End products

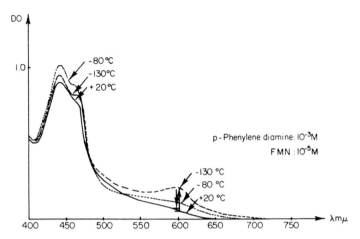

FIG. 6. Effect of temperature on spectra of FMN and *p*-phenylene diamine.

The $(D^+ \ldots A^-)$ term should represent "transient" (semiquinone-like) oxidoreduction processes, which are dependent on oxygen and temperature.

We actually perform the same experiments in the presence and absence of oxygen and try to record eventual free radicals by ESR measurements. At present, concerning the photoreactions of FMN and *p*-phenylene diamine, we think that oxidoreduction processes occur between the reduced and oxidized forms of the *p*-phenylene diamine and also FMN. These different reactions are, of course, oxygen dependent. They occur simultaneously and it is impossible to describe their respective pathways. In order to study the eventual involvement of a photoreduction of FMN using *p*-phenylene diamine as the reducing agent [ethylenediaminetetraacetic acid was previously used by

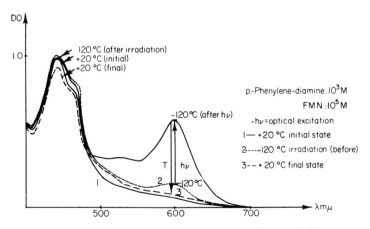

FIG. 7. Enhancement of absorption due to optical excitation.

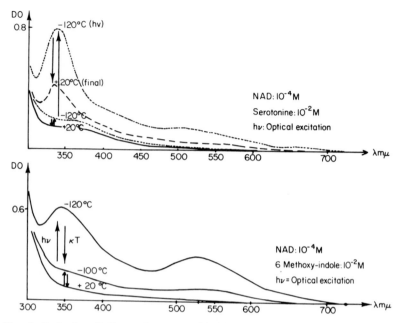

FIG. 8. Optical excitation and spectra of NAD, serotonine, and methoxyindole.

Radda-Calvin on more concentrated (aqueous) solutions at room tempera-
ture], it is necessary to mix and to study our reagents in anaerobic conditions.
At the same time, we could avoid the photooxidation of reduced p-phenylene
diamine which is catalyzed by oxygen. So, in order to determine what the
photoprocesses really do mean, it is necessary to plan some new spectroscopic determinations.

III. Discussion

1. Molecular associations of aromatics (conjugated and planar) are directly
influenced by temperature effects, which act upon Brownian motion, viscosity, and the dielectric constant. This is an essential fact which could be
used for studying many other biologically active compounds and complexes.
    Concerning the extent (resulting molecular weights, sizes, etc.) of such
association processes, we can compare the behavior of the highly polymeriz-
able dye and the electron-donor and electron-acceptor molecules in the same
range of concentrations. It is quite easy to obtain polymer precipitates of the
dye, characterized in solution by an intense and sharp fluorescence at 590–
600 mμ.

Using similar concentrations of D and A molecules of biological interest, such a phenomenon does not occur and when returning to their initial state, the solutions give the same optical density.

So, when we write $(D \ldots A)_n$, we mean that n is probably higher than 2, but we cannot determine its average value and the average size of "polymers." Nevertheless, we use this term because of the hysteresis effect we observe when warming the solutions.

In some cases, it is possible to obtain polymers in more concentrated solutions of aromatic molecules like dyes and even nucleotides. But such a phenomenon depends largely on the solute molecules studied and probably also on some of their characteristics such as dipole moments, chemical composition, and stereochemistry.

2. In considering the other factors (Brownian motion, viscosity) the polarity is an important if not an essential one. This is particularly evident in the case of dyes where the addition of alcohol to aqueous solutions at room temperature inhibits or prevents the normal polymerization. For these same dyes and even for nucleotides, a small decrease in temperature (between 20° and 4°C) increases the dielectric constant and enhances the polymerization. In our experiments on D and A mixtures, the contribution of a charge-transfer process to the overall stability of the molecular association of the D and A compounds seems to be evident, but quite small. Light plays a prominent role on D and A associations by the induction in some cases of thermoreversible photoprocesses in still fluid solutions. But, for the moment, it is difficult to ascertain the nature of such processes and we prefer to limit discussion to the following equilibrium:

$$n \, \mathbf{M} \; \underset{T, \, \eta \searrow \epsilon \searrow}{\overset{1/T, \, \eta' \nearrow \epsilon' \nearrow}{\rightleftharpoons}} \; (\mathbf{M})_n$$

where $n\mathbf{M}$ is the "molecular" compound and $(\mathbf{M})_n$ is the "associated" complex. (M can be a single species or can involve D and A molecules.)

3. In our introduction we defined our experimental techniques based on the variation of the polarity of the hydroalcoholic solvents according to the equation $P = a + b/T$ and we pointed out that we could work in wide ranges of temperature, viscosity, and polarity without changing the global chemical composition of the solvent. We added that reversible changes of polarity could correspond to microscopic changes in the geometric distribution of water and alcohol molecules, presumably equivalent to phase transformations involving the solvent and maybe the solute. A priori the presence of alcohol should involve the "salting-out" of the polymers once they are built.

This is a fact we can verify at room temperature in aqueous solutions of dyes and concentrated solutions of guanylic acid (10^{-2} $1/M$ 10^{-1} M), when we add methanol or ethanol.

But at low temperature, such polymers remain solvated after they are formed. When we raise the temperature we obtain a precipitate, which can be dissolved again by moderate heating (50°C). We studied such a cycle using a hydroalcoholic solution of guanylic acid (in methanol-water, 1 : 1; guanylic acid, 10^{-2} M). Dissolving the solute at 50°C, we can lower the temperature to 0°C and below without getting the classic transparent gel of polyguanylic acid [described and analyzed by Gellert, Lipsett, and Davies for aqueous solutions between 50° and 0°C (18)].

A stable, transparent gel is then obtained between −40° and −60°C depending on solute concentration and composition of solvent mixtures.

Its formation is simply delayed by alcohol. So, there exists a temperature range for which polymerization occurs without "denaturation" or "salting-out" by alcohol. An increase of temperature determines a "salting-out."

In such cases, we obtain macroscopic evidence of the polymerization due to temperature effect, and we can observe that the role of the solvent polarity is prominent since its corresponding geometric distributions and their change (much like phase transformation) are responsible both for polymerization and salting out.

IV. Conclusion

There are several conclusions which seems to spring from the preceding results and discussion.

1. Researchers interested in experiments on organized molecules and molecular complexes or polymers could use the cryometric techniques based upon the relation $\varepsilon = f(1/T)$ which permits simultaneous changes in dielectric constant, Brownian motion, and viscosity. As we saw, such techniques permit many solute molecules to aggregate, and we study their mutual interactions and induce some thermoreversible photoprocesses [till now studied in glassy solutions (rigid media), where solute molecules are not able to dismutate or to change structure].

The reversible formation of polymers without covalent bonds and their isolation after precipitation can also be studied—in particular, 5′-guanylic nucleotide formation.

The study of some "intercalation" processes involving molecules like acridine, other fluorescent derivatives, and aromatic hydrocarbons, could as well be made [previously nucleic acids were used (9)].

2. From a biochemical and a more theoretical point of view concerning the influence and the real meaning of the solvent polarity on molecular asso-

ciation and physicochemical properties of solutes, the results obtained on guanylic acid are of a real interest. They illustrate that polarity, determined by the geometric distribution of solvent molecules, influences the formation and the solvation or the salting out of solute associations. Slight changes of the geometric distribution of water and alcohol molecules make a profound difference in the attachment of water to solute aggregates.

We can understand why as a function of temperature, the alcohols are denaturating or inert solvents toward polynucleotides and also why they permit proteins to dissolve at low temperatures without denaturation when the dielectric constant of the medium reaches the value $\varepsilon \sim 80$ (19–21).

What we mean by an "icelike" structure of "wetted" hydrophilic molecules or chemical groups is a rigidly attached shell of water molecules able to undergo sudden reversible unfreezing. We can see from our experiments that such a mechanistic process should, in fact, correspond to a "phase transformation" according to the views of Hirschfelder (22), since changes in molecular geometry of a binary solvent modify its relations to solute molecules (shift from solvated and nativelike to salted out polymer structures, and conversely).

3. Finally, these observations are still very crude and we have to obtain a great deal of new information about the molecular associations and dissociations due to solvent effects resulting from changes in polarity.

First of all, the mechanism by which changes in polarity of the solvent influence the polar molecules of solute and stabilize the resulting molecular complexes is largely hypothetical. According to current predictions, the higher the dielectric constant of a medium the stronger the complex that two polar molecules can form when they are in suitable orientation (22, 23).

This could explain the aggregation power due to increasing polarity of hydroalcoholic solvents; but we have to learn much more about the eventual influence of dielectric gradients toward isolated, paired, and even stacked polar molecules in order to comprehend eventually how the molecular complex formation or dispersion really occurs. We are now studying this problem using polar aromatic molecules added to media, the dielectric constant of which linearly varies in one direction. From the different experiments we have reported on and those we plan to study we hope to obtain more precise information concerning what polarity really means in biochemistry and biology.

ACKNOWLEDGMENTS

This work was done under contract (67-00-503) with the Comité de biologie moleculaire de la Délégation Generale à la Recherche Scientifique, with the full support of the Centre de Recherches du Service de Santé des Armées. We thank Dr. T. Shiga for helpful suggestions and discussions and also Dr. F. Leterrier and M. C. Balny for technical assistance.

REFERENCES

1. Hirschfelder, J. O., Curtiss, C. F., and Bird, R. B. (1964). "Molecular Theory of Gases and Liquids." Wiley, New York.
2. Pringsheim, P. (1959). "Fluorescence and Phosphorescence," pp. 353–361. Wiley (Interscience), New York.
3. Akerlög, G. (1932). *J. Am. Chem. Soc.* **54**, 4125.
4. Douzou, P. (1967). *J. Chim. Phys.* (in press).
5. Scheibe, G. (1941). *Z. Elecktrochem.* **47**, 73.
6. Levshin, L. W. (1934). *Acta Physicochim. URSS* **1**, 685.
7. Scheibe, G. (1943). *Z. Physik. Chem.* **B53**, 183.
8. Jelley, E. E. (1936). *Nature* **138**, 1009; **139**, 631 (1937). Scheibe, G. *et al.* (1936). *Z. Angew. Chem.* **49**, 563; **50–51**, 212 (1937).
9. Pullman, B. (1965). *In* "Molecular Biophysics" (B. Pullman and M. Weissbluth, eds.), pp. 117–190. Academic Press, New York.
10. Pullman, B., and Pullman, A. (1963). "Quantum Biochemistry," p. 215. Wiley, New York.
11. McGlynn, S. P. (1958). *Chem. Rev.* **58**, 1113.
12. McGlynn, S. P. (1960). *Radiation Res.* Suppl. 2, p. 300.
13. Briegleb, G. (1961). "Electronen-Donator-Acceptor-Komplexe." Springer, Berlin.
14. Szent-Györgyi, A. (1960). "Introduction to a Submolecular Biology." Academic Press, New York.
15. Isenberg, I., and Szent-Györgyi, A. (1958). *Proc. Natl. Acad. Sci. U.S.* **44**, 857.
16. Isenberg, I., and Szent-Györgyi, A. (1959). *Proc. Natl. Acad. Sci. U.S.* **45**, 1229.
17. Radda, G. K., and Calvin, M. (1964). *Biochemistry* **3**, 384.
18. Gellert, M., Lipsett, M. N., and Davies, D. R. (1962). *Proc. Natl. Acad. Sci. U.S.* **48**, 2013.
19. Freed, S., and Sancier, K. M. (1954). *J. Am. Chem. Soc.* **74**, 1273.
20. Freed, S., Turnbull, J. H., and Salmre W. (1958). *Nature* **181**, 1731.
21. Shiga, T., Layani, M., and Douzou, P. (1967). *Bull. Soc. Chim. Biol.* **49**, 507.
22. Hirschfelder, J. O. (1965). *In* "Molecular Biophysics" (B. Pullman and M. Weissbluth, eds.), pp. 325–341. Academic Press, New York.
23. Tsibris, J. C. M., McCormick, D. B., and Wright, L. D. (1965). *Biochemistry* **4**, 504.

The Structure of Antibodies and the Antigen-Antibody Reaction

JEAN SALVINIEN

The Faculty of Sciences, Department of Chemistry
University of Montepellier
Montepellier, France

In 1957, the organizers of the "Quatre Jours de Biochimie" at Montpellier requested me to give a report on the subject of "antigen-antibody reaction from a physical chemical standpoint." As I had just spent 3 months with Linus Pauling at Caltech I had at my disposal first class sources of information. This report was published in *Bulletin de la Societe de Chimie Biologique* (22).

Since that time much important work has been done in the field of immunochemistry. My modest contribution to this symposium will consist of a brief review of those advances in the field most striking to a physical chemist interested in the applications of his field to biology, though unable to devote to them all the time he would desire.

The mechanism of antibody synthesis is still a much debated subject. A priori, it might seem to fall somewhat outside the subject matter here. In fact, antibody synthesis is closely linked to antibody structure and, thus, to the antigenic reactivity of antibody molecules. I shall, therefore, summarize its essential stages.

Before the present-day views on protein synthesis in general had been worked out, two opposing groups of theories were already held: the group known as *instructive theories* or template theories and the group known as *selective theories*.

According to the *instructive theories*, the antigen molecules, when introduced into an organism, modify *by their presence itself* the normal processes of γ-globulin formation, so that the latter acquire the *new* property of reacting specifically with the corresponding antigen molecules. The antigen would thus play the role of a *model* or even a *template*.

The instructive theories are primarily constructed to interpret the remarkable degree of specificity of the antigen-antibody reaction. Thanks to the hapten method of Landsteiner, it was possible to perform numerous experiments proving that every reaction of the antibody is strictly complementary to the homologous antigenic determinant. The correspondence between the two sites is not limited to electrostatic interaction or hydrogen bonds, but must include interactions of a steric order. The reaction site of the antibody is molded closely, *geometrically*, upon the antigen, with the result, in fact, that the van der Waals interactions are maximized. The molding seems to

461

occur not only on the hapten itself, but also on the part of the macromolecule where the hapten is attached. Under these conditions it is very difficult to believe that such a faithful reproduction in *negative* could be obtained without the direct intervention of the template. The physical chemists will always naturally incline toward the instructive theories which for them evoke familiar and very much simpler phenomena such as epitaxy or syncrystallization—which does not mean to be sure, that they would not yield to solid biological arguments.

It is well known that Breinl and Haurowitz in 1930 (5), Alexander in 1931 (2), and Mudd in 1932 (16) proposed the first template theories. In 1940, Pauling (20) thought that the antigen might intervene in such a way as to modify the folding of the polypeptide chain without modifying the amino acid sequence of the molecule. These schemes account for the fact that the antibody is generally bivalent.

Theories of this type, appealing because they are simple and logical, found many supporters among biophysicists.

Their popularity declined, however, when the genetic code controlling protein synthesis was deciphered.

The *normal* polypeptide chains of an organism are synthesized in the cell cytoplasm by the ribosomes. The primary structure of each chain is determined by the base sequence of the DNA belonging to one of the genes of the cell nucleus. This DNA acts through an intermediary. It releases into the cytoplasm an RNA *messenger bearer of the coded message* which has been formed by means of a sort of molding off one of the branches of the DNA. The long filiform molecule of RNA messenger attaches itself to several ribosomes at a time. It then serves as a template for the synthesis of the polypeptide chain formed from the free amino acids, whose subsequent sequence must be in accord with the code message.

Indeed, the primary structure of all proteins synthesized in this way has *already been inscribed* in the genetic code.

At the present time, those who favor the *selective theories* think that antibody synthesis is in no way an exception to the preceding rule. In other words, the pattern of synthesis of an antibody specific for any antigen would already be inscribed in the genetic code. The antigen, then, by its presence would only *stimulate* a selective mechanism either functioning at a reduced rate without it, or else, if not functioning already, functioning at the slightest stimulation. But the antigen has no need to serve as a template since the structure of its antibody would already be preinscribed in the code.

In 1900, well that is, before the discovery of the mechanism of protein biosynthesis, Ehrlich had already suggested the hypothesis of preformed receptors capable, after being stimulated by the corresponding antigen, to multiply and to release antibodies into the circulatory system.

The present-day selective theories enlarge on the ideas of Ehrlich and adapt them to the data furnished by recent discoveries. This is true of the theory of natural selection proposed in 1955 by Jerne (14) and the theory of clones proposed by Burnet (6) in 1959.

I think that the great majority of geneticists favor the selective theories. Watson (27), for instance, in his excellent book on the molecular biology of genes advances very convincing arguments in favor of these theories.

They offer, indeed, undeniable advantages. First of all, they fit antibody biosynthesis into the normal pattern of protein synthesis. They explain, also, the following phenomena better than the instructive theories:

1. The continuation of antibody production by the organism for a certain length of time after the antigen has been eliminated.

2. The more rapid and intense response of the organism to a second injection of antigen. A "recollection" of the first "agression" is therefore retained by the organism.

3. Save for exceptional cases, the organism does not produce antibodies to its own constituents nor against antigens already introduced into the embryo.

The selective theories, on the other hand, explain less well the close complementarity of the antigen to the antibody.

Moreover, these theories inevitably raise a problem of probability. Using the method of Landsteiner or other analogous methods, it is possible to produce an almost unlimited number of antigens. It must be assumed, then, that all the primary structures of all the corresponding antibodies be pre-inscribed in the DNA. In spite of the existence of cross reactions this hypothesis would involve such a complication of the coded information that it would appear, at first view, rather unlikely.

For this reason a certain number of biophysicists remain faithful to the template theory, while attempting, at the same time, to adapt it to the most recent discoveries concerning the structure and production of antibodies. One of the most important of these discoveries from the structural point of view has been the fact that the antibody molecule has been found to consist of two identical heavy chains having a molecular mass of about 60,000 and two identical light chains having a molecular mass of 20,000 (8, 15, 21). These chains are linked together by disulfide bonds. When separated, they are able to rebind again to reform the original antibody molecule. It seems, therefore, that the tertiary structure is already determined by the primary structure of each of the chains.

Analyses of highly purified antibodies, however, have shown that amino-acid sequence of the chains varies *slightly* from one antibody to the other.

Under these conditions, it may be thought that the antibody specificity is acquired at the time that the amino acid sequence is established.

Haurowitz (13) does not consider this hypothesis to be contradictory to the idea of the antigen template, nor even to certain hypotheses of Pauling.

He would agree that the antibody synthesis occurs according to the general mechanism that we have just described briefly—but in the *presence of the antigen*. During this synthesis, the amino acid sequence and the tertiary structure of the antibody molecule are *interdependent*. By a process analogous to the one put forth by Pauling, the antigenic determinants would modify the folding of the antibody molecule in such a way as to keep the complementary sites available at the exterior. Such modifications could not take place without eliminating or changing the position of a few amino acids whose structure or position would interfere with the formation of the antibody reactive sites in the process. *The directives of the coded message are thus slightly transgressed.*

On the other hand, in reply to the probability objections, the adepts of the selective theory have somewhat modified or completed some of their hypotheses.

Finally, although the evolution of biological thinking has been in the direction of the idea of a selectivity of the cells giving rise to antibodies, no final choice between the two groups of theories has been made. The discussion and research which have derived from it are very useful, for, among other things, they are leading us to a better knowledge of antibody structure.

Another group of phenomena, which to its advantage can be studied *in vitro*, brings us also much valuable information on antibody structure. This is the *antigen-antibody reaction* and the possible *precipitation* of the thus formed complexes.

I have called particular attention to this question in my report in 1957. One of the conclusions which could already at that time be drawn from such studies was that the antibody valence is, on the average, in the neighborhood of 2, whereas the valence of the antigen is generally greater than 2, and for proteins containing hapten groups, of the order of 6 to 12.

Thermodynamic studies have been pursued. These studies have confirmed known results and at the same time have rendered these results more precise. For instance, in this Institute (whose fortieth anniversary we are celebrating in 1967), F. Wurmser, R. Wurmser, and their collaborators are continuing to use the system of hemagglutinin which has permitted them to demonstrate the *reversibility* of the first phase of the antigen-antibody reaction and to show that this phenomenon is accompanied by an increase of entropy very likely due to modifications in the tertiary structure of the molecule involved. The authors have recently applied thermodynamics to the attachment of agglutinins to erythrocytes to obtain information about the synthesis of isohemagglutinins (28).

Of the *quantitative* theories seeking to account for the evolution of the reaction toward partial or total precipitation of the molecular complexes, the theory of Goldberg (9–11) seemed very promising. In this brief account I will only attempt to give a general outline of the theory, touching on recent modifications and suggesting what information on the heterogeneity of antibodies the thus perfected theory can furnish us.

Goldberg adopted a method similar to that of Flory and Stockmayer for calculating the most probable distribution of molecular sizes of high polymers with ramified chains. He postulates that due to their multivalence the antigen and the antibody join to form three-dimensional ramified networks containing no cyclical structures. If the initial ratio G/A of the number of molecules G of antigen to the number of molecules A of antibody falls between the appropriate two limits around the point of equivalence, the system may reach a critical point beyond which it passes abruptly from a group of small aggregates to a group of large aggregates, for the most part, bringing about the onset of precipitation. With an ideal system, one should even observe, around the point of equivalence, the formation of an enormous molecular complex incorporating all the antigen and antibody molecules. Outside these limits, for the values of G/A ratios that are either too low or too high, the aggregates are never sufficiently voluminous to become insoluble and precipitate. The two zones of inhibition of precipitation can be interpreted in this way.

The calculations of Goldberg are purely statistical. In order to evaluate its function of distribution, i.e., the number of m_{ik} of the aggregates comprising i molecules of antibodies and k molecules of antigen, the author is obliged to make the following simplifying hypotheses analogous to those of Flory and Stockmayer.

1. During the reaction no closed chains can be formed, i.e., the aggregates never contain cyclic structures.

2. A free reactive site is just as reactive as any other site of the same nature, whatever the shape or the size of the aggregate to which it already belongs may be.

3. The f sites of antigen molecules are identical, as are the g sites of the antibody molecules, g being generally equal to 2.

The first of these hypotheses permits us to count readily the bonds which are set up and the number of free and combined sites in a particle comprising i molecules of antibody and k molecules of antigen. It is difficult to uphold it in any absolute way; but it may be supposed that the error to be derived from not taking the cycles into account is small.

I have already indicated how interesting it would be to test experimentally the structural concepts of Pauling and Goldberg in terms of precipitation phenomena. At the equivalence point Pauling postulates a maximum of regularity, which, in reality, could correspond to a structure that is faintly crystalline and perhaps detectable by X-rays. The anticycle hypothesis of

Goldberg would prohibit any such regularity. It is a hypothesis of disorder. I do not know of any work that has been done on this subject.

The second hypothesis raises objections of a steric kind; and as for the third, it inevitably poses the problem of the homogeneity of the sites.

The theory of Goldberg in its original form, however, is very interesting because it furnishes a very elegant qualitative interpretation of the precipitation phenomenon and the zones of inhibition. It brings us back to the semi-empirical formula of Heidelberger and Kendall.

To explain the absence of an inhibition zone in the presence of an excess of antibodies in the rabbit. Goldberg postulates that the antigen-antibody complexes of this animal, size being equal, are much less soluble than those of the horse.

Immunochemists such as Burtin (7) have found that even at the equivalence point, soluble aggregates can be found in the supernatant above the precipitate. This can be attributed to the fact that the antibodies are not homogeneous and include some molecules which are monovalent and therefore unable to participate effectively in the formation of three-dimensional complexes.

One of the most serious objections to be made to the theory of Goldberg is the discrepancy found between some quantitative results which this theory leads to and those found by experiment. Spiers (25) has done a very clear critical study on this subject and has concluded that the theory is partly unsuccessful.

It was natural, then, that improvements in the theory should be sought. Among recent attempts we may cite the work of Bowman, Palmiter, and Aladjem.

After a first publication of Bowman and Aladjem (4) dealing with equilibrium constants of the hapten-antibody reaction, Palmiter and Aladjem (19) offered a first generalization of Goldberg's theory. Retaining hypotheses (1) and (2) of Goldberg, they supposed that the *f sites of the antigen are different* and, consequently, involve equally as many different equilibrium constants when the f sites unite with one of the two equivalent sites of the antibody molecule. In a second generalization (1) the same two authors extended this hypothesis of the heterogeneity of sites to the antibody. The f sites of the antigen and the two sites of the antibody can thus furnish $2f$ bonds of different energies which correspond to $2f$ equilibrium constants. The heterogeneity of the bonds is described by a probability density function (PDF) of the $2f$ equilibrium constants. A first estimate of the PDF permits one to calculate the distribution of the complexes and to compare the resulting figures with the experimental data. By iterative methods it is possible then to adjust the PDF so that it fits the experimental data as well as possible. In this way we may hope to find out very exactly how heterogeneous the bonds and the sites are.

It should be added that Amano *et al.* (3) did a theoretical analysis rather similar to the one just described.

If the antigen-antibody reaction provides us with a means for learning more about the structures of an antibody and a presumably pure antigen, it also permits us to analyze mixtures of antigens.

Among the analytical methods based on the specificity of the antigen-antibody reaction, two of them seem particularly interesting. They are (1) immunochemical analysis by diffusion introduced by Oudin (18) and Ouchterlony (17) and (2) the doubly selective method of Grabar and Williams (12) in which the electrophoresis of the antigenic mixture is followed by diffusion against antibodies of the separated fractions.

In both methods one uses the diffusion (one against the other in agar-agar gel) of the antigen mixture under study and the antibody mixture collected from the corresponding antiserum. When the homologous couples of antigen-antibody meet, they form a visible precipitate whose front evolves according to a law which depends on the initial and the limiting conditions. If there are no interfering phenomena, as many evolving fronts are formed as there are homologous couples and, therefore, distinct components in the antigenic mixture. Given the extreme specificity of the reaction, one can succeed immediately in obtaining an analysis with very high precision.

This analysis can be made quantitative if one does a mathematical study of the evolution of the precipitation fronts. For this it is necessary that the initial and the limiting conditions be well defined and as simple as possible. In addition it is necessary to formulate simplifying hypotheses which are generally the following: (1) The variations of the diffusion coefficient of a species with concentration are negligible. (2) The antigen-antibody reaction is rapid and practically complete. (3) The precipitate forms almost instantaneously and is immobilized on the spot by the gel. Its possible solubility in the presence of an excess of one of the reactive agents does not appreciably modify the phenomenon. (4) The presence of a zone of precipitation does not appreciably disturb the diffusion process.

It may thus be stated that the flux of reagents (antigen and antibody) through the moving front will always be equal and opposite. If it is a case of monodimensional diffusion following Ox we may state that across this front

$$- D_A \frac{\partial C_A}{\partial x} = D_G \frac{\partial C_G}{\partial x} \tag{1}$$

D_A and D_G designating the diffusion coefficients of the antigen and the antibody, $\partial C_A / \partial x$ and $\partial C_G / \partial x$ their concentration gradients.

It is, of course, to be understood that the concentrations of C_A and C_G must be expressed in gram equivalents per unit volume, the equivalents being determined precisely by the characteristic ratio of the masses of antigen to antibodies at the equivalence point. To use Eq. (1), one then assumes that the precipitate which appears on the diffusion front has the same composition as the precipitate at the equivalence point. This hypothesis is plausible since it is at this point that the precipitate is the least soluble and forms fastest. In addition, the hypothesis has been verified by experiment.

On the diffusion front F the two reagents mutually neutralize each other. Their concentration is zero. We can write

$$C_A = C_G = O/F \tag{2}$$

According to the preceding hypotheses, the precipitation front relative to any given AG couple splits the region of diffusion into two zones: one zone where only antibodies A diffuse to the exclusion of all antigen molecules and another zone where only the antigen diffuses. In each of these zones we may apply the second law of Fick which, for a monodimensional diffusion following Ox can be written as

$$\frac{\partial C_A}{\partial t} = D_A \frac{\partial^2 C_A}{\partial x^2} \qquad \text{(1st zone)} \tag{3}$$

$$\frac{\partial C_G}{\partial t} = D_G \frac{\partial^2 C_G}{\partial x^2} \qquad \text{(2nd zone)} \tag{4}$$

Equations (1–4) and those which translate for each particular case the initial and the limiting conditions make it possible to find the equations of the evolution of the front for a large number of concrete problems.

Thus, in the case of a so-called simple diffusion (setup of Oudin), the abscissa X of the front is a parabolic function of time and has the following form:

$$X^2 = kt \tag{5}$$

This problem has been treated in all its generality by Spiers (26).

In our laboratory, several problems of "double diffusion" have been resolved; especially in collaboration with Moreau (23) and Meffroy-Biget (24).

The advantage of mathematical solutions to these problems is not only to make possible an understanding of the process underlying this phenomenon, but also to make it often possible to obtain the ratio $(C_A)_0/(C_G)_0$ of the concentrations at the outset and of the ratio D_A/D_G of the diffusion coefficients. Antibodies are generally γ-globulins whose molecular mass in man is in the neighborhood of 160,000. Their diffusion coefficient D_A in the gel used is generally known. From the ratio D_A/D_G one deduces D_G. In this way one obtains information as to the size of the antigen molecule.

It may be added that "double diffusions" can lead to stationary conditions characterized by the immobilization of the front.

In this brief discussion, I have only been able to "scratch the surface" of a very rich area, one of the rare areas of biophysics where a physical chemist by formation can dare advance without too many misgivings.

REFERENCES

1. Aladjem, F., and Palmiter, M. T. (1965). *J. Theoret. Biol.* **8**, 8.
2. Alexander, J. (1931). *Protoplasma* **14**, 296.
3. Amano, T., Syozi, I., Tokunaga, T., and Sato, S. (1962). *Biken's J.* **5**, 259.
4. Bowman, J. D., and Aladjem, F. (1963). *J. Theoret. Biol.* **4**, 242.
5. Breinl, F., and Haurowitz, F. (1930). *Z. Physiol. Chem.* **192**, 45.
6. Burnet, F. M. (1959). "The Clonal Selection Theory of Acquired Immunity." Vanderbilt Univ. Press, Nashville, Tennessee.
7. Burtin, P. (1955). *Bull. Soc. Chim. Biol.* **37**, 977.
8. Edelman, G. M., Benacceraf, B., Ovary, Z., and Poulik, M. D. (1961). *Proc. Natl. Acad. Sci. U.S.* **47**, 1751.
9. Goldberg, R. J., and William, J. W. (1952). *Discussions Faraday Soc.* **13**, 224.
10. Goldberg, R. J. (1952). *J. Am. Chem. Soc.* **74**, 5715.
11. Goldberg, R. J. (1953). *J. Am. Chem. Soc.* **75**, 3127.
12. Grabar, P., and Williams, C. A. (1953). *Biochim. Biophys. Acta* **10**, 193.
13. Haurowitz, F. (1965). *Nature,* 847.
14. Jerne, N. K. (1955). *Proc. Natl. Acad. Sci. U.S.* **41**, 849.
15. Karush, F. (1962). *J. Pediat.* **60**, 103.
16. Mudd, S. (1932). *J. Immunol.* **23**, 243.
17. Ouchterlony, Ö. (1949). *Arkiv. Kem., Mineral. Geol.* **B26**, 1.
18. Oudin, J. (1949). Thesis, Sorbonne.
19. Palmiter, M. T., and Aladjem, F. (1963). *J. Theoret. Biol.* **4**, 211.
20. Pauling, L. (1940). *J. Am. Chem. Soc.* **62**, 2643.
21. Porter, R. R. (1963). *Brit. Med. Bull.* **19**, 197.
22. Salvinien, J. (1957). *Bull. Soc. Chim. Biol.* **39**, 11.
23. Salvinien, J., and Moreau, J. J. (1960). *J. Chim. Phys.* **57**, 518.
24. Salvinien, J., and Meffroy-Biget, A. M. (1962). *J. Chim. Phys.* **59**, 501.
25. Spiers, J. A. (1958). *Immunology* **2**, 89.
26. Spiers, J. A., and Augustin, R. (1958). *Trans. Faraday Soc.* **54**, 287.
27. Watson, J. D. (1965). "Molecular Biology of the Gene." Benjamin, New York.
28. Wurmser, S., and Wurmser, R. (1966). *Acta Haematol.* **36**, 239.

A New Allosteric Effect in the Reaction Cycle of Liver Alcohol Dehydrogenase

HUGO THEORELL

Department of Biochemistry
Nobel Medical Institute
Stockholm, Sweden

Ever since my collaborators Bonnichsen and Wassén (1948) produced the first crystalline preparations of alcohol dehydrogenase from horse liver (LADH, E.C. 1.1.1.1.) we have been interested in studies of this enzyme, because its properties make it extremely suitable for experimentation. It is easy to crystallize, it is stable under suitable conditions, it collaborates with a coenzyme, NAD^+-NADH.* The reversible coupling with NADH (Theorell and Bonnichsen, 1951) is accompanied by a shift of the 340 mμ band of free NADH to 325 mμ and an increase in fluorescence intensity (Boyer and Theorell, 1956; Winer and Theorell, 1960) (see Fig. 1). More recently working with Taniguchi *et al.* (1967) we found that the combination of NAD^+ with LADH also gives spectral shifts (see Fig. 2).

Spectrophotometry and spectrofluorimetry in different combinations could, therefore, be used for determinations of the dissociation constants of the enzyme complexes with oxidized (NAD^+, "O") or reduced (NADH, "R") coenzyme, and those of many ternary complexes of enzyme, coenzymes, or fragments of the coenzymes, and different inhibitors, over a pH range from 6 to 10.

It was of interest to have accurate knowledge of these values for defining the working mechanism, both chemically and kinetically, of the enzymes. As a result of 15 years' work both in our institute and in others it may now be regarded as definitely established that the reaction of LADH with its coenzymes and ethanol-aldehyde can from the kinetic point of view be essentially described by the so-called "Theorell-Chance" (1951) mechanism:

$$E + NADH \underset{k_2}{\overset{k_1}{\rightleftharpoons}} E{\cdot}NADH \ [K_{E,R} = k_2/k_1] \tag{1}$$

$$E{\cdot}NADH + H^+ + aldehyde \underset{k_3{}'}{\overset{k_3}{\rightleftharpoons}} E{\cdot}NAD^+ + alcohol \tag{2}$$

$$E{\cdot}NAD^+ \underset{k_1{}'}{\overset{k_2{}'}{\rightleftharpoons}} E + NAD^+ \ [K_{E,O} = k_2{}'/k_1{}'] \tag{3}$$

As seen in Eqs. (1) and (3) the mechanism requires certain simple relations

---

* NAD, nicotinamide adenine dinucleotide; NADH, reduced form.

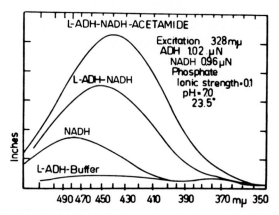

FIG. 1. Relative fluorescence intensities of compounds indicated in the figure. From Winer and Theorell (1960).

between dissociation and rate constants in order to be fulfilled and as Dalziel (1957, 1962, 1963) especially has pointed out there are some other relations between the rate constants that are also required. There will not be time to go into details concerning these relations, but they may be summarized as follows.

1. Equations (1) and (3) are fully valid; the agreement between kinetic determinations and dissociation constants has steadily improved with the increased accuracy of the measurements and the purity of the reagents. The "compulsory order" involving that the enzyme-coenzyme binding must precede the reaction with the substrate means that the reactive binding site

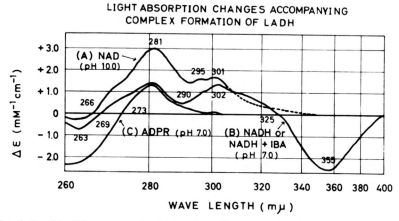

FIG. 2. Double difference spectra. Light absorption increase ($+$) or decrease ($-$) caused by coupling to LADH of NAD^+ (A), NADH (B), or ADPR (C). From Taniguchi et al. (1967).

for the substrate is formed by the attachment of coenzyme to enzyme. Whether this involved a conformational change of the protein molecule remained unknown until recently.

2. Equation (2) is an oversimplification of a more complicated sequence of reactions involving the formation of ternary complexes with the substrates. At moderate concentrations of alcohol, however, these are of no kinetic significance.

The enzyme molecule has *two* binding sites, as already noted in 1951 by Theorell and Chance. Since in all the hundreds of determinations of dissociation constants deviations from monovalent curves have never been observed we draw the conclusion that the binding sites are probably independent of each other and identical. This is supported by X-ray crystallographic data (to be discussed later), according to which the enzyme molecule (molecular weight 83,000) consists of two identical peptide chains. Each molecule contains four rather firmly bound zinc atoms, of which at least one is situated at each of the two coenzyme-binding sites. This is shown by the fact that phenanthroline forms a spectroscopically observable complex with Zn^{2+} and competes with the coenzyme binding. It is the nicotinamide moiety of the coenzyme that has to be bound in the proximity of the zinc, because the rest of the coenzyme molecule, adenosine diphosphate-ribose (ADPR, "A") can be bound to the enzyme molecule simultaneously with phenanthroline (Yonetani, 1963). In this interesting "mosaic" complex "E-Phe-A" which was crystallized by Yonetani, ADPR and phenanthroline exert no influence on each other's dissociation constants, which were found to be the same in the binary E-Phe and E-ADPR complexes as in the "mosaic" complex. This indicates that in the enzyme molecule the binding sites for Phe, as well as ADPR, must be located relatively far away from one another, whereas this cannot be the case in the complexes with the coenzymes, in which the nicotinamide moiety is bound to the Zn^{2+} site and at the same time linked to the ADPR moiety. The attachment of the coenzyme could, therefore, be expected to require a conformational change in the enzyme molecule.

The binary complexes with coenzymes and even more the ternary complexes e.g., with NADH and isobutyramide, are much more stable against various denaturing conditions than the free enzyme (Yonetani and Theorell, 1962). Whereas the reaction of free LADH with *p*-chloromercury sulfonate (PCMS) is practically instantaneous, the reaction with LADH-NADH is measurable on an ordinary recorder and the reaction with LADH-NADH-isobutyramide is much slower (see Fig. 3). This difference very probably depends on the 50 times lower dissociation constant for NADH in the ternary complex as compared with the binary complex. The stability against acid or alkaline reactions is also remarkably enhanced by complex formation (see Figs. 4 and 5) and so is the heat stability (see Fig. 6). These results also speak in favor of a

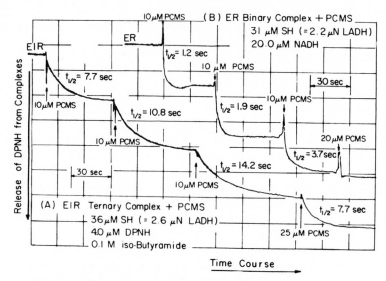

FIG. 3. Release of NADH from LADH·NADH (ER) or LADH·NADH·isobutyramide (IR) after addition of PCMS. From Yonetani and Theorell (1962).

conformational change accompanying the binding of coenzyme to enzyme, and so do the observations of Rosenberg *et al.* (1965) that the optical rotatory dispersion of the ternary complexes is very different from that of the free enzyme.

In recent experiments with Chance and Yonetani (Theorell *et al.*, 1966) we found that the formation of the ternary complexes LADH-NADH-isobutyramide was retarded by a factor of about 1000 when LADH was used as a microcrystalline suspension in 3–3.5 *M* ammonium sulfate, compared with the velocity recorded when the same LADH was dissolved in 2.5 *M*

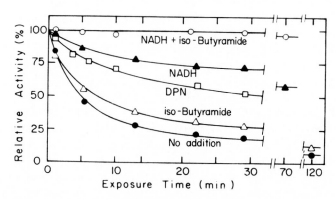

FIG. 4. Inactivation at pH 4.20, 23.5°C. From Yonetani and Theorell (1962).

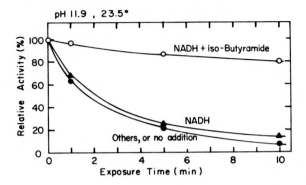

FIG. 5. Inactivation at pH 11.9, 23.5°C. From Yonetani and Theorell (1962).

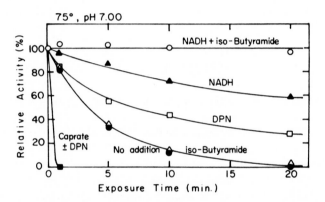

FIG. 6. Inactivation at pH 7, 75°C. From Yonetani and Theorell (1962).

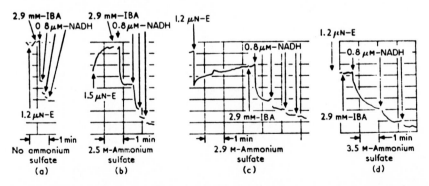

FIG. 7. Fluorescence increase after adding NADH to isobutyramide to dissolved LADH (a); partly dissolved microcrystalline suspensions of LADH in 2.5 M ammonium sulfate (b) or 2.9 M ammonium sulfate (c); all LADH microcrystalline in suspension (d). From Theorell et al. (1966).

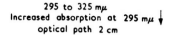

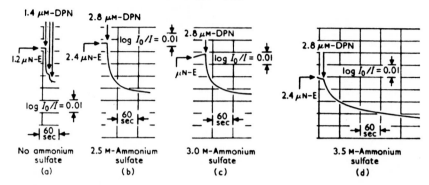

FIG. 8. Light absorption increase at 295 mμ after adding NAD$^+$ to LADH + pyrazole. Conditions similar to Fig. 7. From Theorell *et al.* (1966).

ammonium sulfate (see Figs. 7 and 8). Since the diffusion limitation could be calculated as small or of no importance, these results seemed to indicate that the binding of the coenzyme in solution causes a conformational change in the protein which is prevented by the close packing of the molecules in the protein.

The X-ray crystallographic data of Brändén *et al.* (1965) had already given conclusive evidence of a profound change in the crystal unit cell following complex formation. Crystals of both free enzymes and many complexes were produced either by slowly adding ethanol up to 10–15% to solutions in dilute phosphate buffer (0.05–0.1 μ) at 3°C or by concentrating the solutions by slow evaporation in cellophane tubes.

TABLE I

CRYSTAL DATA ON HORSE LIVER ADH AND SOME COMPLEXES[a]

Property	E	ER$_2$ = E(RI)$_2$ = E(OP)$_2$ = E(OPJ)$_2$
Symmetry	Orthorhombic	Monoclinic
Space group	C222$_1$	P2$_1$
a	56 Å	51 Å
b	75 Å	44 Å
c	181 Å	182 Å
		108°
V	760.000 Å^3	388.000 Å^3
Z	8	2
d	1.2 gm/cm^3	1.2 gm/cm^3

[a] E, Horse liver alcohol dehydrogenase; O, NAD$^+$, nicotinamide adenine dinucleotide; R, NADH; I, isobutyramide; P, pyrazole; and PJ, 3-iodopyrazole (Brändén, 1965).

Already direct inspection showed crystals of two different types: the free enzymes and the complexes with phenanthroline and adenosine diphosphate-ribose gave long prisms, whereas all complexes including coenzyme gave thick parallelepipeds with typical angles (see Figs. 9–12).

The X-ray crystallographic data (see Table I) confirmed the difference. The free enzymes and the three complexes with phenanthroline and/or ADPR gave orthorhombic crystals; complexes with coenzymes and eventually ternary ligands gave monoclinic crystals.* Figure 13 illustrates how Brändén (1965) pictured the relation between the two crystal forms. The X-ray data

FIG. 9. Crystals of LADH. From Yonetani and Theorell (1963).

proved that the molecule contains two identical peptide chains. The work is being continued with the hope of determining the three-dimensional structure of the enzyme and its complexes.

There were many reasons to believe that the coupling of the coenzyme to LADH initiates an allosteric effect on the enzyme. We (Theorell et al., 1967)

---

* LADH·NADH was originally found to be " orthorhombic." It was recently found that this enzyme preparation contained a factor that destroyed NADH, leaving crystals of only LADH. This factor has now been removed so that LADH·NADH crystals can be produced. The crystals were found to be monoclinic.

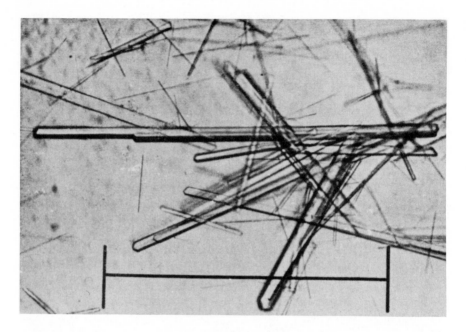

FIG. 10. Crystals of "mosaic" LADH·Phe·ADPR. From Yonetani (1963).

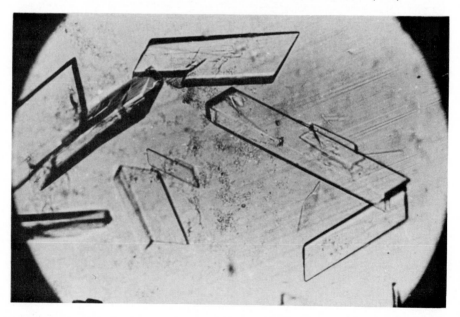

FIG. 11. Crystals of LADH·NADH. From Yonetani and Theorell (1963).

FIG. 12. Crystals of LADH·NADH·isobutyramide. From Yonetani and Theorell (1963)

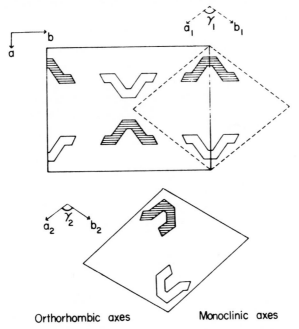

Orthorhombic axes   Monoclinic axes

FIG. 13. Change of unit cell symmetry after conformational change. From Brändén (1965).

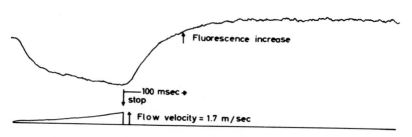

FIG. 14. One of the experimental recorder traces (3 μN E $+$ 3 μM R giving concentrations of 1.5 μM at flow stop). From Theorell *et al.* (1967).

have recently studied this effect by determining the course of the addition reaction between LADH and NADH with the aid of an improved apparatus for "stopped" or "rapid" flow. The fluorescence increase occurring upon mixing these components was recorded. Figure 14 shows a typical curve. It is seen that the fluorescence increase follows an S-shaped curve with a lag period preceding the rapid phase. This proves that the combination reaction takes place in at least two steps. The simplest explanation is that at first a partial coupling between NADH ("R") and LADH ("E") without a fluorescence increase occurs. This presumably involves the ADPR moiety, leaving the nicotinamide free with unchanged fluorescence intensity. It seems highly probable that this reaction should occur first since the association constant of ADPR to E is as high as 10^5 M^{-1} (Yonetani, 1963). Since the association constant of R at pH7 is $10^{6.7}$ it is seen that the coupling of the nicotinamide moiety to E contributes only a factor of $10^{1.7}$. In this primary, low-fluorescent compound (ER) the pyridine ring is presumably restrained from coupling with the active Zn^{2+} site. In a second, intramolecular reaction step a conformational change of the protein (and of NADH?) brings them into proximity, whereupon coupling to the highly fluorescent ER* can occur.

If E equals one-half the LADH molecule the reaction sequence could thus be written

$$E + R \underset{k_{-1}}{\overset{k_{+1}}{\rightleftharpoons}} ER \underset{k_{-2}}{\overset{k_{+2}}{\rightleftharpoons}} ER*$$

This scheme presumes that the two identical peptide chains react independently of each other. Evidence in favor of this assumption was already given by the fact that the dissociation curve between NADH or NAD^+ and LADH has a monovalent shape.

In a series of experiments the concentrations of NADH (micromolar) and enzyme-binding sites (E in "μN" $= 2\ \mu M$) were kept equal to each other, but they varied in total concentration after mixing: 5, 1.5, 0.5, and 0.15 μM. Numerical values for the rate constants k_{+1}, k_{-1}, k_{+2}, and k_{-2} were sought by the aid of a digital computer (IBM 1401) to give the best fit with the experimental values. The amount of computing work was reduced to reasonable dimensions by certain restrictions that followed from earlier kinetic and equilibrium data. For instance, k_{+1} had to be between 15 and 20 μM^{-1} per second, and the total dissociation constant of NADH (K) was known to be 0.20 μM at pH 7 and 23°C, and

$$K = \frac{k_{-1}/k_{+1}}{1 + k_{+2}/k_{-2}} = 0.20\ \mu M$$

The best agreement with the experimental results was obtained by the following values

$$k_{+1} = 17\ \mu M^{-1}/\text{sec}; \qquad k_{-1} = 44\ \text{sec}^{-1}$$
$$k_{+2} = 120\ \text{sec}^{-1}; \qquad k_{-2} = 10\ \text{sec}^{-1}$$

(see Figs. 15–18). The solid-line curves are the computed values for the concentrations of E ($=$ R), ER, and ER*. The points are experimental and are the means of 2–6 experiments. The height of the points is the $\pm$ standard deviation, σ.

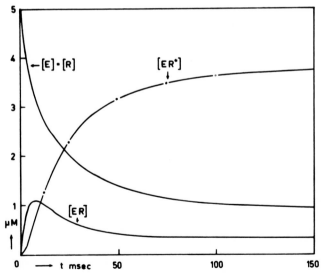

FIG. 15. 10 μN E mixed with 10 μM R to give 5 μM concentration at $t = 0$. Solid-line curves are calculated; points are the average of two experiments: $\sigma < 0.3\%$, 23°C, pH 7.0. From Theorell *et al.* (1967).

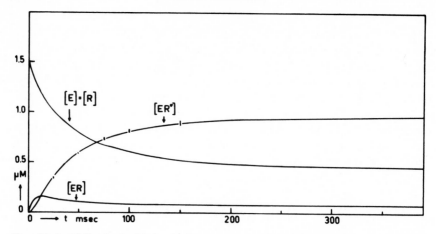

FIG. 16. 3 μN E mixed with 3 μM R to give 1.5 μM concentration at $t = 0$. Solid-line curves are calculated; points are the average of six experiments: height 2σ, 23°C, pH 7.0. From Theorell *et al.* (1967).

It can be seen that the two series with the highest concentrations gave perfect agreement. In the two series with the lowest concentrations some small deviations occurred, as expected from the higher noise level, but they were well within the experimental error.

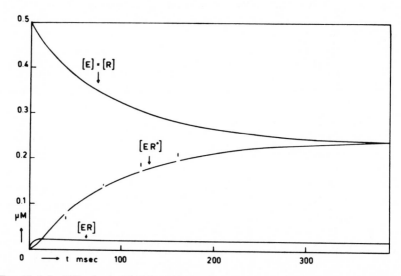

FIG. 17. 1.0 μN E mixed with 1.0 μM R to give 0.5 μM concentration at $t = 0$. Solid-line curves are calculated; points are the average of three experiments: height 2σ, 23°C, pH 7.0. From Theorell *et al.* (1967).

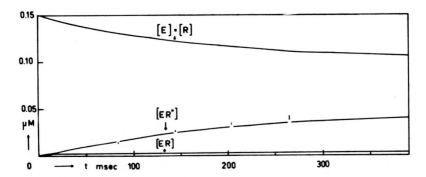

FIG. 18. 0.3 μN E mixed with 0.3 μM R to give 0.15 μM concentration at $t = 0$. Solid-line curves are calculated; points are the average of three experiments: height 2σ, 23°C, pH 7.0. From Theorell *et al.* (1967).

One problem of great interest needs to be discussed: Can the conformational change indicated by our results take place within one of the halves of LADH, leaving the other half intact? In Fig. 19 we have tried to give a schematic representation of different possibilities. The two identical halves are called E_1 and E_2. E_1E_2 (the free enzyme), $E_1PheA \cdot E_2PheA$, and $E_1R \cdot E_2R$ have an open conformation with sulfhydryl groups exposed to PCMS, the protein being labile against heat and extreme pH values. In $E_1R^* \cdot E_2R^*$ the SH groups are fully inaccessible to denaturing agents; in $E_1R \cdot E_2R^*$ half are

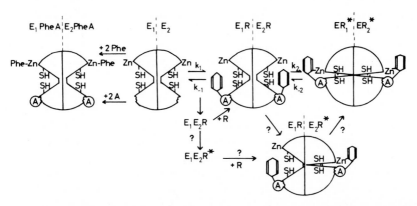

FIG. 19. Schematic representation of possible compounds of LADH with NADH, phenanthroline, and adenosine phosphate-ribose. $E_1 = E_2 = \frac{1}{2}$LADH; R = NADH; R* = NADH, highly fluorescent; Phe = phenanthroline; and A = adenosine diphosphate-ribose.

inaccessible. The possible intermediates in the formation of $ER_1^* \cdot E_2R^*$ may be depicted as follows.

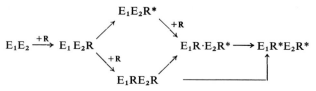

In case the conformational change would require $E_1R \cdot E_2R$ to be formed before it can occur, only the lowest pathway would be followed. This can be ruled out on the basis of experimental evidence.

Two sets of "stopped-flow" experiments were made with concentrations of 3 μN E + 0.5 μM R and 0.5 μN E + 3 μM R after mixing. The results agreed with the computed rate constants: the fluorescence increase curves were equal. If $E_1R \cdot E_2R$ were required for conformational change the rate would be expected to be much slower with a sixfold excess of enzyme, when mainly E_1E_2R would be formed, but it would be faster with excess of R, when mostly E_1RE_2R would be formed.

The possibility that the binding of one R would so greatly favor the attachment of a second R to the other peptide and since mixed complexes of the type E_1E_2R and $E_1E_2R^*$ scarcely exist is compatible with the mono-valent shape of the dissociation curve between R and E.

At present, therefore, all experimental evidence favors the two peptide chains in the LADH molecule operating and even being able to change their conformation independently of one another. Further experiments are, however, planned to ascertain this somewhat surprising conclusion.

In summarizing some previously known facts—(1) the compulsory order reaction kinetics, (2) the stabilization of LADH by complex formation with coenzyme, (3) the slow reactivity of crystalline LADH with coenzyme, and (4) the change in crystal form by complex formation with coenzyme—they all seem to depend on an allosteric effect.

Following a suggestion by Gibson who in similar experiments (cf. Geraci and Gibson, 1967) had not observed any S-shape of the curves, we repeated our experiments, as already reported at the VII International Congress in Tokyo, August, 1967. We then found that because of a technical failure the time constant of the recording system had been considerably greater than believed. With proper time constant the S-shape disappeared. This does not mean that our postulated two-step mechanism is necessarily wrong. An at least twofold binding of the coenzyme to different binding sites of the enzyme cannot reasonably occur simultaneously, and there are, as evident from the text, many reasons to show that a conformational change is induced by the coenzymes coupling. But it means that the rate constants k_2, k_{-1}, and k_{-2}

cannot be resolved by this method, whereas k_1 $(= k_{+1})$ retains its value, $17\mu M^{-1} \times sec^{-1}$. In the formula $K_{ER} = k_2/k_1$ (eq. 1) k_2 will be $= k_{-1}/1 + k_{+2}/k_{-2}) = 3.4 sec^{-1}$, in agreement with $K = 0.20 = (3.4/17)$ μM.

REFERENCES

Bonnichsen, R., and Wassén, A. (1948). *Arch. Biochem. Biophys.* **18**, 361.

Boyer, P. D., and Theorell, H. (1956). *Acta Chem. Scand.* **10**, 447.

Brändén, C.-I. (1965). *Arch. Biochem. Biophys.* **112**, 215.

Brändén, C.-I., Larsson, L.-M., Lindqvist, I., Theorell, H., and Yonetani, T. (1965). *Arch. Biochem. Biophys.* **109**, 195.

Dalziel, K. (1957). *Acta Chem. Scand.* **11**, 1706.

Dalziel, K. (1962). *Biochem. J.* **84**, 244.

Dalziel, K. (1963). *J. Biol. Chem.* **238**, 2850.

Geraci, G. and Gibson, Q. H. (1967). *J. Biol. Chem.* **242**, 4275.

Rosenberg, A., Theorell, H., and Yonetani, T. (1965). *Arch. Biochem. Biophys.* **110**, 413.

Taniguchi, S., Theorell, H., and Åkeson, Å. (1967). *Acta Chem. Scand.* (in press).

Theorell, H., and Bonnichsen, R. (1951). *Acta Chem. Scand.* **5**, 1105.

Theorell, H., and Chance, B. (1951). *Acta Chem. Scand.* **5**, 1127.

Theorell, H., Chance, B., and Yonetani, T. (1966). *J. Mol. Biol.* **17**, 513.

Theorell, H., Ehrenberg, A., and de Zalenski, C. (1967). *Biochem. Biophys. Res. Commun.* (in press).

Winer, A., and Theorell, H. (1960). *Acta Chem. Scand.* **14**, 1729.

Yonetani, T. (1963). *Biochem. Z.* **338**, 300.

Yonetani, T., and Theorell, H. (1962). *Arch. Biochem. Biophys.* **99**, 433.

Yonetani, T., and Theorell, H. (1963). *Arch. Biochem. Biophys.* **100**, 553.

On the Mechanism of Binding Choline Derivatives to an Anticholine Antibody

J. C. METCALFE, A. S. V. BURGEN, AND O. JARDETZKY *

Department of Pharmacology
University of Cambridge, England and
Harvard Medical School
Boston, Massachusetts

I. Introduction

The correlation of chemical structure with biological activity has thus far met with very limited success, despite a century of intense investigation reflected in the extensive literature on the subject accumulated since the original publication of Crum Brown and Fraser (1) in 1869.

The principal reason for this lack of success is that biological activity, however measured, is *at best* a reflection of a thermodynamic parameter, the free energy of association $\Delta F = RT \ln K$, where K is the association constant of the active complex. The ambiguities inherent in the interpretation of thermodynamic parameters in structural terms are well known in chemistry (2). It is therefore not to be hoped that valid correlations between structure and activity will be established until the geometry of the complexes and the relative magnitude of the contributions of different intermolecular forces to the total free energy can be evaluated in detail.

The use of antibodies to antigenic determinants of pharmacological interest as models for drug receptors, introduced by Burgen *et al.* (3, 4), and the demonstration by Jardetzky *et al.* (5–10) that differential changes of the relaxation rates in the high-resolution NMR spectra of both small and large molecules can be used to identify the portions of molecules directly involved in the formation of complexes in solution have made it possible to carry out a systematic experimental investigation of the effects of varying the chemical structure on the structure and dynamic behavior of the complexes formed in a wide range of systems.

In the first paper of this series (11) we have reported an analysis of the stabilization of the tetramethylammonium ion (TMA) and several related compounds by an antibody specific for the choline moiety based on NMR relaxation measurements. It is the purpose of this chapter to present the results of further experiments on the same antibody system using other haptens.

* Present address: Department of Biophysics and Pharmacology, Merck Sharp & Dohme Research Laboratories, Rahway, New Jersey.

II. Materials and Methods

Methods of preparation and measurement have already been reported in detail (9, 11) and will not be repeated here. The spectra were recorded on either a Varian DP 60 or a Varian HA 100 high-resolution NMR spectrometer. Relaxation times were measured with the apparatus described previously (9). The only point bearing emphasis is that all line widths in multiplets were corrected for spin-spin splitting [Eq. 2, Jardetzky and Jardetzky (9)]; this is of particular importance in the case of N-methyl protons, where a latent splitting of ~ 0.6 cps can be detected at extremely high resolution, or more readily, by the wiggle-beat method (12). The usual corrections were applied for field inhomogeneity and viscosity (9, 11).

III. Results and Discussion

For the purpose of mapping the geometry of the antibody binding site the use of mono- and disubstituted tetramethylammonium derivatives is of particular interest, since the relative stabilization of different parts of the substituted molecules will allow some inferences on the orientation of the hapten with respect to the protein surface.

Acetamidophenylcholine ether (AACPE) was studied as a representative of the monosubstituted series and its N-benzyl derivative (NBAACPE) as a representative of the disubstituted series. The structures and NMR spectra of the two compounds are given in Figs. 1 and 2 respectively. The relaxation

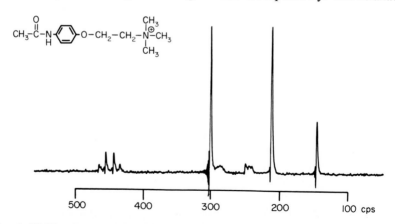

FIG. 1. NMR spectrum of AACPE in D_2O. Chemical shifts are in cycles per second at 60 Mc from hexamethyldisiloxane (HMS). Line assignments are as follows: (*a*) singlet at 140 cps, acetyl methyl; (*b*) singlet at 208 cps, N-methyl (nitrogen splitting unresolved); (*c*) multiplet at 235 cps, N-CH$_2$ of choline; (*d*) multiplet at 290 cps, O-CH$_2$ of choline; (*e*) singlet at 300 cps, HDO; and (*f*) A$_2$B$_2$ pattern at 445 cps, phenyl ring.

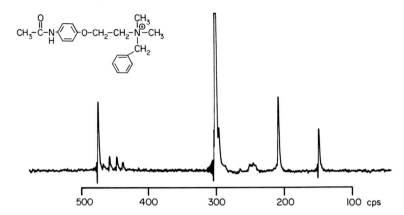

FIG. 2. NMR spectrum of NBAACPE. Shifts and assignments as in Fig. 1, with the exception of the additional singlet at 480 cps due to the benzene ring of the N-benzyl substituent.

rates for different parts of the unbound molecules, calculated from relaxation times as measured by the direct saturation method (9), are given in Table II.

When the spectra of the two compounds are examined in the presence of the specific antibody (11), selective broadening of the lines is observed, whereas in the presence of pooled γ-globulin from nonimmunized rabbits the effects are nonselective and easily accounted for by the viscosity correction. The appearance of single lines for each group and the monotonic dependence of the relaxation rates on the concentration ratio suggest that one is dealing here with a case of rapid exchange between the free and the bound forms of the hapten. The condition for the rapid exchange case is that the dissociation rate of the complex k_D exceeds $(1/T_1)_{bound}$, the relaxation rate in the bound state; k_D may be estimated from the affinity constants if it is assumed that the recombination rate of the complex is of the order of 10^8. Such an estimate is not unreasonable for small molecule-protein interactions, judging from recent measurements of Eigen (13). The values of k_D estimated for the present system and given in Table I make it plausible that the rapid exchange model would apply. In these cases (8)

$$(1/T_1)_{obs} = \alpha(1/T_1)_{bound} + (1 - \alpha)(1/T_1)_{free}$$

where $(1/T_1)_{obs} = (1/T_2)_{obs} = \pi \Delta\nu_{1/2\ obs}$, is the observed relaxation rate, $\Delta\nu_{1/2}$ the observed line width, and α the fraction of hapten bound. In all such cases the exchange process acts as an amplifier for the detection of the relatively small fraction of hapten bound by its effect on the spectrum of the excess free hapten. The limit of amplification is, of course, different in different cases, being determined by the value of $(1/T_1)_{bound}$; but is generally of the order of 100–1000 in proton resonance experiments.

TABLE I

AFFINITIES AND ESTIMATED RATE CONSTANTS FOR
HAPTEN-ANTIBODY COMPLEXES

Hapten	Affinity K_s LM^{-1}	Limiting exchange rate (k_D min)sec^{-1}	Probable	
			k_D sec^{-1}	k_R LM^{-1} sec^{-1}
TMA	1.0×10^4	2×10^2	10^3–10^4	10^7–10^8
AACPE	3.0×10^5	3×10^2	3.3×10^2	$\sim 10^8$
NBAACPE	3.0×10^4	4×10^2	3.3×10^2–10^3	10^7–10^8

A plot of $(1/T_1)_{obs}$ against the molar ratio of antibody combining sites to hapten is shown in Fig. 3 for AACPE and in Fig. 4 for NBAACPE. In both cases it is apparent that the part of the molecule stabilized most is the methyl-ammonium portion of choline and that stabilized least is the acetyl methyl group

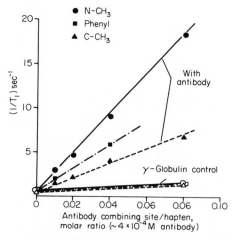

FIG. 3. Concentration dependence of the relaxation rates of the AACPE hapten with the anticholine antibody or nonspecific γ-globulin at constant antibody concentration.

at its opposite end. Stabilization of the phenyl ring is intermediate. A comparison of stabilization factors is given in Table II. It is noteworthy that in the disubstituted NBAACPE the second aromatic moiety is stabilized to a still somewhat greater extent than the N-methyl groups. Both findings are consistent with a pocket-shaped binding site, which allows the first bulky substituent to protrude from its orifice, but requires that the second bulky substituent be forced into its interior. A model of this type of binding site

TABLE II

RELAXATION RATES FOR FREE AND BOUND HAPTEN GROUPS
AND APPROXIMATE STABILIZATION FACTORS[a]

Hapten	Group	$(1/T_1)_{free}$ (sec^{-1})	$(1/T_1)_{bound}$ (sec^{-1})	$S = (1/T_1)_b/(1/T_1)_f$
TMA				
	N-CH$_3$	0.8	130	160
AACPE				
	N-CH$_3$	0.6	230	380
	A$_R$	0.7	120	170
	C-CH$_3$	0.6	80	130
NBAACPE				
	N-A$_R$	0.7	300	430
	N-CH$_3$	0.6	240	400
	A$_R$	(0.7)	(120)	(170)
	C-CH$_3$	0.6	70	120

[a] The stabilization factor S is a convenient, although less relevant measure of stabilization than $(1/T_1)_{bound}$, as it directly reflects the change of correlation times for each group.

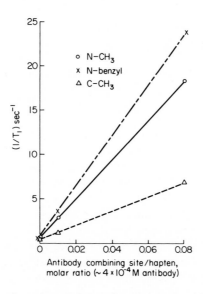

FIG. 4. Concentration dependence of the relaxation rates of NBAACPE hapten with anticholine antibody at constant antibody concentration.

is shown in Fig. 5. The concept derives further support from the additional finding that N-benzyltrimethylammonium ion shows a preferential stabilization of the N-methyl groups over the aromatic moiety. Thus, in effect, when the acetamidophenylether moiety is removed from the NBAACPE, the best fit into the binding site is obtained by reversing the remainder of the molecule, so that the N-benzyl residue protrudes toward the outside.

FIG. 5. Molecular model of the mode of binding of AACPE hapten to anticholine antibody.

Some interaction of the substituent side chain on the N-methyl head with amino acid side chains forming the orifice of the binding site is, of course, not precluded in this model. It may, in fact, contribute to the greater stabilization of the N-methyl head in the substituted compounds, as compared to the unsubstituted TMA (Table II). Such an interpretation must, however, remain uncertain because it is also possible that the lesser stabilization of TMA results from the additional degree of freedom of rotation within the binding site which the TMA may have as a result of its spherical symmetry. The distinction between the two possibilities hinges on a direct comparison of the correlation times of the N-methyl group with those of the amino acid side chains in the walls of the binding site. If these are found to be identical, the complex may be presumed to relax as a unit and the symmetry argument may be ruled out. Comparisons of this type have recently become possible (10), but have not yet been made in the present hapten-antibody system.

It may be hoped that some additional information on the nature of the predominant intermolecular forces can be obtained from a study of the temperature and solvent dependence of the stabilization factor as defined in Table II. In several previous cases it has been found that increase in ionic strength of the solvent produced a marked increase in the stabilization, consistent with a predominantly "hydrophobic" nature of the complexes (8, 9). On the other hand, it might be expected that a predominantly electrostatic complex will be dissociated by an increase in ionic strength.

When the dependence of the stabilization of AACPE by the anticholine antibody on the ionic strength of the medium was examined at the usual operating temperature, 30°C, a progressive dissociation of the complex with an increase in ionic strength was found. In a separate set of experiments, however, the temperature dependence of the stabilization was shown to be anomalous. Its solvent dependence was therefore examined over the entire accessible temperature range.

The results of the combined experiments are presented in Fig. 6 for two ionic strengths, $\mu = 0.1$ and $\mu = 2.0$. At the lower ionic strength the observed relaxation rate of the N-methyl protons goes through a maximum—and so does the $(1/T_2)_{bound}$ calculated from the rapid exchange model, whereas both quantities for the acetyl methyl show a monotonic decline with increasing temperature as expected. The most likely explanation of this finding is as follows.

The relaxation rate of the N-methyl protons, in the bound state $(1/T_2)_{bound}$ being larger than that for the acetyl methyl, is probably of the

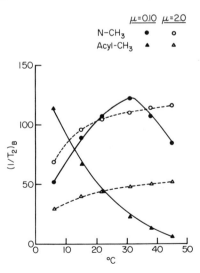

FIG. 6. The effect of temperature and ionic strength on the relaxation rates of the AACPE complex with anticholine antibody.

same order of magnitude as the dissociation rate constant of the complex. Thus, the condition for the rapid exchange model, $k_D > (1/T_2)_{bound}$ is satisfied only for the acetyl methyl, but not for the N-methyl. The observed relaxation rate of the latter thus contains a sizable contribution from an exchange term (12).

Since the rate of exchange might be expected to increase with increasing temperature, so will the observed relaxation rate, until the exchange rate becomes sufficiently rapid to satisfy the rapid exchange condition. In the present case this occurs at about 30°C. If one accepts a value of the order of 10^8 LM^{-1} sec^{-1} as probable for the recombination rate of a hapten-antibody complex, on the basis of temperature-jump measurements by Eigen (13), the stability constant for the AACPE-antibody complex yields a value of 3.3×10^2 sec^{-1} for the dissociation rate constant. Comparison of this value with the value of $(1/T_2)_{bound}$ at 30°C, which is of the same order of magnitude, supports the foregoing argument. At temperatures below 30°C the relaxation of the N—CH$_3$ protons is then increasingly dominated by the exchange process, the contribution of the bound species rapidly becoming negligible; at 5°C the observed line is essentially the line of the free species, slightly broadened by exchange. In contrast, the width of the acetylmethyl line remains on the high-temperature (fast exchange) side of its expected maximum throughout the entire accessible temperature range. At 5°C it may be approximated by a sum of $(1/T_2)_{bound}$ and an exchange term. Extrapolation of the $(1/T_2)_{bound}$ value for the acetyl methyl to 5°C would yield a figure of the order of 50–60 sec^{-1}. If it assumed that the exchange term for the acetylmethyl at 5°C is the same as for the N-methyls (the assumption is equivalent to that of a single exchange process), the total relaxation rate of the acetylmethyl should be of the order of 100–120 sec^{-1}. The findings are quite consistent with this rather crude approximation. The simple model thus accounts for the main features of the temperature curves, including their crossover in the vicinity of 12°C.

The interpretation of the salt effect, on the other hand, is subject to considerably greater ambiguity. If the foregoing model for the temperature effect is correct, and one extrapolates from the high-temperature values of $(1/T_2)_{bound}$ to obtain values of true, i.e., exchange unaffected $(1/T_2)_{bound}$ over the entire temperature range, the predominant effect of high ionic strength at temperatures below 40°C is to destabilize the complex of the N-methyl moeity with the binding site. This is in contrast to the case of TMA (11), in which ionic strength has little, if any, effect on the degree of stabilization and may speak for a somewhat looser binding of the asymmetrically substituted compound. Because of the r^6 dependence of the van der Walls' terms as contrasted to the r^2 dependence of the electrostatic term, the latter will predominate at the relatively larger intermolecular distances in the looser complex. The fact that the observed stabilization of the N-methyls in AACPE is greater than in TMA is not necessarily incompatible with this explanation. The observed stabilization is an average over all degrees of freedom of rotation; it is therefore entirely possible that the greater restriction of the individual modes of rotation in the tighter TMA complex is more than compensated by the additional degrees of freedom resulting from the symmetry of the molecule. It must be borne in mind, however, that changes in the configuration of the binding site or in the conformation of the entire protein could also be conceivably produced by increasing the salt concentration and could contribute to the overall observed effect. A complete analysis of this problem will not be possible until a detailed study of the behavior of the individual protein side chains has been carried out. This applies in particular to the stabilizing effect of salt on the acetyl methyl group, which may reflect the appearance of new binding sites or less specific interactions with the protein surface. Such studies are now in progress (10).

A rather different finding is obtained when one examines the interaction of acetylcholine, or, because of its greater stability in solution, $(-)$-methacholine with the anticholine antibody. The structure and NMR spectrum of methacholine are given in Fig. 7. When the spectrum is examined in the presence of

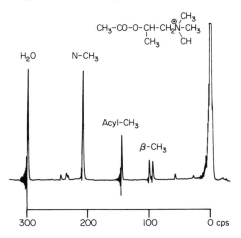

FIG. 7. NMR spectrum of $(-)$-methacholine. Shifts as in Fig. 1, assignments as indicated.

the antibody, it is found that the broadening of all three types of methyl groups is roughly proportional. This, together with the very small temperature coefficient of $(1/T_2)_{bound}$ shown in Fig. 8 strongly suggests that the molecule is bound as a rigid unit, in contrast to the binding of AACPE and NBAACPE. A complete understanding of this phenomenon is again contingent on a detailed study of the behavior of the amino acid side chains which comprise

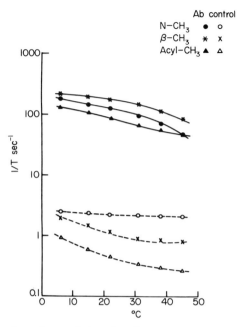

FIG. 8. Temperature dependence of the relaxation rates of (−)-methacholine, free and bound to anticholine antibody.

the binding site. However, one of its possible implications should not be overlooked. The finding that an acetylmethyl group is stabilized to almost the same degree as the N-methyl moiety if it is within the same distance from the latter as in acetylcholine, but not if, as in the case of AACPE or NBAACPE, it is further removed, is rather striking. It suggests that the antibody contains a specific site capable of binding the acetyl moiety, only if it is within a certain critical distance of the N-methyl moeity. Now, it should be borne in mind that the antibody in question was prepared against a phenylcholine ether, which did not have an acetyl group in the required position (11). The finding therefore raises the possibility that the antibody binding site has a structure and conceivably a genetic origin common to other binding sites for acetylcholine, such as may occur in cholinesterase and the acetylcholine receptor. This possibility clearly warrants further investigation.

ACKNOWLEDGMENTS

The authors are indebted to the Smith, Kline and French Foundation and the National Science Foundation for generous support. We are also grateful to Professor Lord Todd for permission to use the NMR spectrometer in the University Chemical Laboratory, Cambridge for part of this work.

REFERENCES

1. Crum Brown, A., and Fraser, T. R., *Trans. Roy. Soc. Edinburgh* **25**, 151 and 693 (1869).
2. Edsall, J. T., and Wyman, J., "Biophysical Chemistry," Vol. 1, Chapter 11. Academic Press, New York, 1958.
3. Metcalfe, J. C., Marlow, H. F., and Burgen, A. S. V., *Nature* **209**, 1142 (1966).
4. Burgen, A. S. V., Marlow, H. F., and Metcalfe, J. C., to be published.
5. Jardetzky, O., Wade, N. G., and Fischer, J. J., *Nature* **197**, 183 (1963).
6. Jardetzky, O., *Advan. Chem. Phys.* **7**, 499 (1964).
7. Wiener, N., and Jardetzky, O., *Arch. Exptl. Pathol. Pharmakol.* **248**, 308 (1964).
8. Fischer, J. J., and Jardetzky, O., *J. Am. Chem. Soc.* **87**, 3237 (1965).
9. Jardetzky, O., and Jardetzky, N. G. W., *Mol. Pharmacol.* **1**, 214 (1965).
10. Meadows, D. H., Markley, J. L., Cohen, J. S., and Jardetzky, O., to be published.
11. Burgen, A. S. V., Jardetzky, O., Metcalfe, J. C., and Wade-Jardetzky, N. G., *Proc. Nat. Acad. Sci. U.S.* **58**, 447 (1967)
12. Pople, J. A., Schneider, W. G., and Bernstein, H. J., "High Resolution Nuclear Magnetic Resonance." McGraw-Hill, New York, 1959.
13. Eigen, H., *Physiol. Chem.* **15**, 344 (1964).

Energy Transfer Methods in the Study of Ligand-Protein Interactions

GREGORIO WEBER

Department of Chemistry and Chemical Engineering
University of Illinois
Urbana, Illinois

I. Introduction

A molecule in an electronic excited state can interact with another molecule in the ground state in such a manner that electronic energy is transferred from the first (donor D) to the second (acceptor A). Symbolically,

$$D + h\nu \longrightarrow D^* \qquad \text{(excitation)}$$
$$D^* + A \longrightarrow D + A^* \qquad \text{(transfer)}$$

Since the acceptor molecule is now in an excited state A^*, emission of light (resulting in sensitized fluorescence), radiationless transition to the ground state $A^* \rightarrow A$ (observed as a quenching of the fluorescence of D by A), or further transfer of energy to another molecule may take place. For electronic energy transfer to occur certain conditions must be fulfilled:

1. Transitions of equal energy D^*–D and A–A^* corresponding to emission in the donor and absorption in the acceptor must exist to permit the transfer of energy. This condition is fulfilled if there is appreciable overlap between the emission spectrum of the donor and the absorption spectrum of the acceptor.

2. The distance between the two molecules must be such that electromagnetic coupling between them is possible. There will be for each system a characteristic distance R at which molecules interacting with optimal orientation (see below) have probability of transfer ν equal to the probability of emission λ. The dependence of the probability of transfer upon the distance r between D and A is, in general, given by the relation,

$$\nu/\lambda = (R/r)^6$$

The values of R calculated from the experimental data of sensitized fluorescence in solution (Förster, 1947; Galanin and Lewschin, 1951) and from the dependence of the polarization of the fluorescence upon the concentration (Pheofilov and Sveshnikoff, 1943; Weber, 1954) vary from 10–50 Å units, depending principally upon the overlap of absorption and emission bands as demanded by a theory due to Förster (1947).

3. The mutual orientation of the two molecules is of importance. Transfer is maximal for parallel oscillators and zero for oscillators at right angles to each other.

We shall be concerned with energy transfer occurring in certain ligand-protein complexes. Since there can be under our conditions up to five moles of ligand bound per mole of protein, two types of energy transfer processes can be expected: (a) protein chromophore-ligand and (b) ligand-ligand. From the previous considerations it would follow that as a result of studies of energy transfer of the types a and b an estimate of the average distance and orientation among the interacting moieties could, in principle, be obtained. This is indeed the case, but it is possible to foresee a more interesting use of such studies: Energy transfer from the protein chromophores to the ligand will proceed with different efficiencies in protein molecules with different numbers of ligands bound. Similarly, ligand-ligand transfer will only take place in molecules with two or more ligands bound and not at all in those with only a single ligand. Thus, some conclusions as to the distribution of ligands among the protein molecules—if not the complete distribution itself—may be reached by such studies. The importance of the knowledge of the distribution of ligands among the protein molecules need no stressing. It is no exaggeration to say that until and unless such distribution is determined binding equilibria are not rigorously defined. Further, we shall introduce another use of energy transfer data. In protein molecules having two distinct relaxation times of rotation (prolate molecules) the transfer of energy among bound ligands may result in an apparent change in average rotational diffusion with increasing transfer, if the ligands have a preferential orientation with respect to the protein geometry. From such an effect the existence of intramolecular orientation of the bound ligands may be inferred.

The study of the transfer from protein chromophores to ligand gives rise to a decrease in yield of protein fluorescence and to an increase in yield of ligand fluorescence and it may be followed by observations of the fluorescence spectrum. The transfer from one ligand chromophore to another does not give rise to changes in spectrum or yield and may be followed by the decrease in the polarization of the fluorescence of the ligand as the number bound increases.

II. The System Anilinonaphthalene Sulfonate (ANS)– Bovine Serum Albumin (BSA)

A description of the binding is given by Daniel and Weber (1966). Five molecules of anilinonaphthalene sulfonate (ANS) are bound with a 150-fold increase in fluorescence yield over that of the free ligand in water.* At pH 7

---

* Equilibrium dialysis experiments show that further molecules of ANS are bound with much less afinity and a quantum yield at least an order of magnitude smaller. There are, apparently, in albumin two physically different types of binding sites available to ANS (Pasby and Weber, 1967).

the binding of these five moles of ANS occurs over a span of free ligand concentration of 1.8 logarithmic units as shown in Fig. 1. The titration curve, apart from an inflection at $\bar{n} = 1$, is characteristic of five sites with a single dissociation constant, which at 25°C equals 2×10^{-6} mole/liter.

As shown schematically in Fig. 2, illumination with λ_1 (275 nm) will preferentially excite the protein, since the molar absorption coefficients of protein and ligand are, respectively, 43,000 and 1500 at this wavelength. The fluorescence emission from the two types (hatched areas in the diagram) can be almost completely resolved from each other, the respective emission maxima being at 344 nm for the protein and at 475 nm for the ligand.

The emission by the ligand can be excited in an entirely independent fashion by the use of a wavelength such as λ_2, situated between 330 and 400 nm, at which there is no appreciable absorption by the protein.

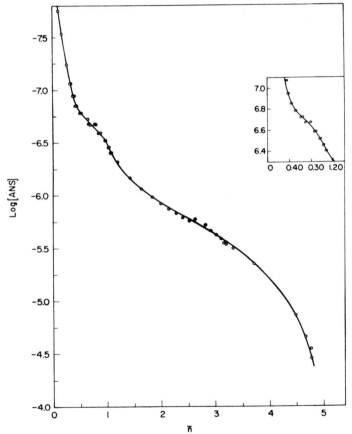

FIG. 1. Plot of logarithm of free anilinonaphthalene sulfonate (log [ANS]) against number bound $\bar{n}$. From Daniel and Weber (1966).

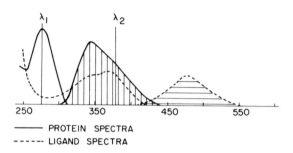

FIG. 2. Schematic of absorption and emission bands in the system ANS-BSA in neutral buffers. Albumin spectra, solid line. ANS spectra, dashed line. Fluorescent emissions are hatched areas.

If protein and ANS are mixed at total concentrations of each much higher than the dissociation constant, virtually stoichiometric binding of ligand to protein takes place provided the ratio of ligand to protein is five or less. In such mixtures containing various ratios of ANS to protein, excitation at 275 nm gives rise to the fluorescence spectra of Fig. 3. In the absence of ligand ($\bar{n} = 0$) only the broad fluorescence band of the protein is seen. As $\bar{n}$ increases a decrease of the 344 nm emission and a corresponding rise of the characteristic ligand fluorescence is seen. There is a single point at which all the curves cross. The significance of this isoemissive point is in every way the same as that of the isobestic point in absorption. We are dealing with two chromophores with unique fluorescence spectra and yield. It follows that ANS must be bound with identical yield of fluorescence at least until $\bar{n} = 2$. Other experiments show that there is no change in yield until at least $\bar{n} = 3.5$ and that the change in yield from $\bar{n} = 3.5$ to $\bar{n} = 5$ must be less than 15%.

III. Ligand Distribution and Protein Fluorescence Yield

In general, if $f_r(0 \leq r \leq 5)$ is the fraction of the protein molecules with r ligands, and Q_r its average fluorescence yield, the observed average yield Q when $\bar{n}$ ligands are bound is

$$Q = \sum_{r=0}^{r=5} f_r Q_r \tag{1}$$

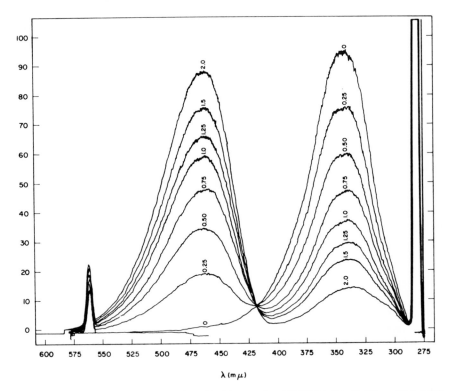

FIG. 3. Spectra of mixtures of ANS and BSA at ratios ANS/BSA 0 to 2. The added ANS is stoichiometrically bound to the protein as the concentrations of both are much higher than the dissociation constant. From Daniel and Weber (1966).

The titration curve shown in Fig. 2, characteristic of multiple binding sites with a unique dissociation constant indicates that the values of f_r must be given by the random distribution. Therefore,

$$ f_r = \binom{5}{r} \left(\frac{\bar{n}}{5}\right)^r \left(1 - \frac{\bar{n}}{5}\right)^{5-r} \tag{2} $$

Moreover, if all sites are equally likely to be occupied

$$ \frac{Q_r}{Q_0} = \frac{\lambda}{\lambda + r\bar{\mu}} = \frac{1}{1 + r\varepsilon} ; \qquad \varepsilon = \bar{\mu}/\lambda \tag{3} $$

where μ is an average probability of transfer from excited tryptophan to a bound ANS molecule. Introducing the values of f_r and Q_r in Eqs. (2) and (3) into Eq. (1), one can determine the initial slope S_i and the final slope S_f in a plot of Q/Q_0 against $\bar{n}$ (Fig. 4) (Weber and Daniel, 1966). As shown by them,

$$S_{i(\bar{n} \to 0)} = \frac{-\varepsilon}{1 + \varepsilon}$$

$$S_{f(\bar{n} \to 5)} = \frac{-\varepsilon}{(1 + 5\varepsilon)(1 + 4\varepsilon)}$$

(4)

We found that $S_i = 0.69$, so that S_f should equal 0.019 in good agreement with the value of 0.021 found experimentally. We have thus used the transfer from tryptophan to ANS to check the validity of the assumption of random distribution of ligands among the protein molecules.

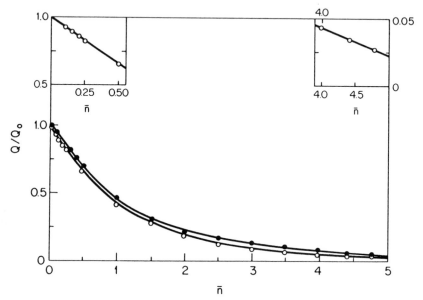

FIG. 4. Plot of Q/Q_0, the relative protein fluorescence yield, as a function of number bound $\bar{n}$ (Weber and Daniel, 1966). The insets show the initial and final slopes in an expanded scale. —O— pH7, —●— pH5.

IV. Fluorescence Polarization Observations

Interpretation of the data of fluorescence polarization requires distinction between two types of depolarizing influences that must be clearly separated—Brownian rotational motion and electronic energy transfer from one ligand molecule to another bound to the same protein molecule. To separate these effects it is necessary to determine the polarization of the fluorescence at different viscosities and to extrapolate to infinite viscosity. The extrapolated value p_0 is then independent of the molecular motions and its variations may be attributed directly to energy transfer. Figure 5 shows plots of the reciprocal of the polarization against the ratio T/η (absolute temperature/viscosity) for various values of $\bar{n}$. In this case the temperature was varied and the change in solvent viscosity was due to the change in temperature. In other studies the viscosity was changed by addition of sucrose at constant temperature with

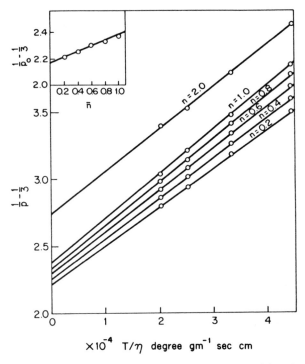

FIG. 5. Plots of reciprocal of polarization against temperature/viscosity (Perrin plots) for ANS/BSA. Mixtures of $\bar{n} = 0.1$ to $\bar{n} = 2$ (Weber and Daniel, 1966).

similar results. The concentrations of protein and ligand used were much greater than the dissociation constant so that stoichiometric binding at all the temperatures used was assured. In accordance with Perrin's law (Perrin, 1926, 1929) the plots are straight lines from which p_0 may be obtained by extrapolation.

At any given temperature, say 25°C, the value of the mean relaxation time ρ of rotation of the macromolecule may be calculated by the use of the equation

$$\frac{1}{p} - \frac{1}{3} = \left(\frac{1}{p_0} - \frac{1}{3}\right)\left(1 + \frac{3\tau}{\rho}\right) \tag{5}$$

when τ (the lifetime of the excited state) is known. We have measured τ by two different direct methods. One consists in illumination with a spark in air at a repetition rate of kilocycles (Koechlin, 1961). The exciting light pulse had a half-width of 4–5 nsec, and the reflected pulse of exciting light or alternatively the excited fluorescence was followed by means of a 100 MHz oscilloscope. The second method (Spencer and Weber, 1967) involved the measurement of the phase shift between photocurrents due to fluorescence and simple scattering, when illumination was done by a sinusoidally modulated source operating at 14.2 MHz. The phase shift and relative modulation of the scattered or fluorescent light were measured by a cross-correlation procedure derived from that of Birks and Little (1953). The first method gave values between 15.7 and 15.0 nsec when $\bar{n}$ varied from 1 to 5. The second method gave values decreasing in the same interval from 16.0 to 15.0 nsec. Using the value of 16.0 nsec and the data of Fig. 5, between $\bar{n} = 0$ and $\bar{n} = 1$, $\bar{\rho}/\tau = 6.5 \mp 0.2$ or $\bar{\rho} = 103 \mp 6$ nsec. The value for $\bar{\rho}$ is 30% smaller than 125–135 nsec obtained by attachment of covalent labels to bovine serum albumin and could conceivably reflect the preferential orientation of the absorbed ANS molecules with respect to the molecular ellipsoid of inertia. These results, except for the absolute values of τ and $\bar{\rho}$, have been reported by Weber and Daniel (1966). They also reported that the value of $\bar{\rho}/\tau$ at $\bar{n} = 2$ equals 7.7 (120 nsec) and is therefore significantly higher than that observed at $\bar{n} = 1$. More recent observations on this phenomenon are due to Dr. S. Anderson, who has shown that the apparent increase in $\bar{\rho}$ with $\bar{n}$ continues beyond $\bar{n} = 2$. At $\bar{n} = 4$ a value of $\bar{\rho}$ equal to 130–140 nsec is observed (Fig. 6). The p_0 values continue to decrease at the same time and a plot of p_0 against $\bar{n}$ for excitation with 366 nm is shown in Fig. 7. We have, therefore, two distinct phenomena revealed by the polarization observations: the decrease in p_0 with $\bar{n}$ and the concomitant increase in $\bar{\rho}$.

V. Significance of Decrease of p_0 with Number Bound

This decrease must be attributed solely to the energy transfer among ligands bound to the same molecule of protein. In attempting to deduce an average distance and orientation among the ligands bound we use the result already obtained from observations of protein-ligand energy transfer that the distribution of ligands among the binding sites does not appreciably differ from the normal distribution. It is further necessary to assume that the depolarization observed is brought about by a single transfer, or in other words, that multiple transfers do not play an important part in the depolarization. This assumption is a sensible one in view of the monotonic decrease in slope found when p_0 is plotted against $\bar{n}$. If multiple transfers were important in this respect, one would expect an increase in slope in the plot of p_0 against $\bar{n}$. With these

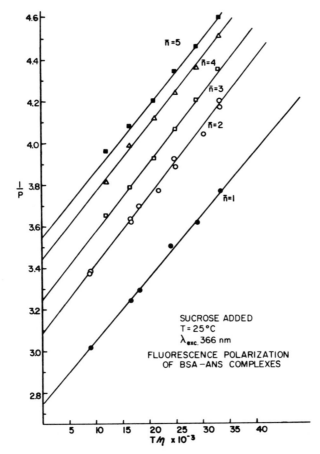

FIG. 6. Similar plots as those of Fig. 5 for $\bar{n} = 1$ to $\bar{n} = 5$ (Anderson and Weber, 1967).

assumptions it is possible to analyze the data in terms of an average probability of transfer between a pair of binding sites and the loss of polarization resulting after a single transfer. From these values an average distance between a pair of sites and an average angle between the exchanging transition moments in emission can be derived (Weber and Daniel, 1966). It must be stressed that such calculations define an equivalent system that would produce the experimental results, but hardly guarantee that the physical situation in the protein is that postulated for the equivalent system, much in the same way as the hydrodynamic ellipsoid associated to the protein molecule need not represent its actual physical shape. In the equivalent system that best fits the experimental data the average distance between a pair of sites is some 20 Å, and the angle defined by the two transition moments is on the average 33°.

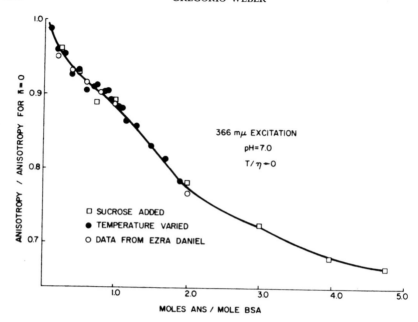

FIG. 7. Plots of $A_0(\bar{n})/A_0(\bar{n} \to 0)$, the ratio of the limiting anisotropies at $\bar{n}$ and 0, for various values of $\bar{n}$ (Anderson and Weber, 1967).

VI. The Increase in Relaxation Time with Number Bound

The simplest way to account for such an increase would be a true change in hydrodynamic parameter. A change in relaxation time of rotation of 30%, however, ought to give rise also to some change in the frictional coefficient for translation, and the molecular weight being constant, to changes in the sedimentation velocity. Careful measurements of $s_{20,w}$ with $\bar{n} = 1$ to $\bar{n} = 5$ revealed constancy within 1%. Alternatively, one can ask the question whether energy transfer among the bound ligands may be a possible cause of apparent increase or decrease in relaxation time, as computed from fluorescence depolarization. This question must be answered in the affirmative. Consider a molecule with cylindrical symmetry: There are two kinds of motion by means of which a change in orientation of the cylinder (Fig. 8) may be brought about. The first is rotation about an arbitrary equatorial diameter and is, therefore, a motion of the axis of revolution in space. Perrin has shown (1934) that as the elongation of a prolate ellipsoidal molecule increases this motion becomes slower. The other type of disorienting motion is a rotation about the axis of revolution (equatorial rotation). This is a fast motion and it does not appreciably lengthen as the prolate ellipsoid increases its elongation. We wish to consider the effects of transfer of electronic energy upon the measured relaxation time. For this purpose we can calculate the effects to be observed in an ellipsoid carrying two oscillators at variable orientation to the axis of revolution. To make matters more specific we assume that in all cases the two oscillators define between them an angle of 33°, a value chosen because as already mentioned the effects produced on p_0 are equivalent to those present in our system. If ligand 1 is the only one directly excited by absorption of light, a polarization p_1 would be observed given by the relation*

$$A_1 = (1 - f)A_{11} + fA_{12}; \qquad A_{ij} = (1/p_{ij} - \tfrac{1}{3})^{-1} \tag{6}$$

MODES OF DISORIENTATION OF CYLINDER

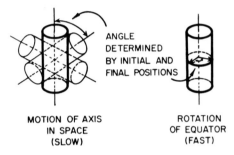

FIG. 8. Modes of disorientation of the cylinder.

where p_{11} is the polarization due to excitation and emission by 1, p_{12} is the polarization when 1 is excited and 2 emits after transfer of the excited state from 1, and f is the fraction of the absorbed quanta transferred from 1 to 2. Similarly, following the excitation of 2 the polarization is given by

$$A_2 = (1 - f)A_{22} + fA_{21} \tag{7}$$

The experimental polarization $\bar{p}$ resulting from equal excitations of 1 and 2 is given by

$$\bar{A} = (1 - f)\left(\frac{A_{11} + A_{22}}{2}\right) + fA_{12} \tag{8}$$

since due to the symmetries operating in the system $p_{12} = p_{21}$. In the calculations of p_{11}, p_{22}, and p_{12} we have employed equations due to Memming (1961) for the depolarization due to the Brownian rotation of prolate ellipsoids. The equations of Memming are equivalent to those of Perrin (1936), but simpler to

---

* The quantity $A = (1/p - \tfrac{1}{3})^{-1}$, the anisotropy of the emission (Jablonski, 1960), derives its usefulness from the fact that the A quantities are additive when weighted with respect to their contribution to the total fluorescent intensity, which is not the case for p itself (Weber, 1952).

manage. We have considered in the calculations—due mainly to Dr. S. Anderson—two types of models. In the first type (I, II, and III in Fig. 9) the two oscillators are contained in a plane together with the axis of revolution, with which they make the various angles shown in the figure though keeping an angle of 33° with each other. Transfer among such oscillators (axial transfer) changes the angle between the excited one and the axis of revolution and is, therefore, equivalent to a diffusion brought about by the slow axial motion. If random orientation of the oscillators in the axial plane was achieved by such transfers the axial mode of diffusion would not further affect the polarization and the effects observed in increasing T/η would be entirely due to the fast equatorial rotation. We would thus expect under conditions of axial transfer an apparent decrease in relaxation time with increasing transfer. Conversely, consider the case in which the oscillators make the same angle with the axis (IV and V in Fig. 9), but determining between them some equatorial angle.

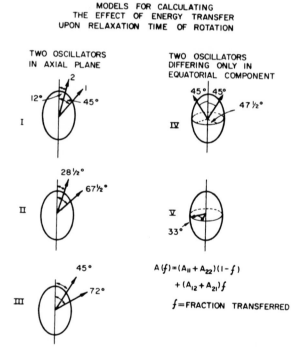

MODELS FOR CALCULATING
THE EFFECT OF ENERGY TRANSFER
UPON RELAXATION TIME OF ROTATION

TWO OSCILLATORS
IN AXIAL PLANE

TWO OSCILLATORS
DIFFERING ONLY IN
EQUATORIAL COMPONENT

$$A(f) = (A_{11} + A_{22})(1-f)$$
$$+ (A_{12} + A_{21})f$$
$$f = \text{FRACTION TRANSFERRED}$$

FIG. 9. Models to which the equations of Memming were applied to calculate the effect of energy transfer upon the apparent value of the rotational relaxation time. I, II, III: Models with oscillators in the axial plane; IV, V: Models with oscillators at a unique angle to the axis. In V the oscillators are at an angle to the equator; in IV they are contained in it. In all models the oscillators make an angle of 33° with each other.

Here transfer does not change the axial position and it is equivalent to equatorial or fast diffusion. In the limit the whole of the depolarization by molecular rotations would be due to the slow axial diffusion and the apparent relaxation time of rotation would be that of the long axis.

In summary axial transfer would show itself in a decrease in apparent

EFFECT OF ENERGY TRANSFER
UPON RELAXATION TIME
OF ROTATION

OSCILLATORS
E_1, E_2 CONTAINED
IN AXIAL PLANE:
TRANSFER REPLACES
SLOW MOTION:
RELAXATION TIME
DECREASES WITH
TRANSFER

OSCILLATORS E_1 AND
E_2 CONTAINED IN
EQUATORIAL PLANE:
TRANSFER REPLACES
FAST MOTION:
RELAXATION TIME
INCREASES WITH
TRANSFER

FIG. 10. Effect of energy transfer upon relaxation time of rotation. The figure intends to convey the principle that transfers in the axial plane eliminate the axial diffusion effects, whereas transfers in an equatorial plane eliminate the effects of equatorial rotation.

relaxation time, whereas equatorial transfer would conversely result in an apparent increase (Fig. 10).

The detailed numerical calculations confirm this qualitative picture and further indicate:

1. The effects of increase or decrease in the apparent relaxation time due to transfer among oscillators is a sensitive function of the elongation. With favorable orientation of the exchanging oscillators an axial ratio of 2 can give rise to appreciable changes.

2. For an ellipsoid of axial ratio 4, estimated for bovine serum albumin from hydrodynamic parameters an increase of 30% in ρ when almost every excitation undergoes one transfer is reasonable if the oscillators are oriented at the same angle to the axis, whether they are contained in the equatorial plane or at an angle to it (Fig. 11).

These considerations show that energy transfer among the oscillators rigidly bound to a protein can furnish evidence regarding their preferential intramolecular orientation, even if it is not possible to fix this unequivocally.

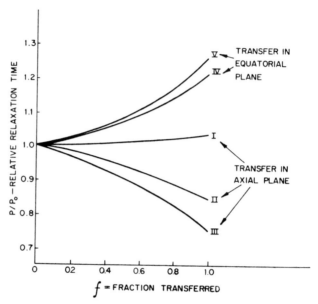

EFFECT OF ENERGY TRANSFER
UPON RELAXATION TIME OF ROTATION
PROLATE ELLIPSOID OF AXIAL RATIO 4.0

FIG. 11. Changes in relative rotational relaxation time ρ/ρ_0 as a function of f in Eq. (8), for the various models of Fig. 9 and a prolate ellipsoid of axial ratio 4. It is seen that only models IV and V could account for the rise in ρ/ρ_0 experimentally observed (Anderson and Weber, 1967).

It may be argued that all the calculations show is that energy transfer is a plausible explanation. Can we, however, furnish direct evidence of the connection between the observed increase in $\bar{p}$ and energy transfer? Such direct evidence would be the disappearance of both effects under suitable conditions. This experiment can, in fact, be performed.

VII. Dependence of Energy Transfer with Excitation Wavelength in ANS

Figure 12 shows the polarization spectrum of the fluorescence of ANS in propylene glycol at $-55°C$. Light of variable wavelength λ, with a bandwidth of 10–20 Å over most of the spectrum was used for the excitation, and the polarization of the emission at wavelengths longer than 500 nm was recorded. The polarization spectrum of the dilute solution (10^{-4} M) shows

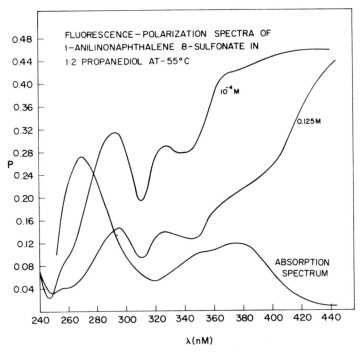

FIG. 12. Polarization spectra of 10^{-4} M and 0.125 M ANS in propylene glycol at $-55°$C. The absorption spectrum at room temperature has been superimposed to show the correspondence of the various electronic transitions with the different polarization values.

certain details of interest such as the degenerate character of the longest wavelength absorption band. In a concentrated solution (0.125 M) the loss of polarization due to energy transfer among the dissolved ANS molecules is clearly seen. At wavelengths longer than 380 nm, however, the extent of energy transfer progressively decreases and at 435 nm hardly any depolarization is observable. The causes of this phenomenon are under study. It seems to be a far from unique occurrence and may be expected in all cases in which the absorption and emission bands are widely separated due to solvent orientation during the excited state (Lippert, 1957). Be it as it may, we need only for our purpose to ascertain the fact that transfer fails to occur after excitation with 435 nm. Observations of p and $\bar{\rho}$ (Figs. 13 and 14) show that in contradistinction to excitation with 366 nm, the change of either quantity with $\bar{n}$ observed upon excitation with 436 nm is negligible. The apparent increase in $\bar{\rho}/\tau$ observed may be due to a shorter lifetime upon excitation with 436 nm. In any case the intrinsic correlation between increase in relaxation time and existence of energy transfer seems demonstrated.

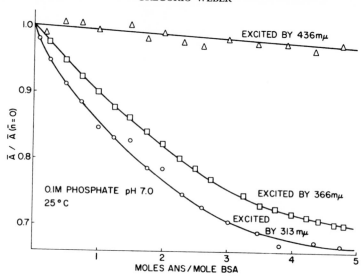

FIG. 13. Ratios $\bar{A}(\bar{n})/\bar{A}(0)$ as functions of $\bar{n}$ for excitation at 313, 366, and 436 nm. The relative inefficiency of energy transfer at the latter wavelength is apparent (Anderson and Weber, 1967).

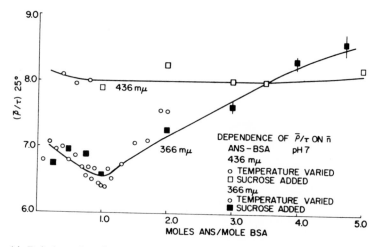

FIG. 14. Relative relaxation times of rotation $\bar{\rho}/\tau$ as a function of $\bar{n}$ for excitation with 366 and 436 nm (Anderson and Weber, 1967).

In summary we wish to stress the several points on which energy transfer methods help us to form a picture of the binding process, namely, the distribution of the ligands among the protein molecules, the approximate orientation and distance among the ligands, and finally the existence of a preferential orientation of the ligands with respect to the protein geometry.

REFERENCES

Anderson, S. R., and Weber, G. (1967). Unpublished data.

Birks, J. B., and Little, W. A. (1953). *Proc. Phys. Soc.* (*London*) **A66**, 921.

Daniel, E., and Weber, G. (1966). *Biochemistry* **5**, 1893.

Förster, T. (1947). *Ann. Physik* [6] **2**, 55.

Galanin, A., and Lewschin, B. (1951). *Zh. Eksperim. i Teor. Fiz.* **21**, 220.

Jablonski, A. (1960). *Bull. Polish Acad., Math. Phys. Ser.* **8**, 259.

Koechlin, Y. (1961). Thesis, Faculté de Sciences, Université de Paris.

Lippert, E. (1957). *Z. Elektrochem.* **61**, 962.

Memming, R. (1961). *Z. Physik. Chem.* (*Frankfurt*) [N.S.] **28**, 168.

Pasby, T. L., and Weber, G. (1967). Unpublished data.

Perrin, F. (1926). *J. Phys.* **7**, 390.

Perrin, F. (1929). *Ann. Phys.* (*Paris*) [10] **12**, 169.

Perrin, F. (1932). *Ann. Phys.* (*Paris*) [10] **17**, 283.

Perrin, F. (1934). *J. Phys.* **5**, 497.

Perrin, F. (1936). *J. Phys.* **7**, 1.

Pheofilov, P., and Sveshnikoff, B. (1943). *J. Phys. USSR* **3**, 493.

Spencer, R. D., and Weber, G. (1967). Unpublished data.

Weber, G. (1952). *Biochem. J.* **51**, 155.

Weber, G. (1954). *Trans. Faraday Soc.* **50**, 554 (1954).

Weber, G., and Daniel, E. (1966). *Biochemistry* **5**, 1900.

Optical Studies of Polypeptide-Solvent Interactions in Sulfuric Acid–Water Mixtures

K. ROSENHECK

Polymer Department
The Weizmann Institute of Science
Rehovot, Israel

I. Introduction

The nature of the interplay between forces governing the conformational properties of biopolymers and those determining the associations of these polymers with small molecules in the environment constitutes a major problem in biophysical research. A difficulty in the optical study of such interactions arises from the possibility that the optical parameters currently used for detecting and assessing certain conformational features may also be subject to medium effects. These parameters have usually been determined under a limited set of specific conditions, and their applicability to a variety of polymer-solvent systems is by no means fully established. Thus, the adequacy of the rotatory dispersion parameter, b_0, in following helix-coil transitions in polypeptide solutions containing strong organic acids was recently questioned (1, 2), after inconsistencies between the results of optical rotatory measurements, on the one hand, and either infrared or nuclear magnetic resonance data, on the other, had been found.

An important illustration of this state of affairs is the recent discovery of highly solvent-dependent far-ultraviolet Cotton effects in a series of diamide models (3), arising from profound changes in the mixing between $\pi-\pi^*$ and $n-\pi^*$ transitions under the influence of various interacting solvents.

The common disadvantage in using organic acids, such as formic, dichloroacetic, and trifluoroacetic acids, as helix-breaking solvents is that, owing to the high absorptivities of these acids, optical measurements in the ultraviolet are, in fact, limited to the determination of b_0 only and that the information residing in the short-ultraviolet absorption spectra, Cotton effects and circular dichroism bands, cannot thus be exploited.

It is for this reason that we have investigated the sulfuric acid–water system as an alternative to the mixed organic solvent system. This system has many attractive features applicable to the kind of work considered here. The main one, of course, is its good optical transmittance down to the short UV limit of present-day spectrometers. Another one is that it dissolves several polypeptides which until now could only be studied in organic solvents. Last but not least, a substantial amount of information has been accumulated in recent years on its mode of interaction with various amides. Also, the ionic equilibria and other interactions, such as H-bonding, existing in sulfuric acid itself and its mixtures with water, are gradually becoming better understood with the aid of experimental data obtained by various physical methods (4).

Using this solvent system we have been studying conformational changes in polypeptides, as well as conformation-independent, medium-induced changes in the optical properties of small amides and oligo-peptides. One of the questions to which these studies are directed is that of the importance of the protonation of the peptide groups constituting the polypeptide chains, relative to that of other interactions, in determining helix stability.

Here we wish to describe observations made on poly-L-glutamic acid and some other water-soluble polypeptides for which comparable aqueous solution studies exist.

II. Experimental Findings

A. The Helix-Coil Transition in Poly-L-Glutamic Acid (PGA)

PGA undergoes a reversible helix-coil transition at moderately high concentrations of sulfuric acid (SA), which by spectral criteria shows great similarity to that occurring in aqueous solutions at varying pH values (Fig. 1). At $\sim 35\%$ SA the spectrum is practically identical with that of the α-helical form of the polymer in water, pH ~ 4. At $\sim 65\%$ SA the absorbance

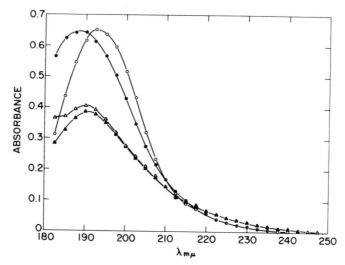

FIG. 1. Absorption spectra of PGA. (▲) $\sim 35\%$ H$_2$SO$_4$ in water; (△) water, pH 4; (●) $\sim 65\%$ H$_2$SO$_4$ in water; (○) water, pH 7.

has increased to that of the random-coil form in water, pH $\sim$ 7. The blue shift of the absorption maximum at the high acid concentration is a common feature of several random polypeptides, as will be shown. This blue shift is usually associated with slight decreases in the intensity of the absorption band.

Figure 2 shows transition profiles for PGA, plotted at three different wavelengths. Figure 3 gives transition profiles for two PGA's having different

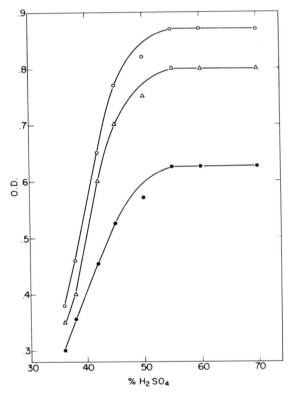

FIG. 2. The variation of optical density of solutions of PGA as a function of H_2SO_4 concentration, measured at different wavelengths. ($\bullet$) 200mμ; ($\triangle$) 195mμ; ($\bigcirc$) 190mμ.

degrees of polymerization. At SA concentrations lower than $\sim$35%, PGA starts to precipitate. The onset of the helix-coil transition must occur somewhat below, but not too far from this concentration, since the absorbance values indicate an almost intact helix. The optical rotatory dispersion (ORD) spectra for the PGA α-helix in water and $\sim$ 35% SA are compared in Fig. 4 and are seen to be practically identical with regard to both location and rotational strength of the Cotton effect.

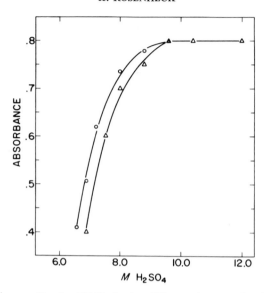

FIG. 3. Transition profiles for PGA's having different degrees of polymerization (DP).
(O) DP ~ 150; (△) DP ~ 475.

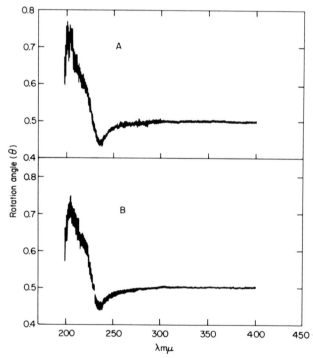

FIG. 4. ORD spectra of helical PGA (DP ~ 475). (A) Water, pH 4; (B) ~ 35% H_2SO_4.

B. Polypeptides in the Random-Coil Form

The effect of varying acidities on poly-L-lysine (PL), which is in the random-coil form throughout the acidity range considered, is shown in Fig. 5 and illustrates the blue shift and intensity change referred to earlier. Practically the same behavior is observed for polysarcosine (PS) (Fig. 6).

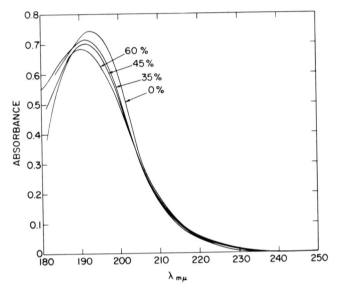

FIG. 5. Absorption spectra of PL (DP ~ 1250) at varying acidities.

The absorption spectral changes in the random polypeptides are thus seen to be relatively minor. More drastic changes are, however, found in the ORD spectra of these materials. Figure 7 shows the Cotton effect for the PGA random-coil and it is seen that its amplitude in ~65% SA has decreased substantially from that exhibited in the aqueous solution. The same result is obtained when the ORD of randomly coiled PL is compared in the same two solvents.

C. The Poly-L-Proline II (PP) Helix

It was somewhat surprising to find similar changes in the absorption spectra of PP (Fig. 8). In Fig. 9 the ORD spectra of PP in the region of the first Cotton effect are shown. Again, decreases in amplitude are seen, as SA concentration increases.

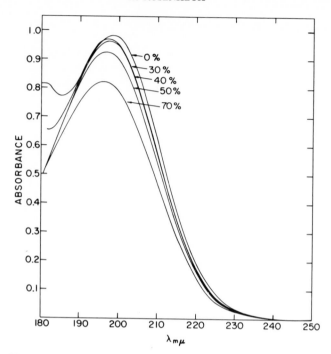

FIG. 6. Absorption spectra of PS (DP $\sim$ 30) at varying acidities.

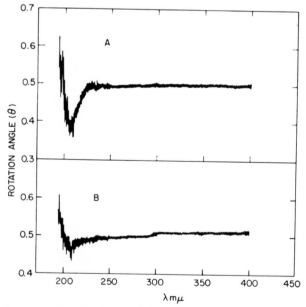

FIG. 7. ORD spectra of randomly coiled PGA (DP $\sim$ 475). (A) water, pH 7; (B) $\sim$ 65% H_2SO_4.

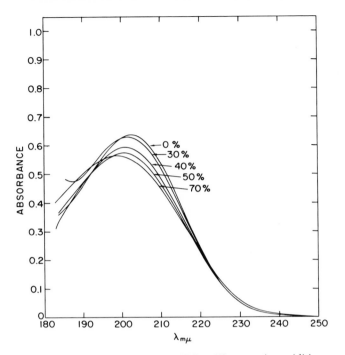

FIG. 8. Absorption spectra of PP (DP ~ 60) at varying acidities.

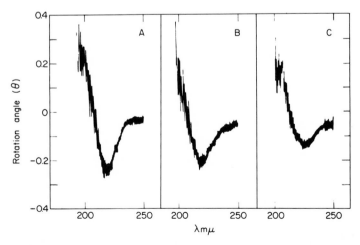

FIG. 9. ORD spectra of PP (DP ~ 60). (A) water; (B) ~ 40% H_2SO_4; (C) ~ 65% H_2SO_4.

III. Discussion

The rather striking similarity in the spectral shifts exhibited by the various randomly coiled polypeptides and also by PP points to a type of polymer-solvent interaction that does not critically depend on the exact chemical structure of the repeating peptide unit, such as degree and kind of substitution on the amide nitrogen. Albeit small, they are rather easily seen, as long as other large-scale spectral changes, such as those due to the α-helix to coil transition in PGA, are absent. A more detailed comparison between the fractional decreases in absorbance induced by varying acidity in these polymers is at present underway. As a first possibility, we can take the ORD data on PGA, PL and PP as an indication of the breakdown of part of an ordered structure. In this case no large changes in the absorption spectra would be expected. This is because the hypo- or hyperchromic effects for the conformational transitions in terms of which these ORD changes could be interpreted are rather insignificant. In the case of PP the decrease of the rotatory strength could signify a partial randomization of the helix, resulting from an increased rate of peptide cis-trans isomerization, occurring at high acidities. In the case of the PGA and the PL random coil it could mean the breakdown of some local ordered regions of polypeptide chains in highly extended form. Such regions have been postulated (5) in order to explain the very existence of the Cotton effect at 200 mμ. It is, however, known (6, 7) that both the formation of the PP helix and the formation of extended chain β-structures do not lead to large intensity changes in the absorption spectra. May it be said in passing that if the decrease in the Cotton effect for random PGA and PL should indeed signal a breakdown of a partly ordered structure, the state of these polymers in SA actually corresponds more closely to a true random-coil than that existing in aqueous solution.

In addition to the possibility of some small-scale conformational transitions occuring in "random-coil" polypeptides and also in the poly-L-proline helix, it may be considered that the spectral changes are a result of a certain amount of protonation of the peptide groups and probably also of hydrogen bonding to one or more of the molecular species present in the SA-water mixtures. Spectral changes in the π-π* band of simple amides occurring on protonation have been found previously, and pK values for the equilibria involved have been calculated from these spectral data (8). The changes taking place in propionamide are shown in Fig. 10 and are seen to be much larger than those found in the polypeptides. The long wavelength edges of the π–π* band in N,N-dimethylacetamide and PS are compared in Figs. 11 and 12, respectively. Their quantitative evaluation indicates that, if the shifts in the small amides are taken as a measure of protonation exclusively, the polymers are protonated to a small degree only—about 10–20% in the range

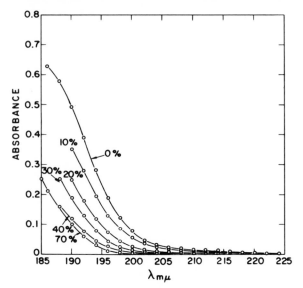

FIG. 10. Absorption spectra of propionamide at varying acidities.

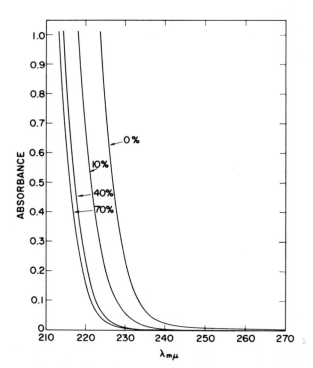

FIG. 11. The onset of the absorption spectrum of *N,N*-dimethylacetamide at varying acidities.

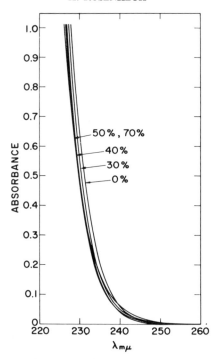

FIG. 12. The onset of the absorption spectrum of PS at varying acidities.

of acidities measured. Similar conclusions have been reached for the poly-L-alanine–chloroform–trifluoroacetic acid system on the basis of NMR measurements (2). Protonation of the peptide group occurs, as is well established, on the carbonyl oxygen (9) and leads to an increase in the barrier to rotation around the amide C—N bond. It thus stabilizes the planarity of the peptide group, and this factor alone should have no weakening effect on a helical polypeptide. There is, however, the possibility that the helix-coil transitions in PGA and other α-helical polypeptides occur in SA as a consequence of electrostatic interactions between these protonated peptide units. Preliminary calculations of the electrostatic energy of a partially protonated α-helix indicate that small degrees of protonation would be sufficient for inducing helix-coil changes even in media of high dielectric screening, such as those considered here. Protonation of the carboxyl groups can probably be neglected, on the account of the much weaker basicity of carboxyl, there being an approximate difference of six orders of magnitude between the respective ionization constants (8).

In conclusion, it may be said that the intrinsic optical properties of polypeptides in the random and helical forms do not seem to be drastically influenced by the medium, on passing from an aqueous solution to a highly acidic solvent, such as the sulfuric acid-water system. The monitoring of the helix-coil transition by the optical parameters currently used is thus not significantly impaired in such media. The small spectral changes, occurring in the absence of a major conformational change, may be interpreted in terms of either or both of two events: (1) The disruption of small ordered regions in the, essentially, randomly coiled polypeptides and the poly-L-proline helix, and (2) a rather limited amount of protonation of the constituent peptide groups.

ACKNOWLEDGMENTS

It is a pleasure to thank Professor A. Katchalsky for very valuable comments, and Miss H. Miller and Miss A. Zacharia for their excellent assistance in carrying out this work.

REFERENCES

1. Hanlon, S., and Klotz, I. M. (1965). *Biochemistry* 4, 37.
2. Stewart, W. E., Mandelkern, L., and Glick, R. E. (1967). *Biochemistry* 6, 143.
3. Schellman, J. A., and Nielsen, E. B. (1967). "Biopolymer Conformations" (G. N. Ramachandran, ed.) Academic Press, New York, 1967.
4. Gillespie, R. J., and Robinson, E. A. (1965). In "Non-aqueous Solvent Systems" (T. C. Waddington, ed.), pp. 117–210. Academic Press, New York.
5. Rosenheck, K., and Sommer, B. (1967). *J. Chem. Phys.* 46, 532.
6. Gratzer, W. B., Rhodes, W., and Fasman, G. D. (1963). *Biopolymers* 1, 319.
7. Rosenheck, K., and Doty, P. (1961). *Proc. Natl. Acad. Sci. U.S.* 47, 1775.
8. Edward, J. T., and Wang, I. C. (1962). *Can. J. Chem.* 40, 966.
9. Fraenkel, G., and Franconi, C. (1959). *J. Am. Chem. Soc.* 82, 4478.

Phospholipid Membranes Are Necessarily Bimolecular

J. F. DANIELLI

Center for Theoretical Biology and
Department of Biophysics
State University of New York at Buffalo
Buffalo, New York

The natural lipids vary from the highly amphipathic, such as lecithin and cephalin, to the weakly amphipathic, such as cholesterol and triglycerides, and to nonpolar compounds, such as carotene and squalene. All lipids contain a substantial nonpolar hydrocarbon moiety, so that the variation in their generalized properties is, in the main, a function of the polar groups, such as the highly polar phosphorylcholine group of lecithin and the weakly polar hydroxyl group of cholesterol.

When these natural lipids are dispersed in aqueous media of physiological ionic strength, the dispersions consist commonly of bimolecular leaflet membranes. Other forms of dispersion are also found; e.g., lysolecithin readily forms micellar structures; most lipids form micelles when long-chain alkyl sulfates or cholic acids are present in large amounts; some form micellar dispersions with certain proteins. But, under near-physiological conditions the prevalence of membranes is striking, and it is rare for these membranes to be thicker than bimolecular. Why is this so?

The capacity to form bilayers requires only that a major component of the system be highly amphipathic. The amphipathic component may be charged, zwitterionic, or uncharged. Variation in ionic strength of the aqueous phase has only second-order effects.

It is not surprising that a pure phospholipid gives bilayers in dilute aqueous suspension. The free energy of transfer of the hydrocarbon moiety of a natural lecithin or cephalin from the bilayer to molecular solution in water is usually of the order of 20,000 calories. Similarly, the free energy of transfer of the polar moiety into the interior of a bilayer from the aqueous surface is of the order of 25,000 calories. Consequently, the only probable structures are bilayers and micelles, and an examination of the geometry of biacyl phospholipids indicates that bilayers will be favored rather than micelles (see, e.g., Vandenheuval, 1966). But why is it that when a relatively nonpolar component, such as cholesterol or triglyceride, or even an aliphatic hydrocarbon is added, the structure remains bimolecular, often with some nonpolar lipid incorporated in the bilayer, and the excess of the nonpolar component appears as droplets in equilibrium with bilayer, rather than constituting membranes greater than bimolecular in thickness?

Clearly, the free energy of a bimolecular membrane must be less than that of thicker membranes, but why? The first clue which led to the solution of this problem was found in the experimental observation, made first I believe by Langmuir and his collaborators about 30 years ago, that there is a significant surface free energy at the interface between random and oriented hydrocarbon chains. This led to the conclusion that it might be possible to calculate the free energy of lipid membranes in a semiempirical manner from a variety of experimental data. This proved to be practicable if electrostatic free energy terms were ignored. This latter assumption was quite reasonable, since it was already well known that the capacity to form bilayers does not depend upon charge. It was then shown that, if the molecules in the two monolayers at the surface of a membrane are well oriented, approximately perpendicular to the plane of the interface, and the molecules in the interior of the membrane are randomly oriented, the free energy of a bilayer is less than that of any thicker (or thinner) membrane by about 20 ergs/cm^2 (Danielli, 1966).

The procedure consisted of two main steps. The first was to consider the surface free energy γ_i at a lipid-water interface as composed of three terms (see Fig. 1), so that

$$\gamma_i = \gamma_{p/w} + \bar{\pi}_c + \gamma_{o/r} \tag{1}$$

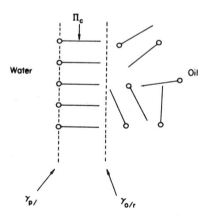

Oil - water interface

$$\gamma_{o/w} = \gamma_d = \left(\gamma_{p/w} + \gamma_{o/r} + \Pi_c \right)$$

FIG. 1. Interface between a lipid droplet and water.

where $\gamma_{p/w}$ is the term for polar group-water interface; $\gamma_{o/r}$ for interface between random and oriented hydrocarbon; and $\bar{\pi}_c$, difference between free energy of hydrocarbon chain oriented in the surface layer and randomly distributed in a bulk phase.

The second step was to make the assumption that, if the surface area of a monolayer was increased, so as to bring CH_2 groups into direct contact with water, the free energy of the system was mainly due to the change in the $\gamma_{p/w}$ term and could be calculated either from the surface pressure-area isotherm of the monolayer or from a simple relationship such as

$$\gamma^*_{p/w} = \alpha\gamma_{p/w} + (1 - \alpha)\gamma_{CH_2/H_2O} \tag{2}$$

where α is the proportion of interface which consists of polar group, $\gamma_{p/w}$ is the free energy appropriate to an interface consisting only of polar groups in contact with water, and γ_{CH_2/H_2O} is the free energy for a surface consisting only of hydrocarbon in contact with water. This assumption is certainly not true, but the value of γ_{CH_2/H_2O} is so large (50 ergs/cm^2) that the calculations remain correct to a first approximation.

I. Calculation of $\gamma_{o/r}$

This quantity has been obtained from: (1) the contact angle observed when a drop of randomly oriented hydrocarbon fluid rests upon a completely oriented hydrocarbon layer; (2) from the interfacial tension critical for the equilibrium: oil dispersed in water $\rightleftarrows$ water dispersed in oil. At this point $\gamma_{p/w} = \gamma_{o/r}$; (3) the Hamaker constant for hydrocarbon-hydrocarbon interactions (this calculation was made by F. Fowkes at the Theoretical Biology Colloquium held at Fort Collins in the summer of 1966); and (4) using the second-order perturbation theory of Hershfelder (Fukuda and Ohki, 1967). In all cases we find that $\gamma_{o/r}$, for a system in which the surface layer is fully oriented, is about 10 ergs/cm^2. Having this value, we can answer the following question.

II. Can a Lipid Membrane Be Thicker Than Bimolecular?

The surface free energy γ_m for a thick membrane ($n \gg 2$, where n equals the number of molecules) is clearly twice that of a lipid-water interface (Fig. 2), i.e., from Eq. (1),

$$\gamma_m = 2\gamma_i \tag{3}$$

If n is reduced, no change in γ_m will occur until $n \simeq 2$, when the interface between random and oriented hydrocarbons will vanish, and we find, for a bilayer,

$$\gamma_b = 2(\gamma_i - \gamma_{o/r}) \simeq 2\gamma_i - 20 \tag{4}$$

Thick membrane

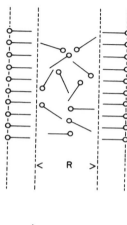

$$\gamma_m = 2\left(\gamma_{p/w} + \gamma_{CH_2^o/CH_2}^r + \Pi_c\right)$$

FIG. 2. Diagram of a membrane more than bimolecular in thickness.

Thus, Eq. (4) shows that the free energy of a bilayer is less than that of any thicker membrane. Consequently, thicker membranes will always thin down to bilayers, and any material which cannot form a bilayer, such as hydrocarbon, will form droplets. Obesity is not a problem for lipid membranes.

We can now ask the next question:

III. Can a Lipid Membrane Be Thinner Than a Bilayer?

Since in a bilayer all the polar groups are already at the water-lipid interface, the increase in area produced by further reduction in thickness (at constant density) must consist of hydrocarbon-water interface, with a high γ_{CH_2/H_2O}. Hence, there is a rapid rise in surface free energy as n falls below 2, and we see that membranes thinner than bimolecular cannot exist unless a constraint is applied additional to those already considered.

In Fig. 3 we plot surface free energy against membrane thickness to illustrate the sharp minimum in free energy for a bilayer. The existence of this minimum indicates that a bilayer is a natural unit of structure in the same sense that the DNA double helix and the protein α-helix are natural units. The bilayer is not the only possible mode of existence of the natural lipids, but under physiological conditions of pH, temperature, and ionic strength, it is the most probable form for phospholipids. The basic reasons for the stability of this structure can be summarized as: (1) the amphipathic nature of the phospholipids; (2) the anisotropic orientation and consequent anisotropy of polarizability of the hydrocarbon chains of the lipids; and (3) the high surface free energy of hydrocarbon-water interfaces.

In developing the theory of bilayers, the consequences of the amphipathic nature of the phospholipids led originally to the postulation of bilayers as a component of cell membranes (Danielli and Davson, 1934; Harvey and Danielli, 1938). The element of novelty which has recently permitted calculation of the free energy (Danielli, 1966) has been the appreciation of the facts that thinning a bilayer can only occur by formation of a new hydrocarbon-water interface of high surface free energy and that, insofar as the hydrocarbon chains of the lipids are oriented at the lipid-water surface, the anisotropy of polarizability of the chains will result in a surface free energy for the interface between the oriented layer and a bulk lipid phase.

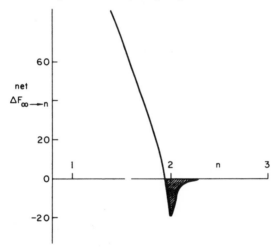

Minimum of free energy for bilayer

FIG. 3. Surface free energy ΔF^s plotted against thickness in molecules.

Naturally, the structure will vary in detail as the constraints on a bilayer are varied. Rise in temperature or in net electrostatic charge will thin the bilayer. Presence of small molecules of hydrocarbon, such as some of the anaesthetic gases, will reduce $\bar{\pi}_c$ and thin the bilayer. Presence of multivalent cations will increase the thickness. Presence of molecules which, in the randomly oriented bulk phase, have a polarizability similar to the anisotropic value (about 2.8) for the oriented hydrocarbon chains will diminish the value of $\gamma_{o/r}$ and so may permit formation of layers with n greater than 2. Some of the halogenated anaesthetic gases may fall into this later category.

As the thickness of a bilayer is reduced, the degree of orientation of the hydrocarbon chains will fall, and so the value of $\gamma_{o/r}$ will diminish. Diminution in thickness will expose more hydrocarbon to water, favoring the transition bilayer → micelle. If we assume that the new hydrocarbon-water interface formed has γ_{CH_2/H_2O} of 50 ergs/cm^2, the approximate free energies of formation of cylindrical micelles and spherical micelles will be 6000 and 12,000 cal, respectively. With more accurate knowledge of these free energies we shall be able to calculate the proportion of material present as bilayer and micelles under any given condition.

An important consequence of the anisotropy of polarizability of bilayer lipid chains is that the dielectric constant and refractive index will show a similar anisotropy. To date, determination of the thickness of artificial bilayers has most commonly been made by optical or capacitance measurements and then the thickness calculated using the bulk values for refractive index or dielectric constant. The use of the bulk values is incorrect, and the anisotropic values must be used. An apparent difficulty is that to obtain the anisotropy values, the degree of orientation of the chains must be known. This difficulty can be avoided by making both optical and capacitance measurements simultaneously, and then finding the degree of anisotropy which gives the same thickness from both sets of data. My colleague S. Ohki is presently engaged in these basic calculations.

IV. The Equilibrium between Bilayers and Droplets

When solid or liquid lipid particles come into contact with aqueous phases, the polar groups will tend to enter the water-lipid interface. If the surface free energy of the lipid-water interface is positive, however, in the absence of agitation the equilibrium state will consist of large lipid droplets dispersed in water or alternatively of water drops dispersed in lipid. If the surface free energy is small or negative, thermal agitation will disperse the system until it consists of bilayers or micelles, with the area per molecule of interface extended until the amount of hydrocarbon-water interface (having high surface free energy) is sufficient to reduce the average surface free energy approximately to zero. Most of the natural phospholipids at physiological temperature, pH, and ionic strength disperse as bilayers under these circumstances. If a relatively nonpolar lipid, such as cholesterol, or a triglyceride is present it may be incorporated in the bilayer in substantial amounts; but it cannot increase the thickness of the membrane so long as the chains remain oriented and $\gamma_{o/r}$ has a significant value. Thus, any lipid in excess to that which can be incorporated in a bilayer will be present as droplets (or solid phase). This excess lipid may or may not be surrounded by a monolayer of the same material as the bilayer. In the former case, the lipid drops will be present in the interior of the bilayer. The condition for this former case is simple.

Let the surface free energy of the lipid droplet for the interface (lipid/aqueous phase) be γ_i. If the droplet is now transferred to the interior of a bilayer, the droplet instead of having an interface with water will have an interface with the oriented hydrocarbon of the bilayer, i.e., the surface free energy of the drop will then be $\gamma_{o/r}$. Then, the condition for entry into the bilayer is

$$\gamma_i > \gamma_{o/r} \qquad (5)$$

Equation (5) is thus the condition which determines whether lipid which cannot be incorporated into a bilayer will exist as an independent droplet or will enter into the interior of a bilayer. Since $\gamma_{o/r}$ cannot be much greater than 10 ergs/cm^2 for bilayers which are fully oriented, a paraffin droplet with $\gamma_i (= \gamma_{CH_2/H_2O})$ of 50 ergs/cm^2 will enter the bilayer. Thus, Eq. (5) gives us a way of measuring the degree of orientation of the chains in a bilayer, for if we determine experimentally the threshold value of γ_i for entry of an oil droplet into a bilayer we obtain the value of $\gamma_{o/r}$. Chambers and Kopac (1937) made studies of entry of droplets into sea urchin egg plasma membranes which show that for this case $\gamma_{o/r}$ is 9.5 ergs/cm^2, so that the lipids of this membrane must be well oriented (Danielli, 1967).

The value of $\gamma_{o/r}$ obtained in this way is a measure of the degree of orientation (or anisotropy) of the hydrocarbon moiety of the bilayer, and the anisotropy should be identical with that calculated from the combination of capacitance and optical measurements suggested earlier.

We can now see why, as in the experiments of Mueller et al. (1964), a thick membrane of phospholipid plus hydrocarbon thins spontaneously to a bilayer plus droplets. The phospholipid is stable only as a bilayer (containing some "dissolved" hydrocarbon) and the excess hydrocarbon must then aggregate as one or more droplets. With the type of equipment used by Rudin and Mueller the droplet is annular, and provides an attachment point for the bilayer (see Fig. 4). The function of the hydrocarbon in this system is thus to

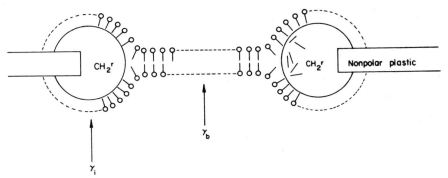

Fig. 4. Diagram of a cross section of a Rudin-Mueller membrane. A bilayer ($\gamma_b \approx 0.5$ ergs/cm^2) is in equilibrium with an annular droplet ($\gamma_i \approx 10$ ergs/cm^2).

provide an annular droplet to support the bilayer and probably also, by incorporation in the bilayer, to raise the value of γ_b from zero to a small positive value. Without the hydrocarbon or its equivalent, the phospholipid would form unattached bilayers.

V. Conclusion

The natural phospholipids spontaneously form bilayers under physiological conditions. The reasons for this can be summarized as follows:

1. The large free energy of removal of hydrocarbon from aqueous solution
2. The large free energy of removal of the polar group from solution in hydrocarbon
3. The existence of a surface free energy at the interface between oriented and random hydrocarbon, arising from the asymmetry of polarizability of hydrocarbon chains
4. The large surface free energy of a hydrocarbon-water interface.

As a result of the existence of these four factors, the bilayer is a natural unit of structure. From these considerations and consideration of the properties of monolayers, we can make the following predictions open to experimental study.

a. At least three and possibly four types of bilayer will exist for a given substance, with different packings of chains and polar groups. Probably only one of these, corresponding to the $L_1 \rightleftharpoons L_2$ transition in expanded monolayers, will be highly sensitive to temperature changes. These different packings will have different degrees of orientation of the hydrocarbons and, consequently, distinct values for $\gamma_{p/w}$, $\bar{\pi}_c$, and $\gamma_{o/r}$ will be found for each type of bilayer.

b. The value of $\bar{\pi}_c$ for the lipids from natural membranes, spread as monolayers on fluids of natural ionic content at physiological temperature, will be similar to that found for dimyristoyl cephalin.

c. The threshold value for the critical surface free energy for penetration of an oil droplet into a bilayer will not exceed approximately 12 ergs/cm² for droplets essentially hydrocarbon in nature and will be lower for droplets having a polarizability of the order of 2.8.

d. Fat droplets, when synthesized inside cells, will appear initially enclosed between the two monolayers of a bilayer.

e. The anaesthetic gases will act principally by reducing the cohesion between chains when the gases are hydrocarbons, but some of the halogenated hydrocarbons may also reduce $\gamma_{o/r}$ and permit formation of lipid membranes thicker than bimolecular.

ACKNOWLEDGMENT

This work was supported by Grant GM 11603 of the National Institutes of Health and Grant NSG 501 from the National Aeronautics and Space Agency. I wish to thank J. Borst, A. Goldup, F. Fowkes, S. Ohki, and F. A. Vandenheuval for helpful comment and discussion.

REFERENCES

Chambers, R., and Kopac, M. J. (1937). *J. Cellular Comp. Physiol.* **9**, 331, 345.

Danielli, J. F. (1966). *J. Theoret. Biol.* **12**, 439.

Danielli, J. F. (1967). *In* "Formation and Fate of Cell Organelles" (K. B. Warren, ed.). Academic Press, New York.

Danielli, J. F., and Davson, H. (1934). *J. Cellular Comp. Physiol.* **5**, 495.

Fukuda, N., and Ohki, S. (1967). *J. Theoret. Biol.* (in press).

Harvey, E. N., and Danielli, J. F. (1938). *Biol. Rev. Cambridge Phil. Soc.* **13**, 319.

Mueller, P., Rudin, D. O., Ti Tien, H., and Wescott, W. C. (1964). *Recent Progr. Surface Sci.* **1**, 379.

Vandenheuval, F. A. (1966). *J. Am. Oil Chemists' Soc.* **43**, 258.

Molecular Associations in Biology —a Brief Summary

PER-OLOV LÖWDIN

Department of Quantum Chemistry
Uppsala University, Uppsala, Sweden and
Quantum Theory Project, University of Florida
Gainesville, Florida

It is evident that it is impossible for anyone to make a proper summary of a symposium which has been so diversified and rich in results as this one—particularly if one has not had the manuscripts of the various authors available in advance. Hence I will only try to comment about some of the main thoughts at this symposium, and I apologize at the very beginning for leaving out many important aspects.

Professor Bernard Pullman started the symposium with an excellent survey of the present situation of molecular biology, considering theory and experience in general and the phenomena of mutagenesis and carcinogenesis in particular. He emphasized the importance of intermolecular forces in discussing molecular associations in biology, and he discussed the nature of the London-van der Waals forces, the formation of charge-transfer complexes, and the hydrogen-bond complementarity of nitrogen bases in the stabilization of certain molecules and molecular complexes. He mentioned also the fundamental role of the electronic structure of certain regions of hydrocarbons—the so-called K and L regions—in the theory of carcinogenesis.

Before proceeding, it may be worthwhile to give a brief summary of the present theoretical background for the various discussions in molecular biology classified after their degree of "sophistication" and in view of the experimental techniques utilized for the investigation. The *macroscopic properties* of molecular systems refer to such concepts as mass, volume, temperature, pressure, pH, etc., and the theory is essentially based on classical thermodynamics and the concept of free energy:

$$F = E + pV - ST \tag{1}$$

One discusses the equilibrium properties as well as the approach to equilibrium and chemical kinetics. Order-disorder phenomena are treated, and one is particularly interested in the melting of biological compounds, e.g., the transition of well-ordered helical structures into random coils. Such problems have been discussed here by Tinoco, Baldwin, and others, and it is remarkable what intriguing results can be obtained by studying the melting process of DNA and RNA as a function of temperature.

On the next level, the properties of biological systems are described by a more detailed *molecular picture*, and thermodynamics is now replaced by statistical mechanics. The analysis of the association of DNA and certain dye molecules by Lifson is an elegant example of this type of approach. On the molecular level, an important experimental tool is rendered by the *molecular spectra*: absorption spectra, fluorescence spectra, optical rotational power, the phenomenon of hypochroism, etc.; and important results of this type have been discussed by Tinoco, Ts'o, and several others.

In connection with order-disorder processes, the *solvent effects* are also of fundamental importance, and they have here been discussed by Sinanoğlu, Rosenheck, Douzou, and others. Sinanoğlu stressed the fact that the phenomenon of hydrophobic bonding is closely related to the problem of surface energy minimization; the fact that, in water, it takes a lot of energy to create free surfaces is one of the reasons for the existence of compact structures, helices, etc.

An elegant discussion of the surface energy was also given by Danielli in his study of *membranes* leading to the result that lipid membranes are necessarily bimolecular, consisting of two monomolecular layers turned structurewise toward each other.

Enzyme kinetics is one of the most complicated phenomena on this level, and, in a brilliant lecture, Theorell spoke about a new allosteric effect in the action cycle of liver alcohol dehydrogenase and the special experimental technique developed in studying the important changes of conformation.

In refining the molecular picture one reaches the *atomic level*, and it is perhaps not surprising that NMR spectroscopy has turned out to be an excellent experimental tool also in biology. Interesting aspects as to these atomic "fingerprints" have here been discussed by Ts'o, Jardetzky, and others.

In looking for finer details in biological system, it is often desirable to avoid the overall effects of random motion, and the use of *crystals* is sometimes a valuable tool in this connection, as reported here by Hoogsteen, Theorell, and others, and crystallographic methods are then of essential value for structure determination. The transformation from "disorder" to "order" is associated with an ordering of the "phases" appearing in the theory to which we will return later.

From the point of view of physics, atoms consist essentially of three *fundamental particles*: electrons, protons, and neutrons—the two latter forming the atomic nuclei of all elements. In chemistry, it is usually sufficient to study the electrons and protons in detail, whereas all atomic nuclei for $Z = 2, 3, \ldots$ are considered as electric point charges. The theoretical treatment is complicated by the fact that these fundamental particles do not obey the laws of classical physics but rather the rules of *modern quantum theory* established in 1925 and summarized in the Schrödinger equation:

$$-\frac{h}{2\pi i}\frac{\partial \Psi}{\partial t} = H\Psi \qquad (2)$$

where $H = T + V$ is the Hamiltonian consisting of the kinetic energy T and the potential energy V. For a system of particles k with the mass m_k and the charge e_k, one has

$$T = \sum_k \frac{p_k^2}{2m_k} = -\frac{h^2}{8\pi^2}\sum_k \frac{1}{m_k}\nabla_k^2 \qquad (3)$$

$$V = \sum_{k<l} \frac{e_k e_l}{r_{kl}} \qquad (4)$$

where V represents the Coulomb interaction of the particles. The basis of the theory is, hence, quite simple.

Ordinary chemistry deals to a large extent with the fundamental particles themselves. Oxidation-reduction reaction is a classic term for the process of *electron transfer*, whereas acid-base reaction deals with *proton transfer*; hence, one can expect that quantum mechanical concepts will be of essential importance in this connection. The electronic structure of stationary states of molecules is also determined by the Schrödinger equation (2), and the associated energy levels E are found by solving the eigenvalue problem

$$H\Psi = E\Psi \qquad (5)$$

which leads to an explanation of atomic and molecular spectra. In certain cases, one should also introduce the electromagnetic field, the spin-orbit coupling, the spin-spin coupling, etc., into the Hamiltonian H. Quantum mechanical aspects have here been discussed by Pullman, Bradley, Claverie, Mantione, and others.

From the point of view of molecular associations, quantum theory is of particular importance as a basis for understanding chemical binding and *intermolecular forces*. The true nature of the Lewis covalent bond, consisting of an electron pair shared between two atoms, was explained in 1927 by Heitler and London as depending on the exchange phenomenon. They showed that, if a and b are the wave functions of two hydrogen atoms at a distance R, a molecular wave function of the type

$$a(1)b(2) + a(2)b(1) \qquad (6)$$

which is invariant under *exchange* of the two electronic coordinates 1 and 2, has a much lower energy than each one of the individual terms; hence, the bond is stabilized by exchange.

A few years later, it was also shown by London that the van der Waals forces between neutral molecules are also an immediate consequence of quantum mechanics and that the main term corresponds to induced dipole-dipole forces. The importance of quantum theory for the understanding of molecular associations is, hence, clearly established.

In 1928, Mulliken started a new development by introducing the *molecular orbitals* (MO) $\varphi = a + b$ and $\chi = a - b$, which are linear combinations (LC) of the atomic orbitals (AO) a and b; the approach is, hence, often denoted by the symbol MO-LCAO. In the molecular orbitals, the electrons move in orbitals associated with the molecule as a whole, and they are of particular importance in discussing molecular symmetry and electronic spectra. It should be observed that the Mulliken wave function $\varphi(1)\varphi(2)$ for the ground state of the hydrogen molecule is not identical to the Heitler-London function and that it actually consists of the so-called "covalent term" [Eq. (6)] and the "ionic terms" $a(1)a(2)$ and $b(1)b(2)$ with both electrons accumulated on the same atom.

In the 1930's, Hückel picked up these ideas in treating aromatic hydrocarbons and conjugated systems, in general having alternating single and double bonds. He showed that, even if the single bonds may be described by localized Lewis electron pairs, it may be more practical to describe the classic double bonds as corresponding to *mobile π electrons* which move around the entire conjugated system in molecular orbitals and give the system its characteristic aromatic properties. This simple picture is now of fundamental importance in the entire field of organic chemistry.

In principle, one finds the molecular orbitals ψ by solving the eigenvalue problem

$$H_{\text{eff}} \psi = \varepsilon \psi \tag{7}$$

where H_{eff} is the "effective Hamiltonian" for the π electrons and ε is the orbital energy. If the atomic orbitals of the conjugated system are denoted by $\phi_\mu (\mu = 1, 2, \ldots, M)$, the best molecular orbitals found by linear combination (MO-LCAO)

$$\psi = \sum_\mu \phi_\mu c_\mu \tag{8}$$

are determined by the following system of linear equations:

$$\sum_v (H_{\mu v} - \varepsilon \Delta_{\mu v}) c_v = 0; \qquad |H_{\mu v} - \varepsilon \Delta_{\mu v}| = 0 \tag{9}$$

where $H_{\mu v} = \langle \phi_\mu | H_{\text{eff}} | \phi_v \rangle$ is the Hamiltonian matrix and $\Delta_{\mu v} = \langle \phi_\mu | \phi_v \rangle = \delta_{\mu v} + S_{\mu v}$ is the metric matrix, containing the overlap matrix $S_{\mu v}$. The molecular orbitals ψ_k may be ordered after increasing values of the orbital energies ε_k and, in the ground state of the molecule, they are occupied from the bottom with two electrons each.

In the independent particle model, all properties of the system are determined by the density matrix:

$$\rho(1, 2) = \sum_{k}^{occ} \psi_k(1)\psi_k{}^*(2) = \sum_{\mu\nu} \phi_\mu(1)R_{\mu\nu}\phi_\nu{}^*(2) \tag{10}$$

where one sums over all occupied orbitals and obtains the charge-order and bond-order matrix

$$R_{\mu\nu} = \sum_{k}^{occ} c_{\mu k} c_{\nu k}^* \tag{11}$$

introduced by Coulson and Longuet-Higgins in the 1940's. The diagonal element $q_\mu = R_{\mu\mu}$ is interpreted as the "charge order" of the atomic orbital ϕ_μ, whereas the nondiagonal element $p_{\mu\nu} = \frac{1}{2}(R_{\mu\nu} + R_{\nu\mu})$ is interpreted as the "bond order" of the association between ϕ_μ and ϕ_ν. The usefulness of these concepts have been demonstrated by several authors.

Even if the molecular orbital picture in the Hückel scheme is an approximation itself to quantum theory, one has, for complicated systems, introduced further approximations. In the *naive Hückel scheme*, one considers only nondiagonal elements $H_{\mu\nu}$ between nearest neighbors ϕ_μ and ϕ_ν, whereas the overlap matrix is neglected completely. The quantities $H_{\mu\mu}$ and $H_{\mu\nu}(\mu \neq \nu)$ are usually determined semiempirically according to certain simple rules.

For the electronic structure of simple hydrocarbons and heterocyclics, the naive Hückel scheme gives surprisingly good results. Starting out from this fact, Alberte and Bernard Pullman have used the naive Hückel scheme to calculate charge and bond orders for the conjugated systems occurring in biochemistry, and the results of their encyclopedic study have been reported in their book [1] on "Quantum Biochemistry." There is little question that this approach gives a first rough picture of the electronic structure of the molecules involved and that it has opened an important path into quantum biology.

The Hückel method is here described in some detail in order to give a background for its use in calculating molecular associations. From the matrix $R_{\mu\nu}$, one can calculate the "*net charges*" associated with the various atoms in conjugated systems, and from this data one can then calculate certain properties—at least in a first approximation.

In the theory of ionic crystals it is well known that the Madelung energy represents the main contribution to the cohesive energy and that the repulsive energy amounts to only about 10%. In his book [2], van Arkel has shown that one may extend these considerations also to comparatively small inorganic molecules and that one gets a rough estimate of the cohesive energy simply by calculating the Coulomb energy [Eq. (4)] for a system of net charges. Even if this approach neglects the virial theorem, one can expect that it should be useful in estimating energy differences.

In molecular biology, Bradley and Pullman have reported at this symposium that they have used extensions of this approach to study interactions also between large conjugated systems under the name of "electrostatic hardsphere approximation" or "monopole approximation" with good results. Using the net charges obtained from the Hückel scheme for two conjugated systems, I and II, separately, they have calculated the Coulomb interaction

$$\sum_i^{I} \sum_j^{II} \frac{e_i e_j}{r_{ij}} \tag{12}$$

and studied its variations as a function of the orientation of the two molecules with respect to each other. Particularly for the hydrogen bonds, they have obtained interesting results in this way.

In classical chemistry, a hydrogen bond was represented by a hydrogen atom bound simultaneously to two electronegative atoms, N, O, F, etc. In quantum chemistry, *a hydrogen bond is essentially a proton shared between two electron lone pairs.* Following Lewis and denoting the proton by H and an electron lone pair by two dots a hydrogen bond has a structure of the type

$$\rangle N\colon\!\!\supset -\!-\!-\!- H\colon\!\!\supset N\langle \tag{13}$$

The hydrogen bonds play an important role in molecular biology in connection with the concept of Watson-Crick complementarity, and it has been stressed by the author that the genetic template is essentially a definite pattern of protons and electron lone pairs.

In connection with chemical bonds, one is accustomed to the fact that their properties are often transferable from one molecule to another and that bond energies are characteristic figures which are additive. That this is not the case for the hydrogen bonds in general has here been clearly stressed by Pullman and Bradley who studied hydrogen-bonded complexes in the monopole approximation based on Eq. (12). They have investigated all possible combinations of purine and pyrimidine bases, and the results indicate that the normal Watson-Crick pairing should be more stable than any other association. They have also investigated the peculiar pairing found by Hoogsteen in crystals and its modifications. Hoogsteen emphasized that there is nothing particular about hydrogen bonds, except that they seem to provide "soft spots" in the van der Waals shells. Perhaps one can summarize the results in the simple statement that *hydrogen bond complementarity is a necessary but not sufficient condition for complex formation.*

In connection with the Watson-Crick pairing in DNA, it is evident that also the *stacking forces* as well as the sugar-phosphate bonds play an important role in the stabilization of the DNA molecule. The stacking forces have also been studied by the Pullmans.

Even if all these results seem rather promising, it is worthwhile to remember that the underlying *quantum theory of intermolecular forces* is still very weak and that one knows very little about the transition from the exchange forces acting on small distances (chemical bonding) to the dispersion forces acting on large distances. A great deal of theoretical work is presently in progress in several countries, and Claverie has presented an interesting contribution to the development by the means of perturbation theory including overlap.

In discussing the dispersion forces, one is often using a terminology related to the so-called *multipole expansion*, but one should remember that this expansion is strictly valid only if the two interacting molecular systems are not spherically connected, i.e., if one can enclose the molecules involved in two spheres which are not overlapping. Since this seldom happens for the large molecules in biology, the terminology and the conceptual basis is from the very beginning rather diffuse.

It is often said that the main problem in *chemical kinetics* is the calculation of the potential energy surfaces related to the intermolecular potentials, but it should be remembered that, in connection with chemical reactions, the exchange forces corresponding to chemical binding are of fundamental importance and that they are of nonlocal character. In spite of all progress during the last few years, quantum theory is still at the starting point as to the deeper understanding of the details of chemical reactions.

A large part of the symposium has been devoted to the properties of the *nucleic acids*, DNA and RNA. The question of the duplication and reading of the genetic code is one of the key problems in molecular biology, which may be illustrated in the following diagram:

$$\begin{array}{c} \text{Stationary state} \\ \diagup \qquad \diagdown \\ \diagup \quad \text{DNA} \quad \diagdown \\ \text{Replication} \text{-----} \text{Transcription} \rightarrow \text{messenger RNA} \rightarrow \text{proteins} \end{array} \qquad (14)$$

The problem of *helix kinetics* has been discussed by Tinoco, Ts'o, Michelson, Geiduschek, Doty, Manago, and others, and a remarkable development of the experimental technique has been demonstrated.

It seems today very likely that both the replication and the transcription processes take place in the *deep groove* of DNA, but it should perhaps be stated that one still does not know the steric details of the opening of the Watson-Crick template in replication. In the transcription process, one still does not know what code is being read in the formation of messenger RNA— whether the Watson-Crick template is opened or whether the code is read on the "copy" of the template situated in the deep groove on the 6-positions of the base pairs, as suggested previously by Stent and Zubay. The discussions indicate that it may be difficult to settle these important problems in the immediate future.

The author has been interested in the problem of the *stability of the genetic code* with respect to incorporation errors and internal errors in DNA associated with proton tunneling. Here it may be sufficient to indicate that all the theoretical results obtained so far strongly support the Watson-Crick model, and interested readers are referred to some recent publications [3].

Even the *shallow groove* of DNA has been discussed at the symposium, and the results by Reich indicate that actinomycin may attack the guanine-cytosine base pair on this side. It should perhaps be observed that the tilt (21°) is in good agreement with flow-dichroism data (23°). Kersten also discussed the interaction between antibiotics and DNA in general.

Hydrocarbons and heterocyclics play a fundamental role in the phenomena of mutagenesis and carcinogenesis, and their interaction with DNA was discussed by several authors. They may be incorporated in the deep groove of DNA or in the shallow groove of DNA, or they may be intercalated between the base pairs. There may also be other possibilities, and there may be several mechanisms working simultaneously.

The interaction between *hydrocarbons* and DNA was discussed by Bergman, who considered several possibilities: the formation of charge-transfer complexes, the stacking of hydrocarbons in nucleic acids depending on dispersion forces, and the interaction with the DNA molecule as a whole. He emphasized the importance of the K and L regions introduced by the Pullmans in their study of carcinogenesis and the fact that these concepts were developed on correlation data before the present knowledge of DNA. Today since we are still ignorant about the ultimate cause of cancer, however, it is impossible to put the interaction between the hydrocarbons and DNA into the picture in a meaningful way.

The interaction between *heterocyclics* and DNA was discussed by Lerman, Jordan, Lifson, Duchesne, and particularly the intercalation mechanism for *acridine dyes* was treated in view of the mutagenic effects. As mentioned previously, Lifson discussed the statistical mechanics of the association process whereas Duchesne studied the transfer of charge and energy.

In connection with intermolecular forces, Mulliken introduced in the early 1950's the concept of *charge-transfer complexes* which were characterized by certain spectral properties in the form of new bands not associated with the parent molecules. If A and D are two molecules (A = acceptor, D = donor), the ground state may be characterized by the wave function $\psi(AD)$, whereas an excited state could be represented by a wave function $\psi(A^-D^+)$ of the charge-transfer type, provided that the two wave functions are orthogonal with respect to each other. In reality, they also have to be noninteracting with respect to the Hamiltonian H. If these conditions are not fulfilled by the original functions, one can easily find two linear combinations

$$\psi = c_1\psi(AD) + c_2\psi(A^-D^+) \tag{15}$$

which satisfy the requirements. The transition from the ground state to the excited state gives rise to a new "charge-transfer absorption band" which is characteristic for the complex, since it is not associated with any of the parent molecules.

The ideas of Mulliken have been of great importance in understanding loosely bound complexes. In tightly bound complexes, the situation is somewhat different, simply because wave functions as in Eq. (15), which are superpositions of a "covalent function" and an "ionic function," are normal for any chemical bond. In discussing charge transfer, the concepts of covalent and ionic character have previously been of essential importance, but Shull showed some years ago that these concepts are by no means "orthogonal" and that functions of the type $\psi(AD)$ and $\psi(A^-D^+)$ may be *overlapping* to a very large extent. This overlap problem constitutes the key difficulty in discussing the existence of real charge-transfer complexes in the sense of Mulliken.

Charge-transfer complexes are believed to be of essential importance in molecular biology, and they have been discussed here by Shifrin, Cilento, Slifkin, Katchalski, Tollin, McCormick, and others. Many interesting results have been reported, but sometimes it has been difficult to follow whether the authors refer to loosely bound complexes in the sense of Mulliken or tightly bound compounds with a large degree of ionic character; perhaps it would be worthwhile for us to try refining out terminology on this point.

An interesting theoretical study of the van der Waals forces in charge-transfer complexes was presented by Mantione; it is evident, that even in this connection, the overlap between the covalent and ionic structures will be of essential importance.

In conclusion, it may perhaps be useful to say a few words about the general connection between experiments and theory. It is clear that, in order to obtain information about nature, experiments come first, and that the gathering of reliable experimental data is the fundamental on which the entire science rests. On the other hand, it is hard to carry out meaningful experiments without having some idea, theory, or model in mind, and there is always an interesting interplay between experiment and theory. In experiment, most concepts are of an operational nature, whereas, in theory, they are usually related to a specific model or a set of axioms. Sometimes, theory provides intuitive concepts of a rather diffuse nature, which become of great importance in the experimental development because of their flexibility and adaptability to various situations. In spite of their diffuse nature, they may often be incorporated in a semiempirical theory to give precise numbers and predictions.

As an example, one can take the classic quantum theory and Bohr's quantum numbers (n, l, m) for the elliptic orbits of the electrons in an atom, which before 1925 led to a basic understanding of the entire periodic system and to a classification of atomic spectra, in spite of the fact that the Bohr theory itself could not be generalized from the hydrogen atom $(Z = 1)$ to the helium atom $(Z = 2)$.

From all points of view, modern quantum theory should be considered as the quintessence of 150 years of experimental experience in physics and chemistry as to the innermost structure of matter. The fact that the eigenvalue problem [Eq. (5)] in terms of atomic units (e, m, h) with proper boundary conditions corresponds to a purely mathematical problem gives rise to *ab initio calculations* which seem to replace experiments. It should be remembered, however, that the stationary states are to a certain extent abstract idealizations and that all phenomena in nature are in reality time-dependent. The main problem is hence to solve the Schrödinger equation (2) when one knows the *initial condition* $\Psi = \Psi_0$ for $t = t_0$, and the latter is, of course, given by the experimental set-up. In this connection, it is also of essential importance to determine the *phase* of the complex wave function Ψ, and this may turn out to be more difficult than anticipated, unless the system has a very high degree of order. The phase problem is closely related to the order-disorder question discussed previously [2]. All in all, this means that, except for the rather small sector referring to energy levels of stationary states, *quantum mechanics is essentially a tool for handling experimental information.*

In molecular biology, experiments concerning molecular spectra refer certainly to energy levels of stationary states, and it should then be possible, at least in principle, to carry out *ab initio* calculations. In reality, however, almost all other phenomena in molecular biology are of a time-dependent nature, and this means that one needs another type of quantum mechanics based on the Schrödinger equation (2) or its solution

$$\Psi(t) = U\Psi_0 \tag{16}$$

where $U(t, t_0) = \exp\{-2\pi i H(t - t_0)/h\}$ is the *evolution operator* [4]. This approach is more suited for treating chemical kinetics and transport phenomena, but it also requires a considerably more detailed knowledge of the experimental situation than one perhaps anticipates in advance. Hence, real progress has to be built on very close cooperation between theory and experiment.

In refining the theory, one should also remember that it is usually comparatively easy to get good results in fair agreement with experience by means of a semiempirical theory based on more or less intuitive concepts and that almost every "improvement" will lead to worse results—which is rather discouraging, of course. This phenomenon was well known in the 1930's and it may be illustrated by the following diagram

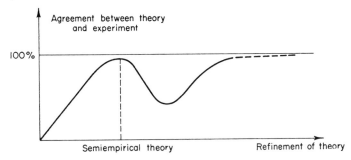

In the first quantum theory of chemical kinetics, one had the "nightmare of the inner shells," and in treating ferromagnetism and cohesive properties of solids, one had the "non-orthogonality catastrophe" referring to the result of the inclusion of the overlap integrals. Similar experiences are shared by those who have tried to refine ligand-field theory and the Hückel scheme; so there is a long way to go. Good agreement between theory and a selected number of experimental data is, therefore, *a necessary but not sufficient criterion* for the accuracy and overall reliability of a specific theoretical approach.

Even if quantum theory has provided chemistry with an entirely new conceptual basis for understanding chemical binding, van der Waals forces, hydrogen bonds, charge-transfer complexes, etc., it is certainly clear that most of the theoretical development, in order to understand molecular associations properly, has yet to come.

In conclusion, on behalf of all participants, I would like to express our sincere gratitude to Professor and Mrs. Bernard Pullman for their excellent organization of this meeting.

REFERENCES

1. Pullman, B., and Pullman, A., "Quantum Biochemistry." Wiley, New York, 1963.
2. van Arkel, A. E., "Molecules and Crystals." Butterworth, London, 1949.
3. Löwdin, P. O. *Advan. Quantum Chem.* **2**, 213 (1965); *Pontif. Acade. Scie. Scripta Varia* No. 31 (1966).
4. Löwdin, P. O. *Biopolymers, Symp.* **1**, 293 (1964); *Advan. Quantum Chem.* **3**, 323 (1967).

Author Index

Numbers in parentheses are reference numbers and indicate that an author's work is referred to although his name is not cited in the text. Numbers in italic show the page on which the complete reference is listed.

Topical Subject Index